Handbook of Hydrometeorology

Handbook of Hydrometeorology

Edited by Frederick Green

SYRAWOOD
PUBLISHING HOUSE
New York

Published by Syrawood Publishing House,
750 Third Avenue, 9th Floor,
New York, NY 10017, USA
www.syrawoodpublishinghouse.com

Handbook of Hydrometeorology
Edited by Frederick Green

International Standard Book Number: 978-1-64740-110-8 (Hardback)

Cataloging-in-Publication Data

Handbook of hydrometeorology / edited by Frederick Green.
p. cm.
Includes bibliographical references and index.
ISBN 978-1-64740-110-8
1. Hydrometeorology. 2. Hydrology. 3. Meteorology. I. Green, Frederick.
GB2805 .H36 2022
551.57--dc23

TABLE OF CONTENTS

PREFACE

Hydrometeorology is a branch of meteorology which focuses on the problems related to water cycle and water budget as well as the rainfall statistics. It also encompasses climatology, hydrology, cloud physics and weather forecasting. Primarily, it is concerned with mitigating the risks of flood and other hydrological threats. It utilizes the data recorded regarding various hydrological processes and observes the risks of natural disasters such as drought, desertification and cyclones on the basis of that data. It also focuses on the transfer of water and energy between the land surface and the lower atmosphere. Hydrometeorology also focuses on water quality and supply. This book outlines the processes and applications of hydrometeorology in detail. It elucidates the concepts and innovative models around prospective developments with respect to this field. Through this book, we attempt to further enlighten the readers about the new concepts in this area.

The researches compiled throughout the book are authentic and of high quality, combining several disciplines and from very diverse regions from around the world. Drawing on the contributions of many researchers from diverse countries, the book's objective is to provide the readers with the latest achievements in the area of research. This book will surely be a source of knowledge to all interested and researching the field.

In the end, I would like to express my deep sense of gratitude to all the authors for meeting the set deadlines in completing and submitting their research chapters. I would also like to thank the publisher for the support offered to us throughout the course of the book. Finally, I extend my sincere thanks to my family for being a constant source of inspiration and encouragement.

Editor

1

Including the dynamic relationship between climatic variables and leaf area index in a hydrological model to improve streamflow prediction under a changing climate

Z. K. Tesemma, Y. Wei, M. C. Peel, and A. W. Western

Department of Infrastructure Engineering, The University of Melbourne, Parkville, Victoria, 3010, Australia

Correspondence to: Y. Wei (ywei@unimelb.edu.au)

Abstract. Anthropogenic climate change is projected to enrich the atmosphere with carbon dioxide, change vegetation dynamics and influence the availability of water at the catchment scale. This study combines a nonlinear model for estimating changes in leaf area index (LAI) due to climatic fluctuations with the variable infiltration capacity (VIC) hydrological model to improve catchment streamflow prediction under a changing climate. The combined model was applied to 13 gauged sub-catchments with different land cover types (crop, pasture and tree) in the Goulburn–Broken catchment, Australia, for the "Millennium Drought" (1997–2009) relative to the period 1983–1995, and for two future periods (2021–2050 and 2071–2100) and two emission scenarios (Representative Concentration Pathway (RCP) 4.5 and RCP8.5) which were compared with the baseline historical period of 1981–2010. This region was projected to be warmer and mostly drier in the future as predicted by 38 Coupled Model Intercomparison Project Phase 5 (CMIP5) runs from 15 global climate models (GCMs) and for two emission scenarios. The results showed that during the Millennium Drought there was about a 29.7–66.3 % reduction in mean annual runoff due to reduced precipitation and increased temperature. When drought-induced changes in LAI were included, smaller reductions in mean annual runoff of between 29.3 and 61.4 % were predicted. The proportional increase in runoff due to modeling LAI was 1.3–10.2 % relative to not including LAI. For projected climate change under the RCP4.5 emission scenario, ignoring the LAI response to changing climate could lead to a further reduction in mean annual runoff of between 2.3 and 27.7 % in the near-term (2021–2050) and 2.3 to 23.1 % later in the century (2071–2100) relative to modeling the dynamic response of LAI to precipitation and temperature changes. Similar results (near-term 2.5–25.9 % and end of century 2.6–24.2 %) were found for climate change under the RCP8.5 emission scenario. Incorporating climate-induced changes in LAI in the VIC model reduced the projected declines in streamflow and confirms the importance of including the effects of changes in LAI in future projections of streamflow.

1 Introduction

Recently, climate changes have been observed in different parts of Australia (Chiew et al., 2011; Cai and Cowan, 2008; Hughes et al., 2012; Lockart et al., 2009; Potter and Chiew, 2011). Specifically, southeastern Australian catchments have experienced changes in streamflow due to fluctuations in climate as observed during the recent "Millennium Drought" (1997–2009) which lasted more than a decade (Chiew et al., 2011; Verdon-Kidd and Kiem, 2009). This drought may be representative of future climatic conditions in this region.

The projected water availability for future climates derived from downscaled outputs from global and regional climate models indicate increases of mean annual runoff by 10–40 % in some parts of the world (high northern latitudes) and 10–30 % reduction elsewhere (southern Europe, Middle East and southeastern Australia) (Milly et al., 2005). More recently, Roderick and Farquhar (2011) examined climate and catchment characteristics for sensitivity to changes in runoff in the Murray–Darling Basin in southeast Australia from a theoretical point of view and estimated that a 10 % change in precipitation would lead to a 26 % change in runoff and a

10 % change in potential evaporation would lead to a 16 % change in runoff with all other variables being constant. In southeastern Australia it has been projected that there will be a reduction in mean annual runoff of 10 % on average when different climate models are used as input to hydrological models (Cai and Cowan, 2008; Chiew et al., 2009; Roderick and Farquhar, 2011; Teng et al., 2012a; Vaze and Teng, 2011). These studies assessed the possible impacts of climate change on total runoff based on rainfall–runoff relationships which only considered first-order effects of changes in precipitation and temperature with subsequent impacts on evaporative demand.

There is evidence that such relationships are not stationary over time (Chiew et al., 2014; Peel and Blöschl, 2011; Vaze et al., 2010), which implies that the studies discussed in the previous paragraph may be missing an important factor. One approach to improving modeling under changing conditions is to use a year-to-year variable monthly leaf area index (LAI) in the hydrological model. Using observed climate variability and streamflow responses, observed monthly LAI has been shown to improve soil moisture prediction (Ford and Quiring, 2013). The improvements are largest under either relatively wet or dry climatic conditions, i.e., in wet and dry years, rather than average years. In most of southeastern Australia, LAI primarily responds to the availability of water and changes in vegetation type, such as conversion of forest to cropland or pasture, but also responds, to a lesser extent, to changes in temperature and rising atmospheric CO_2 concentrations. Most of these LAI responses are expected to be affected by projected climate change. These climate-induced changes in vegetation LAI may impact on evapotranspiration and runoff and hence should be considered when making runoff projections for climate change scenarios.

Dynamic global vegetation models (DGVMs) have been used to assess the vegetation effect of climate change on large-scale hydrological processes and patterns (Murray et al., 2012, 2011). A list of available DGVMs and their process representations (photosynthesis, respiration, allocation, and phenology) can be found in Wullschleger et al. (2014), while Scheiter et al. (2013) provides a review of the possible sources of uncertainty related to representation of plant functional type (PFT) in DGVMs. Most DGVMs overestimate runoff, which is mainly due to model structure problems and operating at low spatial and temporal resolution (Murray et al., 2013). While the relationships between LAI and climatic fluctuation have been modeled (Ellis and Hatton, 2008; O'Grady et al., 2011; Jahan and Gan, 2011; Palmer et al., 2010; Tesemma et al., 2014; White et al., 2010), none of them have been incorporated into hydrological models for the purpose assessing future climate change impacts on streamflow. The poor hydrological sub-models in DGVMs and the static vegetation in most hydrological models mean that importance of the indirect vegetation-related (LAI) effects relative to the direct effects of changes in precipitation and temperature on hydrological response at catchment scale have rarely been studied. This limits understanding of the linkages between climatic fluctuations and vegetation dynamics, and their combined impacts on hydrological processes.

The main objective of this study is to examine the relative effects on mean annual runoff of changes in direct climate forcing (mainly precipitation and temperature) and direct climate forcing combined with climate-induced LAI changes under changed climate scenarios. Comparative analysis of these two cases enables the effect on mean annual runoff of allowing LAI to respond to a changing climate to be identified. Specifically, our study combined the LAI–Climate model developed in Tesemma et al. (2014) with the Variable Infiltration Capacity (VIC) model to assess the impact on catchment runoff of how LAI is modeled (year-to-year constant monthly LAI or year-to-year variable mothly LAI in response to climate) under changing climatic conditions. As noted above, this combined model showed significant improvements in runoff simulations under historical conditions. Here we investigate two sets of changing climatic conditions: (1) the observed Millennium Drought (1997–2009), which is a persistent ($>$ 10 years) large change in climate; and (2) projected climate change for both wet and dry catchments using 38 Coupled Model Intercomparison Project Phase 5 (CMIP5) runs from 15 different global climate models (GCMs) for two future periods, 2021–2050 and 2071–2100, for two emission scenarios, Representative Concentration Pathway (RCP) 4.5 and RCP8.5. The results obtained from this study are expected to demonstrate whether modeling LAI in a way that responds to changing climatic conditions is important for modeling runoff during projected climate change in the study area.

2 Research approach

This section provides details about the data set, the characteristics of the selected catchments and the modeling exercises. The catchment characteristics and data set used in this study are briefly described in Sect. 2.1. The application of multiple outputs from GCMs and emission scenario methods is explained in Sect. 2.2. The relationship between LAI and climatic variables is presented in Sect. 2.3, and the hydrological modeling experiment approach used to assess the impact of changes in climate on runoff is described in Sect. 2.4.

2.1 Catchment characteristics and data set

All the study sub-catchments are located in the Goulburn–Broken catchment which is a tributary of the Murray–Darling Basin, Australia. The Goulburn–Broken catchment extends between 35.8 and 37.7° S and between 144.6 and 146.7° E (Fig. 1a) with a range of altitude from approximately 1790 m on the southern side to 86 m a.m.s.l. on the northern side of the catchment. The mean annual precipitation of the study sub-catchment ranges from 659 (in the north) to 1407 mm year^{-1} (in the south) calculated for the period 1982–2012. The majority of the precipitation (about 60 %) occurs during winter and spring. The reference potential evapotranspiration (PET) calculated using the Food and Agricultural Organization (FAO56) method ranges from 903 (in the north) to 1046 mm year^{-1} (in the south). Hence, the dryness index (mean annual reference potential evapotranspiration divided by mean annual precipitation) varies from 0.64 to 1.6 (Fig. 1b). The dominant land cover type in most of the catchments is forest (mainly tall open *Eucalyptus* forest and *Eucalyptus* woodlands) with some pasture in all catchments. A small amount of cropland is located in some of the catchments (Fig. 1c).

Gridded input data used for the hydrological modeling include the daily precipitation, maximum and minimum temperature, vapor pressure and solar exposure data obtained from the Australian Water Availability Project (AWAP) of the Bureau of Meteorology (Jones et al., 2009) and gridded daily wind run data from McVicar et al. (2008) that were generated from point measurements. All data have a spatial resolution of 0.05° × 0.05° (approximately 5 km × 5 km), and the period from 1982 to 2012 was selected for this study. The daily streamflow data at the outlet of the selected calibration sub-catchments were obtained from the Victorian Water Resources Warehouse (http://data.water.vic.gov.au/monitoring.htm). The missed streamflow data were filled by regressing between neighboring catchments. The elevation data were collected from the GEODATA 9 Second Digital Elevation Model (DEM-9S) Version 3 (Geoscience Australia, 2008). The elevation data were resampled to a resolution of 0.05° × 0.05° using the spatial average. The land cover input data were derived from the National Dynamic Land Cover Data set which provides a land cover map for the whole of Australia at a resolution of 0.00235° × 0.00235° (approximately 250 m × 250 m) and can be accessed at http://www.ga.gov.au/metadata-gateway/metadata/record/gcat_71071. LAI data were collected from the Global Land Surface Satellite (GLASS) product which is available for download from Beijing Normal University (http://www.bnu-datacenter.com). The soil parameters in the VIC model running resolution were derived from the 5' resolution Food and Agriculture Organization data set (FAO, 1995). The root distribution in three soil layers was derived from the global ecosystem root distribution data set (Schenk and Jackson, 2002).

2.2 Applying multiple GCMs and multiple emission scenarios

Outputs from 15 GCM from the CMIP5 (Taylor et al., 2012) were used as input into VIC model. CMIP5 contains model runs for four representative concentration pathways (RCPs), which provide radiative forcing scenarios over the 21st century (Moss et al., 2010; Vuuren et al., 2011). In this study

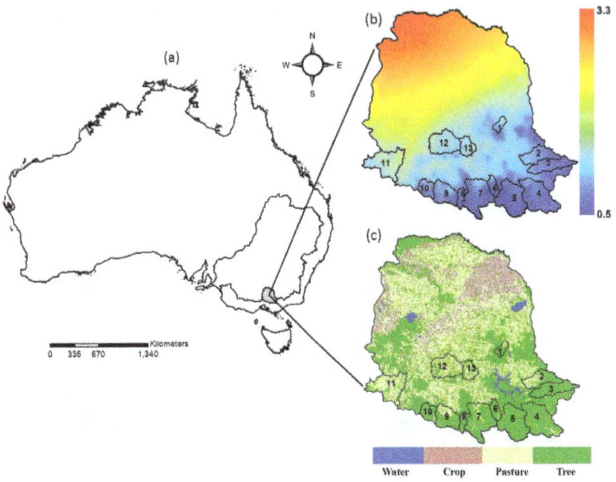

Figure 1. Location map of the study area **(a)**, dryness index (mean annual reference evapotranspiration divided by mean annual precipitation) **(b)** and land cover type **(c)**.

two emission scenarios were chosen: a midrange mitigation scenario, referred to as RCP4.5 and a high emissions scenario RCP8.5 (Meinshausen et al., 2011). RCP4.5 results in a radiative forcing value of 4.5 Wm^{-2} at the end of the 21st century relative to the preindustrial value, while RCP8.5 provides a radiative forcing increase throughout the 21st century to a maximum of 8.5 Wm^{-2} at the end of the century.

CMIP5 GCM data were obtained from http://climexp.knmi.nl (last accessed 28 February 2014). These data were re-sampled to a common grid resolution of 2.5° since each GCM has a different spatial resolution (some are the same, but most are different). A total of 38 RCP4.5 and 38 RCP8.5 runs from 15 different GCMs have been used in this study to include the possible uncertainty among climate models. For each of the 38 runs, daily precipitation, minimum and maximum temperature data were collected for three periods: 1981–2010 (historical run), 2021–2050 and 2071–2100 (future runs). An assessment of the ability of the CMIP5 runs to reproduce the observed baseline seasonality of precipitation, minimum and maximum temperature is shown in Fig. 2. The seasonality in precipitation and temperature were well captured by most CMIP5 runs, with biases which require correction.

Low-spatial-resolution GCM outputs require downscaling for application in catchment hydrology studies. Here the "delta-change" statistical downscaling technique was used to downscale and bias-correct the GCM outputs (Fowler et al., 2007). Delta-change was selected due to its low computational intensiveness and easy applicability to a range of GCMs. We acknowledge the limitations of this method include an assumption of stationarity in change factors, that climate feedbacks are not incorporated and an inability to capture changes in extreme events and year-to-year variability. Dynamic downscaling, which solves some of these prob-

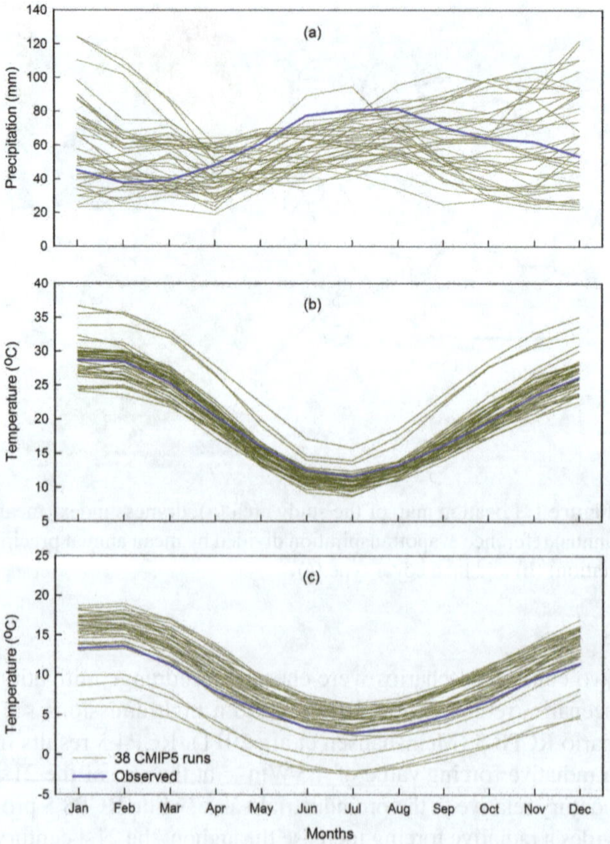

Figure 2. Long-term mean monthly climate observations plotted with the 38 CMIP5 runs during the baseline period (1980–2010) for the Goulburn–Broken catchment **(a)** long-term mean monthly precipitation, **(b)** long-term mean monthly maximum temperature, and **(c)** long-term mean monthly minimum temperature.

lems, was not used as it has high computational demand and is not readily available for a range of GCM runs and scenarios (Fowler et al., 2007). A simple statistical downscaling method was appropriate for this study as we were interested in the impact of including climate-induced LAI change on the runoff results. In the study area, the monthly LAI is strongly related to 6-month and/or 9-month moving average moisture state (precipitation minus reference potential evapotranspiration) (Tesemma et al., 2014). Therefore, so long as the precipitation is consistent between the two runs, we can assess the importance of the change in LAI representation between model runs. It has been suggested that extreme precipitation might change differently to mean precipitation under climate change (Harrold et al., 2005) and the delta-change method does not capture this. Nevertheless, delta-change was used as this study concentrates on average runoff, which is strongly linked to overall catchment wetness, rather than floods, which are linked to a combination of catchment wetness and extreme precipitation. Hence, consideration of extreme precipitation events is less important in this study.

Statistical downscaling was applied to each of the GCM outputs and emission scenarios. Since the study area is covered by four GCM grid cells, the area weighted average precipitation, minimum and maximum temperatures of the GCM grid cells covering the study area were computed. The area-weighted average values were then statistically downscaled using the delta-change approach. Delta changes were calculated separately for each of the 12 months. For temperatures the delta changes were calculated using

$$\Delta_{\mathrm{T}}(j) = \overline{T}_{\mathrm{projn}}(j) - \overline{T}_{\mathrm{baseline}}(j), \tag{1}$$

where $\Delta_T(j)$ is the delta change in the 30-year mean monthly minimum or maximum temperature as simulated by the climate model for the future period and RCP of interest (2021–2050 or 2071–2100, RCP4.5 or RCP8.5), $\overline{T}_{\mathrm{projn}}(j)$, relative to the mean for the baseline period (1981–2010) climate model simulation, $\overline{T}_{\mathrm{baseline}}(j)$. j represents the month. $\Delta_T(j)$ is then applied to the daily baseline (1980–2010) observations, $T_{\mathrm{obs}}(j, i)$, for each pixel of the climate-gridded data (which is the same as in the VIC model grid pixels) to obtain the statistically downscaled minimum or maximum daily temperature, $T\Delta(j, i)$, for month j and day i.

$$T_{\Delta}(j, i) = T_{\mathrm{obs}}(j, i) + \Delta_{\mathrm{T}}(j) \tag{2}$$

For precipitation, the delta-changes value is computed as a proportional change rather than a shift:

$$\Delta_{\mathrm{P}}(j) = \frac{\overline{P}_{\mathrm{projn}}(j)}{\overline{P}_{\mathrm{baseline}}(j)}. \tag{3}$$

And is then applied to the observations using

$$P_{\Delta}(j, i) = P_{\mathrm{obs}}(j, i) \cdot \Delta_{\mathrm{P}}(j). \tag{4}$$

Here $\Delta_P(j)$ is the delta change in 30-year mean monthly precipitation as simulated by the climate model $\overline{P}_{\mathrm{projn}}(j)$ for two future periods (2021–2050 and 2071–2100) relative to the baseline simulation $\overline{P}_{\mathrm{baseline}}(j)$; $P_{\Delta}(j, i)$ is the statistically downscaled daily precipitation for the projected future climate change scenario for month j and day i, $P_{\mathrm{obs}}(j, i)$ is observed daily precipitation for the historical period (1981–2010) for month j and day i for each of the precipitation pixels of the gridded climate data. The delta-change approach maintains a similar (but shifted or scaled) spatial variation of temperature and precipitation as that in the historical, observed gridded data. The daily pattern of weather variation and the relationships between the various weather variables are also maintained. Because historical weather data provides the basis for the temporal patterns, the well-recognized issue of "GCM drizzle" is eliminated. The delta-change method also corrects for differences between the mean elevation of the four GCM grid cells by scaling up or down the historical spatial variation of temperature and precipitation across the catchment.

2.3 Relationship between LAI and climatic variables

Tesemma et al. (2014) showed that monthly LAI of each vegetation type was closely related to changes in moisture state (precipitation minus reference evapotranspiration) of 6-month moving averages for crop and pasture, and 9-month moving averages for trees. Differences in LAI response for the same change in moisture state among the three vegetation types were also observed as differences in model parameters of the LAI–Climate relationship. Tesemma et al. (2014) provide details on the derivation of the LAI–Climate relationship for the Goulburn–Broken catchment. The three LAI models developed for crop, pasture and tree are given below.

$$\mathrm{LAI} = \begin{cases} \dfrac{136.4836}{1+\exp\left(-\left(\frac{(P-\mathrm{PET})-159.4555}{42.5607}\right)\right)}, & \text{if} \quad \text{crop} \\ \dfrac{6.2495}{1+\exp\left(-\left(\frac{(P-\mathrm{PET})-43.6157}{62.8487}\right)\right)}, & \text{if} \quad \text{pasture} \\ \dfrac{4.2091}{1+\exp\left(-\left(\frac{(P-\mathrm{PET})+57.1849}{36.9481}\right)\right)}, & \text{if} \quad \text{tree,} \end{cases} \tag{5}$$

where LAI is the leaf area index of the cover type (tree/pasture/crop), P is the 6-month moving average of precipitation for crop and pasture, and the 9-month moving average for trees, and PET is the respective reference potential evapotranspiration.

The monthly LAI was then simulated for both historical and future climate scenarios using the LAI–Climate model (Eq. 5) driven with the appropriate climatic inputs. In this study monthly average PET (mm day^{-1}) was estimated using the standard FAO Penman–Monteith daily computations (Allen et al., 1998) and then aggregating to monthly values. PET for future climate scenarios was computed using the projected minimum and maximum temperatures,while incoming shortwave radiation and vapor pressure were derived from the daily temperature range using the algorithms of Kimball et al. (1997) and Thornton and Running (1999). The wind speed was kept the same as in the historical observations. A significant literature exists (see discussion in Supplementary Material of McMahon et al., 2015) around the issue of using temperature to drive future changes in PET. We acknowledge this assumption and note that it is likely to have limited impact on our runoff results in the mainly water-limited catchments modeled here. The historical or future precipitation was used in Eq. (5) according to the scenario being modeled. Potential LAI variations in the baseline years (1981–2010) and the two future periods (2021–2050 and 2071–2100), for each of the two future emission scenarios, were simulated using the downscaled outputs from the 38 CMIP5 runs of the 15 GCMs, as input into the LAI–Climate model (Eq. 5). The uncertainty ranges in modeled LAI that come from the difference in climate input were determined by using the downscaled 38 CMIP5 runs individually in Eq. (5).

2.4 Hydrological model and experimental design

In this study we used the three-layers VIC model (version 4.1.2g) to simulate streamflow. The VIC macroscale model is a spatially distributed conceptual hydrological model that balances both water and energy budgets over a grid cell. It simulates soil moisture, evapotranspiration, snowpack, runoff, baseflow and other hydrological properties at daily or sub-daily time steps by solving both the governing water and energy balance equations (Liang et al., 1996). VIC estimates infiltration and runoff using the variable infiltration curve that represents the sub-grid spatial variability in soil moisture capacity (Liang et al., 1994; Zhao et al., 1980) and Penman–Monteith for potential evapotranspiration computation. The ability of the model to incorporate the spatial representation of climate and inputs of soil, vegetation and other landscape properties make it applicable for climate and land use/land cover change impact studies. The VIC model has been widely used for a number of hydrological studies in different climatic zones across the globe (Zhao et al., 2012a, b; Cuo et al., 2013).

The seven most sensitive model parameters (b, Ds, Ws, Dsmax, d2, d3 and exp) in the VIC model (Demaria et al., 2007) were calibrated against observed streamflow from 13 selected sub-catchments with different climate and land cover composition that are representative of the main runoff generating regions of the Goulburn–Broken catchment. The model parameters were calibrated separately for each selected unregulated sub-catchment and applied uniformly within a sub-catchment (Fig. 1). The Multi-Objective Complex Evolution (MOCOM-UA) algorithm (Yapo et al., 1998) was used to calibrate the model. This algorithm was implemented on each of the selected catchments separately to calibrate the model against the observed runoff. The model was first calibrated for the entire period (1982–2012); then, using the calibrated parameters as initial guesses, the model was re-calibrated for the period 1982–1997 and evaluated for the period 1998–2012. During the calibration, VIC ran on a daily basis but the objective function was calculated on a monthly basis. Three criteria (objective functions) were used to evaluate the model's performance during calibration: the Nash–Sutcliffe efficiency (NSE) (Nash and Sutcliffe, 1970) between observed and simulated flow, the logarithm of Nash–Sutcliffe efficiency (logNSE) which penalizes errors at peak flow, and the percentage bias from the observed mean flow (PBIAS).

The VIC model was run at daily time steps with input data at 5 km × 5 km spatial grid resolution for 30 years, from January 1981 to December 2010, to produce the baseline run. Two model experiments were run: the first experiment considered the recent historical climate (Millennium Drought, 1997–2009) and LAI estimates using the simple LAI–Climate model against the relatively normal historical climate period (1983–1995). The second experiment considered the future climate from 38 CMIP5 runs and corresponding LAI derivatives for two periods (2021–2050 and 2071–2100) and two emission scenarios RCP4.5 and RCP8.5 with respect to the historical period (1981–2010). Both sets of simulations were performed over the 13 calibrated sub-catchments within the Goulburn–Broken catchment (Fig. 1b). A flowchart of the modeling method is given in Fig. 3.

To identify the effect on mean annual runoff of allowing LAI to respond to a changing climate, compared with LAI not responding, we used the following steps: (1) the calibrated model was forced with inputs of historical climate data and LAI data modeled from using the historical climate data (1981–2010) to establish baseline streamflow estimates; (2) the model was forced with projected future climate inputs and corresponding modeled LAI to produce projected streamflow for future scenarios; (3) the future climates were input along with the LAI data used in step 1 to produce projected streamflow that ignores projected LAI changes. The difference in mean annual runoff between steps 3 and 1 represents the climate effect (CC effect) on mean annual runoff of only precipitation and temperature. Whereas the difference in mean annual runoff between steps 2 and 1 represents the net effect (CC + LAI effect) on mean annual runoff of allowing LAI to respond to a changing climate in addition to the direct climate forcing (precipitation and temperature). The difference in mean annual runoff between steps 2 and 3 represents the component of the runoff response related to climate-induced changes in LAI. For the Millennium Drought (1997–2009), the above two changes in mean annual runoff were estimated in a similar fashion, taking the 1983–1995 time period as a relatively normal period. The percentage change of mean annual runoff against the historical mean annual runoff for climate change effect (Q_{clim}) (Eq. 6), climate change and LAI effect (Q_{net}) (Eq. 7), and the percentage of CC effect offset by LAI effect (Q_{lai}) (Eq. 8) were estimated as follows:

$$Q_{\text{clim}} = \left[\frac{100 \cdot \left(Q_{\text{historical LAI}}^{\text{future climate}} - Q_{\text{historical LAI}}^{\text{historical climate}} \right)}{Q_{\text{historical LAI}}^{\text{historical climate}}} \right], \tag{6}$$

$$Q_{\text{net}} = \left[\frac{100 \cdot \left(Q_{\text{future LAI}}^{\text{future climate}} - Q_{\text{historical LAI}}^{\text{historical climate}} \right)}{Q_{\text{historical LAI}}^{\text{historical climate}}} \right], \tag{7}$$

$$Q_{\text{lai}} = \left[\frac{100 \cdot (Q_{\text{clim}} - Q_{\text{net}})}{Q_{\text{net}}} \right]. \tag{8}$$

3 Results

This section provides results from the modeling exercises. First the model calibration and evaluation are discussed in Sect. 3.1. The change in climatic variables during (1) the recent observed prolonged drought and (2) future climate change projections for the study catchments are presented in Sect. 3.2. The impact on both LAI (Sect. 3.3) and catchment streamflow (Sect. 3.4) of changes in climate input during the Millennium Drought and future climate change projections are also provided. These results provide readers with a comparison of the anticipated future change in climate with the recently observed drought.

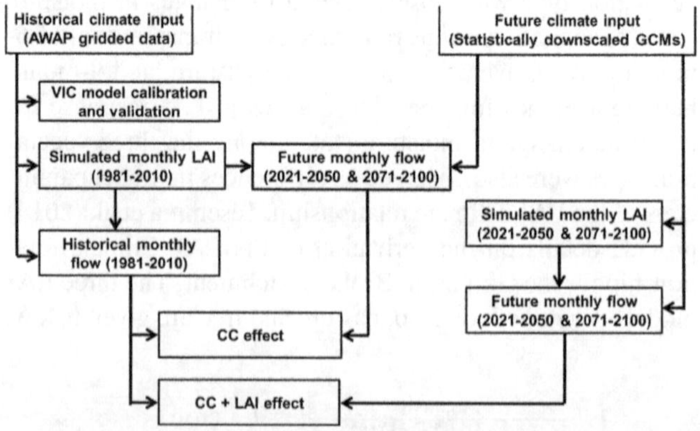

Figure 3. Flowchart showing the modeling experiments and calculation of effects: CC effect indicates the climate change effect of precipitation and temperature with unchanged LAI, CC + LAI effect indicates the climate change effect of precipitation, temperature and leaf area index.

3.1 Model calibration and evaluation results

The calibrated model parameters and model performance during calibration (1982–1997) and evaluation (1998–2012) periods for each sub-catchment are listed in Table 1. Most of the calibrated catchments have a NSE of more than 70 % during both calibration and evaluation periods (Table 1). In most of the selected sub-catchments the simulated runoff for both calibration and evaluation periods met the "satisfactory" criteria according to Moriasi et al. (2007), with a NSE > 50 % and the percentage absolute bias generally less than 25 % during calibration and evaluation periods. Although VIC captured the temporal variability of runoff well, there were some systematic biases in the runoff simulated. The model overestimates peak flow in a few cases and underestimates low flow in most of the sub-catchments. The sources of these biases need to be investigated in order to understand the performance of the model. To do this, the estimated monthly biases are plotted against the monthly climatic inputs: precipitation, temperature and LAI (not shown here). The calibrated sub-catchments showed no relationship between AWAP gridded climate data and simulated runoff biases. The biases are likely related to the model structure (Kalma et al., 1995) rather than the model inputs.

Table 1. Calibrated model parameters and model performance during calibration (1982–1997) and evaluation (1998–2012) periods.

ID	River and station name	Model parameters							Calibration (1982–1997)			Evaluation (1998–2012)		
		b	Ds	Ws	d2	d3	Dsmax	exp	Nash (%)	logNash (%)	Bias (%)	Nash (%)	logNash (%)	Bias (%)
1	Moonee Creek – Lima	0.149	0.598	0.170	1.99	0.47	0.13	2.98	82.7	80.2	2.2	86.1	78.1	8.0
2	Delatite River – Tonga Bridge	0.062	0.014	0.755	0.81	1.88	0.30	2.95	82.7	91.9	6.4	84.2	89.4	−5.4
3	Howqua River – Glen Esk	0.244	0.291	0.006	1.65	0.28	11.60	1.15	90.4	89.4	−2.5	89.3	90.3	−0.8
4	Goulburn River – Dohertys	0.206	0.891	0.035	1.43	0.45	22.01	1.42	95.9	91.0	2.2	92.4	90.8	−2.4
5	Big river – Jamieson	0.183	0.610	0.736	1.70	0.81	0.01	2.19	89.7	86.5	8.9	81.5	85.7	11.9
6	Rubicon River – Rubicon	0.216	0.059	0.200	0.52	1.77	19.29	1.28	93.8	94.9	−2.4	87.4	92.0	3.4
7	Acheron River – Taggerty	0.168	0.030	0.293	1.97	1.84	0.16	2.59	82.6	85.8	9.5	82.4	84.4	−2.4
8	Murrindindi River – above Colwells	0.130	0.801	0.297	1.97	1.89	1.11	2.67	68.9	62.8	14.6	79.7	84.7	3.9
9	Yea River – Devlins Bridge	0.072	0.428	0.646	1.93	1.27	0.05	2.99	79.8	78.3	26.4	68.0	69.3	34.1
10	King Parrot Creek – Flowerdale	0.071	0.041	0.665	0.71	1.95	0.73	2.87	61.5	66.1	45.8	73.0	62.6	41.1
11	Sugarloaf Creek – Ash Bridge	0.001	0.592	0.804	1.31	1.18	0.00	1.39	78.6	73.4	−3.5	59.0	40.0	127.5
12	Hughes Creek – Tarcombe road	0.043	0.215	0.514	1.04	1.88	0.07	3.20	82.5	89.3	9.2	62.7	58.9	39.2
13	Home Creek – Yarck	0.0004	0.415	0.524	0.66	1.91	0.01	2.97	81.7	87.1	−12.7	75.6	64.7	30.7

Table 2. Vegetation type distributions for each sub-catchment and changes in mean annual precipitation, temperature, LAI and streamflow during the Millennium Drought of 1997–2009 relative to 1983–1995.

Variables*	Catchments ID												
	1	2	3	4	5	6	7	8	9	10	11	12	13
Crop cover (%)	0.6	1.0	–	–	–	–	–	–	–	–	1.5	1.2	1.2
Pasture cover (%)	14.4	32.7	3.3	6.4	0.92	5.5	9.94	2.57	25.9	7.62	63.5	56.3	48.8
Tree cover (%)	85.0	66.3	96.7	93.6	99.1	94.5	90.1	97.4	74.1	92.4	35	42.6	50.1
P (%)	−23.2	−23.6	−21.1	−18.0	−17.9	−21.0	−20.1	−20.1	−19.4	−21.7	−19.5	−22.6	−24.1
T (°C)	0.2	0.3	0.3	0.4	0.4	0.3	0.3	0.2	0.3	0.2	0.3	0.3	0.3
LAI crop (%)	−44.2	−48.0	–	–	–	–	–	–	–	–	−38.1	−41.8	−41.4
LAI pasture (%)	−20.5	−21.6	−19.5	−16.9	−16.7	−18.7	−19.0	−19.1	−19.5	−19.7	−19.6	−20.2	−20.8
LAI tree (%)	−11.4	−10.3	−8.2	−6.6	−5.7	−5.9	−7.0	−6.3	−9.1	−9.2	−14.0	−12.5	−13.9
LAI total (%)	−12.9	−14.4	−8.6	−7.3	−5.8	−6.6	−8.2	−6.6	−11.8	−10.0	−17.9	−17.2	−17.6
Q_{clim} (%)	−49.3	−61.5	−43.7	−39.1	−42.9	−29.7	−44.0	−41.2	−55.2	−57.1	−66.3	−61.8	−57.9
Q_{net} (%)	−48.0	−59.7	−42.8	−38.3	−42.3	−29.3	−43.2	−40.6	−53.3	−55.2	−61.4	−56.1	−53.2
Q_{lai} (%)	2.6	3.0	2.1	2.1	1.5	1.3	1.9	1.4	3.6	3.4	8.0	10.2	8.9

* P is the change in mean annual precipitation in percentage (%), T is the change in mean annual temperature in degrees Celsius (°C), Q_{clim} indicates the climate effect on runoff, Q_{net} is the net effect of climate and LAI on runoff and Q_{lai} is the proportion of the climate effect (Q_{clim}) that is offset by the LAI effect.

3.2 Change in the climate variables from change in climate

3.2.1 Millennium Drought

The Millennium Drought brought a decline in the mean annual precipitation over the selected sub-catchments which ranged from 17.9 to 24.1 %, with a mean of 20.9 % when compared with the period 1983–1995. It also brought an increase in mean annual temperature which ranged from 0.2 to 0.4 °C, with an average of 0.3 °C as compared to the temperature in the period 1983–1995. All 13 study sub-catchments experienced a similar change in both precipitation and temperature (Table 2).

3.2.2 Future climate

Averaged over all 38 CMIP5 runs, the mean annual precipitation in 2021–2050 over the selected sub-catchments is projected to decline by 2.9 and 3.7 %, relative to the historical period 1981–2010 under the RCP4.5 and RCP8.5 scenarios respectively. By the end of the century (2071–2100) mean annual precipitation is projected to decline by 5 and 5.2 % under the RCP4.5 and RCP8.5 scenarios respectively (Table 3). The mean annual temperature is also projected to increase in both future periods and emission scenarios (Table 3).

Most precipitation projections showed a shift towards drier climates in all seasons except summer in both emission scenarios and periods. The variability in projected mean monthly precipitation among climate models indicates great uncertainty between GCMs (Fig. 4a–d). The mean monthly temperature of all climate models clearly deviated from the baseline period (1981–2010), underlining the consistent change signal between GCMs (Fig. 4e–h). The median of the 38 CMIP5 mean monthly precipitation data over the Goulburn–Broken catchment in the RCP4.5 emission scenario showed declines in most of the months. The decreases were up to 6 % in 2021–2050 (Fig. 4a) and up to 11 % in

Table 3. Impacts on mean annual precipitation, temperature, LAI and streamflow of projected climate change averaged over 38 CMIP5 runs relative to 1981–2010.

Periods	Variables*	Catchment ID												
		1	2	3	4	5	6	7	8	9	10	11	12	13
2021–2050 RCP4.5	P (%)	−2.9	−2.9	−2.9	−2.9	−2.9	−2.9	−2.9	−2.9	−2.9	−2.9	−2.9	−2.9	−2.9
	T (°C)	0.9	0.9	0.9	0.9	0.9	0.9	0.9	0.9	0.9	0.9	0.9	0.9	0.9
	LAI crop (%)	−12.9	−13.0	–	–	–	–	–	–	–	–	−12.9	−13.0	−12.8
	LAI pasture (%)	−5.9	−5.6	−5.4	−5.6	−5.3	−4.8	−5.4	−5.4	−6.1	−6.1	−6.7	−6.3	−6.3
	LAI tree (%)	−3.9	−2.9	−2.5	−2.4	−2.0	−1.7	−2.1	−1.9	−3.0	−3.0	−5.4	−4.6	−4.8
	LAI total (%)	−4.2	−3.9	−2.6	−2.6	−2.0	−1.8	−2.5	−1.9	−3.8	−3.2	−6.3	−5.6	−5.7
	Q_{clim} (%)	−12.3	−17.6	−11.4	−11.5	−13.5	−6.8	−12.4	−12.6	−17.4	−18.4	−20.3	−18.9	−14.2
	Q_{net} (%)	−11.4	−16.3	−10.9	−11.1	−13.2	−6.6	−11.9	−12.2	−15.8	−17.0	−16.3	−14.8	−11.7
	Q_{lai} (%)	7.9	8.0	4.6	3.6	2.3	3.0	4.2	3.3	10.1	8.2	24.5	27.7	21.4
2021–2050 RCP8.5	P (%)	−3.7	−3.7	−3.7	−3.7	−3.7	−3.7	−3.7	−3.7	−3.7	−3.7	−3.7	−3.7	−3.7
	T (°C)	1.2	1.2	1.2	1.2	1.2	1.2	1.2	1.2	1.2	1.2	1.2	1.2	1.2
	LAI crop (%)	−15.7	−15.7	–	–	–	–	–	–	–	–	−15.7	−15.7	−15.5
	LAI pasture (%)	−7.2	−6.9	−6.7	−6.8	−6.5	−5.9	−6.6	−6.6	−7.4	−7.5	−8.1	−7.7	−7.7
	LAI tree (%)	−4.8	−3.7	−3.1	−3.0	−2.5	−2.1	−2.7	−2.3	−3.7	−3.7	−6.6	−5.6	−5.9
	LAI total (%)	−5.2	−4.8	−3.3	−3.2	−2.5	−2.3	−3.1	−2.4	−4.7	−4.0	−7.7	−6.9	−6.9
	Q_{clim} (%)	−14.6	−20.7	−13.7	−13.8	−16.3	−8.3	−14.8	−15.0	−20.1	−21.3	−23.3	−21.4	−16.1
	Q_{net} (%)	−13.6	−19.2	−13.2	−13.3	−15.8	−8.1	−14.3	−14.5	−18.3	−19.7	−19.0	−17.0	−13.4
	Q_{lai} (%)	7.4	7.8	3.8	3.8	3.2	2.5	3.5	3.4	9.8	8.1	22.6	25.9	20.1
2071–2100 RCP4.5	P (%)	−5.0	−5.0	−5.0	−5.0	−5.0	−5.0	−5.0	−5.0	−5.0	−5.0	−5.0	−5.0	−5.0
	T (°C)	1.6	1.6	1.6	1.6	1.6	1.6	1.6	1.6	1.6	1.6	1.6	1.6	1.6
	LAI crop (%)	−21.1	−21.3	–	–	–	–	–	–	–	–	−20.8	−21.0	−20.7
	LAI pasture (%)	−9.8	−9.5	−9.2	−9.4	−9.0	−8.2	−9.2	−9.2	−10.2	−10.3	−11.0	−10.4	−10.5
	LAI tree (%)	−6.6	−5.1	−4.4	−4.2	−3.5	−3.0	−3.9	−3.4	−5.3	−5.3	−9.2	−7.8	−8.2
	LAI total (%)	−7.2	−6.7	−4.6	−4.5	−3.6	−3.3	−4.4	−3.5	−6.6	−5.7	−10.5	−9.4	−9.5
	Q_{clim} (%)	−19.7	−27.5	−18.6	−18.8	−22.1	−11.5	−20.3	−20.7	−26.9	−28.1	−30.1	−27.7	−21.7
	Q_{net} (%)	−18.3	−25.7	−17.9	−18.1	−21.6	−11.2	−19.6	−20.1	−24.7	−26.2	−25.2	−22.5	−18.6
	Q_{lai} (%)	7.7	7.0	3.9	3.9	2.3	2.7	3.6	3.0	8.9	7.3	19.4	23.1	16.7
2071–2100 RCP8.5	P (%)	−5.2	−5.2	−5.2	−5.2	−5.2	−5.2	−5.2	−5.2	−5.2	−5.2	−5.2	−5.2	−5.2
	T (°C)	2.5	2.5	2.5	2.5	2.5	2.5	2.5	2.5	2.5	2.5	2.5	2.5	2.5
	LAI crop (%)	−28.3	−28.3	–	–	–	–	–	–	–	–	−28.5	−28.5	−28.1
	LAI pasture (%)	−13.6	−13	−12.5	−12.9	−12.2	−11.1	−12.5	−12.5	−14	−14.1	−15.4	−14.6	−14.7
	LAI tree (%)	−9.5	−7.4	−6.3	−6.0	−5.1	−4.3	−5.5	−4.8	−7.6	−7.6	−13.2	−11.2	−11.8
	LAI total (%)	−10.2	−9.4	−6.5	−6.5	−5.2	−4.7	−6.2	−5.0	−9.2	−8.1	−14.9	−13.3	−13.4
	Q_{clim} (%)	−24.0	−33.5	−23.9	−24.2	−27.4	−14.5	−25.0	−25.6	−32.0	−33.0	−35.1	−32.8	−25.3
	Q_{net} (%)	−22.3	−31.3	−23.0	−23.3	−26.7	−14.1	−24.0	−24.8	−29.4	−30.8	−29.2	−26.4	−21.7
	Q_{lai} (%)	7.6	7.0	3.9	3.9	2.6	2.8	4.2	3.2	8.8	7.1	20.2	24.2	16.6

* P is the change in mean annual precipitation in percentage (%), T is the change in mean annual temperature in degrees Celsius (°C), Q_{clim} indicates the climate effect on runoff, Q_{net} is the net effect of climate and LAI on runoff and Q_{lai} is the proportion of the climate effect (Q_{clim}) that is offset by the LAI effect.

2071–2100 (Fig. 4c). Similarly, under the RCP8.5 emission scenario the median monthly precipitation, other than in January and February for both periods, showed decreases of up to 7 % in 2021–2050 (Fig. 4b) and of up to 18 % in 2071–2100 (Fig. 4d). The simulations for January and February showed median increases of up to 4 and 5 % respectively in 2071–2100 from the historical baseline. Some climate models projected very wet future climates while others projected relatively dry climates. There are relatively high uncertainties in the projected mean monthly precipitation results in summer when compared with the mean monthly precipitation in winter among the climates models.

In contrast to precipitation, the projected mean monthly temperatures from all CMIP5 runs showed increases; the median of the mean monthly temperatures of all 38 CMIP5 runs increased by about 0.8 °C in winter and 1 °C in summer in 2021–2050 (Fig. 4e) and by about 1.3 °C in winter and 1.8 °C in summer in 2071–2100 (Fig. 4g) under the RCP4.5 scenario. Under the RCP8.5 emission scenario the temperatures increased by 1 °C in winter and by 1.4 °C in summer during 2021–2050 (Fig. 4f) and by 2 and 3 °C in winter and summer respectively by the end of the 21st century (Fig. 4h). After precipitation the second variable that drives water availability is potential evapotranspiration. Here PET is expected to increase among all CMIP5 runs as it is being driven solely by changes in temperature, given that actual vapor pressure and solar radiation were also simulated as a function of temperature. In the near-future period (2021–2050) the median of all CMIP5 mean monthly PET projections increase by 5–13 % in both emission scenarios, with the largest change in winter and the smallest in summer. In the future period of 2071–2100, the mean monthly PET increased by 7 % in sum-

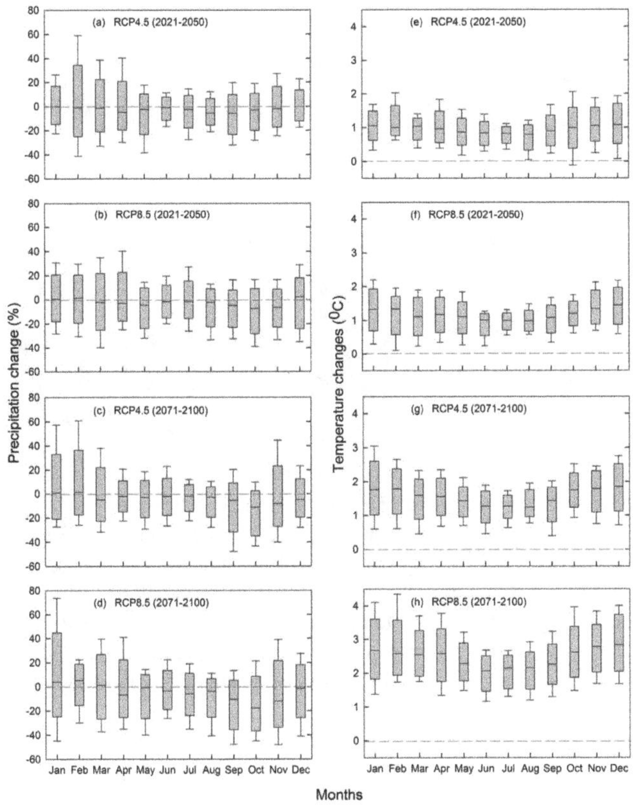

Figure 4. Box plots of percentage changes in the mean monthly precipitation **(a, b, c, d)** and changes in mean monthly temperatures **(e, f, g, h)** in the Goulburn–Broken catchment for the future periods 2021–2050 and 2071–2100 for the 38 CMIP5 runs of climate projections. Changes are relative to the historical (1981–2010) mean monthly precipitation and temperatures. The lower boundary of the box indicates the 25th percentile, a line within the box marks the median, the upper boundary of the box indicates the 75th percentile and the whiskers are delimited by the maximum and minimum.

mer and 25 % in winter under the RCP4.5 emission scenario, and by 10 % in summer and 28 % in winter under the RCP8.5 emission scenario.

3.3 Impact on LAI from change in climate

3.3.1 Millennium Drought

The effects of the Millennium Drought (1997–2009) on modeled crop LAI were very severe with reductions in mean annual LAI between sub-catchments of 38.1–48.0 %, with a mean of 42.7 % (Table 2). The reduction in LAI of pasture was between 16.7 and 21.6 % across the 13 selected sub-catchments, with a spatial average of 19.4 % (Table 2). The LAI of trees responded less than crop and pasture, and reductions were in the range of 5.7–14.0 %, with a spatial mean of 9.2 % (Table 2). A significant reduction in each cover type also brought an overall decline in the areal-weighted sum of all land cover type LAIs in the selected sub-catchments, which ranged from 5.8 to 17.9 % (Table 2), which is similar to the reduction for trees, where trees are the dominant land cover type.

3.3.2 Future climate

The changes in mean monthly LAI of crop, pasture and trees averaged over the whole Goulburn–Broken catchment under future climates vary between the CMIP5 runs and global warming scenarios. Averaged over all 38 CMIP5 runs, the near-future (2021–2050) results for the study sub-catchments showed that the mean annual LAI of cropland, pasture and trees declined by up to 13, 6.7, and 5.4 % under the RCP4.5 scenario, and by up to 16, 8, and 6.6 % under the RCP8.5 scenario (Table 3). A further reduction in the mean annual LAI of each land cover was simulated by the end of the 21st century for both emission scenarios (Table 3).

The effect of projected climate change on monthly total LAI (area-weighted sum of all land cover types LAI) for the study catchment is given in (Fig. 5). The median of the 38 CMIP5-runs-simulated mean-monthly LAI showed declines in all three land cover types. Despite similar percentage changes in mean monthly precipitation and temperature forcing, the mean monthly total LAI across the catchment shows the largest decline in autumn and the smallest decline in spring during both future periods and scenarios. This difference reflects the seasonality of moisture availability influencing plant growth. Based on the median of the 38 CMIP5 runs, the predicted decline in the mean monthly LAI for crop, pasture and trees was 18.1, 10.3, and 7.9 % respectively in the period 2021–2050 (Fig. 5a, e, i) and 27.7, 16.6, and 12.8 % respectively in the period 2071–2100 under RCP4.5 (Fig. 5c, g, k). Larger reductions were simulated under the RCP8.5 emission scenario with 21.4, 12.7, and 9.5 % in the period 2021–2050 (Fig. 5b, f, j) and 36.5, 22.5, and 17.9 % respectively for crop, pasture and trees in the period 2071–2100 (Fig. 5d, h, l).

3.4 Impacts on runoff from change in climate

3.4.1 Millennium Drought

The impact of the Millennium Drought on streamflow due to changes in precipitation and temperature alone and changes in precipitation and temperature and modeled LAI were simulated using the VIC model. The simulated reductions in mean annual streamflow during the Millennium Drought (1997–2009) as compared with the relatively normal period (1983–1995) across the selected sub-catchments due to the change in climate alone ranged from 29.7 to 66.3 %, with a mean of 50 % (Table 2). The reductions in LAI resulting from the decline in precipitation and increase in temperature increased mean annual streamflow by between 1.3 and 10.2 % relative to the direct climate effect above (Table 2, Fig. 6).

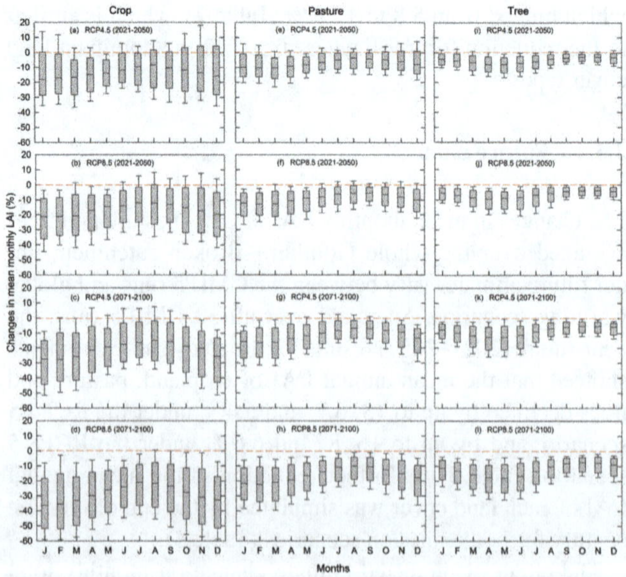

Figure 5. Box plots of changes in mean monthly LAI derived from the 38 CMIP5 runs for climate projections during 2021–2050 and 2071–2100 under RCP4.5 and RCP8.5 scenarios for crop **(a, b, c, d)**, pasture **(e, f, g, h)** and trees **(i, j, k, l)** in the Goulburn–Broken catchment. Changes are relative to LAI calculated using climate time series for the 1981–2010 baseline. The lower boundary of the box indicates the 25th percentile, a line within the box marks the median, the upper boundary of the box indicates the 75th percentile and the whiskers are delimited by the maximum and minimum.

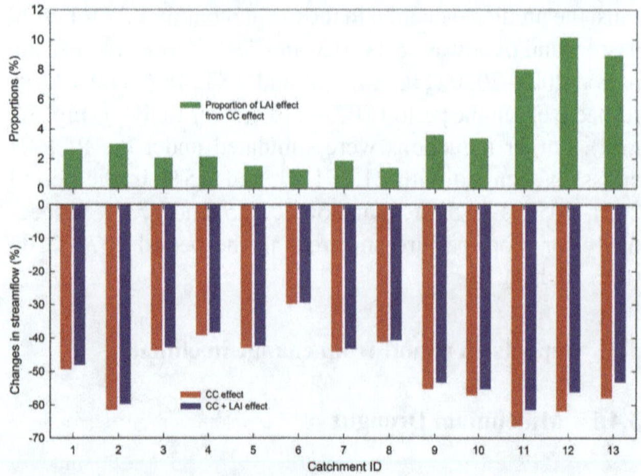

Figure 6. Impacts on catchment mean annual streamflow of the Millennium Drought (1997–2009) relative to the period 1983–1995. CC effect indicates precipitation and temperature effect with unchanged LAI; CC + LAI effect indicates precipitation, temperature and LAI effect. The proportional LAI effect indicates the LAI effect as a percentage of the CC effect.

3.4.2 Future climate

The average of the 38 CMIP5 runs under the RCP4.5 scenario produced declines in mean annual runoff due to the change in precipitation and temperature alone (Q_{clim}), which ranged from 6.8 to 20.3 % in the period 2021–2050 and from 11.5 to 30.1 % for the period 2071–2100 (Table 3 and Fig. 7). For the higher emission scenario (RCP8.5), the reductions were a little larger ranging from 8.3 to 23.3 % in 2021–2050 and from 14.5 to 35.1 % by the end the 21st century (Table 3, Fig. 6). The reductions in runoff due to climate are offset through the LAI effect (Q_{lai}) that ranged from 2.3 to 27.7 % and from 2.3 to 23.1 % in the near- and far-future periods respectively under the RCP4.5 emission scenario. Similar offsets of 2.5–25.9 % and 2.6–24.2 % in the near- and far-future periods respectively were also found under the RCP8.5 emission scenario (Table 3, Fig. 7).

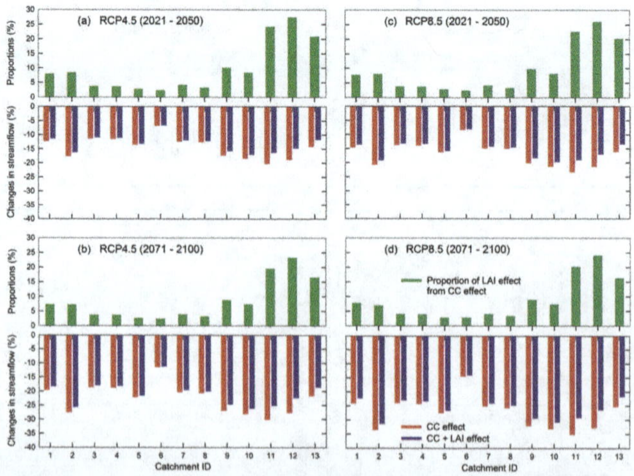

Figure 7. Impact on catchment mean annual streamflow average over the 38 CMIP5 runs of projected climate change for the future periods 2021–2050 and 2071–2100 under RCP4.5 **(a, b)** and RCP8.5 **(c, d)**, relative to the 1981–2010 base period. CC effect indicates precipitation and temperature effect with unchanged LAI; CC + LAI effect indicates precipitation, temperature and LAI effect. The proportional LAI effect indicates the LAI effect as a percentage of the CC effect.

The differences between GCMs in terms of the net climate change impacts (CC + LAI) on mean annual runoff and the LAI contribution to that effect are shown in Figs. 8 and 9 respectively. While large uncertainty exists among the 38 CMIP5 runs, the median between the models showed declines in the net climate change (CC + LAI) projections of mean annual runoff in all catchments (Fig. 8). The median decline in the mean annual runoff due to the net climate change impact was 15.3 and 26.7 % in 2021–2050 and 2071–2100 respectively, under RCP4.5. A larger decline of 21.6 and 31.8 % in 2021–2050 and 2071–2100 respectively occurred under RCP8.5 (Fig. 8). The simulated LAI effects of the climate change showed smaller variation between GCMs than the net climate change (CC + LAI) effect on mean annual runoff. The LAI effect works to offset the reduction in mean annual runoff resulting from lower precipitation and higher temperature. Figure 9 shows the magnitude of the

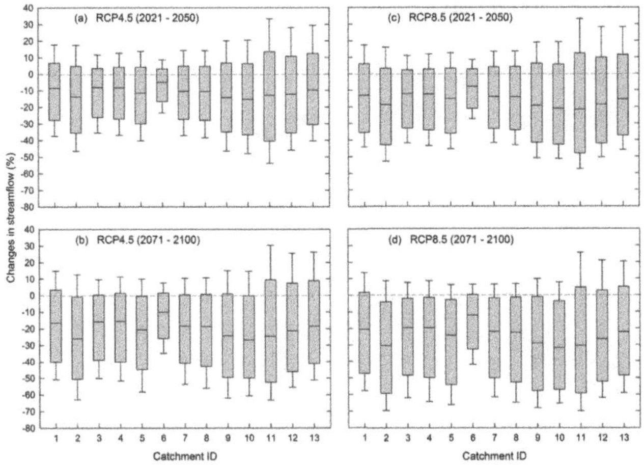

Figure 8. Box plots of the net climate change (CC + LAI) effect on mean annual runoff during 2021–2050 and 2071–2100 under RCP4.5 **(a, b)** and RCP8.5 **(c, d)** emission scenarios from each of the 38 CMIP5 runs. Changes are relative to the historical (1981–2010) period. The lower boundary of the box indicates the 25th percentile, a line within the box marks the median, the upper boundary of the box indicates the 75th percentile and the whiskers are delimited by the maximum and minimum.

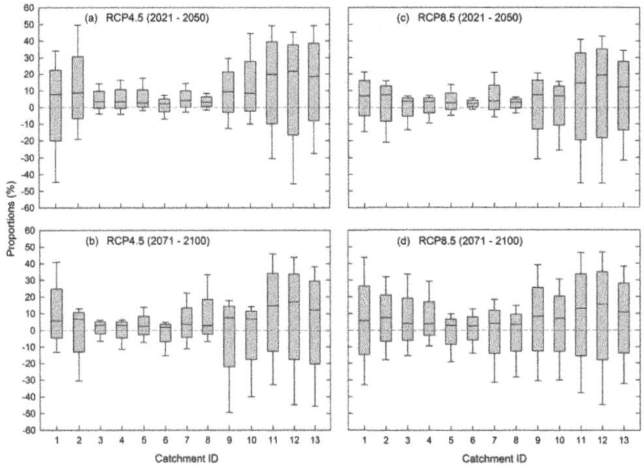

Figure 9. Box plots of contributions of LAI to the climate change effect on mean annual runoff for future (2021–2050, 2071–2100) climate forcings under RCP4.5 **(a, b)** and RCP8.5 **(c, d)** emission scenarios from each of the 38 CMIP5 runs as compared to the historical (1981–2010) period. The LAI effect is normalized by the effect of precipitation and temperature with unchanged LAI (i.e., CC effect) and expressed as a percentage. The lower boundary of the box indicates the 25th percentile, a line within the box marks the median, the upper boundary of the box indicates the 75th percentile and the whiskers are delimited by the maximum and minimum.

LAI effect as a percentage of the magnitude of direct climate change effect (noting they work in opposite directions). The median of this effect across the 38 CMIP5 runs increased up to 20 %, depending on the month. The simulated LAI effect on mean annual runoff showed less variation between GCMs than the net climate change (CC + LAI) effect on mean annual runoff.

The direct climate change (CC) effect, the LAI effect of climate change and the net climate change (CC + LAI) effect on the mean monthly runoff for the selected sub-catchments are given: catchment 6 (Fig. 10a, d, g, j), catchment 10 (Fig. 10b, e, h , k), and catchment 11 (Fig. 10c, f, i , l). Catchments 6 and 10 are located in a high annual precipitation zone with trees as the dominant vegetation cover, whereas catchment 11 is covered mostly with pasture and has relatively lower annual precipitation than catchments 6 and 10. Depending on the month, for the 38 CMIP5 runs in 2021–2050 the median reduction in mean monthly runoff (Q_{net}) was up to 10, 24, and 34 % for catchment 6, 10, and 11 respectively for both the RCP4.5 and RCP8.5 scenarios (Fig. 10). Further reductions projected by the end of the 21st century were up to 17, 37, and 52 % for catchments 6, 10, and 11 respectively under both scenarios (Fig. 10). Catchment 6 showed the lowest seasonality in the climate change effects for both emission scenarios and the LAI-related effects of climate change also showed the smallest seasonal variation. Catchment 11 runoff was the most impacted by projected climate changes and had the greatest benefit from LAI effects of climate change under both emission scenarios and future periods. The seasonal pattern of the LAI effect of climate change is similar under both RCP scenarios. The magnitude of this effect is relatively higher for drier projected future climates.

4 Discussion and conclusion

This study investigated the importance of incorporating the relationship between changing climate, in terms of precipitation and temperature, and vegetation LAI into a hydrological model to estimate changes in mean monthly and mean annual runoff under changing climatic conditions in the Goulburn–Broken catchment, southeastern Australia. A combination of VIC hydrological simulations with a simple model that relates climatic fluctuations with LAI for three different vegetation types revealed that 21st century climate change impacts on LAI significantly influence the projected runoff in the study sub-catchments. LAIs of forest, pasture and crop were predicted to decline in the 21st century due to reductions in precipitation and increases in temperature.

Reduced LAI in response to a drier and warmer climate would reduce transpiration from vegetation and evaporative losses from canopy interception, which leaves the soil relatively wetter than if the LAI response to climate was not included. This is important for the runoff generation process as it promotes saturation excess runoff and sub-surface flow, which are the dominant cause of runoff generation in the study region (Western et al., 1999). Previous studies in the region (Chiew et al., 2009, 2011; Teng et al., 2012a, b) concluded that runoff would decrease due to increases in

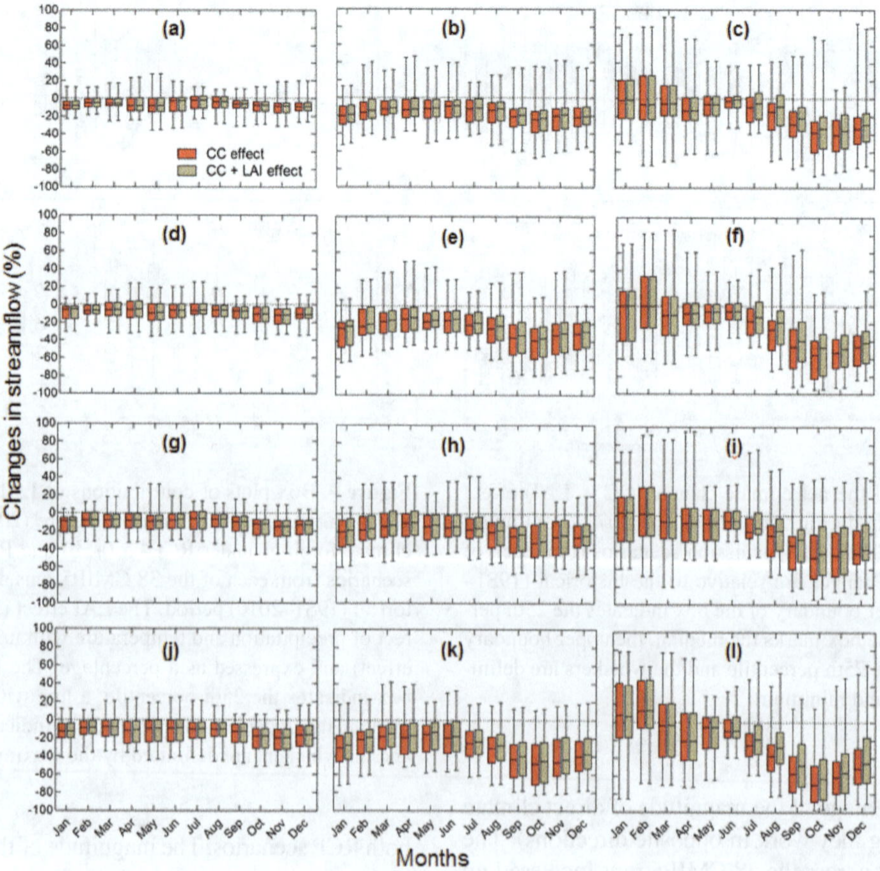

Figure 10. Box plots of impacts on mean monthly streamflow from 38 CMIP5 runs of catchment 6 (**a**, **d**, **g** and **j**), catchment 10 (**b**, **e**, **h** and **k**), and catchment 11 (**c**, **f**, **i** and **l**) of projected climate change for future periods 2021–2050 and 2071–2100 under RCP4.5 and RCP8.5 respectively, relative to the 1981–2010 base period. CC effect indicates precipitation and temperature effect with unchanged LAI; CC + LAI effect indicates precipitation, temperature and LAI effect. The lower boundary of the box indicates the 25th percentile, a line within the box marks the median, the upper boundary of the box indicates the 75th percentile and the whiskers are delimited by the maximum and minimum.

evaporative demand and decreases in precipitation as a result of ongoing warming in the 21st century. However, the relationship between LAI and climatic fluctuations was not taken into account in their modeling experiments. Therefore, in these studies the LAI effect is ignored and there is consequent overestimation of the runoff decline in the range of 2.3–27.7 % (Figs. 6, 7).

Projections of climate-induced vegetation dynamics and their hydrological impacts are influenced by various uncertainties that arise from using downscaled GCM outputs as inputs to the hydrological model. These include large uncertainties in projections for precipitation from the various CMIP5 simulations (Teng et al., 2012b). In addition, the method used to downscale the GCM outputs really only captures changes in the mean; however, any change in variability, which could have an effect on the projected future runoff, is ignored. The ensemble of 38 CMIP5 simulations from 15 GCMs was used to determine the range of uncertainty between GCMs. The results showed that the range of future climate projections from the various GCMs is wide, one climate model could project a very wet future climate while another a relatively dry climate. This suggests future analyses in other catchments should apply downscaled climate change scenarios from several CMIP5 runs from a range of GCM models to the study area in order to get a sense of the possible range of climate change impact on both LAI and streamflow.

The results of this study illustrate that reduction of future precipitation and increase in mean temperature lead to reduction of runoff in a general sense. However, if the hydrological model incorporated dynamic LAI information, as a function of changing climate, it would reduce the impact on runoff that comes from the climate alone. Reduction of LAI due to reduction of precipitation and increase in temperature decreases the evapotranspiration from vegetation and leaves the soil relatively wetter than if climate-induced changes in LAI were not represented in the modeling. The higher catchment moisture contents slightly increased runoff and partially offset the reduction in runoff due to changes in climate.

In interpreting the results presented here it is important to examine the assumptions that were made and the extent to which the results are dependent on those assumptions. Runoff processes can also be triggered by other precipitation characteristics (intensity, duration, inter-storm duration) which have not been considered in this study. If inter-storm durations are expected to increase, this will alter the hydrological fluxes even if the mean precipitation is maintained. However, the Climate–LAI model used in the study area (Tesemma et al., 2014) is related mainly to precipitation and potential evapotranspiration during the previous 6–9 months. This limits the impact of changes in extreme precipitation characteristics in terms of modeling the Climate–LAI relationship. In order to satisfy the aim of this paper, which is to assess the impact of allowing LAI to respond to a changing climate, so long as the precipitation series is consistent between the runs with and without LAI responding to climate, we can assess the importance of the change in LAI on runoff simulation. Hence, in this study, consideration of changing extreme precipitation events is less important, although it would be important for studies with the objective of predicting future floods or reservoir management.

Another assumption of this study was that the impact on runoff of rising atmospheric CO_2 concentrations, via changes in LAI and stomatal conductance, is small relative to the moisture availability effects. Therefore, here we assumed LAI responded only to precipitation and PET changes, not changes in CO_2. Changes in atmospheric CO_2 concentrations could affect vegetation through increasing LAI and narrowing stomata (Ainsworth and Rogers, 2007; Ewert, 2004; Warren et al., 2011). However, increased LAI may be limited by the availability of nutrients, particularly nitrogen (Fernández-Martínez et al., 2014; Körner, 2006). Most of the results on this effect are derived from point experiments which could not be extrapolated to the catchment scale where there is a complex interaction between soil, vegetation and climate. Increasing atmospheric CO_2 could also have two other effects on vegetation dynamics. First, biomass allocation may shift towards more above-ground plant structure (Obrist and Arnone, 2003), which implies more canopy leaf than active rooting area. This change could influence the water balance in either direction by increasing evapotranspiration due to interception losses or by decreasing evapotranspiration through limiting plant water uptake. Second, rising atmospheric CO_2 may favor C_3 species over C_4 species, which could lead to more woody plants compared to some grass species (Yu et al., 2014). This could influence the water balance by increasing evapotranspiration and decreasing runoff. In addition, at the canopy scale, the evapotranspiration effect of increased LAI can be masked by shading among leaves, soil cover and raised canopy humidity (Hikosaka et al., 2005; Bunce, 2004). A study that considered both effects suggested that the fertilization effect of rising CO_2 is larger than the stomatal pore reduction effect, and the net effect is decreases in runoff (Piao et al., 2007). These two effects of increasing atmospheric CO_2 concentrations on vegetation work in opposite directions from a water balance perspective and may offset each other if they are close in magnitude (Gerten et al., 2008). In southeast Australia, it is known that vegetation growth is highly controlled by precipitation (water supply), and is less controlled by temperature and radiation (Nemani et al., 2003). Hence, most vegetation dynamics can be explained by variation in climate, which formed the basis of the LAI–Climate model developed in Tesemma et al. (2014). We acknowledge changing CO_2 levels could influence vegetation growth and water use efficiency and hence runoff, but we expect the impact on runoff to be smaller (Huntington, 2008; Uddling et al., 2008) than that due to changes in moisture state. Hence, exclusion of the fertilization and stomata suppression effects of rising atmospheric CO_2 on vegetation may not change the results significantly. However, the impact on runoff of CO_2 fertilization at the catchment scale remains an important area of ongoing research.

A further assumption was that any effect of climate change on the spatial distribution of plant functional types (PFTs) was ignored. That is, the same spatial distribution of vegetation was used but with changed LAI. We acknowledge that changing climate (i.e increase in temperature) may shift the spatial distribution of PFTs, which has been reported in the Mediterranean climate region (e.g., Lenihan et al., 2003; Crimmins et al., 2011). However, in our study area PFTs are largely determined by historical land use change (human activities), such as forest clearing for agriculture, rather than natural responses of vegetation to changed climatic conditions. Therefore, future changes in the spatial distribution of agricultural crops and pastures are difficult to project as they are not solely due to climate changes. In the forested areas, it is likely that issues that change water use such as changes in fire regime (Heath et al., 2014) and forest age (Cornish and Vertessy, 2001) would dominate over differences between species. *Eucalyptus* species already occupy high-altitude areas of the study catchment, which leaves little room for PFT changes due to upslope migration in a warming climate. Most over-story trees in our study area are *Eucalypts* and while some movement of boundaries between dominant species may be expected, water use characteristics are likely to be relatively similar, and there is insufficient information to represent species-specific details of either migration or water use. Including these effects in the model may improve the results, but there is insufficient understanding of the granularity required to do so at present.

In summary, in this paper we use the VIC hydrological model to assess the impact on mean annual streamflow of ignoring climate-induced changes in LAI for two changing climatic situations: (1) the recently observed Millennium Drought and (2) for downscaled projected future climate change scenarios from 38 CMIP5 runs in the Goulburn–Broken catchment, Australia. In the Millennium Drought (1997–2009), not modeling the response of LAI to changing climatic variables led to further reduction in mean annual runoff, relative to the pre-drought period (1983–1995), of between 1.3 and 10.2 % relative to modeling the dynamic response of LAI to decreased precipitation and increased temperature (Table 2, Fig. 6). For projected climate change under the RCP4.5 emission scenario, ignoring the LAI response to changing climate could lead to a further reduction in mean annual runoff of between 2.3 and 27.7 %, relative to the baseline period (1981–2010), in the near-term (2021–2050) and 2.3–23.1 % later in the century (2071–2100), relative to modeling the dynamic response of LAI to precipitation and temperature changes. Similar results (near-term 2.5–25.9 % and end-of-century 2.6–24.2 %) were found for climate change under the RCP8.5 emission scenario (Table 3, Fig. 7). Due to the strong relationship between climatic variation and LAI, the Climate–LAI interaction should be included in hydrological models for improved climate change impact assessments and modeling under changing climatic conditions, particularly in arid and semi-arid regions where vegetation is strongly influenced by climate.

Acknowledgements. This study was funded by the Australian Research Council (ARC) (project no: ARC LP100100546, ARC FT130100274 and ARC FT120100130), the Natural Science Foundation of China (project no: 91125007) and the Commonwealth of Australia under the Australia China Science and Research Fund (project no: ACSRF800). We would like to thank the University of Melbourne for providing a scholarship to the first author. We thank editor Ciaran Harman and two anonymous reviewers for comments that improved this manuscript.

Edited by: C. Harman

References

Ainsworth, E. A. and Rogers, A.: The response of photosynthesis and stomatal conductance to rising [CO_2]: mechanisms and environmental interactions, Plant Cell Environ., 30, 258–270, doi:10.1111/j.1365-3040.2007.01641.x, 2007.

Allen, R. G., Pereira, L. S., Raes, D., and Smith, M.: Crop evapotranspiration Guidelines for computing crop water requirements, FAO Irrigation and Drainage Paper 56, Food and Agriculture Organization of the United Nations, 1998.

Bunce, J. A.: Carbon dioxide effects on stomatal responses to the environment and water use by crops under field conditions, Oecologia, 140, 1–10, doi:10.1007/s00442-003-1401-6, 2004.

Cai, W. and Cowan, T.: Evidence of impacts from rising temperature on inflows to the Murray–Darling Basin, Geophys. Res. Lett., 35, L07701, doi:10.1029/2008GL033390, 2008.

Chiew, F. H. S., Teng, J., Vaze, J., Post, D. A., Perraud, J. M., Kirono, D. G. C., and Viney, N. R.: Estimating climate change impact on runoff across southeast Australia: method, results, and implications of the modeling method, Water Resour. Res., 45, W10414, doi:10.1029/2008WR007338, 2009.

Chiew, F. H. S., Young, W. J., Cai, W., and Teng, J.: Current drought and future hydroclimate projections in southeast Australia and implications for water resources management, Stoch. Environ. Res. Risk A., 25, 601–612, doi:10.1007/s00477-010-0424-x, 2011.

Chiew, F. H. S., Potter, N. J., Vaze, J., Petheram, C., Zhang, L., Teng, J., and Post, D. A.: Observed hydrologic non-stationarity in far south-eastern Australia: implications for modelling and prediction, Stoch. Environ. Res. Risk A., 28, 3–15, 2014.

Cornish, P. M. and Vertessy, R. A.: Forest age-induced changes in evapotranspiration and water yield in a eucalypt forest, J. Hydrol., 242, 43–63, doi:10.1016/S0022-1694(00)00384-X, 2001.

Crimmins, S. M., Dobrowski, S. Z., Greenberg, J. A., Abatzoglou, J. T., and Mynsberge, A. R.: Changes in climatic water balance drive downhill shifts in plant species' optimum elevations, Science, 331, 324–327, 2011.

Cuo, L., Zhang, Y., Gao, Y., Hao, Z., and Cairang, L.: The impacts of climate change and land cover/use transition on the hydrology in the upper Yellow River Basin, China, J. Hydrol., 502, 37–52, doi:10.1016/j.jhydrol.2013.08.003, 2013.

Demaria, E. M., Nijssen, B., and Wagener, T.: Monte Carlo sensitivity analysis of land surface parameters using the Variable Infiltration Capacity model, J. Geophys. Res., 112, D11113, doi:10.1029/2006JD007534, 2007.

Ellis, T. W. and Hatton, T. J.: Relating leaf area index of natural eucalypt vegetation to climate variables in southern Australia, Agr. Water Manage., 95, 743–747, doi:10.1016/j.agwat.2008.02.007, 2008.

Ewert, F.: Modelling plant responses to elevated CO_2: how important is leaf area index?, Ann. Bot., 93, 619–627, doi:10.1093/aob/mch101, 2004.

Food and Agriculture Organization of the United Nations (FAO): Digital soil map of the world, Version 3.5. FAO, Rome, Italy, 1995.

Fernández-Martínez, M., Vicca, S., Janssens, I., Sardans, J., Luyssaert, S., Campioli, M., Chapin III, F., Ciais, P., Malhi, Y., and Obersteiner, M.: Nutrient availability as the key regulator of global forest carbon balance, Nature Climate Change, 4, 471–476, 2014.

Fowler, H. J., Blenkinsop, S., and Tebaldi, C.: Linking climate change modelling to impacts studies: recent advances in downscaling techniques for hydrological modelling, Int. J. Climatol., 27, 1547–1578, doi:10.1002/joc.1556, 2007.

Ford, T. W. and Quiring, S. M.: Influence of MODIS-Derived Dynamic Vegetation on VIC-Simulated Soil Moisture in Oklahoma, J. Hydrometeorol., 14, 1910–1921, doi:10.1175/JHM-D-13-037.1, 2013.

Geoscience Australia: GEODATA 9 Second Digital Elevation Model (DEM-9S) Version 3, avaialble at: http://www.ga.gov.au/metadata-gateway/metadata/record/gcat_66006, last access: 20 December 2013, 2008.

Gerten, D., Rost, S., von Bloh, W., and Lucht, W.: Causes of change in 20th century global river discharge, Geophys. Res. Lett., 35, L20405, doi:10.1029/2008GL035258, 2008.

Harrold, T. I., Jones, R. N, and Watterson, I. G.: Applying climate changes simulated by GCMs to the generation of fine-scale rainfall scenarios, Hydro 2005, 29th Hydrology and Water Resources Symposium, Canberra, 2005.

Heath, J. T., Chafer, C. J., van Ogtrop, F. F., and Bishop, T. F. A.: Post-wildfire recovery of water yield in the Sydney Basin water supply catchments: An assessment of the 2001/2002 wildfires, J. Hydrol., 519, 1428–1440, doi:10.1016/j.jhydrol.2014.09.033, 2014.

Hikosaka, K., Onoda, Y., Kinugasa, T., Nagashima, H., Anten, N. P. R., and Hirose, T.: Plant responses to elevated CO_2 concentration at different scales: leaf, whole plant, canopy, and population, Ecol. Res., 20, 243–253, doi:10.1007/s11284-005-0041-1, 2005.

Hughes, J. D., Petrone, K. C., and Silberstein, R. P.: Drought, groundwater storage and stream flow decline in southwestern Australia, Geophys. Res. Lett., 39, L03408, doi:10.1029/2011GL050797, 2012.

Huntington, T. G.: CO_2-induced suppression of transpiration cannot explain increasing runoff, Hydrol. Process., 22, 311–314, doi:10.1002/hyp.6925, 2008.

Jahan, N. and Gan, T. Y.: Modelling the vegetation–climate relationship in a boreal mixedwood forest of Alberta using normalized difference and enhanced vegetation indices, Int. J. Remote Sens., 32, 313–335, doi:10.1080/01431160903464146, 2011.

Jones, D. A., Wang, W., and Fawcett, R.: High-quality spatial climate data-sets for Australia, Australian Meteorological and Oceanographic Journal, 58, 233–248, 2009.

Kalma, J. D., Bates, B. C., and Woods, R. A.: Predicting catchment-scale soil moisture status with limited field measurements, Hydrol. Process., 9, 445–467, doi:10.1002/hyp.3360090315, 1995.

Kimball, J. S., Running, S. W., and Nemani, R. R.: An improved method for estimating surface humidity from daily minimum temperature, Agr. Forest Meteorol., 85, 87–98, 1997.

Körner, C.: Plant CO_2 responses: an issue of definition, time and resource supply, New Phytol., 172, 393–411, doi:10.1111/j.1469-8137.2006.01886.x, 2006.

Lenihan, J. M., Drapek, R., Bachelet, D., and Neilson, R. P.: Climate change effects on vegetation distribution, carbon, and fire in California, Ecol. Appl., 13, 1667–1681, 2003.

Liang, X., Lettenmaier, D. P., Wood, E. F., and Burges, S. J.: A Simple hydrologically based model of land surface water and energy fluxes of general circulation models, J. Geophys. Res., 99, 14415–14428, doi:10.1029/94jd00483, 1994.

Liang, X., Wood, E. F., and Lettenmaier, D. P.: Surface soil moisture parameterization of the VIC-2L model: Evaluation and modification, Global Planet. Change, 13, 195–206, doi:10.1016/0921-8181(95)00046-1, 1996.

Lockart, N., Kavetski, D., and Franks, S. W.: On the recent warming in the Murray–Darling Basin: land surface interactions misunderstood, Geophys. Res. Lett., 36, L24405, doi:10.1029/2009GL040598, 2009.

Lockart, N., Kavetski, D., and Franks, S. W.: On the recent warming in the Murray-Darling Basin: Land surface interactions misunderstood, Geophys. Res. Lett., 36, L24405, doi:10.1029/2009GL040598, 2009.

McMahon, T. A., Peel, M. C., and Karoly, D. J.: Assessment of precipitation and temperature data from CMIP3 global climate models for hydrologic simulation, Hydrol. Earth Syst. Sci., 19, 361–377, doi:10.5194/hess-19-361-2015, 2015.

McVicar, T. R., Van Niel, T. G., Li, L. T., Roderick, M. L., Rayner, D. P., Ricciardulli, L., and Donohue, R. J.: Wind speed climatology and trends for Australia, 1975–2006: Capturing the stilling phenomenon and comparison with near-surface reanalysis output, Geophys. Res. Lett., 35, L20403, doi:10.1029/2008GL035627, 2008.

Meinshausen, M., Smith, S. J., Calvin, K., Daniel, J. S., Kainuma, M. L. T., Lamarque, J. F., Matsumoto, K., Montzka, S. A., Raper, S. C. B., Riahi, K., Thomson, A., Velders, G. J. M., and van Vuuren, D. P. P.: The RCP greenhouse gas concentrations and their extensions from 1765 to 2300, Climatic Change, 109, 213–241, doi:10.1007/s10584-011-0156-z, 2011.

Milly, P. C. D., Dunne, K. A., and Vecchia, A. V.: Global pattern of trends in streamflow and water availability in a changing climate, Nature, 438, 347–350, 2005.

Moriasi, D. N., Arnold, J. G., Van Liew, M. W., Bingner, R. L., Harmel, R. D., and Veith, T. L.: Model evaluation guidelines for systematic quantification of accuracy in watershed simulations, T. ASABE, 50, 885–900, 2007.

Moss, R. H., Edmonds, J. A., Hibbard, K. A., Manning, M. R., Rose, S. K., van Vuuren, D. P., Carter, T. R., Emori, S., Kainuma, M., Kram, T., Meehl, G. A., Mitchell, J. F. B., Nakicenovic, N., Riahi, K., Smith, S. J., Stouffer, R. J., Thomson, A. M., Weyant, J. P., and Wilbanks, T. J.: The next generation of scenarios for climate change research and assessment, Nature, 463, 747–756, doi:10.1038/nature08823, 2010.

Murray, S. J., Foster, P. N., and Prentice, I. C.: Evaluation of global continental hydrology as simulated by the Land-surface Processes and eXchanges Dynamic Global Vegetation Model, Hydrol. Earth Syst. Sci., 15, 91–105, doi:10.5194/hess-15-91-2011, 2011.

Murray, S. J., Foster, P. N., and Prentice, I. C.: Future global water resources with respect to climate change and water withdrawals as estimated by a dynamic global vegetation model, J. Hydrol., 448–449, 14–29, doi:10.1016/j.jhydrol.2012.02.044, 2012.

Murray, S. J., Watson, I. M., and Prentice, I. C.: The use of dynamic global vegetation models for simulating hydrology and the potential integration of satellite observations, Prog. Phys. Geog., doi:10.1177/0309133312460072, 2013.

Nemani, R. R., Keeling C. D., Hashimoto, H., Jolly, W. M., Piper, S. C., Tucker, C. J., Myneni, R. B., and Running, S. W.: Climate-driven increases in global terrestrial net primary production from 1982 to 1999, Science, 300, 1560–1563, 2003.

O'Grady, A. P., Carter, J. L., and Bruce, J.: Can we predict groundwater discharge from terrestrial ecosystems using existing eco-hydrological concepts?, Hydrol. Earth Syst. Sci., 15, 3731–3739, doi:10.5194/hess-15-3731-2011, 2011.

Obrist, D. and Arnone, J. A.: Increasing CO_2 accelerates root growth and enhances water acquisition during early stages of development in Larrea tridentate, New Phytol., 159, 175–184, doi:10.1046/j.1469-8137.2003.00791.x, 2003.

Palmer, A. R., Fuentes, S., Taylor, D., Macinnis-Ng, C., Zeppel, M., Yunusa, I., and Eamus, D.: Towards a spatial understanding of water use of several land-cover classes: an examination of relationships amongst pre-dawn leaf water potential, vegetation water use, aridity and MODIS LAI, Ecohydrology, 3, 1–10, doi:10.1002/eco.63, 2010.

Peel, M. C. and Blöschl, G.: Hydrological modelling in a changing world, Prog. Phys. Geogr., 35, 249–261, doi:10.1177/0309133311402550, 2011.

Piao, S., Friedlingstein, P., Ciais, P., de Noblet-Ducoudré, N., Labat, D., and Zaehle, S.: Changes in climate and land use have a larger direct impact than rising CO_2 on global river runoff trends, P. Natl. Acad. Sci. USA, 104, 15242–15247, doi:10.1073/pnas.0707213104, 2007.

Potter, N. J. and Chiew, F. H. S.: An investigation into changes in climate characteristics causing the recent very low runoff in the southern Murray–Darling Basin using rainfall–runoff models, Water Resour. Res., 47, W00G10, doi:10.1029/2010WR010333, 2011.

Roderick, M. L. and Farquhar, G. D.: A simple framework for relating variations in runoff to variations in climatic conditions and catchment properties, Water Resour. Res., 47, W00G07, doi:10.1029/2010WR009826, 2011.

Scheiter, S., Langan, L., and Higgins, S. I.: Next-generation dynamic global vegetation models: learning from community ecology, New Phytol., 198, 957–969, 2013.

Schenk, H. J. and Jackson, R. B.: The global biogeography of roots, Ecol. Monogr., 72, 311–328, 2002.

Taylor, K. E., Stouffer, R. J., and Meehl, G. A.: An overview of CMIP5 and the experiment design, B. Am. Meteorol. Soc., 93, 485–498, doi:10.1175/BAMS-D-11-00094.1, 2012.

Teng, J., Chiew, F. H. S., Vaze, J., Marvanek, S., and Kirono, D. G. C.: Estimation of climate change impact on mean annual runoff across Continental Australia using Budyko and Fu equations and hydrological models, J. Hydrometeorol., 13, 1094–1106, 2012a.

Teng, J., Vaze, J., Chiew, F. H. S., Wang, B., and Perraud, J.-M.: Estimating the relative uncertainties sourced from GCMs and hydrological models in modeling climate change impact on runoff, J. Hydrometeorol., 13, 122–139, doi:10.1175/JHM-D-11-058.1, 2012b.

Tesemma, Z. K., Wei, Y., Western, A. W., and Peel, M. C.: Leaf area index variation for cropland, pasture and tree in response to climatic variation in the Goulburn–Broken catchment, Australia, J. Hydrometeorol., 15, 1592–1606, doi:10.1175/JHM-D-13-0108.1, 2014.

Thornton, P. E. and Running, S. W.: An improved algorithm for estimating incident daily solar radiation from measurements of temperature, humidity, and precipitation, Agr. Forest Meteorol., 93, 211–228, 1999.

Uddling, J., Teclaw, R. M., Kubiske, M. E., Pregitzer, K. S., and Ellsworth, D. S.: Sap flux in pure aspen and mixed aspen-birch forests exposed to elevated concentrations of carbon dioxide and ozone, Tree Physiol., 28, 1231–1243, 2008.

Vaze, J. and Teng, J.: Future climate and runoff projections across New South Wales, Australia: results and practical applications, Hydrol. Process., 25, 18–35, doi:10.1002/hyp.7812, 2011.

Vaze, J., Post, D. A., Chiew, F. H. S., Perraud, J. M., Viney, N. R., and Teng, J.: Climate non-stationarity – validity of calibrated rainfall–runoff models for use in climate change studies, J. Hydrol., 394, 447–457, doi:10.1016/j.jhydrol.2010.09.018, 2010.

Verdon-Kidd, D. C. and Kiem, A. S.: Nature and causes of protracted droughts in southeast Australia: comparison between the Federation, WWII, and Big Dry droughts, Geophys. Res. Lett., 36, L22707, doi:10.1029/2009GL041067, 2009.

Vuuren, D. P., Edmonds, J., Kainuma, M., Riahi, K., Thomson, A., Hibbard, K., Hurtt, G. C., Kram, T., Krey, V., Lamarque, J.-F., Masui, T., Meinshausen, M., Nakicenovic, N., Smith, S. J., and Rose, S. K.: The representative concentration pathways: an overview, Climatic Change, 109, 5–31, 2011.

Warren, J. M., Norby, R. J., and Wullschleger, S. D.: Elevated CO_2 enhances leaf senescence during extreme drought in a temperate forest, Tree Physiol., 31, 117–130, doi:10.1093/treephys/tpr002, 2011.

Western, A. W., Grayson, R. B., and Green, T. R.: The Tarrawarra project: high resolution spatial measurement, modelling and analysis of soil moisture and hydrological response, Hydrol. Process., 13, 633–652, 1999.

White, D. A., Battaglia, M., Mendham, D. S., Crombie, D. S., Kinal, J. O. E., and McGrath, J. F.: Observed and modelled leaf area index in Eucalyptus globulus plantations: tests of optimality and equilibrium hypotheses, Tree Physiol., 30, 831–844, doi:10.1093/treephys/tpq037, 2010.

Wullschleger, S. D., Epstein, H. E., Box, E. O., Euskirchen, E. S., Goswami, S., Iversen, C. M., Kattge, J., Norby, R. J., van Bodegom, P. M., and Xu, X.: Plant functional types in Earth System Models: past experiences and future directions for application of dynamic vegetation models in high-latitude ecosystems, Ann. Bot.-London, 114, 1–16, 2014.

Yapo, P. O., Gupta, H. V., and Sorooshian, S.: Multi-objective global optimization for hydrologic models, J. Hydrol., 204, 83–97, doi:10.1016/S0022-1694(97)00107-8, 1998.

Yu, M., Wang, G., Parr, D., and Ahmed, K.: Future changes of the terrestrial ecosystem based on a dynamic vegetation model driven with RCP8.5 climate projections from 19 GCMs, Clim. Change, 127, 257–271, doi:10.1007/s10584-014-1249-2, 2014.

Zhao, F., Chiew, F. H. S., Zhang, L., Vaze, J., Perraud, J.-M., and Li, M.: Application of a macroscale hydrologic model to estimate streamflow across southeast Australia, J. Hydrometeorol., 13, 1233–1250, doi:10.1175/jhm-d-11-0114.1, 2012a.

Zhao, F. F., Xu, Z. X., and Zhang, L.: Changes in streamflow regime following vegetation changes from paired catchments, Hydrol. Process., 26, 1561–1573, doi:10.1002/hyp.8266, 2012b.

Zhao, R. J., Zhuang, Y. L., Fang, L. R., Liu, X. R., and Zhang, Q. S.: The Xinanjiang Model, Hydrological Forecasting Proceedings Oxford Symposium, IAHS Press, 571 pp., 1980.

2

Responses of runoff to historical and future climate variability over China

Chuanhao Wu[1], **Bill X. Hu**[1,2], **Guoru Huang**[3,4], **Peng Wang**[1], **and Kai Xu**[1]

[1]Institute of Groundwater and Earth Sciences, Jinan University, Guangzhou 510632, China
[2]Department of Earth, Ocean and Atmospheric Sciences, Florida State University, Tallahassee, FL 32306, USA
[3]School of Civil Engineering and Transportation, South China University of Technology, Guangzhou 510640, China
[4]State Key Laboratory of Subtropical Building Science, South China University of Technology, Guangzhou 510640, China

Correspondence: Bill X. Hu (bill.x.hu@gmail.com)

Abstract. China has suffered some of the effects of global warming, and one of the potential implications of climate warming is the alteration of the temporal–spatial patterns of water resources. Based on the long-term (1960–2008) water budget data and climate projections from 28 global climate models (GCMs) of the Coupled Model Intercomparison Project Phase 5 (CMIP5), this study investigated the responses of runoff (R) to historical and future climate variability in China at both grid and catchment scales using the Budyko-based elasticity method. Results show that there is a large spatial variation in precipitation (P) elasticity (from 1.1 to 3.2) and potential evaporation (PET) elasticity (from -2.2 to -0.1) across China. The P elasticity is larger in north-eastern and western China than in southern China, while the opposite occurs for PET elasticity. The catchment properties' elasticity of R appears to have a strong non-linear relationship with the mean annual aridity index and tends to be more significant in more arid regions. For the period 1960–2008, the climate contribution to R ranges from -2.4 to $3.6\,\%\,\mathrm{yr}^{-1}$ across China, with the negative contribution in north-eastern China and the positive contribution in western China and some parts of the south-west. The results of climate projections indicate that although there is large uncertainty involved in the 28 GCMs, most project a consistent change in P (or PET) in China at the annual scale. For the period 2071–2100, the mean annual P is projected to increase in most parts of China, especially the western regions, while the mean annual PET is projected to increase in all of China, particularly the southern regions. Furthermore, greater increases are projected for higher emission scenarios. Overall, due to climate change, the arid regions and humid regions of China are projected to become wetter and drier in the period 2071–2100, respectively (relative to the baseline 1971–2000).

1 Introduction

Climate change has become increasingly significant (IPCC, 2013), and numerous studies have reported that climate warming is likely leading to the alteration of the hydrological cycle (Oki and Kanae, 2006; Jung et al., 2010). The dynamic properties of the hydrological cycle are governed by the interactions and feedbacks between atmospheric and land surface hydrologic processes on a catchment scale. The potential consequences of anthropogenic climate change on the hydrological cycle have received significant attention over the last 2 decades (Wang et al., 2012; IPCC, 2013).

Runoff (R), as a commonly adopted indicator of the hydrologic cycle, is critical to human lives and economic activities (Milly et al., 2005). There is a great deal of previous work exploring the impact of climate variations on R, with the motivation stemming from the region's vast resources (Christensen et al., 2004; Guo et al., 2009; Piao et al., 2010; Chen et al., 2012; Harding et al., 2012; Wang et al., 2012; Y. P. Xu et al., 2013), dangers of flooding (Kay et al., 2006, 2009, Kay and Jones, 2012; Raff et al., 2009; Liu et al., 2013; Xiao et al., 2013; Wang et al., 2013; Smith et al., 2014; Wu et al., 2014, 2015), and agricultural water uses (Vano et al., 2010). The most common practices in these previous studies are to use the hydrological models driven by the output from global climate models (GCMs) to simulate the hydrological process (e.g. R) under future climate change scenarios. However, the key issue faced by such studies is the need to convert coarse-resolution GCM outputs to local catchment-scale climatic variables at a higher spatial resolution to serve as the input to a hydrological model (Vano et al., 2015; Wu et al., 2015). The impact assessments are resource-intensive and usually subject to uncertainties related to the choice of hydrological model, GCMs, emissions scenarios, and downscaling techniques (Vano et al., 2014, 2015).

With the uncertainty in R due to climate change, simple tools able to provide robust estimates of this impact are essential to support policy and planning decisions. Climate elasticity, as an important indicator, provides a measure of sensitivity of the changes in R due to the changes in climate. Schaake (1990) made the first attempt to introduce the concept of elasticity and related the climate elasticity of R to precipitation (P). Since then numerous climate elasticity methods have been developed for evaluating the hydrologic response to climate change all over the world (Schaake, 1990; Dooge et al., 1999; Sankarasubramanian et al., 2001; Milly and Dunne, 2002; Fu et al., 2007; Zheng et al., 2009; Ma et al., 2010; Yang and Yang, 2011; Yang et al., 2014; Vano et al., 2015). Sankarasubramanian et al. (2001) provided a detailed category of climate elasticity methods for modelling climate change impacts. One of the most common methods is to analytically derive the sensitivity of R based on the Budyko hypothesis, due to its clear theory and the fact that it does not rely on a large amount of data (Yang and Yang, 2011). More importantly, the Budyko-based elasticity method can derive the climate elasticity and can also represent the impact of the catchment characteristics through the parameters of the Budyko model. Accordingly, it is widely applied for the assessment of the hydrologic impacts of climate change (Dooge et al., 1999; Zheng et al., 2009; Yang and Yang, 2011; Yang et al., 2014).

China is a vast land, spanning many degrees of latitude with complicated terrain, which results in a large regional variation in climate elasticity. The investigation of the P elasticity of R has been reported in many regions of China, such as the Miyun Reservoir basin (Ma et al., 2010), Luan River basin (X. Y. Xu et al., 2013), the headwater catchments of the Yellow River basin (Zheng et al., 2009), Poyang Lake basin (Sun et al., 2013), and Hai River and Yellow River basins (Yang and Yang, 2011; Liu and McVicar, 2012). Recently Yang et al. (2014) investigated the climate elasticity of R for the 210 catchments of China based on the Budyko-based elasticity approach. The results indicated that the P elasticity exhibits a large regional variation, with a small range in southern China, the Songhua River basin, and the northwest and a large range in the Hai River basin, the Yellow River basin, and the Liao River basin. Although the aforementioned studies have certainly made advances in understanding the climate elasticity of R in China, our knowledge about the responses of R to climate change over various temporal and spatial scales remains rather limited due to the large regional variation in climate types and catchment characteristics. The question of how climate change will affect R over China in the future is also an important problem to be addressed. Developing a more accurate and quantitative understanding of the changing water resources over various temporal and spatial scales under a changing environment is therefore a high priority for China.

Based on the unique long-term (1960–2008) land surface dataset of China and the climate projections from 28 GCMs of the Coupled Model Intercomparison Project Phase 5 (CMIP5), the objectives of this research are (1) to investigate the changes of R and climate variables and their relationship at an interannual scale; (2) to quantitatively estimate the climate elasticity and catchment properties' elasticity of R across China at both grid and catchment scales; and (3) to predict climate change and the changes in R due to future climate change for China from the CMIP5 projections at both grid and catchment scales.

2 Data and methodology

2.1 Datasets

Monthly data of potential evaporation (PET) covering the period 1960–2008 over China are provided by the Terrestrial Hydrology Research Group of Princeton University (Sheffield et al., 2006, 2012). The PET is estimated by the Penman equation (Penman, 1948; Shuttleworth, 1993), using the updated meteorological dataset obtained from Sheffield et al. (2006, 2012). A long-term (1960–2008) daily land surface dataset over China, including P, surface runoff (RS), and baseflow (BS), with a 0.25° spatial resolution, was obtained from the Land Surface Processes and Global Change Research Group (Zhang et al., 2014). In this dataset, P is driven by interpolating gauged daily precipitation from 756 meteorological stations of the Chinese Meteorological Administration (CMA). RS and BS are derived from the Variable Infiltration Capacity (VIC) model forced by the gridded daily climate forcings (i.e. P, maximum and minimum temperature, and wind speed). VIC model parameters, including the infiltration shape parameter, the second and third soil layer depths, and the three parameters in the base flow scheme, were estimated by using an optimization algorithm of the multi-objective complex evolution of the University of Arizona (Zhang et al., 2014). The simulated monthly RS and BS match well with the observations at the large river basins in China (Zhang et al., 2014). Compared with the global product of a similar nature, this dataset provides a more reliable estimate of land surface variables over China (Nijssen et al., 2001; Adam et al., 2006; Rodell et al., 2004; Sheffield et al., 2006; Sheffield and Wood, 2007; Pan et al., 2012). In this study, the data of P, RS, and BS are ini-

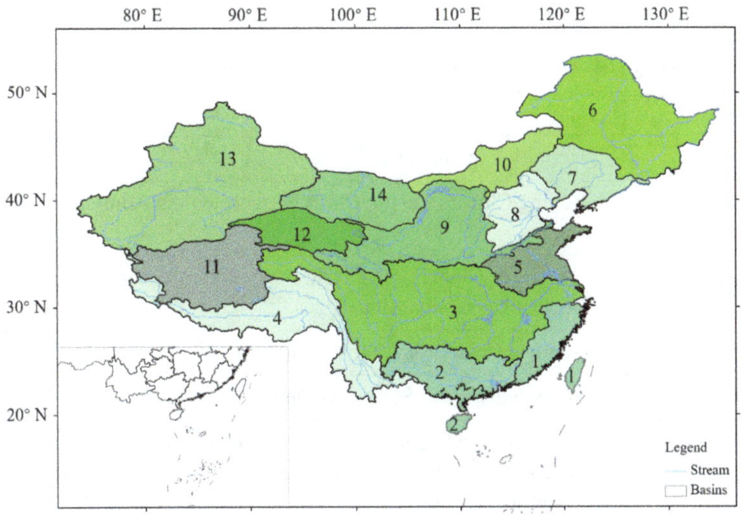

Figure 1. Location of the main river basins in China. The numbers denote the river basins with increasing aridity index: (1) Southeast Drainage (0.52); (2) Pearl River (0.64); (3) Yangtze River (0.81); (4) Southwest Drainage (1.19); (5) Huaihe River (1.19); (6) Heilongjiang River (1.43); (7) Liaohe River (1.71); (8) Haihe River (2.14); (9) Yellow River (2.38); (10) Inner Mongolia River (4.41); (11) Qiangtang River (4.70); (12) Qinghai River (6.68); (13) Xinjiang River (8.09), (14) Hexi River (8.63). The numbers in the parentheses indicate the 1960–2008 mean aridity index.

tially regridded onto 0.5° grids over China using the linear interpolation method. All the daily data (P, RS, and BS) and monthly data (PET) are then aggregated temporally for the annual scale over China. The R was calculated by the sum of RS and BS for each of the 0.5° grid points.

Climate projections from 28 CMIP5 GCMs (as shown in Table 1) are provided by the Canadian Climate Data and Scenarios (CCDS, http://climate-scenarios.canada.ca/?page=main). These data, including simulations of surface air temperature (T), P, sea ice thickness, sea ice concentration, snow depth, and near-surface wind speed, are statistically downscaled and regridded onto a common $1^\circ \times 1^\circ$ global grid by the CCDS. In this study, monthly P and monthly T over China, including one historical simulation for the period 1971–2000 and three emission scenarios (RCP2.6, RCP4.5, and RCP8.5) for the future period 2071–2100 from each of the 28 CMIP5 models and the multi-model ensemble of 28 CMIP5 models, are used for the projections of climate change. The data are initially disaggregated to 0.5° grids over China then corrected by using a "delta change" method (Wu and Huang, 2016), on the basis of the observed data of P and T as provided by the Climatic Research Unit (CRU) of the University of East Anglia (Harris et al., 2014).

Figure 2 shows the comparison of observed mean annual T and P and the corresponding simulations from 28 CMIP5 models before and after bias correction for the 14 basins in China. The basin number is consistent with that given in Fig. 1. As shown, the uncorrected model simulations tend to underestimate T and overestimate P for most of the basins, with more uncertainties for the simulation of P than for T. Compared to the uncorrected model results, the bias correction results represent large improvements and show a good agreement with the observed values for these basins. Therefore, the bias correction model simulations are acceptable for the investigation of climate change projections in this study.

As the GCM data used only consist of P and T, the PET of GCM is estimated by the Thornthwaite method (Thornthwaite, 1948) and then corrected by a multiplicative bias correction method as follows:

$$\mathrm{PET}_{\mathrm{cor,GCM},i} = \mathrm{PET}_{\mathrm{Th,GCM},i} \times \frac{\overline{\mathrm{PET}}_{\mathrm{Pen,obs},i}}{\overline{\mathrm{PET}}_{\mathrm{Th,obs},i}}, \tag{1}$$

where $\mathrm{PET}_{\mathrm{cor,GCM},i}$ and $\mathrm{PET}_{\mathrm{Th,GCM},i}$ are bias-corrected annual PET and the PET calculated from the Thornthwaite method, respectively, for the ith grid point of the GCMs. $\overline{\mathrm{PET}}_{\mathrm{Pen,obs},i}$ and $\overline{\mathrm{PET}}_{\mathrm{Th,obs},i}$ are the 49-year (1960–2008) averages of PET calculated from the Penman and Thornthwaite methods, respectively, for the ith grid point.

Based on the T data from the CRU, the Thornthwaite method is used to calculate PET to test the applicability of Eq. (1). Figure 3 shows a comparison of annual PET calculated from the Penman method and that from the Thornthwaite method corrected by Eq. (1) during the period 1960–2008. It is clear that the corrected PET agrees well with the PET from the Penman method, with the correlation coefficients of 0.94 and 0.958 at the catchment and grid scales, respectively. This suggests that Eq. (1) can be acceptable for the bias correction of PET in the GCMs.

Table 1. CMIP5 GCMs used in this study. The GCM data were statistically downscaled and regridded onto a common 1° × 1° global grid from the Canadian Climate Data and Scenarios (CCDS).

No.	Model	Institution (country)	Resolution
1 2	BCC-CSM1-1 BCC-CSM1-1-m	Beijing Climate Center, China Meteorological Administration, China	1° × 1°
3	BNU-ESM	College of Global Change and Earth System Science, Beijing Normal University, China	1° × 1°
4	CCSM4	National Center for Atmospheric Research, USA	1° × 1°
5	CESM1-CAM5	Community Earth System Model Contributors, USA	1° × 1°
6	CNRM-CM5	Centre national de Recherches Météorologiques/Centre Européen de Recherche et Formation Avancée en Calcul Scientifique, France	1° × 1°
7	CSIRO-Mk3-6-0	Commonwealth Scientific and Industrial Research Organization in collaboration with Queensland Climate Change Centre of Excellence, Australia	1° × 1°
8	CanESM2	Canadian Centre for Climate Modelling and Analysis, Canada	1° × 1°
9	EC-EARTH	EC-EARTH consortium	1° × 1°
10	FGOALS-g2	LASG, Institute of Atmospheric Physics, Chinese Academy of Sciences and CESS, Tsinghua University, China	1° × 1°
11	FIO-ESM	The First Institute of Oceanography, SOA, China	1° × 1°
12 13 14	GFDL-CM3 GFDL-ESM2G GFDL-ESM2M	NOAA Geophysical Fluid Dynamics Laboratory, USA	1° × 1°
15 16	GISS-E2-H GISS-E2-R	NASA Goddard Institute for Space Studies, USA	1° × 1°
17	HadGEM2-AO	National Institute of Meteorological Research/Korea Meteorological Administration, South Korea	1° × 1°
18	HadGEM2-ES	Met Office Hadley Centre (additional HadGEM2-ES realizations contributed by Instituto Nacional de Pesquisas Espaciais), UK	1° × 1°
19 20	IPSL-CM5A-LR IPSL-CM5A-MR	Institut Pierre-Simon Laplace, France	1° × 1°
21 22	MIROC-ESM MIROC-ESM-CHEM	Japan Agency for Marine-Earth Science and Technology, Atmosphere and Ocean Research Institute (the University of Tokyo), and National Institute for Environmental Studies, Japan	1° × 1°
23	MIROC5	Atmosphere and Ocean Research Institute (The University of Tokyo), National Institute for Environmental Studies, and Japan Agency for Marine-Earth Science and Technology, Japan	1° × 1°
24 25	MPI-ESM-LR MPI-ESM-MR	Max-Planck-Institut für Meteorologie (Max Planck Institute for Meteorology), Germany	1° × 1°
26	MRI-CGCM3	Meteorological Research Institute, Japan	1° × 1°
27 28	NorESM1-M NorESM1-ME	Norwegian Climate Centre, Norway	1° × 1°

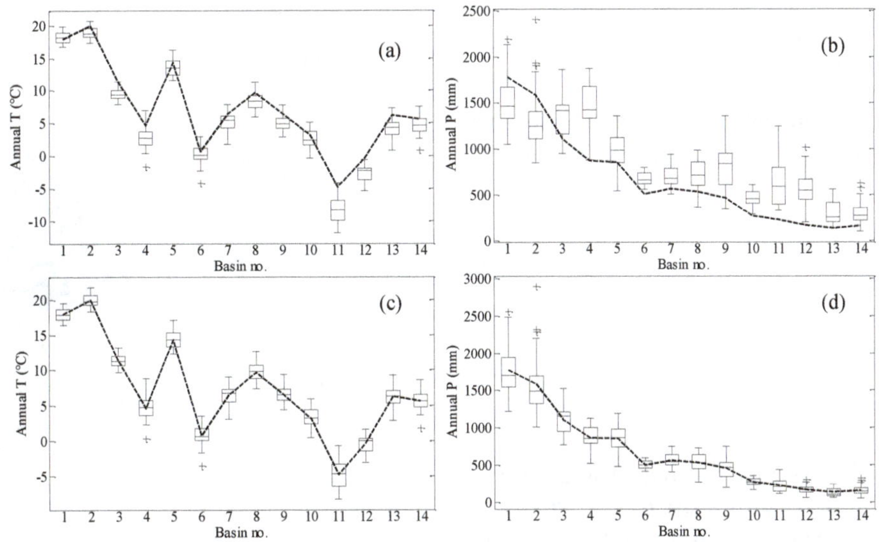

Figure 2. Box plots of the simulation results of **(a)** mean annual T and **(b)** mean annual P and the bias correction results of **(c)** mean annual T and **(d)** mean annual P from 28 GCMs for the period 1971–2000 in the 14 river basins. The boxes denote the interquartile model spread (range between the 25th and 75th quantiles), with the horizontal line indicating the ensemble median and the whiskers showing the extreme range of the 28 CMIP5 model simulations. The blue dotted lines denote the observed results of mean annual T and mean annual P for the period 1971–2000. The basin number is consistent with that given in Fig. 1.

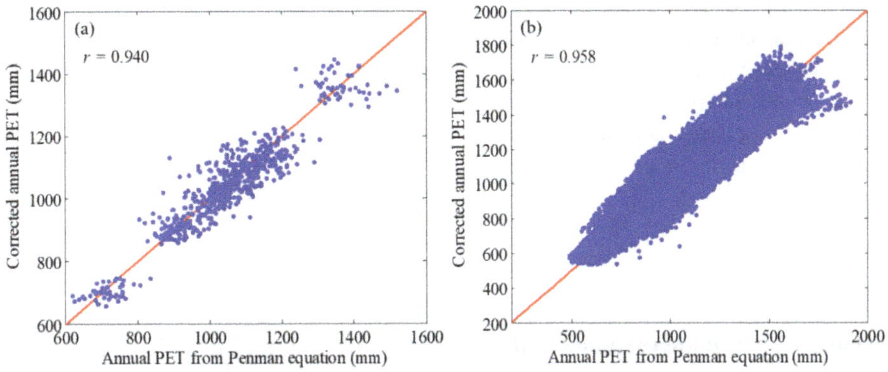

Figure 3. Comparison of annual PET calculated from the Penman method and the Thornthwaite method corrected by Eq. (1) during the period 1960–2008 for **(a)** the 14 river basins and **(b)** all 0.5° grid points over China.

2.2 Sensitivity of runoff to climate and catchment properties

The Budyko framework has been widely used to study basin-scale water and energy balances. Two of the one-parameter formulations of the Budyko curve proposed by Choudhury (1999) (Eq. 2; see also Yang et al., 2008) and Fu (1981) (Eq. 3; see also Zhang et al., 2004) are expressed as

$$E = P\frac{\mathrm{PET}}{(P^n + \mathrm{PET}^n)^{1/n}},\ n \in (0, \infty) \tag{2}$$

$$E = P + \mathrm{PET} - \left(P^\omega + \mathrm{PET}^\omega\right)^{1/\omega},\ \omega \in (1, \infty), \tag{3}$$

where n and ω are empirical parameters, representing the effects of other factors (e.g. land surface characteristics, the average slope, vegetation type or land use, and climate seasonality) on the water–energy balance (Yang et al., 2008, 2014; Roderick and Farquhar, 2011; D. Li et al., 2013). Yang et al. (2008) calibrated the parameters n and ω using long-term water balance data from 108 catchments from the non-humid regions of China and found that these two empirical parameters are linearly correlated.

Based on the Budyko hypothesis and assuming steady state conditions, Roderick and Farquhar (2011) and Yang and Yang (2011) derived the elasticity method to estimate the contribution to R from the changes in climate (represented by P and PET) and catchment properties as follows:

$$\frac{\mathrm{d}R}{R} = \varepsilon_{\mathrm{P}} \cdot \frac{\mathrm{d}P}{P} + \varepsilon_{\mathrm{PET}} \cdot \frac{\mathrm{dPET}}{\mathrm{PET}} + \varepsilon_n \cdot \frac{\mathrm{d}n}{n}, \tag{4}$$

where ε_{P}, $\varepsilon_{\mathrm{PET}}$, and ε_n represent the elasticity coefficients of P, PET, and catchment properties respectively and are expressed as

$$\varepsilon_{\mathrm{P}} = \frac{P}{R}\left(1 - \frac{\partial E}{\partial P}\right) \tag{5}$$

$$\varepsilon_{\mathrm{PET}} = -\frac{\mathrm{PET}}{R}\frac{\partial E}{\partial \mathrm{PET}} \tag{6}$$

$$\varepsilon_n = -\frac{n}{R}\frac{\partial E}{\partial n}, \tag{7}$$

where $\frac{\partial E}{\partial P}$, $\frac{\partial E}{\partial \mathrm{PET}}$, and $\frac{\partial E}{\partial n}$ denote the first-order partial derivatives of the Budyko equation with respect to P, PET, and the parameter n. In this study, both Eqs. (2) and (3) are used for the estimation of the elasticity of P, PET, and catchment properties over China.

2.3 Trend estimate method

The Mann–Kendall (M–K) non-parametric test (Mann, 1945; Kendall, 1975) is an effective tool for detecting the statistical significance of trends in the time series of meteorological and hydrological variables (Yang et al., 2014; Wu and Huang, 2015). In this study, the M–K method is used to detect the significance of monotonic trends in hydroclimatic time series. The non-parametric trend slope estimator developed by Sen (1968) is used for the magnitude estimation of the trends in a hydroclimatic time series.

3 Results

3.1 Interannual variability of climatic variables and runoff

The standard deviations for annual P, PET, and R are computed for each of the 0.5° grids in China, and the PET deviation ratio ($\sigma_{\mathrm{PET}}/\sigma_{\mathrm{P}}$) and the R deviation ratio ($\sigma_{\mathrm{R}}/\sigma_{\mathrm{P}}$) are calculated. The spatial distributions of PET deviation ratio and R deviation ratio across China are displayed in Fig. 4a and b. As shown, the PET deviation ratio is rather small in most parts of China, especially the southern regions, while a larger value is observed mainly in the Xinjiang region, where there are greater aridity indices. Generally, atmospheric water is enough to accommodate the limited PET in humid climates, which would lead to a limited response of PET to P variability. Specifically, the interannual variability of PET is more sensitive to that of P in arid climates (with water limits) than in humid climates (with energy limits). In contrast to the PET deviation ratio, the R deviation ratio tends to increase from arid climates to humid climates. The reason for this is that, in arid climates, the catchment water supply is very limited and gives priority to evaporation and soil storage capability, which leads to little variation in R.

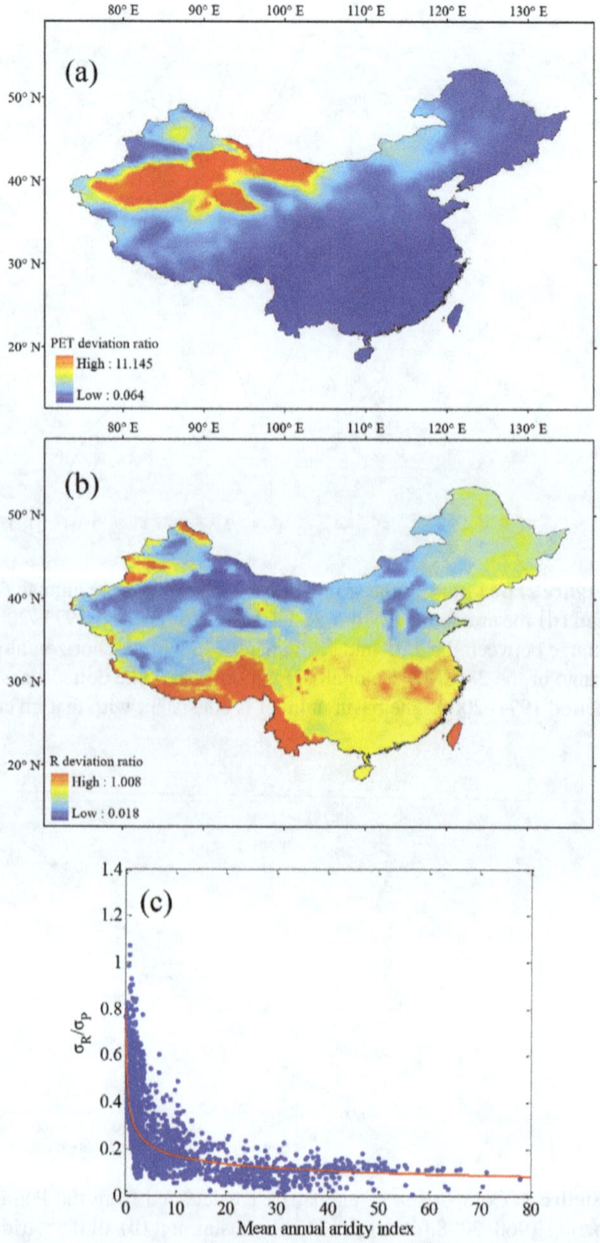

Figure 4. Spatial distributions of **(a)** PET deviation ratio and **(b)** R deviation ratio and **(c)** the relationship between R deviation ratio and mean annual aridity index ($\overline{\phi}$) for all 0.5° grid points in China.

Figure 4c shows the relationship between the R deviation ratio and mean annual aridity index ($\overline{\phi}$) for all 0.5° grids in China. As indicated, $\overline{\phi}$ is a major control for the deviation ratio under not very dry conditions (e.g. $\overline{\phi} < 10$); that is, the R deviation ratio decreases with increased $\overline{\phi}$. However, under very dry conditions (e.g. $\overline{\phi} > 10$) the R deviation ratio becomes insensitive to $\overline{\phi}$, since in this case, other factors, such as soil storage capacity, can also have a large impact on the variation of R.

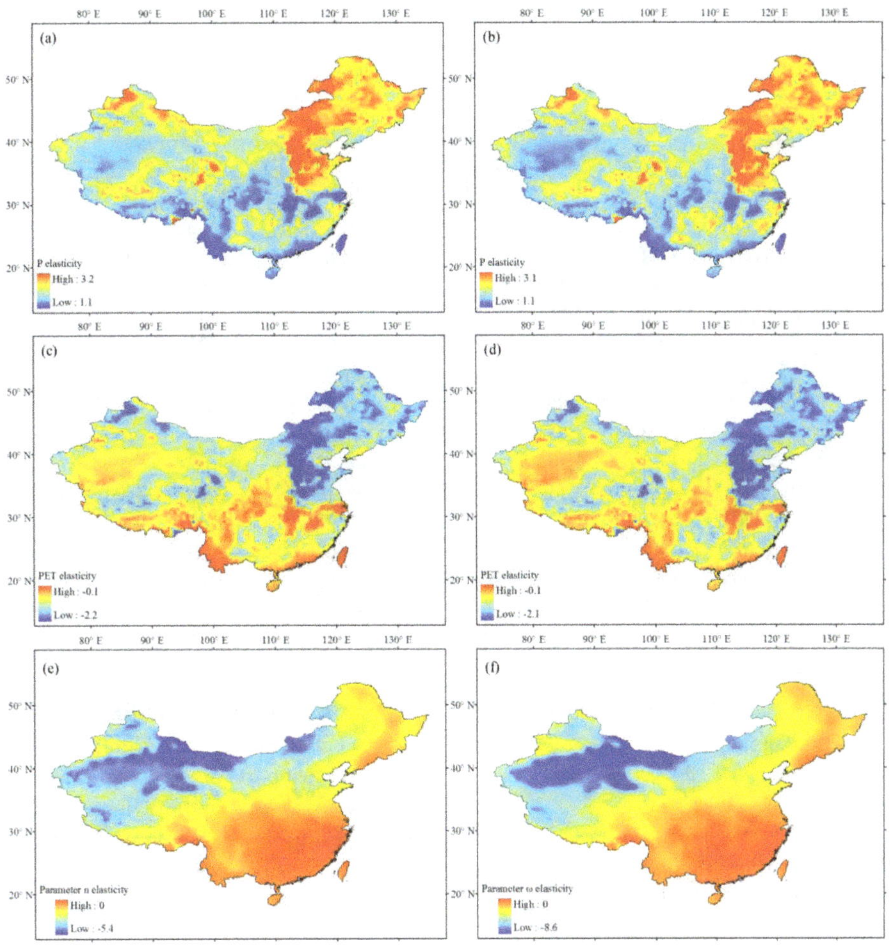

Figure 5. Spatial distributions of the P elasticity of R across China from **(a)** Eq. (2) and **(b)** Eq. (3). Spatial distributions of the PET elasticity of R across China from **(c)** Eq. (2) and **(d)** Eq. (3). Spatial distributions of the parameter elasticity of R across China from **(e)** Eq. (2) and **(f)** Eq. (3).

3.2 Sensitivity of runoff to climate and catchment properties

3.2.1 Climate elasticity

The P elasticity and PET elasticity of R based on Eqs. (2) and (3) are estimated at each of the 0.5° grids in China. As shown in Fig. 5, the spatial patterns of P elasticity and PET elasticity from Eqs. (2) and (3) are almost the same in all regions of China. There is a large spatial variation in P elasticity and PET elasticity, i.e. ranging from 1.1 to 3.2 and from −2.2 to −0.1 across China, respectively. In particular, P elasticity is more significant in the north-eastern and western areas than in southern China, which is in contrast to PET elasticity. Figure 6 shows the relationship between $\overline{\phi}$ and climate (P and PET) elasticity. As shown, the P (PET) elasticity first increases (decreases) and then decreases (increases) with the increase of $\overline{\phi}$ under not very dry conditions (i.e. $\overline{\phi} < 10$). However, when $\overline{\phi}$ becomes large enough (e.g. $\overline{\phi} > 10$), both P and PET elasticity become insensitive to $\overline{\phi}$.

The climate elasticity estimated for each of the 14 large basins is shown in Table 2. The values of P elasticity are in the range of 1.39–2.28, with a larger (∼ smaller) elasticity in the Haihe River and Inner Mongolia River (Southwest Drainage). A similar phenomenon is found for PET elasticity, which suggests that Haihe River (Southwest Drainage) is the most (least) sensitive to PET among the 14 basins. Overall the values of P elasticity and PET elasticity derived by Eq. (2) are very close to those from Eq. (3), but the difference between them tends to be larger for dry basins with increasing aridity indices.

By using the estimates of climate elasticity derived by Eq. (2), the change in R as a function of the percentage change in P and PET is calculated for the 14 basins (Fig. 7). The R is positively related to P and negatively related to PET, and the magnitudes and patterns of the response of R to changes in P and PET vary on different scales. Generally, the R is more sensitive to climate in the Haihe River and Inner

Table 2. The estimations of P elasticity, PET elasticity, and catchment properties' elasticity of R in the 14 river basins of China based on Eqs. (2) and (3). The basin number is consistent with that given in Fig. 1. The numbers in the parentheses indicate the 1960–2008 mean aridity index.

Basin	ε_P		ε_{PET}		ε_n or ε_ω	
no.	Eq. (2)	Eq. (3)	Eq. (2)	Eq. (3)	Eq. (2)	Eq. (3)
1 (0.52)	1.64	1.65	−0.64	−0.65	−0.24	−0.33
2 (0.64)	1.63	1.64	−0.62	−0.63	−0.41	−0.61
3 (0.81)	1.55	1.56	−0.55	−0.55	−0.57	−0.93
4 (1.19)	1.40	1.39	−0.40	−0.39	−0.73	−1.44
5 (1.19)	2.09	2.08	−1.08	−1.07	−1.03	−1.47
6 (1.43)	2.06	2.04	−1.05	−1.02	−1.25	−1.83
7 (1.71)	1.92	1.88	−0.91	−0.87	−1.35	−2.10
8 (2.14)	2.28	2.21	−1.29	−1.22	−1.89	−2.70
9 (2.38)	1.78	1.72	−0.79	−0.73	−1.53	−2.54
10 (4.41)	2.23	2.11	−1.22	−1.10	−2.78	−4.16
11 (4.70)	1.81	1.72	−0.82	−0.72	−2.17	−3.67
12 (6.68)	1.72	1.62	−0.73	−0.63	−2.28	−4.08
13 (8.09)	1.66	1.56	−0.65	−0.55	−2.26	−4.27
14 (8.63)	1.63	1.53	−0.64	−0.54	−2.26	−4.30

Mongolia River, while relatively weak sensitivity is found in the Southwest Drainage and Yangtze.

3.2.2 Catchment properties' elasticity

The spatial distributions of catchment properties' elasticity from Eqs. (2) and (3) are displayed in Fig. 5e and f. As shown, the catchment properties' elasticities for these two equations are rather similar across China, and the values of Eq. (3) are generally smaller than those from Eq. (2). Regarding the spatial pattern, the catchment properties' elasticity is very weak (approximately equal to 0) in southern China and some regions of north-eastern China, but it tends to be more significant in some water-limited regions of north-west China. Figure 6c shows the relationship between $\overline{\phi}$ and the parameter elasticity for all 0.5° grids in China. It suggests that $\overline{\phi}$ is a major control for catchment properties' elasticity across China; i.e. the catchment properties' elasticity would become stronger with increasing aridity indices. The catchment properties' elasticities estimated for the 14 large basins are shown in Table 2. The catchment properties' elasticity shows a large spatial variation, ranging from −2.78 to −0.24 for Eq. (2) and from −4.3 to −0.33 for Eq. (3). Overall, the changes in R are more sensitive to catchment properties in arid basins with larger aridity indices, which is consistent with the findings at the grid scale.

3.3 Climate change during 1960–2008

The annual trend magnitudes in P, R, PET, and aridity index during the period 1960–2008 are shown in Fig. 8a–d. As indicated, both P and R show an increasing trend mainly in the north-west and south-east regions and a decreasing trend mainly in the central region and the North China Plain. A significant increasing trend in PET is detected mainly in north-eastern China and eastern China, while the decreases mainly occur in most parts of western China. The aridity index tends to show an increasing trend in most parts of China, indicating an increasing risk of meteorological drought in these regions during the past several decades. In contrast, the decrease of aridity index is only found in some parts of western China.

3.4 Changes in runoff due to climate change during 1960–2008

Using the estimates of climate elasticity from Eq. (2), the contributions of P, PET, and climate (i.e. P and PET) to R in China for the period 1960–2008 are calculated (as shown in Fig. 8e–g). A positive contribution (up to 3.7 % yr^{-1}) from P to R is mainly recorded in western China, while a negative contribution is found mainly in north-eastern China and the North China Plain. Negative and positive contributions of PET to R mainly occur in north-eastern China and western China, respectively. The contribution of climate, i.e. the sum of the contributions from P and PET, ranges from −2.4 to 3.6 % yr^{-1} across China. The spatial pattern of climate is rather similar to that of P, showing a negative contribution in north-eastern China and a positive contribution in western China and some parts of the south-east. In particular, the largest positive contribution of climate occurs in the Tibetan Plateau. The contributions of P, PET, and climate (i.e. P and PET) to R in the 14 river basins for the period 1960–2008 are shown in Table 3. A positive contribution of P is detected in Southeast Drainage, Southwest Drainage, Qiangtang, Qinghai, Xinjiang and Hexi, while an opposite contribution is found in other basins. In contrast, a negative contribution of PET is found in most of the basins (except for

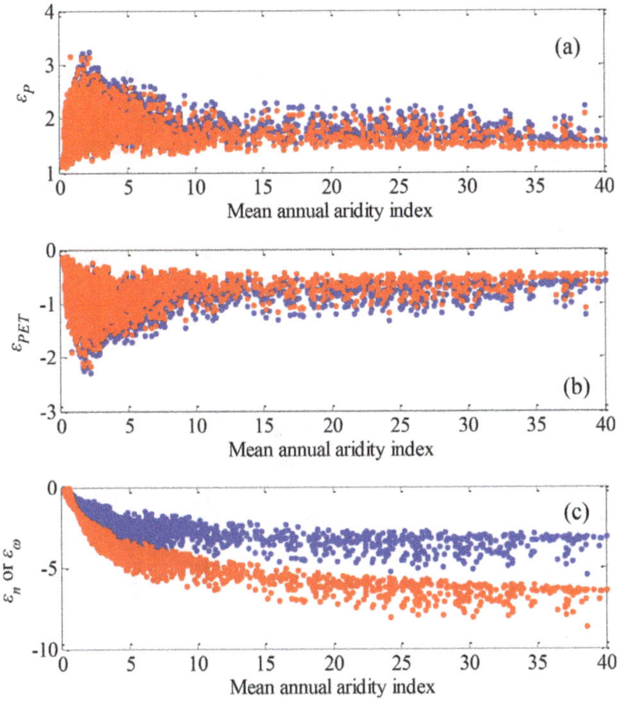

Figure 6. The relationship between mean annual aridity index and **(a)** P elasticity, **(b)** PET elasticity, and **(c)** parameter elasticity. The blue points represent the case of Eq. (2), and the red points represent the case of Eq. (3).

Qiangtang and Hexi). In general, there is an increased R in Southeast Drainage, Southwest Drainage, Qiangtang, Qinghai, Xinjiang, and Hexi (from 0.06 to 1 % yr^{-1}) and a decreased R in other basins (from -1.12 to -0.12 % yr^{-1}).

3.5 Future climate change

Figure 9 shows the uncertainty range of the relative change in mean annual P and PET in the basins for the period 2071–2100 under the RCP2.6, RCP4.5, and RCP8.5 scenarios as predicted by 28 CMIP5 models (relative to the baseline 1971–2000). As shown, there is a large difference between different GCMs and emission scenarios, which highlights the uncertainty inherent in projections of climate change. However, overall P is projected to increase in most of the basins, and greater increases are projected for higher emission scenarios. Meanwhile, greater increases tend to be projected for more arid basins, suggesting a decreasing risk of meteorological drought in the future. The average changes (red dotted lines) of mean annual P for the 14 basins range from 2.4 to 11.0 % in RCP2.6, from 4.2 to 16.0 % in RCP4.5, and from 3.1 to 23.7 % in RCP8.5. The largest increase in the RCP2.6 and RCP8.5 scenarios is found for the Qinghai River, while the largest increase in the RCP4.5 scenario is projected for the Hexi River. For PET, there is an increase projected in all basins due to climate warming, with the largest and smallest increases in the RCP8.5 and RCP2.6 scenarios, respectively. However, a large uncertainty exists among the GCMs, which is similar to that for P. Furthermore, the uncertainty range tends to be larger with higher emission scenarios. The average changes (red dotted lines) of PET for the basins range from 7.0 to 12.0 % in RCP2.6, from 13.5 to 22.2 % in RCP4.5, and from 27.9 to 49.8 % in RCP8.5. The largest and smallest average increases are projected for the Pearl River and Qiangtang River, respectively.

Figure 10 displays the multi-model ensemble median relative change in mean annual P and PET in China for the period 2071–2100 (relative to the baseline 1971–2000). The projected changes in P (or PET) have a similar spatial pattern for the three emission scenarios; that is, P is projected to show an increase in western China and the north-east, and PET is projected to increase significantly in southern China and some parts of the Tibetan Plateau, especially for the RCP8.5 scenario. In addition, note that there are small changes in P and significant increases in PET projected for southern China. This would result in an increasing risk of meteorological drought in the future.

3.6 Future changes in runoff due to climate change

Based on the estimates of elasticity from Eq. (2), the percentage changes in the contributions of annual P and PET, as well as climate, to R from the 28 GCMs for the period 2071–2100 are calculated for each of the 14 basins (relative to the baseline 1971–2000). As shown in Fig. 11, the changes in P contribution mainly follow the changes in P itself (Fig. 9). A positive contribution from P is projected for most of the basins, and larger contributions occur in more arid basins, as well as in higher emission scenarios. Negative contributions of PET to R are projected for all basins due to the negative coefficients of PET elasticity. Smaller contributions of PET are mainly found in the Southwest Drainage. In contrast, larger contributions are projected mainly in the Huaihe River, Haihe River, and Inner Mongolia River, where the percentage decreases from the 28 models can be up to 25, 35, and 90 % in the RCP2.6, RCP4.5 and RCP8.5 scenarios, respectively.

Climate change is projected to reduce the R in some humid basins, such as the Southeast Drainage and Pearl River, where the average changes in the three emission scenarios range from -22.83 to -3.0 % and from -23.6 to -3.5 %, respectively (Fig. 11g–i). For other basins, particularly for arid basins, the R is projected to increase due to climate change. The largest average changes in R under the RCP2.6 and RCP4.5 scenarios are found in the Qinghai River (12.85 and 16.18 %, respectively). For the RCP8.5 scenario, they are found in the Qiangtang River (18.59 %). Note that there is an obvious decrease in R (-17.59 %) projected for the Huaihe River under the RCP8.5 scenario, which is mainly caused by the larger negative contribution of PET.

Figure 12 shows the spatial distributions of the relative changes in the contributions of annual P and PET as well as climate to R in China for 2071–2100. This is based on the

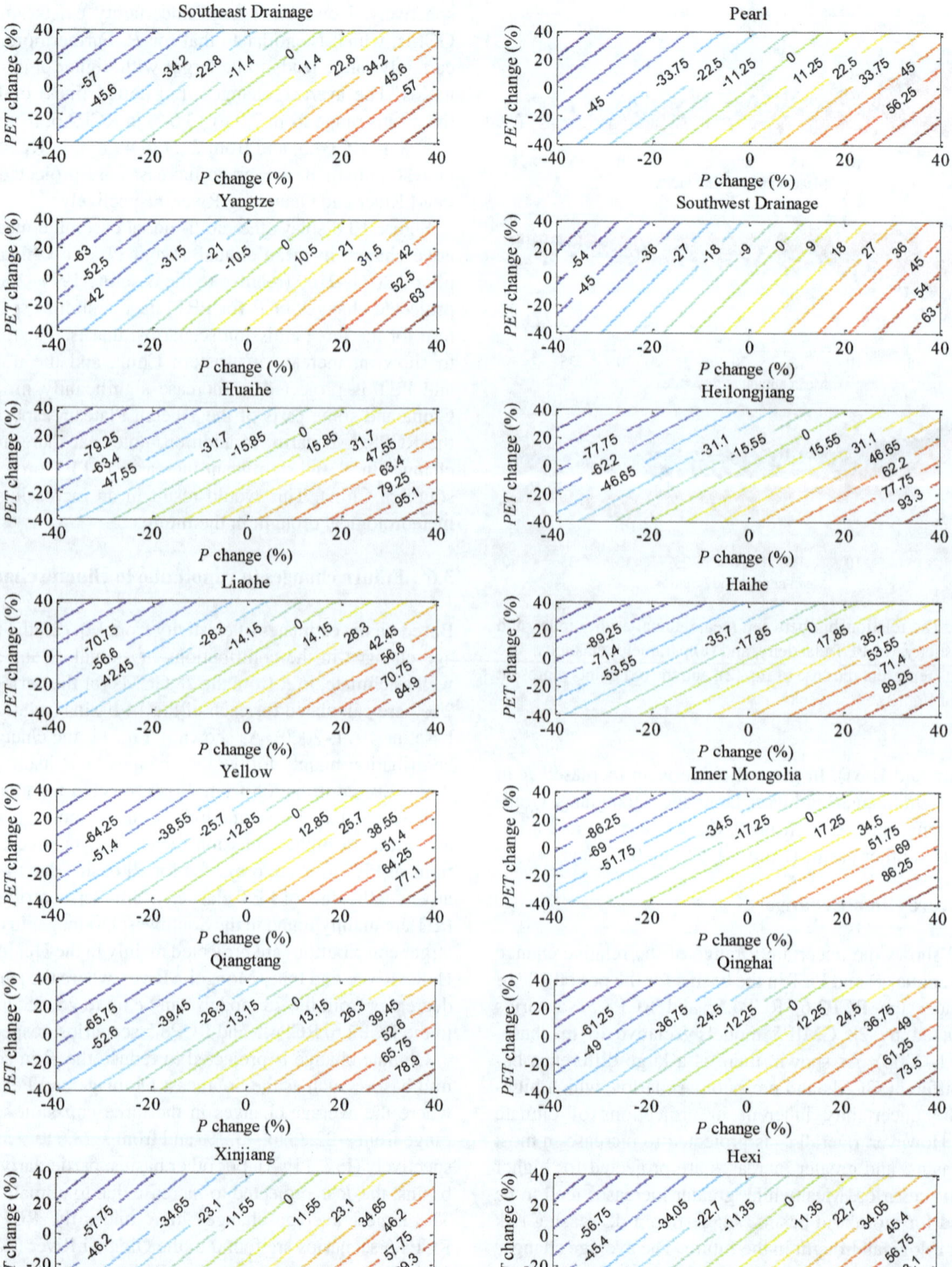

Figure 7. Contour plot of percentage R change due to the changes in P and PET for the 14 river basins. The P elasticity and PET elasticity of R are estimated based on Eq. (2).

CMIP5 multi-model ensemble medians. Compared with the baseline 1971–2000, the increases in R due to the changes in P are projected in western China and some parts of northern China, and this phenomenon is particularly significant in

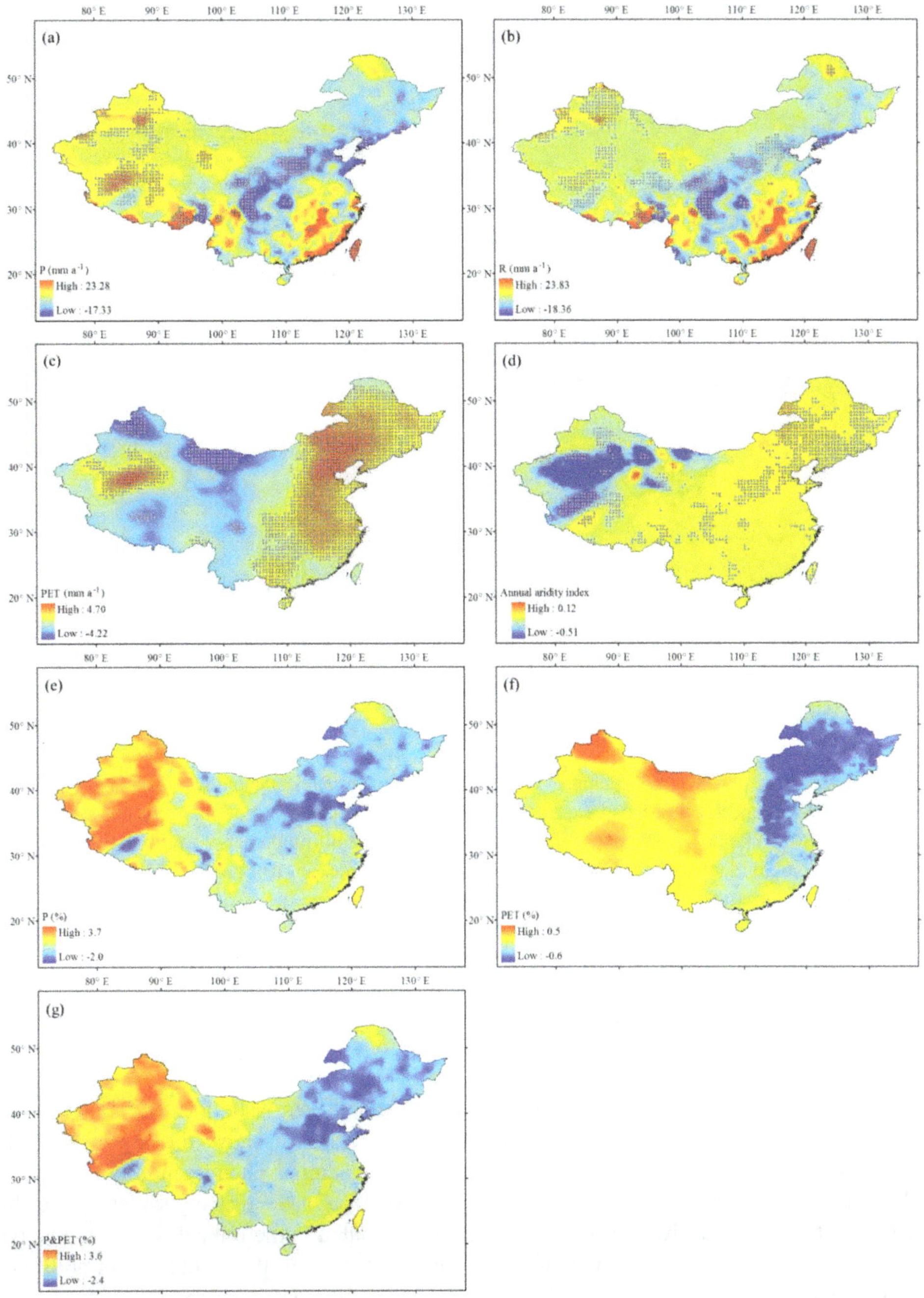

Figure 8. Trend magnitudes in annual time series of **(a)** P, **(b)** R, **(c)** PET, and **(d)** aridity index for the period 1960–2008 and spatial distributions of the contributions (unit: $\%\,\mathrm{yr}^{-1}$) of **(e)** P, **(f)** PET, and **(g)** climate (i.e. P and PET) to R in China for the period 1960–2008. The trend magnitudes are estimated by Sen's method. Grey dots are shown as statistically significant positive/negative trends ($p < 0.05$).

the RCP8.5 scenario (up to 60.3 %). In contrast, the changes in PET are projected to reduce the R in all of China, with the larger decreases occurring mainly in the North China Plain, north-eastern China, and some parts of western China. Overall, climate change is projected to cause an obvious increase (decrease) of R in western China (southern China) under any emission scenario (Fig. 12g–i). This suggests that the arid regions (humid regions) in China will become wetter (drier) in the future.

4 Discussion

4.1 The estimation of elasticity

The Budyko-based elasticity method is applied to quantify sensitivity of runoff to climate and catchment properties across China. Two Budyko models proposed by Choudhury (1999) and Fu (1981) are used for the comparison of the estimation of the climate elasticity of R. The results suggest that the climate elasticity is insensitive to the Budyko equa-

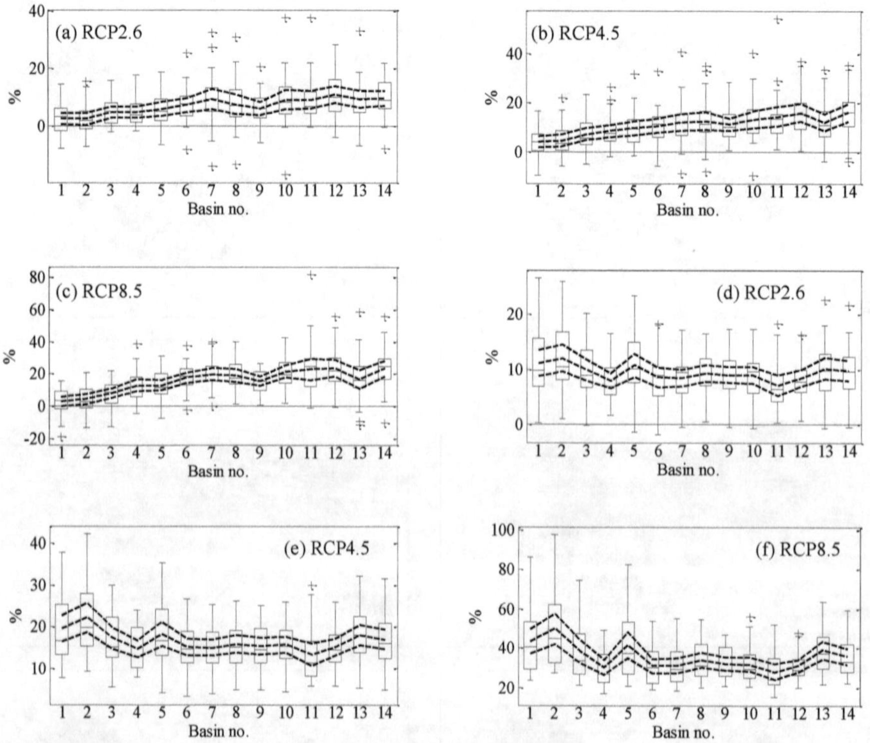

Figure 9. Box plots of relative change (%) in mean annual P under **(a)** RCP2.6, **(b)** RCP4.5, and **(c)** RCP8.5 scenarios and in mean annual PET under **(d)** RCP2.6, **(e)** RCP4.5, and **(f)** RCP8.5 scenarios calculated from 28 CMIP5 models in 14 basins for the period 2071–2100 (relative to the baseline 1971–2000). The boxes denote the interquartile model spread (range between the 25th and 75th quantiles), with the horizontal line indicating the ensemble median and the whiskers showing the extreme range of the 28 CMIP5 model simulations. Red dotted lines denote the average values of the multi-model ensemble. Blue dotted lines denote the 95 % significance levels' range of the average values of the multi-model ensemble. The basin number is consistent with that given in Fig. 1.

tions. The climate elasticity of R has been estimated in many regions of China. For example, the values of P elasticity are estimated as 2.4 for the Miyun Reservoir basin (Ma et al., 2010), 2.6 for the Luan River basin (X. Y. Xu et al., 2013), 2.1 for the headwater catchments of the Yellow River basin (Zheng et al., 2009), 1.4–1.7 for the Poyang Lake basin (Sun et al., 2013), 1.7–3.1 for the Hai River basin (Xu et al., 2014), 1.1–2.0 for southern China, the Songhua River basin, and the north-west, 2.1–4.8 for the Hai River basin, the Yellow River basin, and the Liao River basin (Yang et al., 2014), and 1.6–3.8 for the 63 catchments of China (Yang and Yang, 2011). In addition, the PET elasticity is estimated as -1.04 for the headwater catchments of the Yellow River basin (Zheng et al., 2009) and from -1 to -0.2 for the Poyang Lake basin (Sun et al., 2013). Those results are close to our results for P elasticity ranging from 1.1 to 3.2 and for PET elasticity ranging from -2.2 to -0.1 in China. It is worth noting that the values of P elasticity tend to be larger in the north-eastern and some parts of western China that are located in arid climates. This is in good agreement with the findings by Sankarasubramanian et al. (2001), which indicated that a larger P elasticity occurs in more arid regions. However, some parts of Xinjiang, which is more arid than southern China, have smaller P elasticity. Meanwhile, some parts of southern China, which is more humid than other regions in China, have larger P elasticity (Fig. 5). In addition, the Haihe River basin, located in a less arid climate than that of the north-west, shows the largest P elasticity in China (Table 2). A similar phenomenon is also introduced in Yang et al. (2014). One of the major reasons for this difference may be attributed to the impacts of human activities that alter the patterns of R in these regions. In addition, uncertainties in water budget data, such as the errors in the simulation of R and in the estimation of PET, may also contribute to this difference.

The comparisons for the estimates of ε_n and ε_ω suggest that although the values of ε_n and ε_ω are mainly dependent on the parameters of Budyko models, the spatial pattern of ε_n is consistent with that of ε_ω at the 0.5° grid points over China (Fig. 5e and f). Yang et al. (2008) indicated that the parameters n and ω from Eqs. (2) and (3) are linearly correlated. We also conducted a regression analysis of ε_n and ε_ω for all 0.5° grid points over China and found a strong linear correlation between ε_n and ε_ω($\varepsilon_\omega = 1.7061\varepsilon_n + 0.0986$, $r^2 = 0.96$). In addition, our results show that R is more sensitive to catchment properties (ε_n and ε_ω) in the more arid regions (Fig. 5e

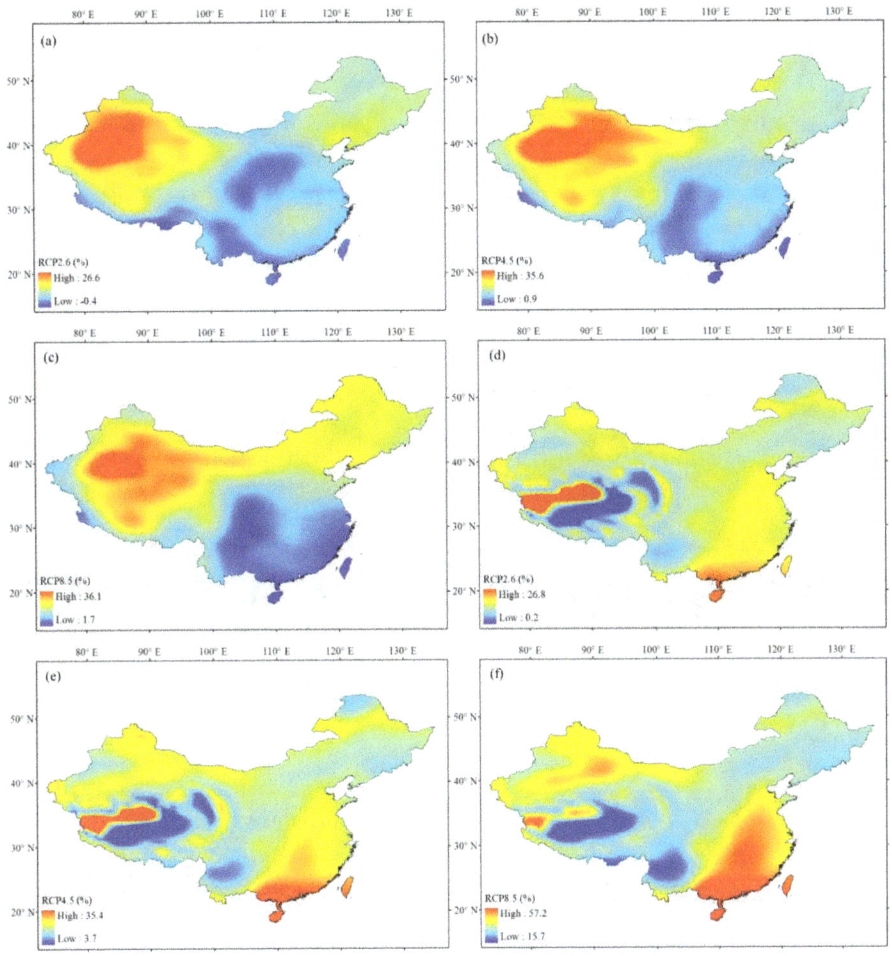

Figure 10. The CMIP5 multi-model ensemble median relative change (%) in mean annual P under **(a)** RCP2.6, **(b)** RCP4.5, and **(c)** RCP8.5 scenarios and in mean annual PET under **(d)** RCP2.6, **(e)** RCP4.5, and **(f)** RCP8.5 scenarios in China for the period 2071–2100 (relative to the baseline 1971–2000).

and f). The possible internal connection is that the arid regions with less vegetation coverage and stronger evaporation do not effectively hold the rainfall water that will be evaporated, leading to the smaller proportion of rainfall for R.

4.2 Sensitivity analysis for PET calculation methods

We compare four PET calculation methods, including the Penman method, the Thornthwaite method, the FAO-56 Penman–Monteith method (Allen et al., 1998), and the Thornthwaite method corrected by Eq. (1), to test the robustness of the PET elasticity result subject to PET uncertainties. In terms of mean annual PET as shown in Fig. 13a, the Thornthwaite method gives relatively low PET among the four methods, especially in arid basins (e.g. Qiangtang, Qinghai, Xinjiang, and Hexi). This is in agreement with previous studies, which indicated that the Thornthwaite method tends to underestimate PET in the arid areas (Hashemi and Habibian, 1979; Malek, 1987; Garcia et al., 2004). In contrast, the mean annual PET values from the other three methods are quite consistent, especially the Penman method and the Thornthwaite method corrected by Eq. (1). A similar result was also reported by Zeng and Cai (2016), which indicated that estimations of water balance at both annual and month scales are generally robust under various PET calculation methods (not including the Thornthwaite method). The PET elasticity calculations from the four different PET data for the 14 river basins are shown in Fig. 13b. The Thornthwaite method yields stronger PET elasticity than the other three methods in most of the basins, mainly due to the underestimation of PET. However, the other three methods give very similar results in all 14 basins. In summary, the estimation of PET elasticity is robust according to the PET calculations from the Penman method, the FAO-56 Penman–Monteith method, and the Thornthwaite method corrected by Eq. (1) but is not acceptable for the Thornthwaite method itself.

In general, the Thornthwaite method corrected by Eq. (1) significantly improves the accuracy of PET (Figs. 3 and 13a).

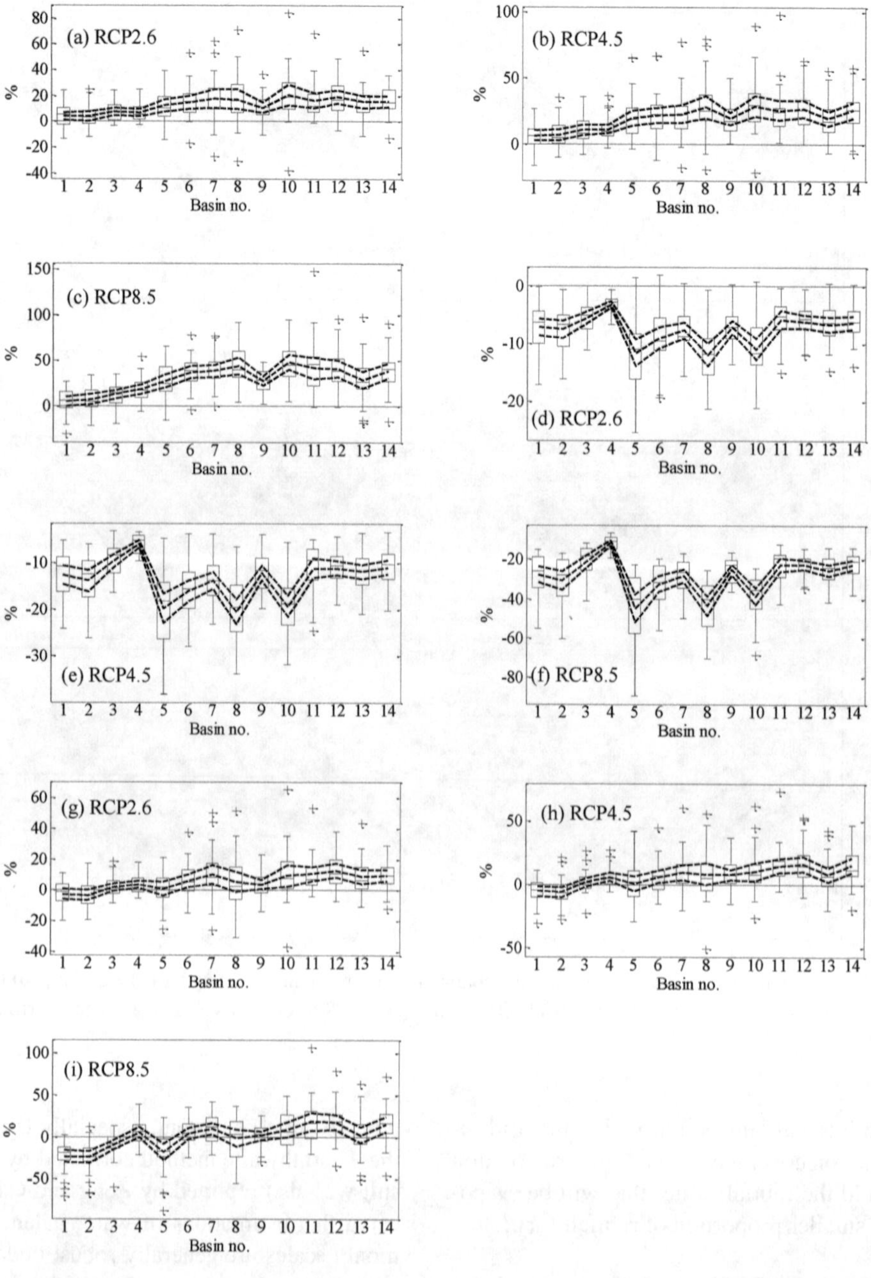

Figure 11. Box plots of relative change (%) in the contributions of annual P to R under **(a)** RCP2.6, **(b)** RCP4.5, and **(c)** RCP8.5 scenarios, in the contributions of annual PET to R under **(d)** RCP2.6, **(e)** RCP4.5, and **(f)** RCP8.5 scenarios, and in the contributions of climate to R under **(g)** RCP2.6, **(h)** RCP4.5, and **(i)** RCP8.5 scenarios calculated from 28 CMIP5 models in 14 basins for the period 2071–2100 (relative to the baseline 1971–2000). The boxes denote the interquartile model spread (range between the 25th and 75th quantiles), with the horizontal line indicating the ensemble median and the whiskers showing the extreme range of the 28 CMIP5 model simulations. Red dotted lines denote the average values of the multi-model ensemble. Blue dotted lines denote the 95 % significance levels' range of the average values of the multi-model ensemble. The basin number is consistent with that given in Fig. 1.

However, it should be emphasized that the Thornthwaite method is an empirical equation that neglects the effects of atmospheric conditions, such as wind speed, humidity and radiation (McVicar et al., 2012). In addition, Eq. (1) used for the bias correction of PET is part of a delta method (Graham et al., 2007; Sperna Weiland et al., 2010), which only considers the average change but ignores the differences in the standard deviation and the coefficient of variation between the projection and baseline periods (Watanabe et al., 2012). Therefore, a more physically based PET calculation method

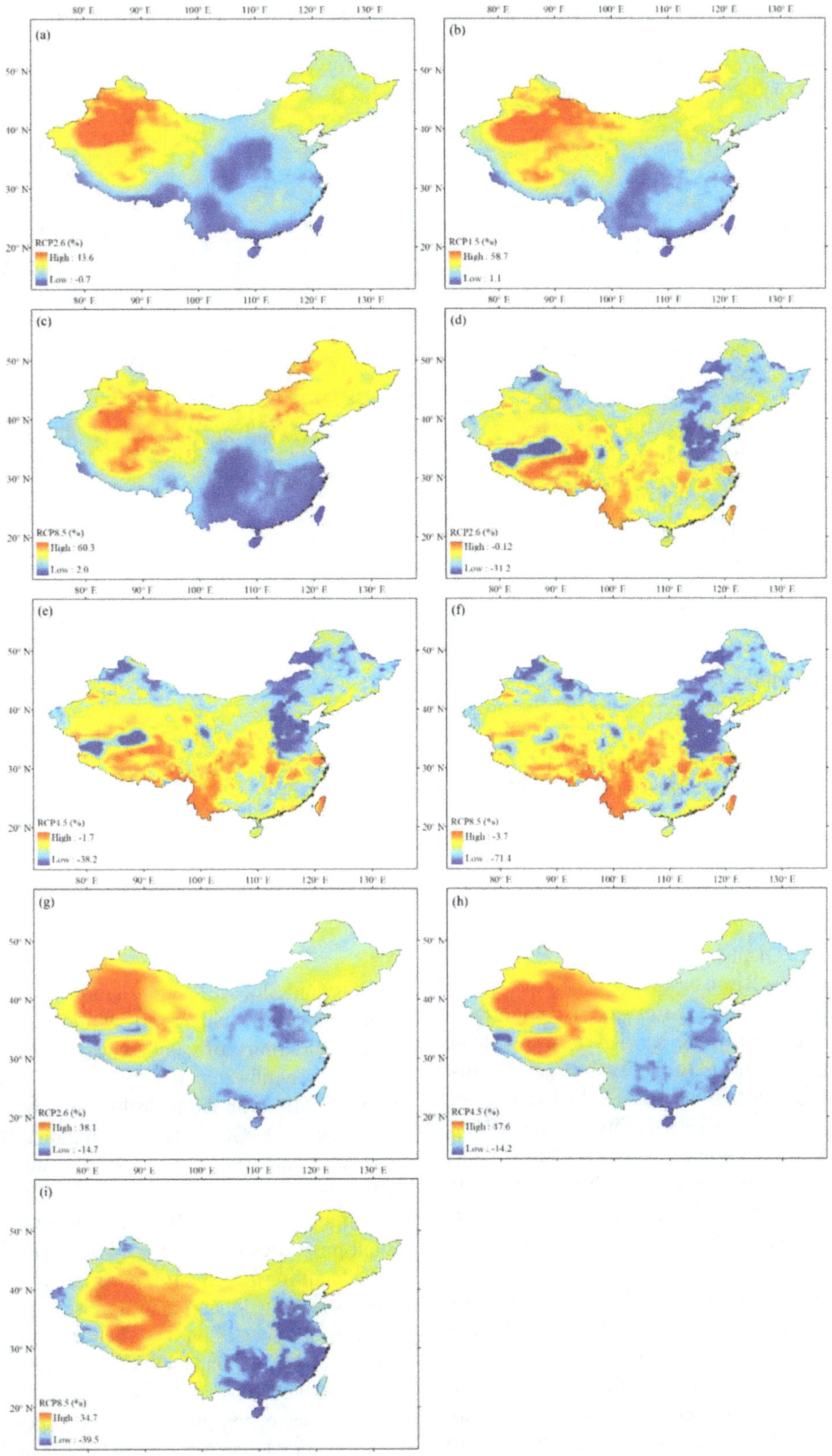

Figure 12. The CMIP5 multi-model ensemble median relative change (%) in the contributions of annual P to R under **(a)** RCP2.6, **(b)** RCP4.5, and **(c)** RCP8.5 scenarios, in the contributions of annual PET to R under **(d)** RCP2.6, **(e)** RCP4.5, and **(f)** RCP8.5 scenarios, and in the contributions of climate to R under **(g)** RCP2.6, **(h)** RCP4.5, and **(i)** RCP8.5 scenarios in China for the period 2071–2100 (relative to the baseline 1971–2000).

Table 3. The contributions of P, PET, and climate (i.e. P and PET) to R in the 14 basins of China for the period 1960–2008. The basin number is consistent with that given in Fig. 1. The numbers in the parentheses indicate the 1960–2008 mean aridity index.

Basin no.	P ($\% \, a^{-1}$)	PET ($\% \, a^{-1}$)	P and PET ($\% \, a^{-1}$)
1 (0.52)	0.19	−0.13	0.06
2 (0.64)	−0.03	−0.09	−0.12
3 (0.81)	−0.07	−0.07	−0.14
4 (1.19)	0.14	−0.01	0.13
5 (1.19)	−0.18	−0.27	−0.45
6 (1.43)	−0.35	−0.31	−0.66
7 (1.71)	−0.57	−0.34	−0.91
8 (2.14)	−0.74	−0.38	−1.12
9 (2.38)	−0.38	−0.04	−0.42
10 (4.41)	−0.40	−0.26	−0.66
11 (4.70)	0.99	0.01	1.00
12 (6.68)	0.43	−0.01	0.42
13 (8.09)	0.84	−0.02	0.82
14 (8.63)	0.11	0.08	0.19

(such as the Penman method) needs to be considered to fully understand the PET calculation uncertainties in the projections of climate change.

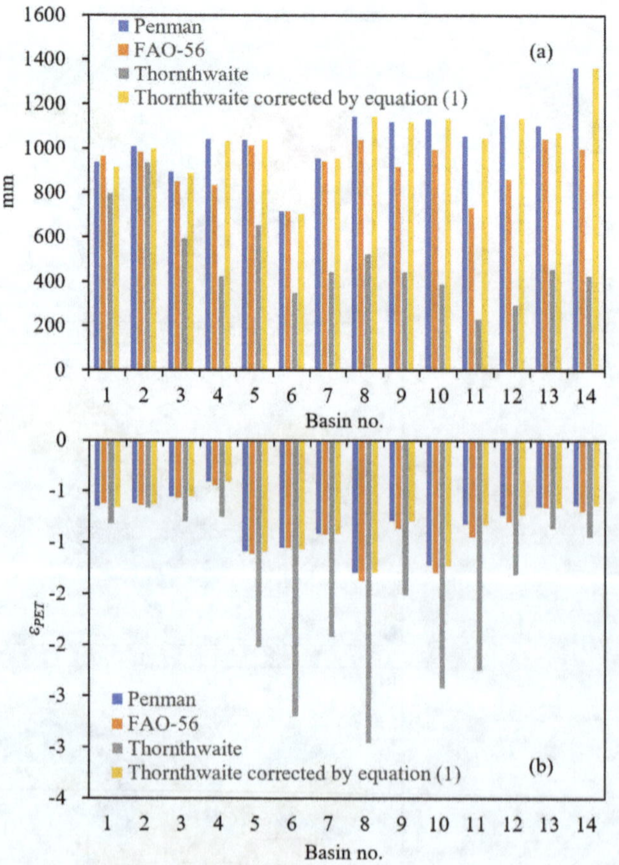

Figure 13. (a) Mean annual PET calculated from the four methods for the 14 river basins of China during the period 1960–2008. **(b)** PET elasticity calculated from Eq. (2) based on the four PET data for the 14 river basins of China during the period 1960–2008. The basin number is consistent with that given in Fig. 1.

4.3 The projections of climate change and runoff

The hydrological impacts of climate change have been investigated in many regions of China, such as the Hanjiang basin (Chen et al., 2007; Guo et al., 2009), the catchment of the Loess Plateau (Wang et al., 2013), the Qingjiang River basin (Chen et al., 2012), the Qiantang River basin (Y. P. Xu et al., 2013), the Songhuajiang River basin (Su et al., 2015), the south-eastern Tibetan Plateau (F. Li et al., 2013), the Pearl River basin (Yan et al., 2015), the Xin River basin (Zhang et al., 2016), the subcatchments of the Yangtze and Yellow River basins (Xu et al., 2011), the Huang-Huai-Hai region (Lu et al., 2012), and 10 major river basins in China (Wang et al., 2012). There is a large uncertainty involved in these impact studies, which results in a large difference in climate projections. For example, Wang et al. (2012) indicated that the prevailing pattern of "north dry and south wet" in China will likely be exacerbated under future climate warming. However, the results of most GCMs in this study suggest that the arid regions and humid regions of China are projected to become wetter and drier in the future, respectively. The main difference between the two studies is the use of different climate models, emission scenarios, and time periods. This also demonstrates that the results of climate projections should be taken with caution, since the regional climate simulations (especially of precipitation) from the GCMs are still not robust at the present stage.

This study focuses on the hydrological change due to climate change (i.e. changes in P and PET), while the effects of the variability of catchment properties (e.g. land cover change, groundwater and river water extraction, urbanization, irrigation) on the hydrology are overlooked here. Most of the available GCMs lack key regional feedback processes involving land use, such as forest plantations, irrigation, and urbanization feedbacks that are critically important throughout China (Piao et al., 2010). The projected changes in catchment properties therefore need to be involved in the GCMs to account for their hydrological impacts. In addition, recent studies indicated that plant responses to increasing CO_2 tend to keep more water on land, hence resulting in a greater increase in R (Milly and Dunne, 2016; Swann et al., 2016); that is to say, the hydrological models (e.g. VIC model), without the schemes of the plant stomatal responses to CO_2, would lead to an underestimation of R under high CO_2. Therefore, the implications of plants needing less water under high CO_2 should be included in the assessment of hydrological impacts of climate change.

4.4 Uncertainties

Generally, a multitude of sources of uncertainty are involved in the impact assessment of climate change. In this study, uncertainty mainly comes from the GCMs, emission scenarios, the elasticity method, and the estimation error of the water budget data. To highlight the uncertainty from the GCMs, the 28 GCMs, as produced by different research institutes around the world, are used for the comparison of climate change projections. There is a large difference in the projections of P and PET among the 28 GCMs. In particular, the uncertainty range of P tends to be larger for more arid regions, while the uncertainty range of PET tends to be larger for more humid regions (Fig. 9). This highlights the impact of potential misleading conclusions if only one climate model were to be used for the impact assessments. The large uncertainty driven by the GCMs in relation to the hydrological impacts of climate change has been reported in many previous studies (Kay et al., 2009; Prudhomme and Davies, 2009; Chen et al., 2011; Teng et al., 2012; Liu et al., 2013; Wu et al., 2014, 2015). It is worth noting that although the projected ranges of P and PET show large variability in various GCMs, most project a consistent change (i.e. increase) in P and PET for the future period (Fig. 9). In contrast, the uncertainty from the emission scenarios is smaller than that from the GCMs, since the projected changes in P (or PET) show a similar pattern under all emission scenarios (Fig. 9). The main difference is that the projected changes tend to be more significant in higher emission scenarios.

The elasticity equation (i.e. Eq. 4) used in this study is driven from the linear approximation of the Budyko equation (Eqs. 2 and 3) by neglecting the higher order. Such approximation would possibly lead to uncertainty in the estimation of climate elasticity. Yang et al. (2014) indicated that the error in the estimation of elasticity tends to increase with increasing changes in P and PET, as well as the increased parameter of the Budyko equation. Future research is needed to quantify the effects of the errors on the estimation of elasticity under various climate conditions.

In addition to uncertainty in PET calculation (as discussed in Sect. 4.2), there are also uncertainties associated with the estimates of other water budget components, such as R. As shown in Fig. 14, the sensitivity of climate (i.e. P and PET) elasticity to R varies considerably between basins and tends to be larger in more humid basins. Moreover, PET elasticity is more sensitive to changes in R compared with P elasticity for all 14 basins. As indicated by Zhang et al. (2014), although the R is realistically estimated for most of the basins (especially for humid basins) in China with a small relative error, there is still a large relative error for few arid basins in western China due to the lack of meteorological observations. Therefore, the large errors in simulated R of the VIC model may result in large uncertainties in the elasticity calculation, particularly in western China. Also note that some other natural water sources, such as snow and glaciers, which

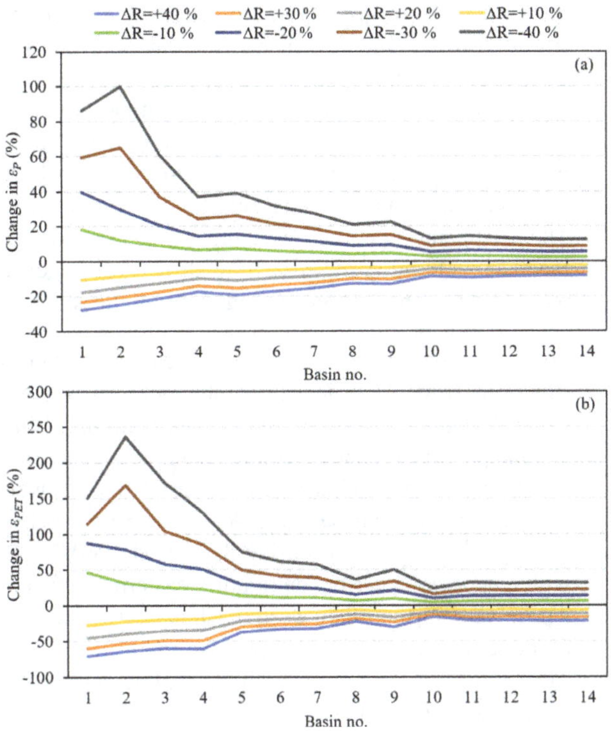

Figure 14. Comparison of changes in **(a)** P elasticity and **(b)** PET elasticity in response to changes in R for the 14 river basins of China. The basin number is consistent with that given in Fig. 1.

may contribute to R, are overlooked in this study. Lute and Abatzoglou (2014) highlighted the importance of extreme snowfall events in shaping the interannual variability of the water balance. The melting of snow and glaciers is generally significant at a seasonal timescale in some high-altitude regions of China. Neglecting the effects of snow and glaciers would lead to a bias in the modelling of R for these regions.

5 Conclusion

In this study, the Budyko-based elasticity method was used to investigate the responses of runoff to historical and future climate variability over China at both grid and catchment scales. The climate and catchment properties' elasticities of runoff were estimated based on the long-term (1960–2008) land surface data from Zhang et al. (2014). Twenty-eight GCMs with three emission scenarios from the CMIP5 were collected for the projections of climate change and its contribution to runoff in China during the period 2071–2100. The uncertainties associated with the estimates of PET, R, and climate elasticity, as well as climate projections, are discussed in detail. The main findings are summarized as follows.

1. The interannual variability of PET is more sensitive to that of P in more arid regions, while the opposite occurs in the response of interannual variability of R to

that of P. A large spatial variation exists in P elasticity (from 1.1 to 3.2) and PET elasticity (from -2.2 to -0.1) across China. The P elasticity is larger in north-eastern and western China than in southern China, which is opposite to that of PET elasticity. Among the 14 river basins, the Haihe River and Southwest Drainage have the largest and smallest climate elasticities, respectively. The catchment properties' elasticity of R is sensitive to mean annual aridity indices and tends to be stronger in more arid regions with increasing aridity indices.

2. For the period 1960–2008, the positive (negative) contributions from P to R are mainly found in western China (north-eastern China and the North China Plain), and the positive (negative) contributions of PET mainly occur in western China (north-eastern China). Overall, the climate contribution to R ranges from -2.4 to $3.6\,\%\,\mathrm{yr}^{-1}$ across China during the period 1960–2008, with a negative contribution in north-eastern China and a positive contribution in western China and some parts of the south-west. The largest positive and negative contributions of climate occur in the Qiangtang and Haihe River basins, respectively.

3. There is a large uncertainty in climate projections among the 28 GCMs. Moreover, the uncertainty range of the P (PET) projection tends to be larger for more arid (humid) regions. However, most of the GCMs project a consistent change in annual P or annual PET. For the period 2071–2100, the P is projected to increase in most parts of China, especially the western regions, and the PET is projected to increase in all of China, particularly the southern regions. Furthermore, greater increases are projected for higher emission scenarios. Due to future climate warming, the arid regions and humid regions of China are projected to become wetter and drier in the period 2071–2100, respectively (relative to the baseline 1971–2000).

The results of this study (especially of the climate change projections) should be taken with caution, since uncertainties in the results exist because of several issues, including the different simulations of GCMs, the estimation error of climate elasticity, and the estimation error in the water budget components. A thorough investigation of the uncertainty involved in the hydrologic effects of climate change in China should be considered in future research.

Data availability. The PET data are provided by the Terrestrial Hydrology Research Group of Princeton University (http://hydrology.princeton.edu/data.php; Sheffield et al., 2006, 2012). The land surface data over China, including P, RS, and BS, are provided by the Land Surface Processes and Global Change Research Group (http://hydro.igsnrr.ac.cn/resources.html; Zhang et al., 2014). The CMIP5 GCM data are obtained from the Canadian Climate Data and Scenarios (CCDS, http://climate-scenarios.canada.ca/?page=main).

Competing interests. The authors declare that they have no conflict of interest.

Acknowledgements. This research was supported by the Fundamental Research Funds for the Central Universities (grant no. 21617301) and partly supported by funding from the National Natural Science Foundation of China (grant no. 41530316) and the National Key Research and Development Program of China (grant no. 2016YFC0402805). We thank the editors and two reviewers for their valuable comments that considerably helped to improve this paper.

References

Adam, J. C., Clark, E. A., Lettenmaier, D. P., and Wood, E. F.: Correction of global precipitation products for orographic effects, J. Climate., 19, 15–38, 2006.

Allen, R., Pereira, L., Raes, D., and Smith, M.: Crop evapotranspiration – Guidelines for computing crop water requirements, FAO, Rome, 1998.

Chen, H., Guo, S. L., Xu, C. Y., and Singh, V. P.: Historical temporal trends of hydro-climatic variables and runoff response to climate variability and their relevance in water resource management in the Hanjiang basin, J. Hydrol., 344, 171–184, https://doi.org/10.1016/j.jhydrol.2007.06.034, 2007.

Chen, H., Xiang, T. T., Zhou, X., and Xu, C. Y.: Impacts of climate change on the Qingjiang watershed's runoff change trend in China, Stoch. Environ. Res. Risk A., 26, 847–858, https://doi.org/10.1007/s00477-011-0524-2, 2012.

Chen, J., Brissette, F. P., Poulin, A., and Leconte, R.: Overall uncertainty study of the hydrological impacts of climate change for a Canadian watershed, Water Resour. Res., 47, W12509, https://doi.org/10.1029/2011WR010602, 2011.

Choudhury, B. J.: Evaluation of an empirical equation for annual evaporation using field observations and results from a biophysical model, J. Hydrol., 216, 99–110, 1999.

Christensen, N. S., Wood, A. W., Voisin, N., Lettenmaier, D. P., and Palmer, R. N.: The effects of climate change on the hydrology and water resources of the Colorado River basin, Climatic Change, 62, 337–363, 2004.

Dooge, J. C., Bruen, M., and Parmentier, B.: A simple model for estimating the sensitivity of runoff to long-term changes in precipitation without a change in vegetation, Adv. Water Resour., 23, 153–163, 1999.

Fu, B. P.: On the calculation of the evaporation from land surface, Sci. Atmos. Sin., 5, 23–31, 1981.

Fu, G., Charles, S. P., and Chiew, F. H.: A two-parameter climate elasticity of streamflow index to assess climate change effects on annual streamflow, Water Resour. Res., 43, W11419, https://doi.org/10.1029/2007WR005890, 2007.

Garcia, M., Raes, D., Allen, R., and Herbas, C.: Dynamics of reference evapotranspiration in the Bolivian highlands (Altiplano), Agr. Forest Meteorol., 125, 67–82, 2004.

Graham, L. P., Andréasson, J., and Carlsson, B.: Assessing climate change impacts on hydrology from an ensemble of regional climate models, model scales and linking methods – A case study on the Lule River basin, Climatic Change, 81, 293–307, https://doi.org/10.1007/s10584-006-9215-2, 2007.

Guo S., Guo, J., Zhang, J., and Chen, H.: VIC distributed hydrological model to predict climate change impact in the Hanjiang basin, Sci. China Ser. E, 52, 3234–3239, 2009.

Harding, B. L., Wood, A. W., and Prairie, J. R.: The implications of climate change scenario selection for future streamflow projection in the Upper Colorado River Basin, Hydrol. Earth Syst. Sci., 16, 3989–4007, https://doi.org/10.5194/hess-16-3989-2012, 2012.

Harris, I., Jones, P. D., Osborna, T. J., and Lister, D. H.: Updated high-resolution grids of monthly climatic observations – the CRU TS3.10 Dataset, Int. J. Climatol., 34, 623–642, 2014.

Hashemi, F. and Habibian, M. T.: Limitations of temperature-based methods in estimating crop evapotranspiration in arid-zone agricultural projects, Agr. Forest Meteorol., 20, 237–247, 1979.

IPCC: Climate Change 2013: The Physical Science Basis, in: Contribution of Working Group I to the Fifth Assessment Report of the Intergovernmental Panel on Climate Change, edited by: Stocker, T. F., Qin, D., Plattner, G.-K., Tignor, M., Allen, S. K., Boschung, J., Nauels, A., Xia, Y., Bex, V., and Midgley, P. M., Cambridge University Press, Cambridge, UK and New York, NY, USA, 1535 pp., 2013.

Jung, M., Reichstein, M., Ciais, P., Seneviratne, S. I., Sheffield, J., Goulden, M. L., Bonan, G., Cescatti, A., Chen, J. Q., deJeu, R., Dolman, A. J., Eugster, W., Gerten, D., Gianelle, D., Gobron, N., Heinke, J., Kimball, J., Law, B. E., Montagnani, L., Mu, Q. Z., Mueller, B., Oleson, K., Papale, D., Richardson, A. D., Roupsard, O., Running, S., Tomelleri, E., Viovy, N., Weber, U., Williams, C., Wood, E., Zaehle, S., and Zhang, K.: Recent decline in the global land evapotranspiration trend due to limited moisture supply, Nature, 467, 951–954, 2010.

Kay, A. L. and Jones, D. A.: Transient changes in flood frequency and timing in Britain under potential projections of climate change, Int. J. Climatol., 32, 489–502, 2012.

Kay, A. L., Jones, R. G., and Reynard, N. S.: RCM rainfall for UK flood frequency estimation. II. Climate change results, J. Hydrol., 318, 163–172, https://doi.org/10.1016/j.jhydrol.2005.06.013, 2006.

Kay, A. L., Davies, H. N., Bell, V. A., and Jones, R. G.: Comparison of uncertainty sources for climate change impacts: flood frequency in England, Climatic Change, 92, 41–63, 2009.

Kendall, M. G.: Rank Correlation Methods, 4th Edn., Charles Grifin, London, 1975.

Li, D., Pan, M., Cong, Z., Zhang, L., and Wood, E.: Vegetation control on water and energy balance within the Budyko framework, Water Resour. Res., 49, 969–976, https://doi.org/10.1002/wrcr.20107, 2013.

Li, F., Zhang, Y., Xu, Z., Teng, J., Liu, C., Liu, W., and Mpelasoka, F.: The impact of climate change on runoff in the southeastern Tibetan Plateau, J. Hydrol., 505, 188–201, 2013.

Liu, L. L., Fischer, T., Jiang, T., and Luo, Y.: Comparison of uncertainties in projected flood frequency of the Zhujiang River, South China, Quatern. Int., 304, 51–61, 2013.

Liu, Q., and McVicar, T. R.: Assessing climate change induced modification of Penman potential evaporation and runoff sensitivity in a large water-limited basin, J. Hydrol., 464, 352–362, 2012.

Lu, G. H., Xiao, H., Wu, Z. Y., Zhang, S. L., and Li, Y.: Assessing the impacts of future climate change on hydrology in Huang-Huai-Hai region in China using the PRECIS and VIC models, J. Hydrol. Eng., 18, 1077–1087, 2012.

Lute, A. C. and Abatzoglou, J. T.: Role of extreme snowfall events in interannual variability of snowfall accumulation in the western United States, Water Resour. Res. 50, 2874–2888, 2014.

Ma, H. A., Yang, D. W., Tan, S. K., Gao, B., Hu, Q. F.: Impact of climate variability and human activity on streamflow decrease in the Miyun Reservoir catchment, J. Hydrol., 389, 317–324, 2010.

Malek, E.: Comparison of alternative methods for estimating ETP and evaluation of advection in the Bajah area, Iran, Agr. Forest Meteorol., 39, 185–192, 1987.

Mann, H. B.: Non-parametric tests against trend, Econometrica, 13, 245–259, 1945.

McVicar, T. R., Roderick, M. L., Donohue, R. J., Li, L. T., Van Niel, T. G., Thomas, A., Grieser, J., Jhajharia, D., Himri, Y., and Mahowald, N. M.: Global review and synthesis of trends in observed terrestrial near-surface wind speeds: implications for evaporation, J. Hydrol., 416–417, 182–205, 2012.

Milly, P. C. D. and Dunne, K. A.: Macroscale water fluxes, 2. Water and energy supply control of their interannual variability, Water Resour. Res., 38, 1206, https://doi.org/10.1029/2001WR000760, 2002.

Milly, P. C. D. and Dunne, K. A.: Potential Evapotranspiration and Continental Drying, Nat. Clim. Change, 6, 946–949, 2016.

Milly, P. C. D., Dunne, K. A., and Vecchia, A. V.: Global pattern of trends in streamflow and water availability in a changing climate, Nature, 438, 347–350, 2005.

Nijssen, B., Schnur, R., and Lettenmaier, D. P.: Global retrospective estimation of soil moisture using the Variable Infiltration Capacity land surface model, 1980–93, J. Climate, 14, 1790–1808, 2001.

Oki, T. and Kanae, S.: Global hydrological cycles and world water resources, Science, 313, 1068–1072, 2006.

Pan, M., Sahoo, A. K., Troy, T. J., Vinukollu, R. K., Sheffield, J., and Wood, E. F.: Multisource estimation of long-term terrestrial water budget for major global river basins, J. Climate, 25, 3191–3206, 2012.

Penman, H. L.: Natural evaporation from open water, bare soil and grass, P. Roy. Soc. Lond., 193, 120–145, 1948.

Piao, S., Ciais, P., Huang, Y., Shen, Z., Peng, S., Li, J., Zhou, L., Liu, H., Ma, Y., Ding, Y., Friedlingstein, P., Liu, C., Tan, K., Yu, Y., Zhang, T., and Fang, J.: The impacts of climate change on water resources and agriculture in China, Nature, 467, 43–51, 2010.

Prudhomme, C. and Davies, H. N.: Assessing uncertainties in climate change impact analyses on river flow regimes in the UK. Part 2: future climate, Climatic Change, 93, 197–222, 2009.

Raff, D. A., Pruitt, T., and Brekke, L. D.: A framework for assessing flood frequency based on climate projection information, Hydrol. Earth Syst. Sci., 13, 2119–2136, https://doi.org/10.5194/hess-13-2119-2009, 2009.

Rodell, M., Houser, P. R., Jambor, U., Gottschalck, J., Mitchell, K., Meng, C. J., Arsenault, K., Cosgrove, B., Radakovich, J., Bosilovich, M., Entin, J. K., Walker, J. P., Lohmann, D., and Toll, D.: The Global Land Data Assimilation System, B. Am. Meteorol. Soc., 85, 381–394, 2004.

Roderick, M. L. and Farquhar, G. D.: A simple framework for relating variations in runoff to variations in climatic conditions and catchment properties, Water Resour. Res., 47, W00G07, https://doi.org/10.1029/2010WR009826, 2011.

Roderick, M. L., Sun, F., Lim, W. H., and Farquhar, G. D.: A general framework for understanding the response of the water cycle to global warming over land and ocean, Hydrol. Earth Syst. Sci., 18, 1575–1589, https://doi.org/10.5194/hess-18-1575-2014, 2014.

Sankarasubramanian, A., Vogel, R. M., and Limbrunner, J. F.: Climate elasticity of streamflow in the United States, Water Resour. Res., 37, 1771–1781, 2001.

Schaake, J. C.: From climate to flow, in: Climate Change and U.S. Water Resources, edited by: Waggoner, P. E., John Wiley, New York, 177–206, 1990.

Sen, P. K.: Estimates of the regression coeffcient based on Kendall's tau, J. Am. Stat. Assoc., 63, 1379–1389, 1968.

Sheffield, J. and Wood, E. F.: Characteristics of global and regional drought, 1950–2000: Analysis of soil moisture data from off-line simulation of the terrestrial hydrologic cycle, J. Geophys. Res., 112, D17115, https://doi.org/10.1029/2006JD008288, 2007.

Sheffield, J., Goteti, G., and Wood, E. F.: Development of a 50-year high-resolution global dataset of meteorological forcings for land surface modeling, J. Climate, 19, 3088–3110, https://doi.org/10.1175/JCLI3790.1, 2006.

Sheffield, J., Wood, E. F., and Roderick, M. L.: Little change in global drought over the past 60 years, Nature, 491, 435–438, https://doi.org/10.1038/nature11575, 2012.

Shuttleworth, W. J.: Evaporation, in: Handbook of Hydrology, edited by: Maidment, D. R., McGraw-Hill, Sydney, 1993.

Smith, A., Bates, P., Freer, J., and Wetterhall, F.: Investigating the application of climate models in flood projection across the UK, Hydrol. Process., 28, 2810–2823, https://doi.org/10.1002/hyp.9815, 2014.

Sperna Weiland, F. C., van Beek, L. P. H., Kwadijk, J. C. J., and Bierkens, M. F. P.: The ability of a GCM-forced hydrological model to reproduce global discharge variability, Hydrol. Earth Syst. Sci., 14, 1595–1621, https://doi.org/10.5194/hess-14-1595-2010, 2010.

Su, B., Zeng, X., Zhai, J., Wang, Y., and Li, X.: Projected precipitation and streamflow under sres and rcp emission scenarios in the songhuajiang river basin, China, Quatern. Int., 380, 95–105, 2015.

Sun, S. L., Chen, H. S., Ju, W. M., Song, J., Zhang, H., Sun, J., and Fang, Y. J.: Effects of climate change on annual streamflow using climate elasticity in Poyang Lake Basin, China, Theor. Appl. Climatol., 112, 169–183, 2013.

Swann, A. L., Hoffman, F. M., Koven, C. D., and Randerson, J. T.: Plant responses to increasing CO_2 reduce estimates of climate impacts on drought severity, P. Natl. Acad. Sci. USA, 113, 10019–10024, 2016.

Teng, J., Vaze, J., Chiew, F. H. S., Wang, B., and Perraud, J.: Estimating the Relative Uncertainties Sourced from GCMs and Hydrological Models in Modeling Climate Change Impact on Runoff, J. Hydrometeorol., 13, 122–139, 2012.

Thornthwaite, C. W.: An approach toward a rational classification of climate, Geogr. Rev., 38, 55–94, 1948.

Vano, J. A., Scott, M., Voisin, N., Stockle, C. O., Hamlet, A. F., Mickelson, K. E. B., Elsner, M. M., and Lettenmaier, D. P.: Climate change impacts on water management and irrigated agriculture in the Yakima River basin, Washington, USA, Climatic Change, 102, 287–317, https://doi.org/10.1007/s10584-010-9856-z, 2010.

Vano, J. A., Udall, B., Cayan, D. R., Overpeck, J. T., Brekke, L. D., Das, T., Hartmann, H. C., Hidalgo, H. G., Hoerling, M., McCabe, G. J., Morino, K., Webb, R. S., and Lettenmaier, D. P.: Understanding uncertainties in future Colorado River streamflow, B. Am. Meteorol. Soc., 95, 59–78, 2014.

Vano, J. A., Nijssen, B., and Lettenmaier, D. P.: Seasonal hydrologic responses to climate change in the Pacific Northwest, Water Resour. Res., 51, 1959–1976, 2015.

Wang, G. Q., Zhang, J. Y., Jin, J. L., Pagano, T. C., Calow, R., Bao, Z. X., Liu, C. S., Liu, Y. L., and Yan, X. L.: Assessing water resources in China using PRECIS projections and a VIC model, Hydrol. Earth Syst. Sci., 16, 231–240, https://doi.org/10.5194/hess-16-231-2012, 2012.

Wang, G. Q., Zhang, J. Y., Xuan, Y. Q., Liu, J. F., Jin, J. L., Bao, Z. X., He, R. M., Liu, C. S, Liu, Y. L., and Yan, X. L.: Simulating the impact of climate change on runoff in a typical river catchment of the Loess Plateau, China, J. Hydrometeorol., 14, 1553–1561, 2013.

Watanabe, S., Kanae, S., Seto, S., Yeh, P. J. F., Hirabayashi, Y., and Oki, T.: Intercomparison of bias-correction methods for monthly temperature and precipitation simulated by multiple climate models, J. Geophys. Res., 117, D23114, https://doi.org/10.1029/2012JD018192, 2012.

Wu, C. H. and Huang, G. R.: Changes in heavy precipitation and floods in the upstream of the Beijiang River basin, South China, Int. J. Climatol., 35, 2978–2992, 2015.

Wu, C. H. and Huang, G. R.: Projection of climate extremes in the Zhujiang River basin using a regional climate model, Int. J. Climatol., 36, 1184–1196, 2016.

Wu, C. H., Huang, G. R., Yu, H. J., Chen, Z. J., and Ma, J. G.: Impact of climate change on reservoir flood control in the upstream area of the Beijiang River Basin, South China, J. Hydrometeorol., 15, 2203–2218, https://doi.org/10.1175/JHM-D-13-0181.1, 2014.

Wu, C. H., Huang, G. R., and Yu, H. J.: Prediction of extreme floods based on CMIP5 climate models: a case study in the Beijiang River basin, South China, Hydrol. Earth Syst. Sci., 19, 1385–1399, https://doi.org/10.5194/hess-19-1385-2015, 2015.

Xiao, H., Lu, G. H., Wu, Z. Y., and Liu, Z. Y.: Flood response to climate change in the Pearl River basin for the next three decades, J. Hydraul. Eng., 12, 1409–1419, 2013.

Xu, H., Taylor, R. G., and Xu, Y.: Quantifying uncertainty in the impacts of climate change on river discharge in sub-catchments of the Yangtze and Yellow River Basins, China, Hydrol. Earth Syst. Sci., 15, 333–344, https://doi.org/10.5194/hess-15-333-2011, 2011.

Xu, X. Y., Yang, H. B., Yang, D. W., and Ma, H.: Assessing the impacts of climate variability and human activities on annual runoff in the Luan River basin, China, Hydrol. Res., 44, 940–952, 2013.

Xu, X. Y., Yang, D. W., Yang, H. B., and Lei, H. M.: Attribution analysis based on the Budyko Hypothesis for detecting the dominant cause of runoff decline in Haihe basin, J. Hydrol., 510, 530–540, 2014.

Xu, Y. P., Zhang, X., Ran, Q., and Tian, Y.: Impact of climate change on hydrology of upper reaches of Qiantang River Basin, East China, J. Hydrol., 483, 51–60, 2013.

Yan, D., Werners, S. E., Ludwig, F., and Huang, H. Q.: Hydrological response to climate change: The Pearl River, China under different RCP scenarios, J. Hydrol.: Reg. Stud., 4, 228–245, 2015.

Yang, H. and Yang, D.: Derivation of climate elasticity of runoff to assess the effects of climate change on annual runoff, Water Resour. Res., 47, W07526, https://doi.org/10.1029/2010WR009287, 2011.

Yang, H., Yang, D., Lei, Z., and Sun, F.: New analytical derivation of the mean annual water-energy balance equation, Water Resour. Res., 44, W03410, https://doi.org/10.1029/2007WR006135, 2008.

Yang, H., Qi, J., Xu, X., Yang, D., and Lv, H.: The regional variation in climate elasticity and climate contribution to runoff across China, J. Hydrol., 517, 607–616, 2014.

Zeng, R. and Cai, X.: Climatic and terrestrial storage control on evaporation temporal variability: Analysis of river basins around the world, Geophys. Res. Lett., 43, 185–195, https://doi.org/10.1002/2015GL066470, 2016.

Zhang, L., Hickel, K., Dawes, W. R., Chiew, F. H. S., Western, A. W., and Briggs, P. R.: A rational function approach for estimating mean annual evapotranspiration, Water Resour. Res., 40, W02502, https://doi.org/10.1029/2003WR002710, 2004.

Zhang, X., Tang, Q., Pan, M., Tang, Y.: A Long-Term Land Surface Hydrologic Fluxes and States Dataset for China, J. Hydrometeorol., 15, 2067–2084, https://doi.org/10.1175/JHM-D-13-0170.1, 2014.

Zhang, Y., You, Q., Chen, C., and Ge, J.: Impacts of climate change on streamflows under RCP scenarios: A case study in Xin River Basin, China, Atmos. Res., 178, 521–534, 2016.

Zheng, H., Zhang, L., Zhu, R., Liu, C., Sato, Y., and Fukushima, Y.: Responses of streamflow to climate and land surface change in the headwaters of the Yellow River Basin, Water Resour. Res., 45, W00A19, https://doi.org/10.1029/2007WR006665, 2009.

3

Vegetation dynamics and climate seasonality jointly control the interannual catchment water balance in the Loess Plateau under the Budyko framework

Tingting Ning[1,2], **Zhi Li**[3], **and Wenzhao Liu**[1,2]

[1]State Key Laboratory of Soil Erosion and Dryland Farming on the Loess Plateau, Institute of Soil and Water Conservation, CAS & MWR, Yangling, Shaanxi 712100, China
[2]University of the Chinese Academy of Sciences, Beijing 100049, China
[3]College of Natural Resources and Environment, Northwest A&F University, Yangling, Shaanxi 712100, China

Correspondence to: Wenzhao Liu (wenzhaoliu@hotmail.com)

Abstract. Within the Budyko framework, the controlling parameter (ω in the Fu equation) is widely considered to represent landscape conditions in terms of vegetation coverage (M); however, some qualitative studies have concluded that climate seasonality (S) should be incorporated in ω. Here, we discuss the relationship between ω, M, and S, and further develop an empirical equation so that the contributions from M to actual annual evapotranspiration (ET) can be determined more accurately. Taking 13 catchments in the Loess Plateau as examples, ω was found to be well correlated with M and S. The developed empirical formula for ω calculations at the annual scale performed well for estimating ET by the cross-validation approach. By combining the Budyko framework with the semi-empirical formula, the contributions of changes in ω to ET variations were further decomposed as those of M and S. Results showed that the contributions of S to ET changes ranged from 0.1 to 74.8 % (absolute values). Therefore, the impacts of climate seasonality on ET cannot be ignored, otherwise the contribution of M to ET changes will be estimated with a large error. The developed empirical formula between ω, M, and S provides an effective method to separate the contributions of M and S to ET changes.

1 Introduction

The water cycle has been influenced greatly by human activities and climate change since the 1960s, and considerable variability in hydrological processes has been observed in many basins around the world; this has led to a series of problems concerning essential water resources (Stocker et al., 2014). Analyses of the mechanisms of the interactions among the water balance, climate, and catchment surface conditions are important for understanding these complex processes at different spatio-temporal scales (Zhang et al., 2008), and such work has practical significance in regard to the improvement of water resources and land management (Rodriguez-Iturbe, 2000; Xu et al., 2014).

Budyko (1948, 1974) postulated that precipitation (P, represents the water supply from the atmosphere) and potential evapotranspiration (ET_0, represents the demand by the atmosphere) are the two dominant variables that control the long-term average water balance. The Budyko framework is considered one of the most abiding frameworks linking climatic conditions to the runoff (R) and actual evapotranspiration (ET) of a catchment (Donohue et al., 2007), and it has been used successfully to investigate interactions between hydrological processes, climate variability, and landscape characteristics (e.g. Milly, 1994; Woods, 2003; Yokoo et al., 2008; Yang et al., 2009; Liu et al., 2012). A series of empirical formulas have been developed for the Budyko curve based on theoretical research and case studies of regional water balance over the past 50 years. Among them, the Fu (Fu, 1981; Zhang et al., 2004) and Choudhury–Yang equations (Choudhury, 1999; Yang et al., 2008) have been used widely; furthermore, the controlling parameters ω (in the Fu equation) and n (in the Choudhury–Yang equation) are related linearly (Yang et al., 2008).

Deviations from the Budyko curve have been detected in previous studies, which indicates that in addition to climate conditions, other variables can also influence the variability of regional water balances (Yang et al., 2007; Wang and Alimohammadi, 2012). Two kinds of factors have been identified to be responsible for the deviations. The first type of factor is related to land surface conditions, and includes vegetation dynamics (Donohue et al., 2007, 2010; Yang et al., 2009; Li et al., 2013; Zhang et al., 2016), soil properties, and topography (Yang et al., 2007; Peel et al., 2010). The second type of factor includes seasonal climate variability (in addition to P and ET_0), such as storm depth (Donohue et al., 2012; Shao et al., 2012; Li et al., 2014), frequency of daily rainfall (Milly, 1994), and differences in the timing of P and ET_0 (Budyko, 1961; Potter et al., 2005). All of these factors can be encoded into the controlling parameter of the Budyko equations (e.g. ω in the Fu equation and n in the Choudhury–Yang equation). So far, a great deal of attention has been paid to the relationships between land surface conditions and the controlling parameter. Based on satellite products of vegetation such as the normalized difference vegetation index (NDVI), vegetation has been found to correlate well with the controlling parameter, and some empirical relationships have been successfully developed (Yang et al., 2009; Li et al., 2013). In particular, the controlling parameter can be better represented by vegetation when higher spatiotemporal resolution products are used. Therefore, the impacts of dynamic changes in vegetation on hydrology can be effectively quantified.

Many current studies attribute any effects of the controlling parameter to landscape characteristics (Roderick and Farquhar, 2011; Zhou et al., 2015; Zhang et al., 2016). However, both empirical evidence and modelling tests have demonstrated the important function of climate seasonality on catchment water yield, and thereby, evidence exists that climate seasonality also strongly affects the controlling parameter in the Budyko equations (Berghuijs and Woods, 2016). Some indices and models have thus been developed to address this issue, and several potential solutions have been discussed (Milly, 1993, 1994; Potter et al., 2005; Yokoo et al., 2008; Feng et al., 2012; Li, 2014). Yang et al. (2012) introduced the climate seasonality index into the Budyko framework and proposed an empirical equation to include its effect in the estimation of the long-term controlling parameters; however, by focusing on the mean annual scale, the effects of vegetation dynamics were not considered. Therefore, how the vegetation dynamics and climate seasonality jointly control the interannual variability in the controlling parameters needs further interpretation.

Therefore, the primary motivation behind this study was to detect the potential linkages between the controlling parameter and surface condition change, as well as climate seasonality at the annual scale. The specific objectives were to derive an appropriate analytic formula between parameter ω in the Fu equation and the above two factors for typical catchments in the Loess Plateau, and then quantify the impacts of vegetation change and climate seasonality variability on the catchment water balance.

2 Methods

2.1 Annual water balance definition

The Budyko framework assumes that the long-term average water balance is in a steady state (Wang and Alimohammadi, 2012), and the water storage change in a catchment can be negligible. The interannual variability of the water balance in individual basins can also be studied by overlooking the interannual variation of the catchment water storage (Sankarasubramanian and Vogel, 2002; Yang et al., 2007; Potter and Zhang, 2009). However, water storage change can be great when analysing the interannual variability of the water balance (Wang, 2012). To minimize the potential errors introduced by neglecting water storage variation, the hydrological year (Sivapalan et al., 2011; Carmona et al., 2014) and moving windows (Jiang et al., 2015) were introduced to the time series of annual hydrological variables. Similar to Sivapalan et al. (2011) and Carmona et al. (2014), the hydrological year rather than the calendar year is introduced to calculate the annual ET in our study, and this is called the "measured" ET in the subsequent discussion. Specifically, as the study area has a continental monsoon climate with most rainfall occurring in summer and autumn (July–September), a hydrological year is defined as July to June of the following year. In this way, the water input occurs mainly at the beginning of the year and the water is consumed within that year.

2.2 Identification of factors determining parameter ω in Fu's equation

The Fu equation is used in this study with the following expressions:

$$\frac{\mathrm{ET}}{P} = 1 + \frac{\mathrm{ET}_0}{P} - \left[1 + \left(\frac{\mathrm{ET}_0}{P}\right)^{\omega}\right]^{1/\omega}$$

or

$$\frac{\mathrm{ET}}{\mathrm{ET}_0} = 1 + \frac{P}{\mathrm{ET}_0} - \left[1 + \left(\frac{P}{\mathrm{ET}_0}\right)^{\omega}\right]^{1/\omega}, \tag{1}$$

where ω is the controlling parameter of the Budyko curve. ET_0 is calculated by using the equation of Priestley and Taylor (1972).

The important issue regarding the parameterization of ω in Fu's equation is to choose factors with physical meanings. According to the results from related studies, land surface conditions can be mainly represented by vegetation, which was also true in this study. Although soil properties and topography also influence the variation of catchment water balance, they are relative stable for a catchment. Therefore, vegetation dynamics (i.e. vegetation coverage) were chosen to represent the variations in surface conditions and the effects of soil and topography were not considered in our study. The vegetation coverage (*M*) was estimated by the following equation (Yang et al., 2009):

$$M = \frac{\mathrm{NDVI} - \mathrm{NDVI_{min}}}{\mathrm{NDVI_{max}} - \mathrm{NDVI_{min}}}, \tag{2}$$

where $\mathrm{NDVI_{max}}$ and $\mathrm{NDVI_{min}}$ are the NDVI values of dense forest (0.80) and bare soil (0.05) in the previous studies (Yang et al., 2009; Li et al., 2013), respectively. The maximum NDVI in our study is 0.85, thus the value of $\mathrm{NDVI_{max}}$ in Eq. (2) was defined as 0.85.

The effects of seasonal variations in coupled water and energy are important for better understanding the water cycle for different climatic regions. For example, two catchments with the same annual *P* and $\mathrm{ET_0}$ may have different evapotranspiration partitioning or runoff if the balance between *P* and $\mathrm{ET_0}$ within a year are different, Solar radiation was considered as the dominant factor that controls the climate seasonality and thus the seasonality of $\mathrm{ET_0}$ can be expressed by sine function. Similarly, the seasonality of *P* also exhibited the characteristic of sine distribution (Milly, 1994; Woods, 2003; Yang et al., 2012):

$$P(t) = \overline{P}(1 + \delta_P \sin\vartheta t) \tag{3a}$$

$$\mathrm{ET_0}(t) = \overline{\mathrm{ET_0}}(1 + \delta_{\mathrm{ET_0}} \sin\vartheta t), \tag{3b}$$

where δ_P and $\delta_{\mathrm{ET_0}}$ are the ratios of the amplitudes of the monthly harmonics to the monthly averages of precipitation and potential evapotranspiration, respectively. Larger absolute values of δ_P and $\delta_{\mathrm{ET_0}}$ mean larger variability of climate seasonality. ϑ is the duration of the seasonal cycle, $2\pi/\vartheta$ equal to 1 year. Woods (2003) summarized the modelled climate of Eq. (3a) and (b) in dimensionless form and defined the climate seasonality index (*S*) and here it was used to reflect the non-uniformity in the intra-annual distribution of water and energy in our study:

$$S = \left|\delta_P - \delta_{\mathrm{ET_0}} \varnothing\right|, \tag{4}$$

where $\varnothing$ is the dryness index, $\varnothing = \overline{\mathrm{ET_0}}/\overline{P}$. If $S = 0$, there is no seasonal fluctuation of the difference between *P* and $\mathrm{ET_0}$. Larger values of *S* indicate that the larger changes in the balance between *P* and $\mathrm{ET_0}$ during the seasonal cycle.

2.3 Evaluating the contributions of climate change and surface condition alterations to ET changes

Based on the climate elasticity method, which was introduced by Schaake (1990) and improved by Sankarasubramanian et al. (2001), the contribution of change for each factor to runoff was defined as the product of the sensitivity coefficient and the variation of the factor (Roderick and Farquhar, 2011):

$$\mathrm{d}R = \frac{\partial R}{\partial P}\mathrm{d}P + \frac{\partial R}{\partial \mathrm{ET_0}}\mathrm{dET_0} + \frac{\partial R}{\partial \omega}\mathrm{d}\omega. \tag{5}$$

However, due to ignoring the higher orders of the Taylor expansion in Eq. (5), this method will result in high errors (Yang et al., 2014b). Recently, Zhou et al. (2016) proposed a new method to partition climate and catchment effect on the mean annual runoff based on the Budyko complementary relationship, called "the complementary method". The algebraic identities in their work can ensure that the change in runoff can be decomposed into two components precisely without any residuals. Here, we extend "the complementary method" to conduct attribution analysis of ET changes for each basin by further incorporating the effects of vegetation coverage and climate seasonality:

$$\begin{aligned}\Delta\mathrm{ET} = {} & \alpha\left[\left(\frac{\partial \mathrm{ET}}{\partial P}\right)_1 \Delta P + \left(\frac{\partial \mathrm{ET}}{\partial \mathrm{ET_0}}\right)_1 \Delta\mathrm{ET_0} + P_2\Delta\left(\frac{\partial \mathrm{ET}}{\partial P}\right) + \mathrm{ET_{0,2}}\Delta\left(\frac{\partial \mathrm{ET}}{\partial \mathrm{ET_0}}\right)\right] \\ & + (1-\alpha)\left[\left(\frac{\partial \mathrm{ET}}{\partial P}\right)_2 \Delta P + \left(\frac{\partial \mathrm{ET}}{\partial \mathrm{ET_0}}\right)_2 \Delta\mathrm{ET_0} + P_1\Delta\left(\frac{\partial \mathrm{ET}}{\partial P}\right) + \mathrm{ET_{0,1}}\Delta\left(\frac{\partial \mathrm{ET}}{\partial \mathrm{ET_0}}\right)\right],\end{aligned} \tag{6}$$

where α is a weighting factor that varies from 0 to 1, which can determine the upper and lower bounds of the climate and the controlling parameter effect. In this study, we defined $\alpha = 0.5$ according to the recommendation of Zhou et al. (2016). The difference operator (Δ) refers to the difference of a variable from period I (1981 to the changing point detected by Pettitt's test; Pettitt, 1979) to period II (period-I end to 2012), e.g. $\Delta\mathrm{ET_0} = \mathrm{ET_{0,2}} - \mathrm{ET_{0,1}}$. Then the contributions of *P*, $\mathrm{ET_0}$, and ω changes to the ET change can be expressed as follows:

$$C\_(P) = \alpha\left[\left(\frac{\partial \mathrm{ET}}{\partial P}\right)_1 \Delta P\right] + (1-\alpha)\left[\left(\frac{\partial \mathrm{ET}}{\partial P}\right)_2 \Delta P\right] \tag{7a}$$

$$C\_(\mathrm{ET_0}) = \alpha\left[\left(\frac{\partial \mathrm{ET}}{\partial \mathrm{ET_0}}\right)_1 \Delta\mathrm{ET_0}\right] + (1-\alpha)\left[\left(\frac{\partial \mathrm{ET}}{\partial \mathrm{ET_0}}\right)_2 \Delta\mathrm{ET_0}\right] \tag{7b}$$

$$C\_(\omega) = \alpha\left[P_2\Delta\left(\frac{\partial \mathrm{ET}}{\partial P}\right)+\mathrm{ET}_{0,2}\Delta\left(\frac{\partial \mathrm{ET}}{\partial \mathrm{ET}_0}\right)\right] + (1-\alpha)\left[P_1\Delta\left(\frac{\partial \mathrm{ET}}{\partial P}\right)+\mathrm{ET}_{0,1}\Delta\left(\frac{\partial \mathrm{ET}}{\partial \mathrm{ET}_0}\right)\right]. \quad (7c)$$

After obtaining the contribution of parameter ω to the ET change, the contributions of vegetation coverage (M) and climate seasonality (S) changes to the ET change can be further decomposed as follows.

First, the contributions of M and S changes to the variation of parameter ω are calculated by using the total differential method similar to Eq. (5) based on the relationship between ω and M as well as S we built:

$$\mathrm{d}\omega = \frac{\partial \omega}{\partial M}\mathrm{d}M + \frac{\partial \omega}{\partial S}\mathrm{d}S \quad (8a)$$

Ignoring the higher orders of the Taylor expansion, Eq. (8a) can be rewritten as

$$\Delta\omega = \frac{\partial \omega}{\partial M}\Delta M + \frac{\partial \omega}{\partial S}\Delta S \quad (8b)$$

Furthermore, the individual relative contribution (RC) of M and S to ω can be calculated. Then, the contributions of M and S changes to ET variation can be obtained as follows:

$$C\_(M) = C\_(\omega) \times \mathrm{RC}\_(M) \quad (9a)$$

$$C\_(S) = C\_(\omega) \times \mathrm{RC}\_(S). \quad (9b)$$

3 Study area and data

The Loess Plateau, which is located in the upper and middle reaches of the Yellow River in China, experiences a sub-humid and semiarid continental monsoon climate (Ning et al., 2016). Frequent heavy summer storms, sparse vegetation coverage, easily erodible wind-deposited loess soil, and a long agricultural history have all contributed to severe drought and soil erosion problems in this region (Li et al., 2012). To recover and preserve the ecosystem, the Chinese government has launched numerous soil and conservation measures since the 1950s, and these include biologic measures ("Grain to Green" project) and engineering measures (building terraces and sediment trapping dams) (Mu et al., 2007). As a result, the hydrological processes of this area have undergone significant changes (Huang and Zhang, 2004; Zhang et al., 2008). Thirteen catchments on the Loess Plateau were selected as our study area (Fig. 1).

Monthly runoff data for the 13 catchments were supplied by the Yellow River Conservancy Commission. Detailed information about the catchment characteristics and data durations are shown in Table 1. Daily meteorological data (1960–2012) comprised of precipitation, daily maximum and minimum temperatures, atmospheric pressures, wind speeds, mean relative humidity values, and sunshine durations, which were recorded at 96 stations, were provided by the China Meteorological Administration. The new NDVI third-generation (NDVI3g) dataset was used to represent the vegetation characteristics of the study area, and detailed information about this dataset was presented earlier by Fensholt and Proud (2012). The maximum value composite (MVC) procedure (Holben, 1986) was applied to produce the annual NDVI values.

4 Results

4.1 The variability of parameter ω

The Budyko framework is usually used for analyses of long-term average catchment water balance; however, in this study, it was employed for the interpretation of the interannual variability of the water balance by using the hydrological year approach described earlier. To validate the feasibility of using Fu's equation for interannual variability, the evapotranspiration ratio (ET/P) and dryness index (ET_0/P) on the annual scale for 13 basins are presented in the supporting information (Fig. S1 in the Supplement), and it can be seen that almost all points are focused on Fu's curves in each basin. Therefore, Fu's equation was considered adequate for the analysis of the interannual variability of the water balance.

If the controlling parameter ω on the annual scale can reflect the combined impacts of vegetation change and climate seasonality, it should also exhibit interannual variability with the seasonal variation in vegetation and climate, especially in those basins affected significantly by climate change and human activities. Obviously, this is true for basins in Loess Plateau (Fig. 2). During 1961–2012, ω values in all 13 basins had an upward trend. Along with such a changing trend in ω, ET should increase for the same levels of P and ET_0. Before the 1980s, the variation in ω for each basin was relatively gentle; however, since that time, it has increased dramatically.

4.2 Development of the semi-empirical formula for parameter ω

The relationships between the annual parameter ω and vegetation coverage M as well as the climate seasonality index S were first explored in each study basin during the period 1981–2012, and the results are shown in Figs. S2 and S3. We can see that the parameter ω generally had a positive correlation with M, which implies that evapotranspiration increased with improvements in the vegetation conditions. However, ω was correlated negatively with S, which means that larger seasonal variations of coupled water and energy resulted in less evapotranspiration in this area. The relationships between ω and M as well as S imply that the annual variation in parameter ω can be estimated by the changes in vegetation dynamics and climate seasonality.

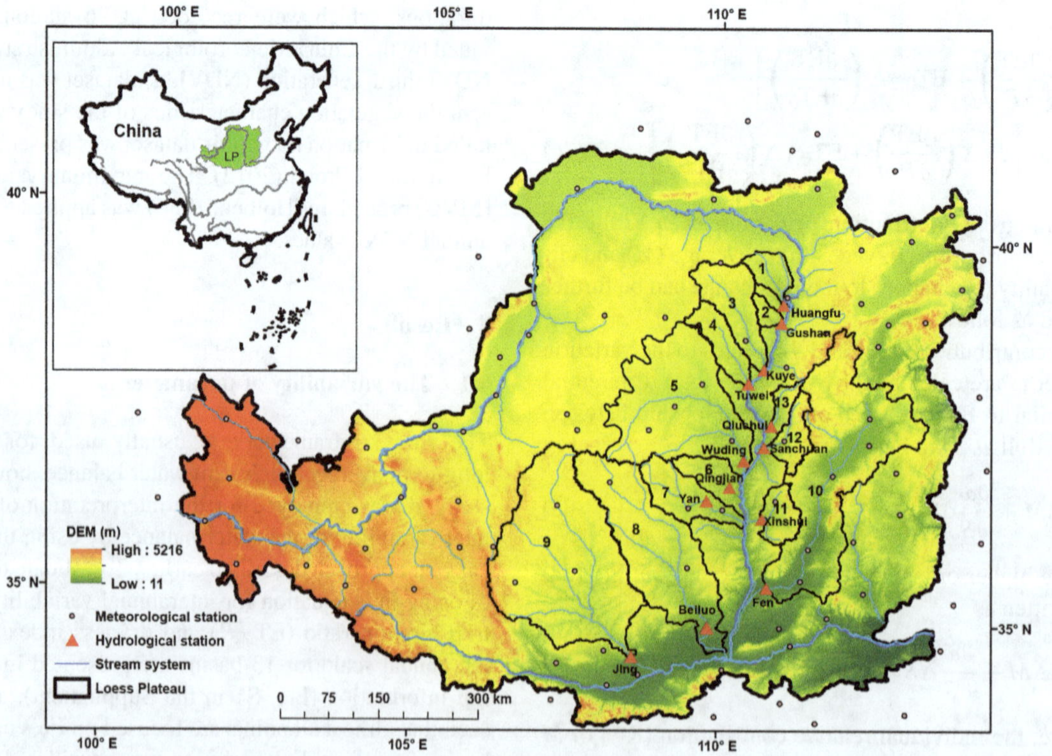

Figure 1. Locations of the study area and hydrometeorological stations.

Table 1. Long-term hydrometeorological characteristics and vegetation coverage (1981–2012). Because a few runoff data points were missing for several basins, the data length in these basins was less than 32. Each item represents the mean annual value.

ID	Basin	Data length, year	P, mm	ET_0, mm	ET, mm	ω	M	S
1	Huangfu	32	372	972	347	3.15	0.40	0.94
2	Gushan	25	394	975	359	2.75	0.45	0.79
3	Kuye	32	375	1018	333	2.45	0.40	0.99
4	Tuwei	32	383	1031	308	1.99	0.38	0.95
5	Wuding	32	385	1045	356	2.68	0.43	0.95
6	Qingjian	32	451	1009	417	3.00	0.56	0.60
7	Yan	30	462	984	433	3.21	0.66	0.51
8	Beiluo	28	502	960	475	3.76	0.82	0.34
9	Jing	32	529	936	497	3.74	0.64	0.51
10	Fen	30	465	982	452	4.21	0.82	0.43
11	Xinshui	30	478	992	458	3.77	0.82	0.45
12	Sanchuan	24	444	998	397	2.70	0.53	0.58
13	Qiushui	23	442	1006	418	3.33	0.64	0.60

To expand the sample size and span a wider range of climate conditions, as well as to make the derived semi-empirical formula of parameter ω more representative, relationships were then developed based on the combined dataset from the 13 basins (Fig. 3). These results also indicate a good relationship between ω and M ($R^2 = 0.43$, $p < 0.01$) as well as S ($R^2 = 0.24$, $p < 0.01$).

To develop the semi-empirical formula of parameter ω, the limiting conditions of the two variables were considered as follows.

In Eq. (1), when $\omega \to 1$, $\mathrm{ET} \to 0$. After further considering the range of M and S, we can get that if $\mathrm{ET} \to 0$, which indicates that $T \to 0$ (transpiration), and thus $M \to 0$, or implies that $R \to P$, the match of P and ET_0 tends to be worst, i.e. $S \to \infty$.

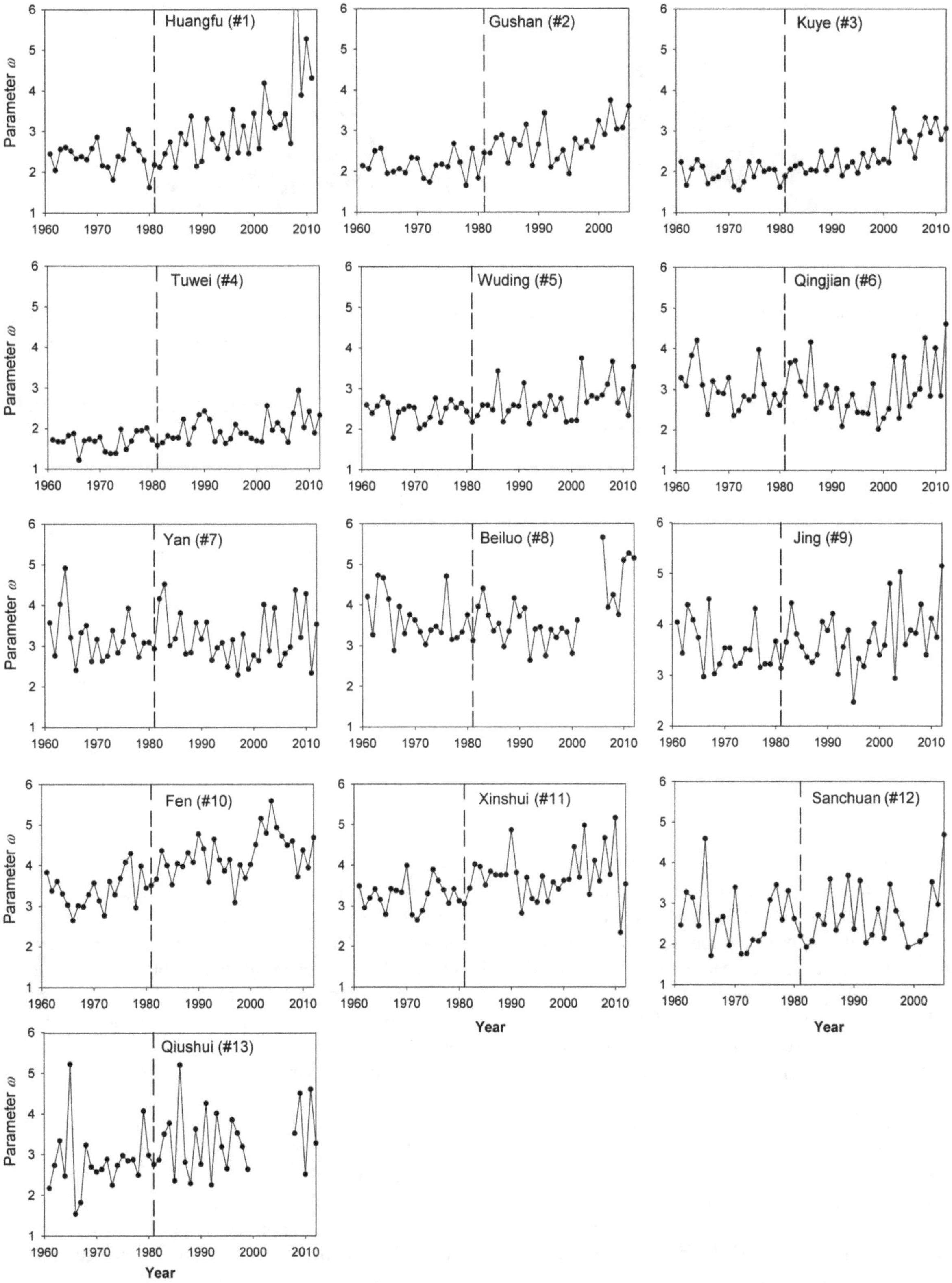

Figure 2. The variability of parameter ω for 13 basins between 1961 and 2012.

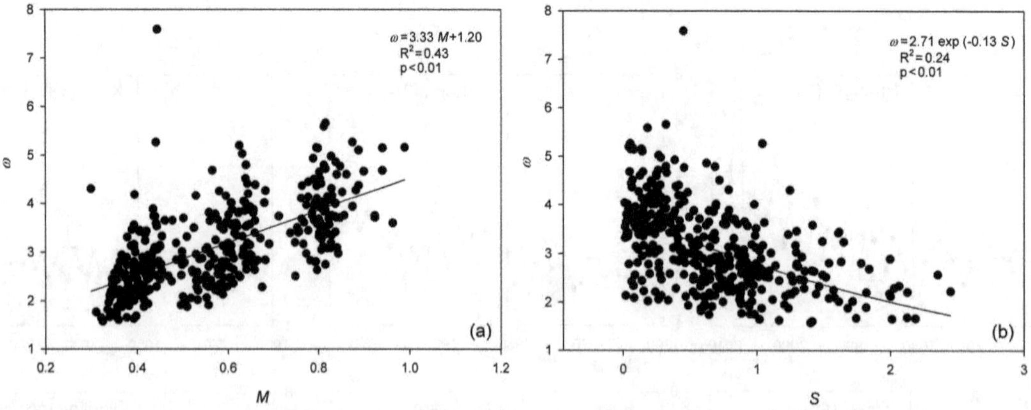

Figure 3. Relationship between the **(a)** annual ω and vegetation coverage (*M*) and **(b)** ω and climate seasonality index (*S*) based on the combined dataset from 13 basins.

Considering the relationships shown in Fig. 3 and given the above limiting conditions, the general form of parameter ω can be expressed as follows:

$$\omega = 1 + a \times M^b \times \exp(cS), \tag{10}$$

where *a*, *b*, and *c* are constants. Using the least linear squares regression method, the semi-empirical formula of parameter ω is derived as follows:

$$\omega = 1 + 3.683 \times M^{0.798} \times \exp(-0.246S). \tag{11}$$

The coefficient of determination R^2 and the statistics for the *F* test of the modelled ω were 0.53 and 214.54, respectively.

A cross-validation approach was chosen to calibrate and test the above semi-empirical formula for parameter ω. Specifically, the dataset for the 13 basins in our study was separated into two groups. One was applied to build the semi-empirical formula, and it consisted of 12 basins for each time; the other was used for testing the performance of the semi-empirical formula, and it consisted of the remaining basin. In total, the cross-validation process was conducted 13 times. After building the semi-empirical formula by using the vegetation coverage and climate seasonality index data for the 12 basins, the parameter ω for the validated basin was modelled by using this fitted formula, and the annual ET for the validated basin was evaluated with the modelled ω, which is referred to as the "modelled" ET. Then, the "modelled" ET was compared with the "measured" ET.

Table 2 shows the cross-validation results for each basin. Except for basins 4 and 12, the MAE (mean absolute error) and RMSE (square root of the mean square error) values for each cross-validation process were relatively low, with mean values of 14.6 and 18.8 mm, respectively. The NSE coefficient (Nash–Sutcliffe coefficient of efficiency) for each process was greater than 0.85, thus suggesting that vegetation changes and climate seasonality can well explain the variation in the controlling parameter of the catchment water balance on the shorter timescale.

4.3 Quantitative attribution of the variation in ET

The impacts of vegetation changes on ET have been widely studied with the Budyko framework by assuming surface conditions can be represented by the controlling parameter. However, according to the developed relationships in our study, the controlling parameter is not only related to surface condition change, but also to climate seasonality. The contributions of changes in climate (*P*, ET_0, and *S*) and vegetation (*M*) to the ET change were thus estimated by using the semi-empirical formula for parameter ω in the context of Fu's framework.

Trends in hydrometeorological variables and vegetation coverage were first analysed for each basin (Table 3). ET_0, *M*, and *S* in almost all basins exhibited an upward trend, though with different significances. Based on the sensitivity coefficients of ET (Table S1 in the Supplement) and the changes in mean annual *P*, ET_0, ω, *M*, and *S* from period I to period II (Table 4), the changes in ET due to those in *P*, ET_0, *M*, and *S* were estimated using the method described in Sect. 2.3. The contributions of four variables to ET change for each basin were presented in Table 4. In basins 1, 3–4, and 6, the ET changes were controlled by vegetation improvement; however, in the other basins, the dominant factor was precipitation. Except for basins 6, 9, and 12, elevated vegetation in most basins positively contributed to ET changes, which is consistent with Feng et al. (2016). ET in several basins showed a downward trend even though *M* positively contributed to ET changes; which is due to the offsetting effect of the other factors.

It should be noted that the climate seasonality (represented by *S*) played an important role in the catchment ET variation. The contributions of *S* to ET changes ranged from 0.1 to 74.8 % (absolute values). Besides basins 6, 9, and 12, the climate seasonality had a negative effect on ET variation in most of the basins, which means that larger seasonality differences between seasonal water and energy will lead

Table 2. Cross-validation results for each basin.

ID	Validated basin	Model coefficients			ET estimation accuracy		
		a	*b*	*c*	MAE	RMS	NSE
1	Huangfu	2.773	0.583	−0.120	19.7	22.3	0.90
2	Gushan	2.707	0.508	−0.116	14.8	19.8	0.91
3	Kuye	2.604	0.435	−0.100	19.0	24.7	0.86
4	Tuwei	3.350	0.627	−0.224	31.4	35.2	0.85
5	Wuding	3.525	0.803	−0.211	10.0	15.1	0.96
6	Qingjian	2.705	0.470	−0.143	15.4	19.6	0.95
7	Yan	3.560	0.808	−0.210	11.2	14.3	0.98
8	Beiluo	3.773	0.832	−0.241	9.7	11.3	0.97
9	Jing	3.456	0.814	−0.188	17.7	23.3	0.91
10	Fen	3.550	0.746	−0.253	6.9	9.5	0.98
11	Xinshui	3.561	0.805	−0.216	6.6	10.0	0.99
12	Sanchuan	3.287	0.964	−0.115	27.0	31.3	0.88
13	Qiushui	2.617	0.482	−0.093	17.2	21.2	0.93

Table 3. Trend analysis for the hydrometeorological variables and vegetation coverage.

ID	Basin	Period	ET, mm yr^{-2}	ET_0, mm yr^{-2}	*P*, mm yr^{-2}	*M*, yr^{-1}	*S*, yr^{-1}
1	Huangfu	1981–2011	1.89	1.17	0.61	0.002*	0.001
2	Gushan	1981–2005	1.28	1.99	0.49	0.003**	0.009
3	Kuye	1981–2012	2.34*	2.04*	0.53	0.004**	0.006
4	Tuwei	1981–2012	1.87	2.33**	0.53	0.005**	0.006
5	Wuding	1981–2012	0.88	1.17	0.31	0.005**	0.004
6	Qingjian	1981–2012	−0.45	1.78*	−0.94	0.007**	0.006
7	Yan	1981–2006	3.43	3.43**	−3.56	0.001	0.014
8	Beiluo	1981–2001	−5.4*	4.6*	−6.2*	0.001	0.017
9	Jing	1981–2012	−0.03	1.23*	−0.75	0.003**	−0.001
10	Fen	1981–2006	−0.14	2.04*	−0.62	0.0004	0.003
11	Xinshui	1981–2006	0.01	3.08*	−0.37	0.0001	0.011
12	Sanchuan	1981–2004	1.49	1.84	0.09	−0.0004	0.005
13	Qiushui	1981–1999	−0.50	1.79	−0.83	0.002	0.008

* and ** indicate the trend is significant at the level of $p = 0.05$ and $p = 0.01$ by the Mann–Kendall test, respectively.

to smaller amounts of evapotranspiration. Accordingly, if ω is supposed to only represent the landscape condition, the effects of landscape condition change on ET variation will be underestimated in basins 1, 3, 6–7, 9, and 11. Except for basin 9, the area of these basins is relative smaller; while its effects will be overestimated in the other basins, and the error would be equal to the contributions of *S* to ET changes.

5 Discussion

Although the controlling parameter ω showed a good relationship with the vegetation change and climate seasonality index, two groups of deviations around the regressed curves were detected (Fig. 3). The deviation points for the relationship between ω and *M* were mainly located at the top of the curve, i.e. corresponding to the same *M* values, where ω values were greater. We checked those points and found that precipitation and vegetation coverage in those years were normal, but runoff was very low compared to normal years. Excluding abrupt climate change, possible reasons for the extremely low runoff in those years include dam and reservoir operations, as well as irrigation diversions. A study conducted by Liang et al. (2015) on the same basins that we investigated in the Loess Plateau showed that check-dams increased continuously starting from the 1960s. By the year 2006, the numbers of dams along basins 10 and 5 reached 482 and 181, respectively. Dams can intercept stormwater runoff for a short period during flood seasons and allow more time for infiltration (Polyakov et al., 2014). A total of 21 large and 136 medium-sized reservoirs were installed along the Yellow River by 2001. Such infrastructure can also influence the runoff change by controlling the flooding, regulating the water discharge, and diverting the water to other regions (Chen et al., 2005). Agricultural production is heavily

Table 4. Attribution analysis for ET changes for each basin. The relative contribution of a certain variable to the ET change ($\varphi(x)$) was calculated as follows: $\varphi(x) = (C\_(x)/\Delta\text{ET}) \times 100\,\%$, where $C\_(x)$ represents the contribution of each variable.

ID	Basin	Break point of ET	Change from period I to period II					$\text{ET}_0/P/M/S$-induced ET change (mm)					Relative contribution to ET change (%)			
			ΔET	ΔET_0	ΔP	ΔM	ΔS	$C\_(\text{ET}_0)$	$C\_(P)$	$C\_(\omega)$	$C\_(M)$	$C\_(S)$	$\varphi\_(\text{ET}_0)$	$\varphi\_(P)$	$\varphi\_(M)$	$\varphi\_(S)$
1	Huangfu	2001(ns)	41.7	7.0	22.2	0.02	0.01	0.28	18.67	22.70	22.73	−0.04	0.7	44.8	54.6	−0.1
2	Gushan	2000(ns)	33.6	64.9	20.6	0.04	−0.10	2.81	17.01	13.77	8.87	4.90	8.4	50.6	26.4	14.6
3	Kuye	2000(**)	51.4	32.0	17.3	0.05	0.05	1.54	13.34	36.48	57.15	−20.67	3.0	26.0	111.3	−40.2
4	Tuwei	2000(**)	43.2	39.6	24.0	0.07	−0.03	2.57	15.28	25.35	21.74	3.61	5.9	35.4	50.3	8.4
5	Wuding	2000(*)	35.2	17.6	26.9	0.08	−0.12	0.77	21.82	12.64	8.13	4.51	2.2	61.9	23.1	12.8
6	Qingjian	1988(**)	−50.1	32.0	−48.0	0.08	0.19	2.06	−37.80	−14.31	−51.75	37.44	−4.1	75.5	103.4	−74.8
7	Yan	1985(**)	−82.3	44.6	−86.9	0.05	0.30	3.19	−69.52	−15.96	20.28	−36.24	−3.9	84.5	−24.6	44.0
8	Beiluo	1985(**)	−65.1	49.4	−79.8	0.02	0.19	4.33	−62.9	−6.75	3.47	−10.22	−6.6	96.3	−5.3	15.7
9	Jing	1990(**)	−33.7	43.0	−47.8	0.03	0.11	4.1	−37.2	−0.61	−13.96	13.35	−12.2	110.3	41.4	−39.6
10	Fen	2005(ns)	23.1	8.5	21.2	0.07	−0.20	0.33	19.00	3.81	2.10	1.71	1.4	82.1	9.1	7.4
11	Xinshui	1990(**)	−19.1	39.7	−24.7	0.02	0.09	2.06	−21.08	−0.14	0.35	−0.49	−10.8	110.1	−1.8	2.6
12	Sanchuan	1996(ns)	−27.0	45.4	−43.4	−0.01	0.22	3.01	−32.52	2.56	0.19	2.37	−11.2	120.6	−0.7	−8.8
13	Qiushui	1996(ns)	−80.3	77.5	−103.5	−0.01	0.68	3.76	−83.68	−0.40	−0.02	−0.37	−4.7	104.2	0.1	0.5

* and ** indicate the significance level of $p = 0.05$ and $p = 0.01$, respectively; ns indicates that the significance level exceeds 0.05.

dependent on irrigation throughout the entire Yellow River basin, and it has been reported that water consumption by agricultural irrigation accounted for nearly 80.0 % of the entire water consumed from 1998 to 2011 (Wang et al., 2014). Thereby, water withdrawn for irrigation also plays an important role in the changing trends in runoff. In this study, the deviation points around the relationship curve between the annual ω and S fell in the upper left, and they were likely influenced by the low runoff. However, separation of the impacts on runoff from vegetation change, climate seasonality, and engineering works will have to await future work.

The relationships of parameter ω with vegetation dynamics and climate seasonality in some single basins were not significant in this study. Similarly, Yang et al. (2014a) also found a weak relationship between parameter n and vegetation coverage in 201 basins in China. This implies that the parameter ω might represent the combined effects of some other factors. For example, strong interactions among vegetation, climate, and soil conditions will lead to specific hydrologic partitioning at the catchment scale. In dry years, with low soil water contents, plants are trying to adapt by making use of hydrological processes, e.g. ground water dynamics and plant water storage mechanisms (Soylu et al., 2011). Therefore, the relationship between the parameter ω and vegetation dynamics can be influenced by climate and soil conditions. However, it is difficult to separate the climatic and soil components from the vegetation change. Moreover, Zhang et al. (2001) reported that the impact of different vegetation types on catchment water balance can be vastly different, and the plant-available water coefficient in their function, which is similar to parameter ω in Fu's equation, is related to vegetation type. Therefore, the vegetation type may also be an important variable that influences the parameter ω.

Despite that catchment-scale water storage changes are usually assumed to be zero on a long-term scale, the interannual variability of water storage can be an important component in annual water budget during dry or wet years (Wang and Alimohammadi, 2012), and cannot be ignored. However, the Loess Plateau has a sub-humid to semiarid climate, the water storage and its annual variation are relatively small compared with humid regions (see Fig. 5 from Mo et al., 2016). For example, using GRACE (Gravity Recovery and Climate Experiment), the water storage variations in the Yangtze, Yellow, and Zhujiang from 2003 to 2008 were analysed by Zhao et al. (2011), and the values for the Yangtze and Zhujiang basins were 37.8 and 65.2 mm, while no clear annual variations are observed in the Yellow River basin (3.0 mm). Furthermore, Mo et al. (2016) found that the water storage in Yellow River kept decreasing from 2004 to 2011, whereas it was changing slowly with a rate of 1.3 mm yr^{-1}. Therefore, considering the small water storage change in study area, ignoring water storage change in a period of hydrologic year is reasonable.

There were still errors in the attribution analysis of ET changes. As the changes in evapotranspiration were decomposed without residual by the complementary method (Eqs. 6–7), the errors were induced from the developed empirical formula for ω (Eq. 11). This suggested that ω cannot be completely explained by M and S, and it might include some other factors. Therefore, further discussion of factors influencing ω remains a goal for future work.

6 Conclusions

This study explored the concomitant effects of vegetation dynamics and climate seasonality on the variation in interannual controlling parameter ω from Fu's equation in the Loess Plateau. First, to reduce the impact of ignoring the water storage change on annual catchment water balance, the hydrological year approach was introduced to examine the interannual variability of the controlling parameter ω for the 13 basins in the Loess Plateau from 1961 to 2012. The findings showed that parameter ω in all these basins presented an increasing trend, especially after the 1980s. Furthermore, we checked the relationship between ω and vegetation dynamics (represented by the annual vegetation coverage, M) as well as climate seasonality (represented by the climate seasonality index, S). The interannual changes of parameter ω were found to be related strongly to M and S. As such, a semi-empirical formula for the annual value of ω was developed based on these two parameters, and it has proven superior for estimating the actual evapotranspiration (ET) by a cross-validation approach. Finally, based on the proposed semi-empirical formula for parameter ω, the contributions of changes in climate (P, ET_0, and S) and vegetation (M) to ET variations were estimated. The results showed that the improved vegetation conditions in almost all basins made a positive contribution to the ET change, but these effects were largely offset by other variables in some basins. The contribution of landscape condition changes to ET variation will be estimated with a large error if the effects of climate seasonality are ignored.

Competing interests. The authors declare that they have no conflict of interest.

Acknowledgements. This study was supported by the National Natural Science Foundation of China (no. 41571036), the National Key Research and Development Program of China (no. 2016YFC0501602), and the Public Welfare Industry (Meteorological) Research Project of China (no. GYHY201506001).

Edited by: Y. Chen

References

Berghuijs, W. R. and Woods, R. A.: Correspondence: Space-time asymmetry undermines water yield assessment, Nature Communications, 7, 11603, doi:10.1038/ncomms11603, 2016.

Budyko, M. I.: Evaporation under natural conditions, Gidrometeorizdat, Leningrad, English translation by IPST, Jerusalem, 1948.

Budyko, M. I. and Zubenok, L. I.: Determination of evaporation from the land surface, Izv. Akad.Nauk SSR, Ser. Geogr., 6, 3–17, 1961 (in Russian).

Budyko, M. I.: Climate and life, Academic, New York, 1974.

Carmona, A. M., Sivapalan, M., Yaeger, M. A., and Poveda, G.: Regional patterns of interannual variability of catchment water balances across the continental US: A Budyko framework, Water Resour. Res., 50, 9177–9193, doi:10.1002/2014wr016013, 2014.

Chen, J. S., Wang, F. Y., Meybeck, M., He, D. W., Xia, X. H., and Zhang, L. T.: Spatial and temporal analysis of water chemistry records (1958–2000) in the Huanghe (Yellow River) basin, Global Biogeochem. Cy., 19, GB3016, doi:10.1029/2004gb002325, 2005.

China Meteorological Data Service Center: Dataset of Daily Climate Data from Chinese Surface Stations for Global Exchange (V3.0), available at: http://data.cma.cn/data/detail/dataCode/SURF_CLI_CHN_MUL_DAY_CES_V3.0.html (last access: 9 March 2017), 2012.

Choudhury, B. J.: Evaluation of an empirical equation for annual evaporation using field observations and results from a biophysical model, J. Hydrol., 216, 99–110, doi:10.1016/s0022-1694(98)00293-5, 1999.

Donohue, R. J., Roderick, M. L., and McVicar, T. R.: On the importance of including vegetation dynamics in Budyko's hydrological model, Hydrol. Earth Syst. Sci., 11, 983–995, doi:10.5194/hess-11-983-2007, 2007.

Donohue, R. J., Roderick, M. L., and McVicar, T. R.: Can dynamic vegetation information improve the accuracy of Budyko's hydrological model?, J. Hydrol., 390, 23–34, doi:10.1016/j.jhydrol.2010.06.025, 2010.

Donohue, R. J., Roderick, M. L., and McVicar, T. R.: Roots, storms and soil pores: Incorporating key ecohydrological processes into Budyko's hydrological model, J. Hydrol., 436, 35–50, doi:10.1016/j.jhydrol.2012.02.033, 2012.

Feng, X., Vico, G., and Porporato, A.: On the effects of seasonality on soil water balance and plant growth, Water Resour. Res., 48, W05543, doi:10.1029/2011wr011263, 2012.

Feng, X., Fu, B., Piao, S., Wang, S., and Ciais, P.: Revegetation in China's Loess Plateau is approaching sustainable water resource limits, Nature Climate Change, 6, 1019–1022, doi:10.1038/NCLIMATE3092, 2016.

Fensholt, R. and Proud, S. R.: Evaluation of Earth Observation based global long term vegetation trends – Comparing GIMMS and MODIS global NDVI time series, Remote Sens. Environ., 119, 131–147, doi:10.1016/j.rse.2011.12.015, 2012.

Fu, B.: On the calculation of the evaporation from land surface, Scientia Atmospherica Sinica, 5, 23–31, 1981 (in Chinese).

Holben, B. N.: Characteristics of maximum-value composite images from temporal AVHRR data, Int. J. Remote Sens., 7, 1417–1434, doi:10.1080/01431168608948945, 1986.

Huang, M. and Zhang, L.: Hydrological responses to conservation practices in a catchment of the Loess Plateau, China, Hydrol. Process., 18, 1885–1898, doi:10.1002/hyp.1454, 2004.

Jiang, C., Xiong, L. H., Wang, D. B., Liu, P., Guo, S. L., and Xu, C. Y.: Separating the impacts of climate change and human activities on runoff using the Budyko-type equations with time-varying parameters, J. Hydrol., 522, 326–338, doi:10.1016/j.jhydrol.2014.12.060, 2015.

Li, D., Pan, M., Cong, Z., Zhang, L., and Wood, E.: Vegetation control on water and energy balance within the Budyko framework, Water Resour. Res., 49, 969–976, doi:10.1002/wrcr.20107, 2013.

Li, D.: Assessing the impact of interannual variability of precipitation and potential evaporation on evapotranspiration, Adv. Water Resour., 70, 1–11, doi:10.1016/j.advwatres.2014.04.012, 2014.

Li, Z., Zheng, F. L., and Liu, W. Z.: Spatiotemporal characteristics of reference evapotranspiration during 1961–2009 and its projected changes during 2011–2099 on the Loess Plateau of China, Agr. Forest Meteorol., 154, 147–155, doi:10.1016/j.agrformet.2011.10.019, 2012.

Liang, W., Bai, D., Wang, F. Y., Fu, B. J., Yan, J. P., Wang, S., Yang, Y. T., Long, D., and Feng, M. Q.: Quantifying the impacts of climate change and ecological restoration on streamflow changes based on a Budyko hydrological model in China's Loess Plateau, Water Resour. Res., 51, 6500–6519, doi:10.1002/2014wr016589, 2015.

Liu, X., Liu, C., Luo, Y., Zhang, M., and Xia, J.: Dramatic decrease in streamflow from the headwater source in the central route of China's water diversion project: Climatic variation or human influence?, J. Geophys. Res.-Atmos., 117, D06113, doi:10.1029/2011JD016879, 2012.

Milly, P. C. D.: An analystic solution of the stochastic storage problem applicable to soil-water, Water Resour. Res., 29, 3755–3758, doi:10.1029/93wr01934, 1993.

Milly, P. C. D.: Climate,soil-water storage, and the average annual water-balance, Water Resour. Res., 30, 2143–2156, doi:10.1029/94wr00586, 1994.

Mo, X., Wu, J. J., Wang, Q., and Zhou, H.: Variations in water storage in China over recent decades from GRACE observations and GLDAS, Nat. Hazards Earth Syst. Sci., 16, 469–482, doi:10.5194/nhess-16-469-2016, 2016.

Mu, X., Zhang, L., McVicar, T. R., Chille, B., and Gau, P.: Analysis of the impact of conservation measures on stream flow regime in catchments of the Loess Plateau, China, Hydrol. Process., 21, 2124–2134, doi:10.1002/hyp.6391, 2007.

NASA Earth Exchange: NASA GIMMS AVHRR Global NDVI (3g) (Normalized Difference Vegetation Index-3rd generation), available at: https://nex.nasa.gov/nex/projects/1349/ (last access: 9 March 2017), 2014.

Ning, T., Li, Z., Liu, W., and Han, X.: Evolution of potential evapotranspiration in the northern Loess Plateau of China: recent trends and climatic drivers, Int. J. Climatol., 36, 4019–4028, doi:10.1002/joc.4611, 2016.

Peel, M. C., McMahon, T. A., and Finlayson, B. L.: Vegetation impact on mean annual evapotranspiration at a global catchment scale, Water Resour. Res., 46, W09508, doi:10.1029/2009wr008233, 2010.

Pettitt, A.: A non-parametric approach to the change-point problem, Appl. Stat., 28, 126–135, doi:10.2307/2346729, 1979.

Polyakov, V. O., Nichols, M. H., McClaran, M. P., and Nearing, M. A.: Effect of check dams on runoff, sediment yield, and retention on small semiarid watersheds, J. Soil Water Conserv., 69, 414–421, doi:10.2489/jswc.69.5.414, 2014.

Potter, N. J. and Zhang, L.: Interannual variability of catchment water balance in Australia, J. Hydrol., 369, 120–129, doi:10.1016/j.jhydrol.2009.02.005, 2009.

Potter, N. J., Zhang, L., Milly, P. C. D., McMahon, T. A., and Jakeman, A. J.: Effects of rainfall seasonality and soil moisture capacity on mean annual water balance for Australian catchments, Water Resour. Res., 41, W06007, doi:10.1029/2004wr003697, 2005.

Priestley, C. and Taylor, R.: On the assessment of surface heat flux and evaporation using large-scale parameters, Mon. Weather Rev., 100, 81–92, 1972.

Roderick, M. L. and Farquhar, G. D.: A simple framework for relating variations in runoff to variations in climatic conditions and catchment properties, Water Resour. Res., 47, W00G07, doi:10.1029/2010wr009826, 2011.

Rodriguez-Iturbe, I.: Ecohydrology: A hydrologic perspective of climate-soil-vegetation dynamics, Water Resour. Res., 36, 3–9, doi:10.1029/1999wr900210, 2000.

Sankarasubramanian, A. and Vogel, R. M.: Annual hydroclimatology of the United States, Water Resour. Res., 38, 19-11–19-12, doi:10.1029/2001WR000619, 2002.

Sankarasubramanian, A., Vogel, R. M., and Limbrunner, J. F.: Climate elasticity of streamflow in the United States, Water Resour. Res., 37, 1771–1781, doi:10.1029/2000wr900330, 2001.

Schaake, J. C.: From climate to flow, in: climate change and U.S. water resources, edited by: Waggoner, P. E., chap. 8, 177–206, John Wiley, NY, 1990.

Shao, Q. X., Traylen, A., and Zhang, L.: Nonparametric method for estimating the effects of climatic and catchment characteristics on mean annual evapotranspiration, Water Resour. Res., 48, W03517, doi:10.1029/2010wr009610, 2012.

Sivapalan, M., Yaeger, M. A., Harman, C. J., Xu, X. Y., and Troch, P. A.: Functional model of water balance variability at the catchment scale: 1. Evidence of hydrologic similarity and space-time symmetry, Water Resour. Res., 47, W02522, doi:10.1029/2010wr009568, 2011.

Soylu, M. E., Istanbulluoglu, E., Lenters, J. D., and Wang, T.: Quantifying the impact of groundwater depth on evapotranspiration in a semi-arid grassland region, Hydrol. Earth Syst. Sci., 15, 787-806, doi:10.5194/hess-15-787-2011, 2011.

Stocker, T., Qin, D., Plattner, G.-K., Tignor, M., Allen, S. K., Boschung, J., Nauels, A., Xia, Y., Bex, V., and Midgley, P. M.: Climate change 2013: The physical science basis, Cambridge University Press Cambridge, UK, and New York, 2014.

Wang, D.: Evaluating interannual water storage changes at watersheds in Illinois based on long-term soil moisture and groundwater level data, Water Resour. Res., 48, W03502, doi:10.1029/2011wr010759, 2012.

Wang, D. and Alimohammadi, N.: Responses of annual runoff, evaporation, and storage change to climate variability at the watershed scale, Water Resour. Res., 48, W05546, doi:10.1029/2011wr011444, 2012.

Wang, F., Zhao, G. J., Mu, X. M., Gao, P., and Sun, W. Y.: Regime Shift Identification of Runoff and Sediment Loads in the Yellow River Basin, China, Water, 6, 3012–3032, doi:10.3390/w6103012, 2014.

Woods, R.: The relative roles of climate, soil, vegetation and topography in determining seasonal and long-term catchment dynamics, Adv. Water Resour., 26, 295–309, doi:10.1016/s0309-1708(02)00164-1, 2003.

Xu, X., Yang, D., Yang, H., and Lei, H.: Attribution analysis based on the Budyko hypothesis for detecting the dominant cause of runoff decline in Haihe basin, J. Hydrol., 510, 530–540, doi:10.1016/j.jhydrol.2013.12.052, 2014.

Yang, D., Sun, F., Liu, Z., Cong, Z., Ni, G., and Lei, Z.: Analyzing spatial and temporal variability of annual water-energy balance in nonhumid regions of China using the Budyko hypothesis, Water Resour. Res., 43, W04426, doi:10.1029/2006WR005224, 2007.

Yang, D., Shao, W., Yeh, P. J. F., Yang, H., Kanae, S., and Oki, T.: Impact of vegetation coverage on regional water balance in the nonhumid regions of China, Water Resour. Res., 45, W00A14, doi:10.1029/2008wr006948, 2009.

Yang, H., Yang, D., Lei, Z., and Sun, F.: New analytical derivation of the mean annual water-energy balance equation, Water Resour. Res., 44, W03410, doi:10.1029/2007wr006135, 2008.

Yang, H., Lv, H., Yang, D., and Hu, Q.: Seasonality of precipitation and potential evaporationanditsimpact on catchment water-energy balance , Journal of Hydroelectric Engineering, 31, 54–59+93, 2012 (in Chinese).

Yang, H. B., Qi, J., Xu, X. Y., Yang, D. W., and Lv, H. F.: The regional variation in climate elasticity and climate contribution to runoff across China, J. Hydrol., 517, 607–616, doi:10.1016/j.jhydrol.2014.05.062, 2014a.

Yang, H. B., Yang, D. W., and Hu, Q. F.: An error analysis of the Budyko hypothesis for assessing the contribution of climate change to runoff, Water Resour. Res., 50, 9620–9629, doi:10.1002/2014wr015451, 2014b.

Yokoo, Y., Sivapalan, M., and Oki, T.: Investigating the roles of climate seasonality and landscape characteristics on mean annual and monthly water balances, J. Hydrol., 357, 255–269, doi:10.1016/j.jhydrol.2008.05.010, 2008.

Zhang, L., Dawes, W. R., and Walker, G. R.: Response of mean annual evapotranspiration to vegetation changes at catchment scale, Water Resour. Res., 37, 701–708, doi:10.1029/2000wr900325, 2001.

Zhang, L., Hickel, K., Dawes, W. R., Chiew, F. H. S., Western, A. W., and Briggs, P. R.: A rational function approach for estimating mean annual evapotranspiration, Water Resour. Res., 40, W02502, doi:10.1029/2003wr002710, 2004.

Zhang, L., Potter, N., Hickel, K., Zhang, Y., and Shao, Q.: Water balance modeling over variable time scales based on the Budyko framework – Model development and testing, J. Hydrol., 360, 117–131, doi:10.1016/j.jhydrol.2008.07.021, 2008.

Zhang, S., Yang, H., Yang, D., and Jayawardena, A. W.: Quantifying the effect of vegetation change on the regional water balance within the Budyko framework, Geophys. Res. Lett., 43, 1140–1148, doi:10.1002/2015gl066952, 2016.

Zhao, Q. L., Liu, X. L., Ditmar, P., Siemes, C., Revtova, E., Hashemi-Farahani, H., and Klees, R.: Water storage variations of the Yangtze, Yellow, and Zhujiang river basins derived from the DEOS Mass Transport (DMT-1) model, Science China-Earth Sciences, 54, 667–677, doi:10.1007/s11430-010-4096-7, 2011.

Zhou, G., Wei, X., Chen, X., Zhou, P., Liu, X., Xiao, Y., Sun, G., Scott, D. F., Zhou, S., Han, L., and Su, Y.: Global pattern for the effect of climate and land cover on water yield, Nature Communications, 6, 5918, doi:10.1038/ncomms6918, 2015.

Zhou, S., Yu, B., Zhang, L., Huang, Y., Pan, M., and Wang, G.: A new method to partition climate and catchment effect on the mean annual runoff based on the Budyko complementary relationship, Water Resour. Res., 52, 7163–7177, doi:10.1002/2016wr019046, 2016.

4

Comparative analyses of hydrological responses of two adjacent watersheds to climate variability and change using the SWAT model

Sangchul Lee[1,2], **In-Young Yeo**[3,4], **Ali M. Sadeghi**[2], **Gregory W. McCarty**[2], **Wells D. Hively**[5], **Megan W. Lang**[4,6], **and Amir Sharifi**[7]

[1]Department of Environmental Science and Technology, University of Maryland, College Park, MD 20742, USA
[2]US Department of Agriculture-Agricultural Research Service, Hydrology and Remote Sensing Laboratory, Beltsville, MD 20705, USA
[3]School of Engineering, the University of Newcastle, Callaghan NSW 2308, Australia
[4]Department of Geographical Sciences, University of Maryland, College Park, MD 20742, USA
[5]US Geological Survey, Eastern Geographic Science Center, Reston, VA 20192, USA
[6]US Fish & Wildlife Service, National Wetlands Inventory, Falls Church, VA 22041, USA
[7]Department of Energy and Environment, Government of the District of Columbia, Washington, DC 20002, USA

Correspondence: Sangchul Lee (sangchul.lee84@gmail.com)

Abstract. Water quality problems in the Chesapeake Bay Watershed (CBW) are expected to be exacerbated by climate variability and change. However, climate impacts on agricultural lands and resultant nutrient loads into surface water resources are largely unknown. This study evaluated the impacts of climate variability and change on two adjacent watersheds in the Coastal Plain of the CBW, using the Soil and Water Assessment Tool (SWAT) model. We prepared six climate sensitivity scenarios to assess the individual impacts of variations in CO_2 concentration (590 and 850 ppm), precipitation increase (11 and 21 %), and temperature increase (2.9 and 5.0 °C), based on regional general circulation model (GCM) projections. Further, we considered the ensemble of five GCM projections (2085–2098) under the Representative Concentration Pathway (RCP) 8.5 scenario to evaluate simultaneous changes in CO_2, precipitation, and temperature. Using SWAT model simulations from 2001 to 2014 as a baseline scenario, predicted hydrologic outputs (water and nitrate budgets) and crop growth were analyzed. Compared to the baseline scenario, a precipitation increase of 21 % and elevated CO_2 concentration of 850 ppm significantly increased streamflow and nitrate loads by 50 and 52 %, respectively, while a temperature increase of 5.0 °C reduced streamflow and nitrate loads by 12 and 13 %, respectively. Crop biomass increased with elevated CO_2 concentrations due to enhanced radiation- and water-use efficiency, while it decreased with precipitation and temperature increases. Over the GCM ensemble mean, annual streamflow and nitrate loads showed an increase of ~ 70 % relative to the baseline scenario, due to elevated CO_2 concentrations and precipitation increase. Different hydrological responses to climate change were observed from the two watersheds, due to contrasting land use and soil characteristics. The watershed with a larger percent of croplands demonstrated a greater increased rate of 5.2 kg N ha^{-1} in nitrate yield relative to the watershed with a lower percent of croplands as a result of increased export of nitrate derived from fertilizer. The watershed dominated by poorly drained soils showed increased nitrate removal due do enhanced denitrification compared to the watershed dominated by well-drained soils. Our findings suggest that increased implementation of conservation practices would be necessary for this region to mitigate increased nitrate loads associated with predicted changes in future climate.

1 Introduction

Located in the Mid-Atlantic region, Chesapeake Bay (CB) is the largest and most productive estuary in the United States (US). The Chesapeake Bay Watershed (CBW) covers an area of 166 000 km^2 and is home to more than 18 million people and 3600 species of plants and animals (Chesapeake Bay Program, 2016). Despite significant restoration efforts, the health of the Bay has continued to deteriorate, primarily due to excessive nutrient and sediment loads from agricultural lands (Rogers and McCarty, 2000). Najjar et al. (2010) suggested that the current water quality problems in the bay are expected to worsen under climate variability and change. General circulation models (GCMs) have projected increases in temperature and precipitation of up to 5.0 °C and 21 %, respectively, by the end of this century in the CB region (Najjar et al., 2009), which could lead to substantial changes in hydrology and nitrogen (N) cycling. For instance, Howarth et al. (2006) reported that greater precipitation is anticipated to increase N loads to CB by ~ 65 %. With precipitation and temperature changes, elevated CO_2 concentrations affecting stomatal conductance have also been viewed as one of the decisive factors modifying watershed hydrological processes (Chaplot, 2007; Wu et al., 2012a, b).

Numerous studies have been conducted to demonstrate the impacts of changes in CO_2 concentrations, precipitation, and temperature on streamflow and N loads. Elevated CO_2 concentrations are predicted to increase streamflow by reduction of evapotranspiration (ET) that results from a decrease in plant stomatal conductance (Field et al., 1995; Jha et al., 2006; Wu et al., 2012a, b). Jha et al. (2006), for example, showed that a doubling of CO_2 concentration increased water loads by ~ 36 % in the upper Mississippi River basin. Precipitation increase/decrease has been found to directly affect the rise/fall of streamflow levels (Jha et al., 2006; Ficklin et al. 2009; Wu et al., 2012a; Praskievicz, 2014; Uniyal et al., 2015). Ficklin et al. (2009) found that a change in precipitation of +20 and −20 % led to changes in water loads by nearly +17 and −14 %, respectively, in the San Joaquin River watershed, California. Temperature increase was reported to reduce streamflow during summer seasons due to the intensified ET values, and to increase streamflow during winter seasons due to increased snowmelt (Jha et al., 2006; Ficklin et al., 2009, 2013; Wu et al., 2012a; Praskievicz, 2014). Interestingly, in most studies, the response of N loads to climate variability was found to be similar to the response of streamflow (Ficklin et al., 2009; Wu et al., 2012a; Praskievicz, 2014; Gombault et al., 2015). According to the projected climatic conditions (e.g., elevated CO_2 concentrations, precipitation and temperature increases) illustrated in Najjar et al. (2009), substantial variations in streamflow and N loads are anticipated in the CBW. Therefore, it is important to investigate potential climate change impacts on watershed hydrological processes to efficiently mitigate potential water quality degradation.

Climate change impacts on hydrological processes have not been fully investigated in the CBW region. Howarth et al. (2006) attempted to quantify N loads under modified climate conditions, but their projections relied on the statistical relationships between river discharge/precipitation and N loads. Lee et al. (2015) predicted changes in streamflow and nitrate loads at the outlet of the watershed in response to climate variability (e.g., elevated CO_2 concentrations, precipitation and temperature increase). To cope with climate change-driven modifications, it is imperative to have an understanding of a wide range of changes in hydrological processes (Najjar et al., 2010). A simple projection of aggregated watershed responses (i.e., water quality variables at the outlet of the watershed) would be limited to suggesting conservation practices to reduce climate change impacts. An understanding of internal watershed processes (i.e., water and nutrient transport mechanisms) within a watershed can guide site-specific management plans to aid conservation decision making. In addition, climate impacts on agriculture are extremely important for the CB region because agriculture is the single largest nutrient source to CB and crop growth modified by climate change can substantially impact internal watershed processes (Najjar et al., 2010). However, previous studies have not fully demonstrated climate change impacts on internal watershed processes while considering detailed agricultural management practices.

Moreover, responses of watershed hydrological processes to climate variability and change can vary by watershed characteristics (e.g., land use and soil drainage conditions). For example, several studies showed that watersheds with a greater area of croplands released a higher amount of nitrate than watersheds with less cropland, mainly due to increased input of agricultural N (Jordan et al., 1997; Hively et al., 2011; McCarty et al., 2014). Thus, climate change can lead to greater nitrate export from watersheds with a larger percentage of cropland area, due to increased export of N from fertilizer application. Additionally, different soil characteristics can also lead to different responses in watershed-scale water and N cycles under climate change. A study by Chiang (1971) showed that well-drained soils with a high infiltration rate promote water percolation, increasing groundwater contribution to streamflow. Nitrate leaching is also found to frequently occur in well-drained soils (Lee et al., 2016a). In contrast, poorly drained soils with a low infiltration rate provide anaerobic conditions favorable to denitrification, resulting in nitrate removal in soils and groundwater (Denver et al., 2010; Lee et al., 2016a; Sharifi et al., 2016). For example, prior converted croplands, which are also known as "currently farmed historical wetlands", and are commonly associated with poorly drained soils, were also shown to have prominent impacts on reducing agrochemical loadings in the CBW region during the winter season, when ET is low and the groundwater table is high (Tiner and Burke, 1995; Denver et al., 2014; McCarty et al., 2014; Sharifi et al., 2016). Artificial drainage systems in agricultural lands are widely developed on poorly drained soils in this region, resulting in an increase in water and nutrient transport to nearby streams through surface runoff (McCarty et al., 2008; Fisher et al., 2010). Therefore, water and nitrate fluxes in watersheds with different soil characteristics are expected to show distinctive responses to climate variability and change.

This study aimed at evaluating the impacts of potential climate variability and change on water and nitrate budgets in two adjacent watersheds on the Coastal Plain of the CBW, using the Soil and Water Assessment Tool (SWAT) model. This process-based water quality model has been widely used to predict climate change impacts on numerous watersheds (Gassman et al., 2007; Uniyal et al., 2015). We prepared six climate sensitivity scenarios to assess the individual impacts of changes in CO_2 concentration (590 and 850 ppm), precipitation (11 and 21 %), and temperature (2.9 and 5.0 °C) increase. This sensitivity analysis was prepared to develop in-depth knowledge and understanding of how each climate factor affects internal watershed processes and crop growth. Then, SWAT simulations were conducted using five GCM projections (referred to as the GCM scenario) to evaluate watershed internal processes and crop growth under foreseeable climate conditions that consider simultaneous changes in CO_2, precipitation, and temperature. We used the GCM projections to describe foreseeable changes, as the combination of climate factors and their interactions could not provide complete climate change/variability information including seasonal and inter-decadal variability (Mearns, 2001). We first assessed climate change impacts on water and nitrate loads by analyzing internal watershed processes and crop growth, and then analyses comparing the two watersheds were conducted to identify critical landscape characteristics that affected nitrate loads. Finally, suggestions were provided regarding conservation practice implementation to improve the resilience of coastal watersheds to future climate change in the CBW region.

2 Materials and methods

2.1 Study area

This study was undertaken on two adjacent watersheds, Tuckahoe Creek Watershed (TCW, $\sim$220.7 km^2) and Greensboro Watershed (GW, $\sim$290.1 km^2). They are subwatersheds of the Choptank River Watershed located in the Coastal Plain of the CBW (Fig. 1). The Choptank River Watershed is one of the Conservation Effects Assessment Project (CEAP) Benchmark watersheds of the US Department of Agriculture (USDA)-Natural Resources Conservation Service (NRCS). The US Environmental Protection Agency (USEPA) has listed this watershed as "impaired" under Section 303(d) of the 1972 Clean Water Act, primarily due to the excessive nutrient and sediment loads (McCarty et al., 2008). The two adjacent sub-watersheds have distinctive characteristics considering the distribution of land use and soil drainage conditions (Fig. 2 and Table 1). The TCW is dominated by agricultural lands (54 %) and forest (32.8 %) with well-drained soils, classified as hydrologic soil groups

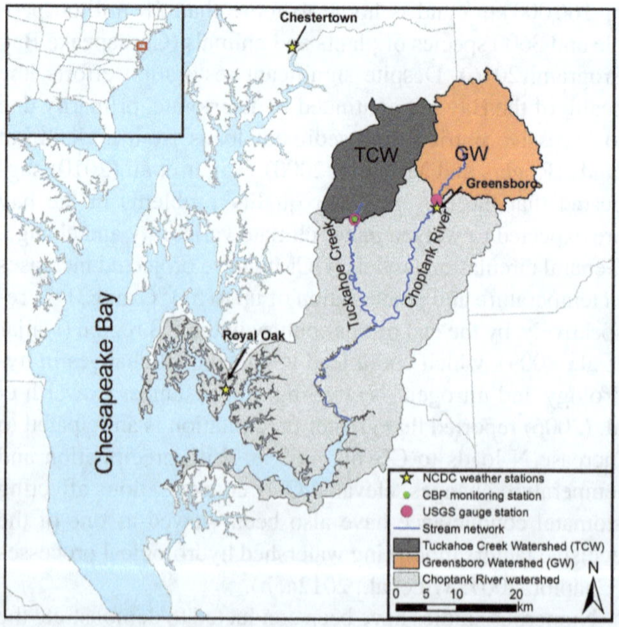

Figure 1. The location of the Tuckahoe Creek Watershed (left) and Greensboro Watershed (right) (adapted from Lee et al., 2016a).

(HSGs) A or B. These soils account for 56 % of the total watershed and 69.5 % of the agricultural lands (Fig. 2). Thus, water and nitrate fluxes tend to be easily percolated and leached into soils and groundwater, and groundwater flow is considered to be a major water pathway for nutrient fluxes to streams in the TCW (Lee et al., 2016a). In comparison, forest (48.3 %) is the major land use type in the GW, followed by agriculture (36.1 %). Soils that are poorly drained (HSGs C or D) occupy 75 % of the total area and 67.2 % of agricultural lands, which results in low infiltration rates and high denitrification potential.

2.2 SWAT

The SWAT is a process-based watershed model, developed to assess the impact of human activities and land use on water and nutrient cycles within agricultural watersheds (Neitsch et al., 2011). The SWAT divides a watershed into subwatersheds using a digital elevation model (DEM), and each sub-watershed is further divided into hydrologic response units (HRUs) based on a unique combination of land use, soil type, and slope. Model simulation is performed at the HRU level, and the simulated outputs aggregated at the subwatershed and then further at the watershed level through routing processes. The amounts of surface runoff and infiltration are calculated based on the Soil Conservation Service (SCS) curve number (CN) method, and the CN values are updated daily based on soil permeability, land use type, and antecedent soil water conditions. Water infiltrated into soils is either delivered to streams through lateral flow or further percolated into groundwater, when soil water content exceeds

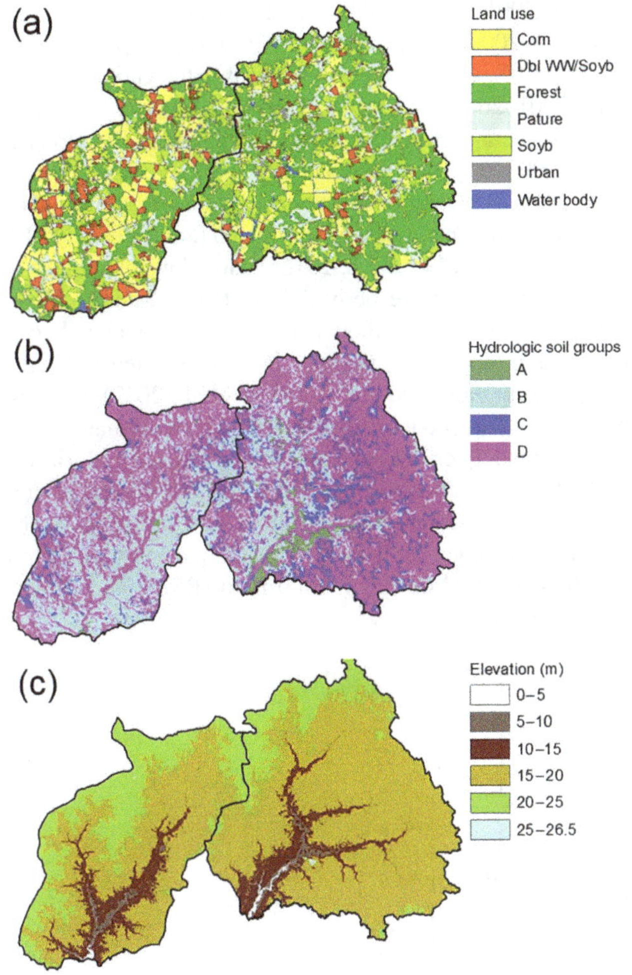

Figure 2. The physical characteristics of the Tuckahoe Creek Watershed (left) and Greensboro Watershed (right); **(a)** land use, **(b)** hydrologic soil groups, and **(c)** elevation (adapted from Lee et al., 2016a). Note: Dbl WW/Soyb stands for double crops of winter wheat and soybean in a year. Hydrologic soil groups (HSGs) are characterized as follows: Type A – well-drained soils with a 7.6–11.4 mm h^{-1} water infiltration rate; Type B – moderately well-drained soils with 3.8–7.6 mm h^{-1}; Type C – moderately poorly drained soils with 1.3–3.8 mm h^{-1}; Type D – poorly drained soils with 0–1.3 mm h^{-1} (Netisch et al., 2011).

field capacity. The groundwater portion is then transported to streams through groundwater flow, percolated into the deep groundwater aquifer, or discharged to the soil profile. The amount of nitrate in soils is increased by nitrification, mineralization of soil organic and crop residue, biological N fixation, and fertilization, and decreased through denitrification and plant uptake (Neitsch et al., 2011). Nitrate fluxes move via surface runoff, lateral flow, percolated water from soil to groundwater, and groundwater flow. Nitrate concentration in the mobile water (i.e., surface runoff, lateral flow, and percolated water) is first determined and then nitrate fluxes in the mobile water are calculated based on the nitrate concentra-

Table 1. Soil properties and land use distribution of the Tuckahoe Creek Watershed (TCW) and Greensboro Watershed (GW) (adapted from Lee et al., 2016a).

Land use	TCW	GW
Agriculture	54.0 %	36.1 %
	(69.5/30.5 %)	(32.8/67.2 %)
Forest	32.8 %	48.3 %
Pasture	8.4 %	9.3 %
Urban	4.2 %	5.6 %
Water body	0.6 %	0.7 %
Hydrologic soil groups (HSGs)	TCW	GW
A	0.3 %	3.1 %
B	55.8 %	22.4 %
C	2.2 %	4.2 %
D	41.7 %	70.3 %

Note: values in parentheses – "( )" – denote the proportion of well-drained soils (HSG-A&B) and poorly drained soils (HSG-C&D) used for agricultural lands, respectively.

tion and the amount of mobile water. Nitrate in groundwater is re-distributed in four ways: remains in the groundwater, recharges to deep groundwater, moves to streams, or discharges to the soils. Nitrate removal by biological and chemical processes in groundwater is simulated by the first-order kinetics. Refer to Netisch et al. (2011) for further details.

The SWAT model has the capability of simulating the impacts of CO_2 concentration on ET and biomass accumulations. The Penman–Monteith method used for this study considers CO_2 effects on ET based on the relationship between plant stomatal conductance and CO_2 concentration:

$$g_{l,CO_2} = g_l \times [1.4 - 0.4 \times (CO_2/330)], \quad (1)$$

where g_{l,CO_2} is the leaf conductance modified to reflect CO_2 effects, and g_l is the leaf conductance without the effect of CO_2. The equation shows the linear reduction of the leaf conductance with increasing CO_2 and results in a 40 % reduction in leaf conductance for all plants when CO_2 concentration is doubled. According to Eq. (1) elevated CO_2 concentrations decrease plant stomatal conductance and canopy resistance, subsequently reducing ET. Refer to Neitsch et al. (2011) for details on the Penman–Monteith method.

The simulation of crop growth in the SWAT is based on potential heat unit theory. The model considers the impacts of CO_2 concentration on crop biomass growth by modifying the radiation-use efficiency (RUE) of the plant as follows:

$$\mathrm{RUE} = \frac{100 \cdot CO_2}{CO_2 + \exp(r_1 - r_2 \cdot CO_2)}, \quad (2)$$

where RUE is the radiation-use efficiency of a plant, and $r_1$and r_2 are coefficients.

$$\Delta \mathrm{bio} = \mathrm{RUE} \cdot H_{\mathrm{phosyn}}, \quad (3)$$

where Δbio is a potential increase in plant biomass on a given day and H_{phosyn} is the amount of intercepted photosynthetically active radiation on a given day.

2.3 Baseline SWAT input data

Climate and geospatial data needed for the SWAT simulation are summarized in Table 2. Daily precipitation and temperature were obtained from three meteorological stations operated by the National Oceanic and Atmospheric Administration (NOAA) National Climate Data Center (NCDC) at Chestertown, Royal Oak, and Greensboro, Maryland (USC00181750, USC00187806, and US1MDCL0009, respectively). Due to data unavailability, humidity, wind speed, and solar radiation were generated using the SWAT built-in weather generator (Neitsch et al., 2011). Monthly streamflow data were downloaded from US Geological Survey (USGS) gauge stations on Tuckahoe Creek near Ruthsburg (USGS no. 01491500) and the Choptank River near Greensboro (USGS no. 01491000) (Fig. 1). The USGS LOAD ESTimator (LOADEST, Runkel et al., 2004) was used to generate continuous monthly nitrate loads from nitrate grab sample data (133 samples over the simulation period) that were obtained from the Chesapeake Bay Program (CBP, TUK no. 0181) for the TCW and from USGS gauge station data (USGS no. 01491000) for the GW. The LOADEST is commonly used to generate continuous data from discrete data, and it has been shown to accurately generate water quality variables (Jha and Jha, 2013; Lee et al., 2016b). The land use, soil maps, and DEM were prepared as shown in Table 2.

We identified representative agricultural practices for this region using multiple geospatial data (Lee et al., 2016a). Major crop rotations and their year-to-year placement were derived through analysis of the USDA-National Agricultural Statistics Service (NASS) Cropland Data Layer (CDL) for the period of 2008–2012. We assumed that crop rotation and land use did not change over the simulation period so that agricultural N input did not vary for the baseline and GCM scenarios. Detailed agricultural management information (e.g., the amount, type, and application timing of fertilizer, and planting and harvesting timings of individual crops) was developed through literature review and communications with local experts (Table A1). Detailed information about the development of crop rotation and land management is available in Lee et al. (2016a).

2.4 Baseline SWAT calibration and validation

The SWAT model simulations were performed at a monthly time step for 16 years; this included 2-year warm-up (1999–2000), 8-year calibration (2001–2008), and 6-year validation (2009–2014) periods. The SWAT model was run at a daily time step based on daily climate input data, and daily outputs were aggregated to monthly outputs. It should be noted that due to unavailability of water quality observations prior to 2001, model calibration and validation were initiated from 2001. Compared to past 30-year precipitation data (1981–2010), the climate condition over the calibration period (2001–2008) was shown to include representative wet, dry, and average climate conditions, while the validation period (2009–2014) was dominated by wet conditions. Critical parameters used for model calibration were selected based on previous studies conducted in this region (Sexton et al., 2010; Yeo et al., 2014; Lee et al., 2016a) and allowable ranges of these parameters were derived from the literature as indicated in Table 3. Streamflow parameters were manually calibrated and then nitrate parameters were adjusted following SWAT calibration guidelines (Arnold et al., 2012). A set of parameters that produced the best model performances and fulfilled model performance criteria suggested by Moriasi et al. (2007) were chosen for model validation. Model performance was evaluated using the following statistics: Nash–Sutcliffe efficiency coefficient (NSE), root mean square error (RMSE) standard deviation (SD) ratio (RSR), and percent bias (P-bias).

$$\mathrm{NSE} = 1 - \left[\frac{\sum_{i=1}^{n} (O_i - S_i)^2}{\sum_{i=1}^{n} (O_i - \overline{O})^2} \right], \tag{4}$$

$$\mathrm{RSR} = \frac{\mathrm{RMSE}}{\mathrm{SD_{obs}}} = \frac{\left[\sqrt{\sum_{i=1}^{n} (O_i - S_i)^2}\right]}{\left[\sqrt{\sum_{i=1}^{n} (O_i - \overline{O})^2}\right]}, \tag{5}$$

$$P\text{-bias} = \left[\frac{\sum_{i=1}^{n} (O_i - S_i) \times 100}{\sum_{i=1}^{n} O_i} \right], \tag{6}$$

where O_i is the observed data at time step i, S_i is the simulated output at time step i, $\overline{O}$ is the mean of observed data over all time steps, and n is the total number of observed data. We also calculated NSE for the natural logarithm of streamflow to evaluate model performance for low flows (Kiptala et al., 2014). In addition, the 95 % prediction uncertainty (95 PPU) band was represented to evaluate model uncertainty (Singh et al., 2014). The 95 PPU was computed based on all simulated outputs generated during the calibration process. The 95 PPU was represented as the range of values between the 2.5 and 97.5 percentiles of the cumulative distribution of simulated outputs.

2.5 Climate sensitivity and GCM scenarios

To evaluate the impacts of climate variability and change on watershed hydrological processes, climate sensitivity and GCM scenarios were prepared as illustrated below (see Sect. 2.5.1 and 2.5.2). The calibrated SWAT model was sim-

Table 2. List of the SWAT model input data.

Data	Source	Description	Year
DEM	MD-DNR	LiDAR-based 2 m resolution	2006
Land use	USDA-NASS	Cropland Data Layer (CDL)	2008–2012
	MRLC	National Land Cover Database (NLCD)	2006
	USDA-FSA-APFO	National Agricultural Imagery Program digital Orthophoto quadimagery	1998
	US Census Bureau	TIGER road map	2010
Soils	USDA-NRCS	Soil Survey Geographical Database (SSURGO)	2012
Climate	NCDC	Daily precipitation and temperature	1999–2014
Streamflow	USGS	Monthly streamflow	2001–2014
Water quality	USGS and CBP	Daily grab nitrate samples	2001–2014

Note: MD-DNR: Maryland Department of Natural Resources, USDA-NASS: USDA-National Agricultural Statistics Service, MRLC: Multi-Resolution Land Characteristics Consortium, USDA-FSA-APFO: USDA-Farm Service Agency-Aerial Photography Field Office, TIGER: Topologically Integrated Geographic Encoding and Referencing, and USDA-NRCS: USDA-Natural Resources Conservation Service.

Table 3. List of calibrated parameters.

Parameter	Variable	Description (unit)	Range	Calibrated value	
				TCW	GW
CN2[1]	Streamflow	Curve number	−50–50 %	−30 %	0 %
ESCO[1]		Soil evaporation compensation factor	0–1	1	0.95
SURLAG[1]		Surface runoff lag coefficient	0.5–24	0.5	0.5
SOL_AWC[1]		Available water capacity of the soil layer (m H_2O mm soil^{-1})	−50–50 %	−10 %	−1 %
SOL_K[1]		Saturated hydraulic conductivity (mm h^{-1})	−50–50 %	50 %	49 %
SOL_Z[1]		Depth from soil surface to bottom of layer (mm)	−50–50 %	−20 %	−31 %
ALPHA_BF[1]		Baseflow recession constant (1 days^{-1})	0–1	0.07	0.051
GW_DELAY[1]		Groundwater delay time (days)	0–500	120	45
GW_REVAP[1]		Groundwater "revap" coefficient	0.02–0.2	0.10	0.02
RCHRG_DP[1]		Deep aquifer percolation fraction	0–1	0.01	0.05
GWQMN[1]		Threshold depth of water in the shallow aquifer required for return flow to occur (mm)	0–5000	1.9	1.0
CH_K2[1]		Effective hydraulic conductivity (mm h^{-1})	0–150	0	20
CH_N2[1]		Manning coefficient	0.01–0.3	0.29	0.021
NPERCO[2]	Nitrate	Nitrogen percolation coefficient	0.01–1	0.5	0.2
N_UPDIS[2]		Nitrogen uptake distribution parameter	5–50	50	50
ANION_EXCL[2]		Fraction of porosity from which anions are excluded	0.1–0.7	0.59	0.6
ERORGN[2]		Organic N enrichment ratio for loading with sediment	0–5	4.92	4.1
BIOMIX[2]		Biological mixing efficiency	0.01–1	0.01	0.01
SOL_NO3[3]		Initial NO_3 concentration in soil layer (mg N kg^{-1})	0–100	11.23	0
CDN[4]		Denitrification exponential rate coefficient	0–3.0	0.3	1.8
SDNCO[4]		Denitrification threshold water content	0.1–1.1	1.0	1.0

* refers to a default value. The ranges of parameters with superscripts (1–4) were adapted from Gitau and Chaubey (2010), Yeo et al. (2014), Seo et al. (2014), and Neitsch et al. (2011), respectively.

ulated using the climate sensitivity and GCM scenarios for comparison with baseline water and nitrate budgets.

2.5.1 Climate sensitivity scenarios

A climate sensitivity analysis aids in identifying the degree or threshold of responses of hydrologic variables to climate-induced modifications, and a sensitivity scenario generally assumes constant changes throughout the year (Mearns, 2001). Following the approach in Mearns (2001), six climate sensitivity scenarios were prepared by modifying the baseline data (1999–2014) to assess individual effects of elevated CO_2 concentrations, precipitation, and temperature on watershed hydrological processes (Table 4). Sensitivity scenarios were designed to change one variable while holding other variables constant throughout the simulations. Baseline precipitation and temperature were modified by percent

Table 4. Climate sensitivity scenarios developed by modifying baseline values.

Scenario	Percent increase in precipitation (%)	Absolute increase in temperature (°C)	Replacement of CO_2 (ppm)
Baseline	0	0	330
1	0	0	590
2	0	0	850
3	11	0	330
4	21	0	330
5	0	2.9	330
6	0	5.0	330

and absolute changes using anomaly and absolute data, respectively, as illustrated in Najjar et al. (2009). They reported mean temperature and precipitation changes over CB for three future periods (2010–2039, 2040–2069, and 2070–2099) relative to the baseline period (1971–2000) based on GCM outputs (Najjar et al., 2009). We used the maximum increase rate (and value) for 2040–2069 (precipitation: 11 % and temperature: 2.9 °C) and 2070–2099 (precipitation: 21 % and temperature: 5.0 °C) to set the precipitation and temperature sensitivity scenarios. For example, baseline precipitation increased by 11 and 21 % for Scenarios 3 and 4, respectively, and 2.9 and 5.0 °C were added to the baseline temperature for Scenarios 5 and 6, respectively (Table 4). The baseline CO_2 concentration was set as the default value (330 ppm) for simulations. For the first and second scenarios, the baseline CO_2 concentration was replaced with 590 and 850 ppm, respectively. The upper value of 850 ppm was used because GCMs used for temperature and precipitation sensitivity scenarios were forced with the assumption of CO_2 concentration of 850 ppm (Najjar et al., 2009). The lower value of 590 ppm (the average of 330 and 850 ppm) was considered to be the level of CO_2 concentration around the middle of the 21st century.

2.5.2 GCM scenario

A GCM-based scenario is the most commonly used method for assessing future climate change impacts (Mearns, 2001). We downloaded projected climate data (e.g., daily precipitation and maximum and minimum temperature) from the World Climate Research Program's (WCRP's) Coupled Model Intercomparison Project5 (CMIP5) archive (Brekke et al., 2013). Five GCM data under the Representative Concentration Pathway (RCP) 8.5 scenario were downloaded (Table A2), because the RCP 8.5 indicates the highest value of CO_2 concentration in the CMIP5. To be consistent with the period of the baseline data (1999–2014), 16-year future data (2083–2098) were used in this study. We further refined GCM data using the delta change method because spatially downscaled data are consistent with historical observations at the global scale, but could be significantly inconsistent at fine spatial scales, such as a watershed (Wang et al., 2014). The delta change method was calculated as follows:

$$P_{\text{delta}} = \text{GCM}_{P\text{-future, monthly}} / \text{GCM}_{P\text{-baseline, monthly}}, \tag{7}$$

$$T_{\text{delta}} = \text{GCM}_{T\text{-future, monthly}} - \text{GCM}_{T\text{-baseline, monthly}}, \tag{8}$$

$$\text{DGCM}_{P\text{-future, daily}} = \text{OBS}_{P\text{-baseline, daily}} \times P_{\text{delta}}, \tag{9}$$

$$\text{DGCM}_{T\text{-future, daily}} = \text{OBS}_{T\text{-baseline, daily}} + T_{\text{delta}}, \tag{10}$$

where P_{delta} and T_{delta} indicate precipitation (P) and temperature (T) biases in GCM data, respectively, $\text{GCM}_{\text{future,monthly}}$ and $\text{GCM}_{\text{baseline,monthly}}$ indicate the monthly average of GCM data for the future (2083–2098) and baseline (1999–2014) periods, respectively, $\text{OBS}_{\text{baseline,daily}}$ indicates observed daily climate, and $\text{DGCM}_{\text{future,daily}}$ indicates unbiased future climate data. We calculated the ensemble mean of delta-change values from the five GCMs, because substantial variations existed among the GCM projections (Shrestha et al., 2012; Van Liew et al., 2012). Then, the SWAT model was simulated using the ensemble mean to predict hydrological processes under future climate conditions. Similar to the baseline scenario, humidity, wind speed, and solar radiation values were generated using the SWAT built-in weather generator owing to data unavailability. We assumed the CO_2 concentration for the GCM scenario to be 936 ppm, as the specified CO_2 concentration under the RCP8.5 scenario (Meinshausen et al., 2011).

2.6 Analyses of simulation outputs

Simulated outputs were summarized at multiple temporal scales (e.g., monthly, seasonal, and annual). Annual averages of streamflow, ET, and nitrate loads were calculated to investigate changes in water and nitrate budgets in response to climate sensitivity and GCM scenarios. The response of crop growth to climate variability and change was also analyzed to show the effects of modified crop biomass on hydrology and the N cycle. For comparative analyses between two watersheds, water and nitrate yields were summarized seasonally for climate sensitivity scenarios (i.e., summer (April–September) and winter (October–March)) and monthly for the GCM scenario. Note that water and nitrate yields indicate the summations of water and nitrate fluxes transported from lands to streams by surface runoff, lateral flow, and groundwater flow. All simulation outputs were normalized by total watershed size.

We conducted a statistical analysis to test whether the simulation results under climate sensitivity and GCM scenarios were statistically different from those under the baseline scenario using parametric (paired t-test) and nonparametric (Wilcoxon signed rank) methods. Note that we used monthly outputs (168 samples over 14 years) for this analysis. The statistical significance for the difference was indicated by the p-value.

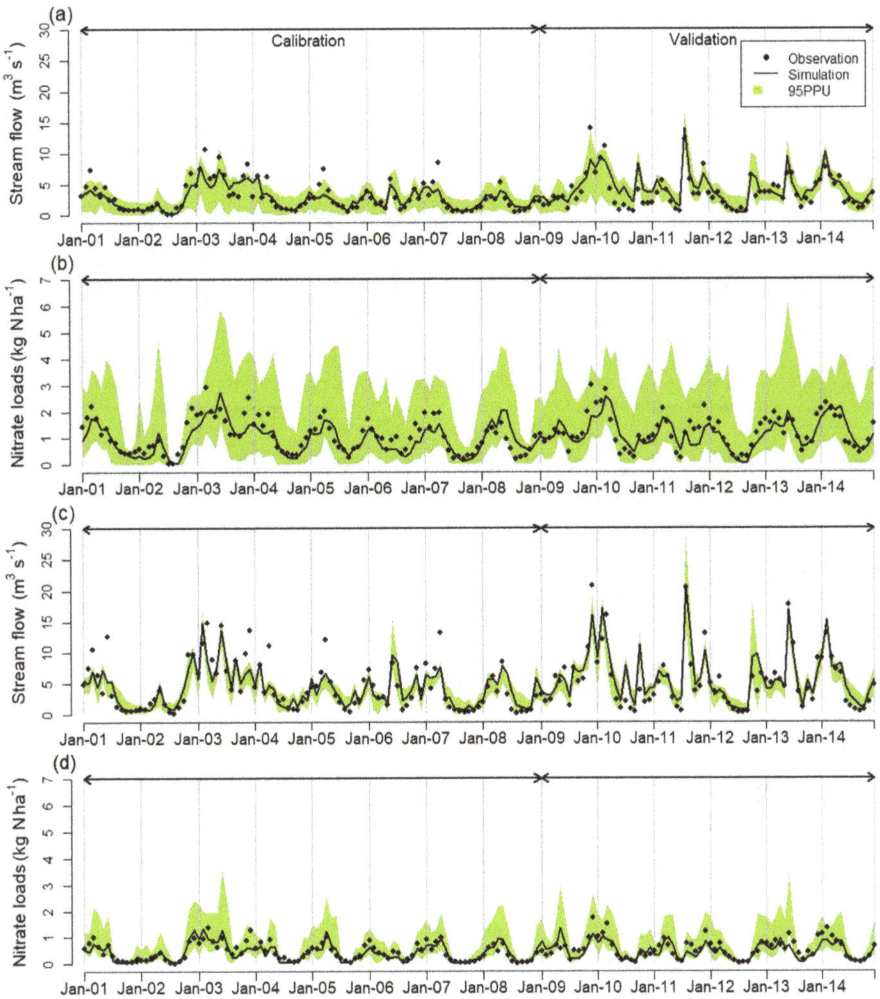

Figure 3. Simulated and observed monthly streamflow and nitrate loads for the **(a, b)** TCW and **(c, d)** GW during calibration and validation periods. Note: 95 PPU stands for 95 % prediction uncertainty.

3 Results and discussions

3.1 Model calibration and validation

Monthly simulations for streamflow and nitrate loads were compared with corresponding observations (Fig. 3). Results show that simulated values for monthly streamflow were in good agreement with observations, but simulated peak streamflows were underestimated relative to observations. This underestimation was attributed to the inherent limitations of the SWAT model and limited climate data to capture local storm effects as it does not account for the intensity and duration of the precipitation (Qiu et al., 2012). Previous studies conducted in this region showed similar results, though the overall simulation results accurately replicated the observations (Yeo et al., 2014; Lee et al., 2016a). Simulated nitrate loads were also well matched with actual observations and the uncertainty band (shown as green in Fig. 3) captured most observations in the two watersheds. Overall, model performance measures fulfilled "good" (e.g., $0.65 < \text{NSE} \leq 0.75$) or "very good" ($0.75 < \text{NSE}$) criteria for streamflow and at least "satisfactory" ($0.5 < \text{NSE} \leq 0.65$) for nitrate loads (Table 5). The model performance measures for low flows (NSE for the natural logarithm of streamflow) also indicated "satisfactory" to "very good" (Table 5). These results demonstrated that the calibrated model replicated actual conditions reasonably well (Moriasi et al., 2007; Arnold et al., 2012).

3.2 Responses to climate sensitivity scenarios

3.2.1 Water and nitrate budgets

The 14-year averages of annual hydrologic variables under the baseline and climate sensitivity scenarios are presented in Fig. 4. Elevated CO_2 concentrations (590 and 850 ppm) and precipitation increases (11 and 21 %) led to significant increases in annual streamflow and nitrate loads by 50 and 52 % for the TCW and 43 and 33 % for the GW, respectively,

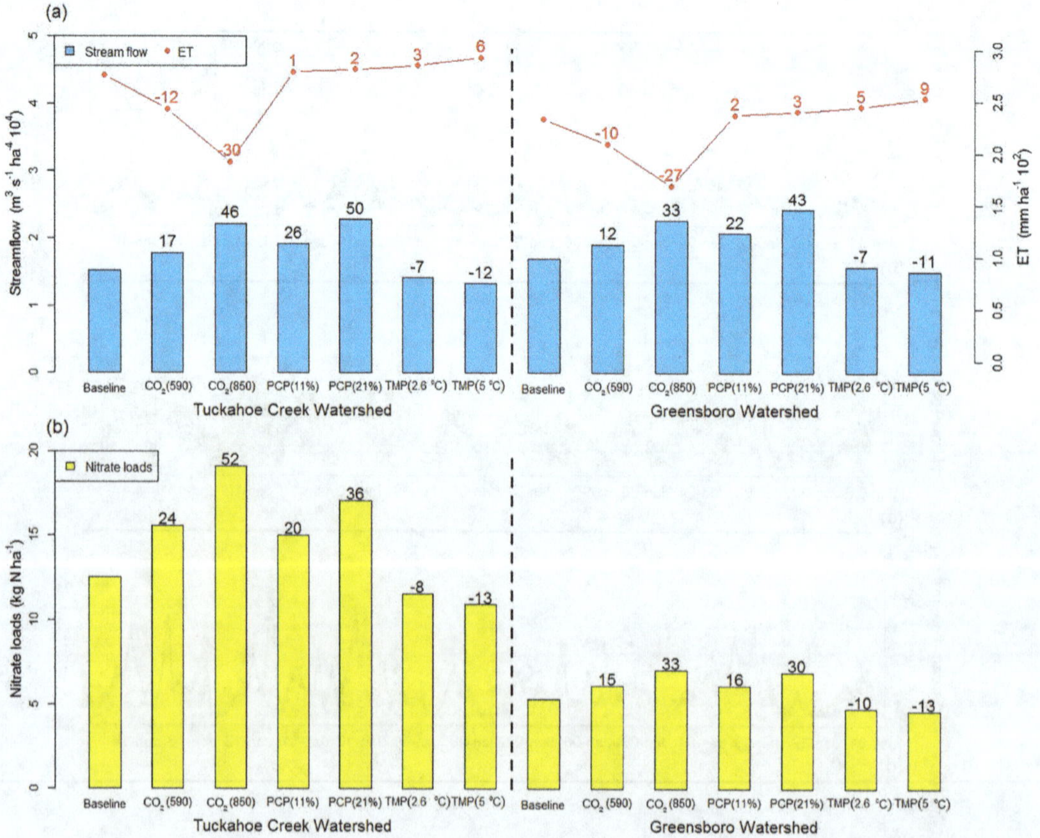

Figure 4. The 14-year average of annual hydrologic variables under the baseline and climate sensitivity scenarios at the watershed scale: **(a)** streamflow and evapotranspiration (ET), and **(b)** nitrate loads. Note: the red and black numerical values above the bar and dot graphs, respectively, indicate the relative changes (%) in hydrologic variables for climate sensitivity scenarios relative to the baseline scenario (relative change (%) = (sensitivity scenarios − baseline) / baseline × 100). PCP and TMP stand for precipitation and temperature, respectively.

Table 5. Model performance measures for monthly streamflow and nitrate loads.

Period	Variable	Streamflow		Nitrate loads	
		TCW	GW	TCW	GW
Calibration	NSE	0.723[b] (0.828[c])	0.686[b] (0.719[b])	0.623[a]	0.702[b]
	RSR	0.523[b]	0.556[b]	0.610[a]	0.542[b]
	P-bias (%)	−5.8[c]	−3.2[c]	−9.8[c]	−4.1[c]
Validation	NSE	0.674[b] (0.556[a])	0.790[c] (0.727[b])	0.604[a]	0.567[a]
	RSR	0.566[b]	0.454[c]	0.624[a]	0.652[a]
	P-bias (%)	17.8[b]	13[c]	−5.6[c]	−12.1[c]

Model performances were rated based on the criteria of Moriasi et al. (2007): [a] satisfactory, [b] good, and [c] very Good.
[a] Satisfactory (0.5 < NSE ≤ 0.65, 0.6 < RSR ≤ 0.7, and ±15 ≤ *P*-bias < ±25),
[b] Good (0.65 < NSE ≤ 0.75, 0.5 < RSR ≤ 0.6, and ±10 ≤ *P*-bias < ±15), and
[c] Very Good (0.75 < NSE ≤ 1.0, 0.0 < RSR ≤ 0.5, *P*-bias < ±10). A value in parentheses indicates the NSE for the natural logarithm of streamflow.

relative to the baseline scenario (p-value < 0.01) (Fig. 4). Elevated CO_2 concentrations lowered plant stomatal conductance, resulting in a decrease in ET of 30 % and thereby increased streamflow and corresponding increases in nitrate loads (Fig. 4). The reduced rate of ET (driven by CO_2 concentrations of 850 ppm) demonstrated in this study is supported by previous studies using SWAT, such as Ficklin et al. (2009, −40 %; 970 ppm) and Pervez et al. (2015, −12 %; 660 ppm). Precipitation increase resulted in a direct increase in streamflow, leading to increased nitrate loads. Compared to the baseline scenario, a temperature increase of 5 °C significantly reduced annual streamflow and nitrate loads by 12 and 13 % for the TCW and 11 and 13 % for the GW (p-value < 0.01), respectively, due to intensified ET (Fig. 4).

It should be noted that the standard version of SWAT tends to overestimate the impact of CO_2 on reduction of ET (Eckhardt and Ulbrich, 2003). Maximum leaf area index (LAI) is assumed to be constant regardless of variation in CO_2 concentration in SWAT. However, maximum LAI is known to increase with increasing CO_2 concentration (Eckhardt and Ulbrich, 2003). In addition, the degree of reduction in stomatal conductance varies by plant species, which also is not taken into account in the SWAT model. Another model simplification, which increases uncertainty, is the application of the same reduction rate to all plants. For example, C3 crops (soy-

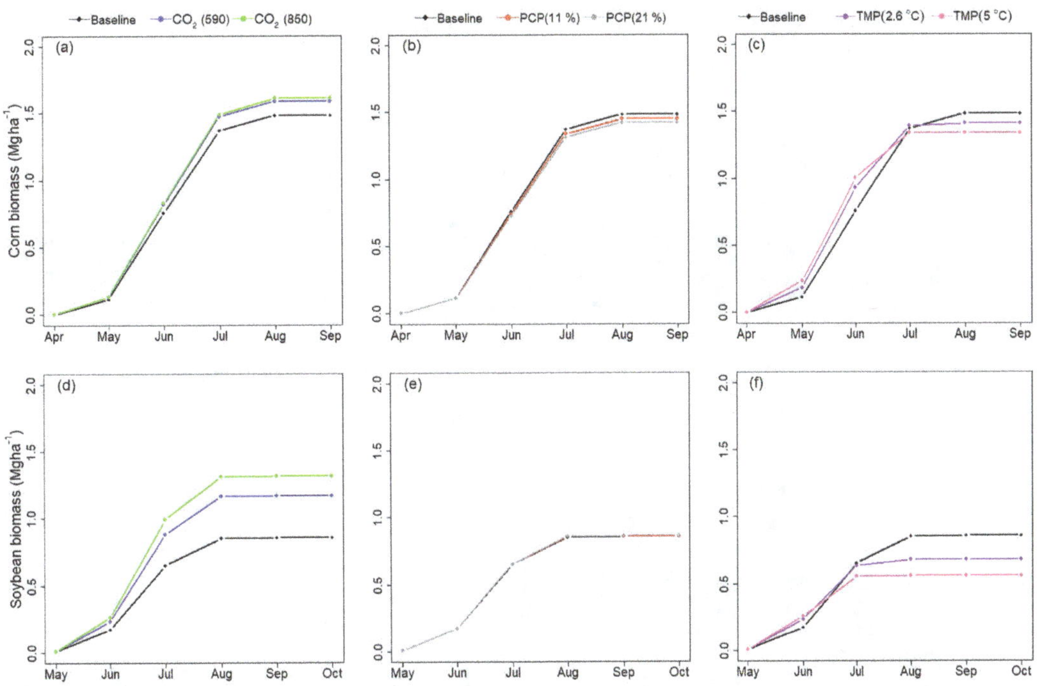

Figure 5. The responses of crop biomass growth to the climate sensitivity scenario: **(a–c)** corn and **(d–f)** soybean. Note: PCP and TMP in the legend stand for precipitation and temperature, respectively.

bean and wheat) are known to have less reduction in stomatal conductance with rising CO_2 concentration compared to C4 crops (corn) (Ainsworth and Rogers, 2007). Both factors could contribute to overestimating reduction of ET and resultant increase in streamflow and nitrate loads (Eckhardt and Ulbrich, 2003).

Changes in crop growth under climate sensitivity scenarios had great impacts on water and nitrate budgets. Although precipitation increase resulted in the greatest increase in annual streamflow, annual nitrate loads were greater under elevated CO_2 concentrations (Fig. 4), due to increased crop biomass and high N availability from mineralization of crop residues (Fig. 5a, b). Elevated CO_2 concentrations stimulated crop growth by decreasing water demand and increasing radiation-use efficiency (Abler and Shortle, 2000; Parry et al., 2004). For example, simulated corn and soybean biomass increased from 1.5 and 0.9 Mg ha^{-1} (baseline concentration of 330 ppm) to 1.6 and 1.3 (CO_2 concentration of 850 ppm) Mg ha^{-1}, respectively (Fig. 5a, b). Increased crop biomass left greater amounts of crop residue following crop harvest (winter seasons: October–March), which contributed to increasing nitrate in soils through mineralization (Lee et al., 2016a). Our simulation results indicated that mineralized nitrate under elevated CO_2 concentrations increased by 27 % for the TCW and 23 % for the GW during winter seasons, compared to the baseline values (Fig. A1). Increased crop residue resulted in greater nitrate loads under elevated CO_2 concentrations than under conditions of increased precipitation. In contrast, temperature increase led to lower crop biomass than the baseline value, due to increased heat stress (Fig. 5c, f). Lower biomass reduced remaining crop residue and subsequently reduced mineralized nitrate by 22 % during winter seasons, compared to the baseline value (Fig. A1). Reduction of mineralized nitrate contributed to decreased nitrate loads in conjunction with intensified ET. Precipitation increase slightly decreased corn biomass because increased precipitation reduced the availability of nutrients for crops (Fig. 5b), leading to increased nutrient stress. However, soybean biomass did not change in response to precipitation increase (Fig. 5e) since soybean crops can generate N through fixation as needed.

3.2.2 Comparative analyses

For the purpose of comparing the two watersheds in response to climate sensitivity scenarios, 14-year averages of seasonal water and nitrate yields were calculated (Fig. 6). Both elevated CO_2 concentrations and precipitation increase led to greater water and nitrate yields for the two watersheds during winter and summer seasons, compared to the baseline scenario. However, the seasonal pattern of nitrate yield differed between the two watersheds. Wintertime water yield was greater than summertime water yield for both watersheds, which was consistent with the seasonal pattern of nitrate yield for the GW. However, summertime nitrate yield increases were greater than wintertime increases for the TCW, apparently due to the difference in percent agricultural lands between the TCW (54.0 %) and GW (36.1 %). Increased wa-

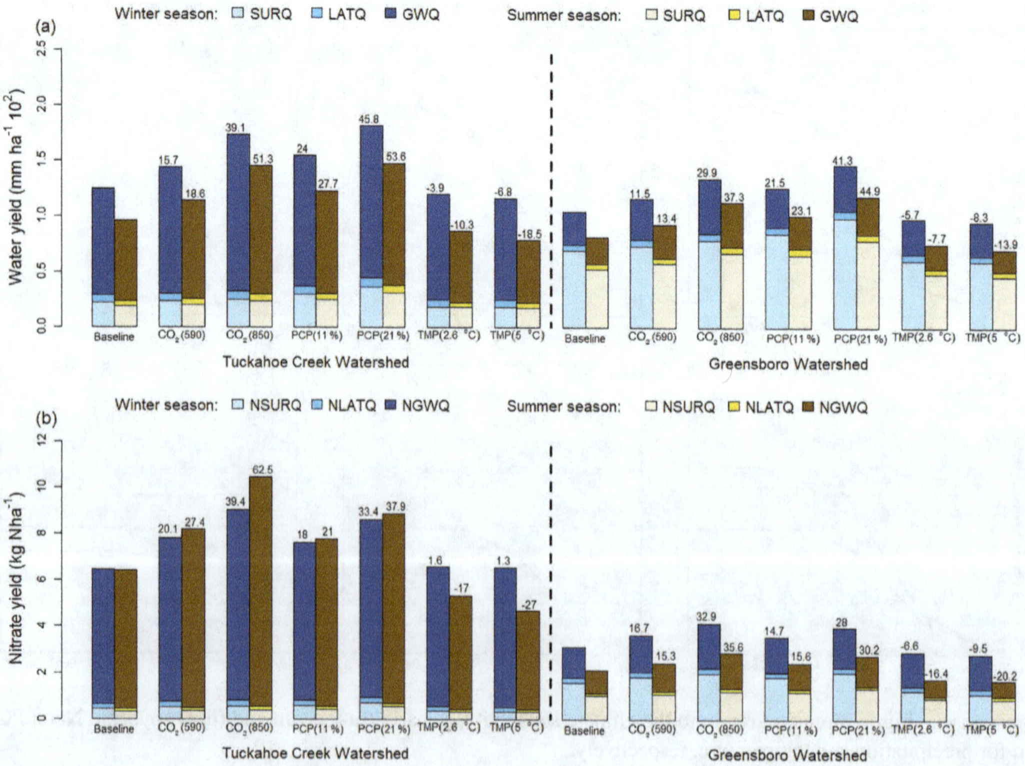

Figure 6. The 14-year average of seasonal hydrologic variables under the baseline and climate sensitivity scenarios at the watershed scale: **(a)** water and **(b)** nitrate yields. Note: the number on the bar graph indicates the relative changes (%) in hydrologic variables for climate sensitivity scenarios relative to the baseline scenario. Water and nitrate yields indicate the summations of water and nitrate fluxes transported from lands to streams by surface runoff, lateral flow, and groundwater flow. PCP and TMP stand for precipitation and temperature, respectively. SURQ, LATQ, and GWQ indicate water fluxes transported by surface runoff, lateral flow, and groundwater flow, respectively. NSURQ, NLATQ, and NGWQ indicate nitrate fluxes transported by surface runoff, lateral flow, and groundwater flow, respectively.

ter yield could accelerate the export of nitrate added to the watersheds through fertilizer activities, which mainly occurs during summer seasons. Accordingly, increased water yield caused by elevated CO_2 concentrations and precipitation increase induced considerable increase in summertime nitrate yield by $\sim 62.5\,\%$ for the TCW, while only moderately increasing yield by $\sim 35.6\,\%$ for the GW, which is dominated by forest instead of croplands.

Temperature increase reduced summertime water and nitrate yields by 18.5 and 27 % for the TCW and 13.9 and 20.2 % for the GW, respectively, mainly due to increased water loss by ET (Table A3). Wintertime water yield also decreased for the two watersheds, but changes in wintertime nitrate yield differed between the two watersheds. A decrease of 9.5 % in wintertime nitrate yield was found for GW, but wintertime nitrate yield increased by 1.6 % for the TCW (Fig. 6b), due to modified crop growth patterns and contrasting soil characteristics between the two watersheds. Temperature increase can drive crops to reach maturity earlier while exerting increased heat stress on crops, leading to lower biomass compared to the baseline (Fig. 5c, f). These two factors collectively reduced soil water and nitrate consumption by crops at the end of the growth stage under temperature increase scenarios, subsequently increasing soil water content and nitrate leaching compared to the baseline (Fig. A2). Nitrate leached into groundwater was discharged to streams through groundwater flow during winter seasons. The TCW showed increased nitrate leaching of $1.0\,\mathrm{kg\,N\,ha^{-1}}$ compared to the GW, due to a larger percentage of well-drained soils with a high infiltration rate. Different leaching rates between the TCW and GW soils led to a greater increase in wintertime nitrate flux transported by groundwater flow (NGWQ) for the TCW ($0.21\,\mathrm{kg\,N\,ha^{-1}}$) compared to the GW ($0.16\,\mathrm{kg\,N\,ha^{-1}}$) (Fig. 6b). However, intensified ET reduced wintertime water and nitrate fluxes transported by surface runoff (SURQ and NSURQ, respectively) for the two watersheds (Table A3), while water fluxes transported by lateral and groundwater flow (LATQ and GWQ, respectively) were rarely changed. Because the majority of water flux was transported by groundwater flow for the TCW and surface runoff for the GW (Fig. 6a), a decrease in SURQ led to a substantial reduction of wintertime NSURQ for GW ($0.45\,\mathrm{kg\,N\,ha^{-1}}$) with less reduction shown in the TCW ($0.12\,\mathrm{kg\,N\,ha^{-1}}$), compared to the baseline (Fig. 6b). There-

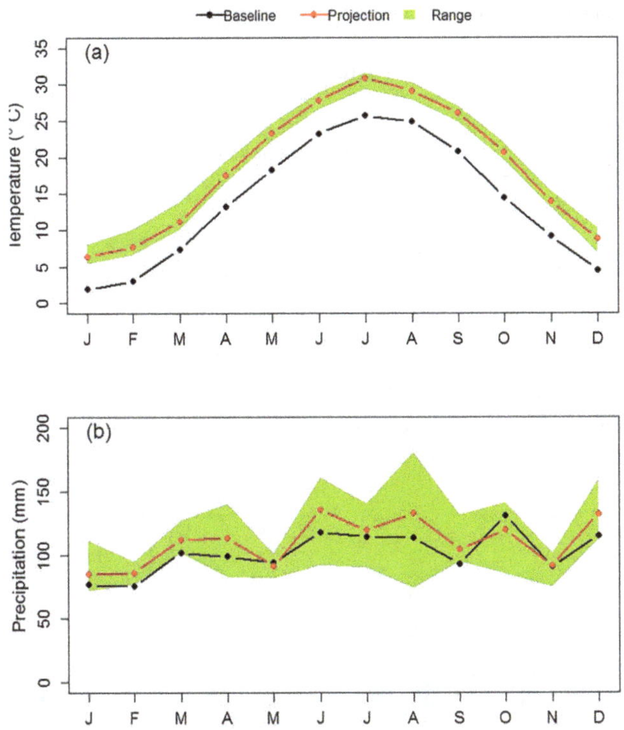

Figure 7. Monthly average of **(a)** mean temperature and **(b)** cumulative precipitation for the baseline (2001–2014) and future (2085–2098) periods. Note: "Projection" stands for the ensemble mean of five GCM data, and the range stands for the interval between the maximum and minimum values of five GCM data.

fore, both increased NGWQ and decreased NSURQ during winter seasons collectively led to an increasing pattern of wintertime nitrate yield for the TCW and a decreasing pattern for the GW, compared to the baseline scenario. Note that denitrification was rarely affected by temperature increase because reduced soil water content resulting from increased ET at higher temperatures decreased denitrification.

3.3 Responses to the GCM scenario

3.3.1 Comparison of climate data

The monthly averages of mean temperature and cumulative precipitation under the baseline scenario were compared with the ensemble means of five GCMs (Fig. 7). Projected temperature was constantly higher than the baseline value throughout the year by 3.8–6.2 °C (Fig. 7a). Compared to the baseline, projected precipitation was greater except for May and October (Fig. 7b). Monthly cumulative precipitation was up to 19 mm greater in August and up to 11 mm lower in October, in comparison to the baseline values. Note that the annual average of mean temperature increased from 13.9 °C (baseline) to 18.6 °C (projection), and the annual average of cumulative precipitation also increased from 1221 mm (baseline) to 1322 mm (projection).

3.3.2 Water and nitrate budgets

Baseline hydrologic variables (e.g., streamflow, ET, and nitrate loads) are compared with the simulated outputs in Table 6. Relative to the baseline scenario, annual streamflow and nitrate loads significantly increased by 70 and 66 % for the TCW and 50 and 56 % for the GW, respectively (p-value < 0.01). These increasing patterns were mainly caused by two factors: (1) increased precipitation and (2) decreased ET resulting from an elevated CO_2 concentration of 936 ppm. Annual precipitation increased by 8 % and elevated CO_2 concentrations reduced ET by 32 % for the TCW and 26 % for the GW (Table 6).

3.3.3 Comparative analyses

Responses of the two watersheds to the GCM scenario were compared using the monthly averages of water and nitrate yields as shown in Fig. 8. Relative to the baseline, projected water and nitrate yield was greater over the year. The greatest increase in water yield was observed in August and September when the increased rate of precipitation was greatest. However, the increased rate of nitrate yield was higher in April than other months, due to a significant export of nitrate from fertilizer applications.

The increased rate of nitrate yield (under the GCM scenario relative to the baseline scenario) was 5.2 kg N ha^{-1} greater overall in the TCW compared to the GW, mainly due to the difference in watershed characteristics (Fig. 8b, d). First, a larger percentage of croplands in TCW led to greater nitrate export from fertilizer application compared to GW with a smaller percent of croplands. This was because increased water yield resulting from an elevated CO_2 concentrations and increased precipitation promoted the export of nitrate in the soil profile (Suddick et al., 2013). For example, nitrate yield increased by 1.4 kg N ha^{-1} for the TCW and 0.9 kg N ha^{-1} for the GW in April, when fertilizer application occurred, compared to the baseline. Second, a larger percentage of poorly drained soils in the GW contributed to reducing nitrate yield via greater potential of denitrification, compared to the TCW dominated by well-drained soils, under the GCM scenario. Increased soil water content resulting from an elevated CO_2 concentration of 936 ppm provided anaerobic conditions for denitrification. Compared to the baseline, the GW and TCW showed increased nitrate removal by denitrification of 3.9 and 0.5 kg N ha^{-1} under the GCM scenario, respectively. Eventually, the GW lost 8.7 kg N ha^{-1} more nitrate via denitrification than the TCW, which likely led to a lower nitrate yield for the GW.

4 Implications and limitations

The key results of this study suggest important future research for improving our understanding of climate change impacts on nutrient loads into the CBW. Analysis of cli-

Table 6. The 14-year average of hydrologic variables under the baseline and GCM scenarios.

Variables	TCW			GW		
	Baseline	GCM scenario	Relative change (%)	Baseline	GCM scenario	Relative change (%)
Streamflow ($m^3 s^{-1} ha^{-1} 10^4$)	1.5	2.5 (2.3–2.8)	70	1.7	2.5 (2.3–2.8)	50
ET ($mm\,ha^{-1}$)	2.7	1.8	−32	2.3	1.7	−26
Nitrate loads ($kg\,N\,ha^{-1}$)	12.5	20.8 (19.8–22.0)	66	5.3	8.2 (7.8–8.9)	56

Note: the numbers within parentheses indicate the maximum and minimum values of simulations with five GCM data. Relative change indicates the percent changes in the ensemble mean relative to the baseline value.

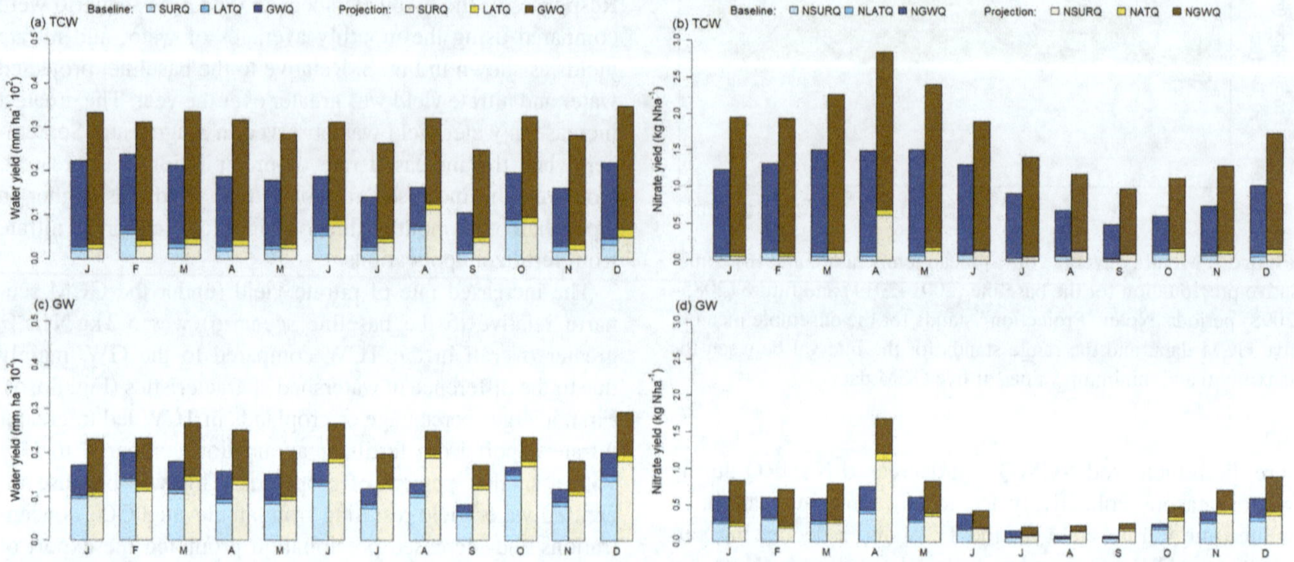

Figure 8. The 14-year average of monthly water and nitrate yields under the baseline and GCM scenarios. Note: the descriptions of abbreviations are available in the caption of Fig. 6.

mate variability and change impacts on watershed hydrological processes illustrate the close relationship between agricultural activities and future nitrate export in the watershed dominated by croplands, due to excessive export of nitrate from springtime fertilizer application. Changes in crop growth resulting from climate change are likely to alter agricultural activities and associated nitrate loads. Fertilizer application might increase in the future because increased extreme climate conditions (e.g., high-intensity rainfall and flooding) might lead to increased risk of nutrient loss to leaching and runoff, reducing the fertilizer use efficiency of field crops (Suddick et al., 2013). Our simulation indicated considerable increases in nitrate transported by surface runoff (NSURQ) due to increased precipitation in April, when the vast majority of fertilizers were applied (Fig. 8b, d). As a result, projected corn biomass appeared to be 0.03 $Mg\,ha^{-1}$ lower than the baseline value, likely due to increased nutrient stress (Fig. 9a). Conversely, soybean biomass increased under the GCM scenario because soybean could accumulate N through biological fixation and elevated CO_2 concentrations contributed to biomass growth (Fig. 9b). To adapt to warmer temperatures, early planting of summer crops could be suggested to increase crop production while reducing heat stress (Woznicki et al., 2015). For example, when planting dates were shifted 10 days earlier, soybean yield increased on average by 0.03 $Mg\,ha^{-1}$ (Fig. 9b). Contrary to our expectation, corn yield decreased under the earlier planting date, due to increased nutrient stress resulting from intensified precipitation. Lastly, irrigation patterns could be changed due to decreased ET resulting from elevated CO_2 concentrations. However, there are limited studies investigating projected future agricultural practices. There-

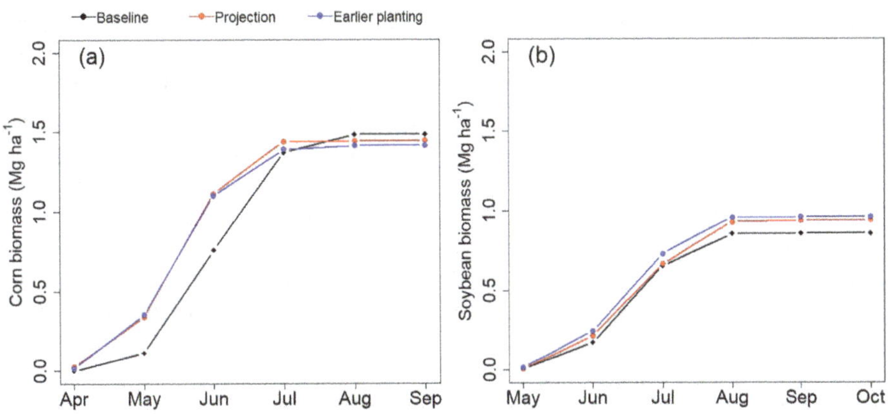

Figure 9. Crop biomass growth under the baseline and GCM scenarios: **(a)** corn and **(b)** soybean. Note: "Projection" stands for the simulated biomass planted on the original planting dates under the GCM scenario. Earlier planting indicates the simulated biomass planted 10 days earlier than the original planting dates under the GCM scenario.

fore, it is crucial to investigate potential agricultural activities under climate change scenarios and their projected effects on nitrate loads.

Climate change-driven modifications indicated a potential overall increase in nitrate export. Therefore, the importance of conservation practices aimed at N mitigation would be even more critical in the future. Comparative analyses of two watersheds can provide a practical guideline and have implications for agricultural watersheds in coastal areas in the CBW because our analyses considered climate change impacts on croplands (crop growth, water and nutrient cycling) and nutrient transport mechanisms in the context of detailed agricultural management practices. In addition, the two watersheds showed the typical site characteristics in the coastal watershed, in terms of topographic and soil characteristics, and the agricultural practices commonly used in the southern CBW. Hence, the findings from this study can be applicable to other catchments in the CBW region in preparing climate change adaptation strategies. For example, the effective management of nutrients applied in manure or fertilizer would be even more critical for reducing nitrate export from a watershed dominated by croplands. Winter cover crops, which are widely implemented in this region, would likely show increased value in mitigating agricultural nitrate loss during winter seasons, considering increased N availability and increased wintertime precipitation. In a watershed dominated by poorly drained soils, wetland restoration would be well positioned to enhance denitrification (McCarty et al., 2014), as would be the use of drainage control structures on ditches and tiles draining prior converted croplands (poorly drained areas of the farm landscape).

Note that although forest litterfall can have a significant impact on nutrient cycles (Zhang et al., 2014), the current version of the SWAT model is limited in representing forest impacts (Yang et al., 2016). In our simulation, growth of deciduous trees was simulated for forest areas with the default setting. This setting allowed tree growth to affect water and nutrient cycling via ET and uptake, but simulated tree growth was considerably underestimated compared to actual growth and litterfall was rarely considered (Yang et al., 2016). Hence, our simulation might poorly represent the ecological responses of forests to climate change. Future work should accurately consider forest ecosystems through model improvement.

5 Summary and conclusion

Water quality degradation by human activities on agricultural lands is a great concern on the Coastal Plain of the CBW. This degradation is expected to increase in the future due to changes in climate variability and conditions. Currently, there is limited information about how climate change will influence hydrology and nutrient cycles. This study used the SWAT model to simulate the impacts of potential climate variability and change on two adjacent watersheds in the Coastal Plain of the CBW. The climate sensitivity and GCM scenarios were prepared to assess the individual and combined impact of three climate factors (e.g., increases in CO_2 concentration, precipitation, and temperature). We performed comparative analyses between the two watersheds to demonstrate how key landscape characteristics influence the watershed level response to climate variability and change.

Our simulation results showed that water and nitrate budgets in two watersheds on the Coastal Plain of the CBW were significantly sensitive to climate variability and change. Compared to the baseline scenario, a precipitation increase of 21 % and elevated CO_2 concentration of 850 ppm resulted in increases in streamflow and nitrate loads of 50 and 52 %, respectively. A temperature increase of 5.0 °C reduced streamflow and nitrate loads by 12 and 13 %, respectively. Under the GCM scenario, annual streamflow and nitrate loads increased by 70 and 66 %, respectively, compared to the baseline scenario. Contrasting land use and soil characteristics led to different patterns of nitrate yield between two watersheds. The watershed with a larger percent of cropland showed a 5.2 kg N ha^{-1} greater increase in the rate of nitrate yield (under the GCM scenario relative to the baseline scenario) compared to the watershed with a lower percent of cropland under the GCM scenario, due to increased export of nitrate derived from fertilizer. Increased nitrate loss by denitrification also contributed to smaller increases in nitrate yield in the watershed dominated by poorly drained soils compared to the watershed dominated by well-drained soils. Based on our results, we suggest that increased implementation of conservation practices, such as nutrient management planning, winter cover crops, and wetland restoration and enhancement, is necessary to mitigate increased nitrate loads facilitated by climate change. These findings may help watershed managers and decision makers establish climate change adaptation strategies for mitigating water quality degradation in areas impaired by excessive agricultural nutrient loadings.

Appendix A

Table A1. Management schedules for the baseline scenario (adapted from Lee et al., 2016a).

Baseline scenario (no winter cover crop)			
Crop	Planting	Fertilizer	Harvest
Corn (after corn)	30 April (no-till)	157 kg N ha^{-1} of poultry manure on 20 April 45 kg N ha^{-1} of sidedress 30 % UAN on 7 June	3 October
Corn (after soybean and double-crop soybean)	30 April (no-till)	124 kg N ha^{-1} of poultry manure on 20 April 34 kg N ha^{-1} of sidedress 30 % UAN on 7 June	3 October
Soybean	20 May (no-till)		15 October
Double-crop winter wheat (Dbl WW)	10 October	34 kg N ha^{-1} of sidedress 30 % UAN on 8 October 45 kg N ha^{-1} of sidedress 30 % UAN on 1 March 67 kg N ha^{-1} of sidedress 30 % UAN on 5 April	27 June
Double-crop soybean (Dbl Soyb)	29 June		1 November

Note: UAN stands for urea-ammonium nitrate. The typical nitrogen content for poultry manure is assumed to be 2.8 % (Glancey et al., 2012).

Table A2. Five general circulation models (GCMs) used to construct the GCM scenario.

No.	Model	Full name	Modeling group
1	BCC-CSM1-1.1	Beijing Climate Center (BCC) – Climate System Model (CSM)	Beijing Climate Center, China Meteorological Administration
2	CCSM4.1	Community Climate System Model (CCSM) 4.1	National Center for Atmospheric Research
3	GFDL-ESM2G.1	Geophysical Fluid Dynamics Laboratory (GFDL) – Earth System Model (ESM)	NOAA Geophysical Fluid Dynamics Laboratory
4	IPSL-CM5A-LR.1	Institut Pierre-Simon Laplace (IPSL) – Climate Model(CM)5A-Low Resolution	Institut Pierre-Simon Laplace
5	MIROC-ESM-CHEM.1	An atmospheric chemistry coupled version of Model for Interdisciplinary Research on Climate (MIROC) – Earth System Model (ESM)	Japan Agency for Marine-Earth Science and Technology, Atmosphere and Ocean Research Institute (The University of Tokyo), and National Institute for Environmental Studies

Table A3. Seasonal evapotranspiration (ET, $\mathrm{mm\,ha^{-1}\,10^2}$) under climate sensitivity scenarios (CO_2: carbon dioxide concentration; PCP: precipitation; and TMP: temperature).

Scenario	Corn		Soybean	
	Winter	Summer	Winter	Summer
Baseline	0.74	2.00	0.68	1.64
CO_2 (590 ppm)	0.65	1.77	0.59	1.49
CO_2 (850 ppm)	0.50	1.40	0.46	1.22
PCP (11 %)	0.75	2.03	0.68	1.68
PCP (21 %)	0.75	2.05	0.69	1.71
TMP (2.6 °C)	0.81	2.03	0.77	1.67
TMP (5.0 °C)	0.87	2.05	0.83	1.70

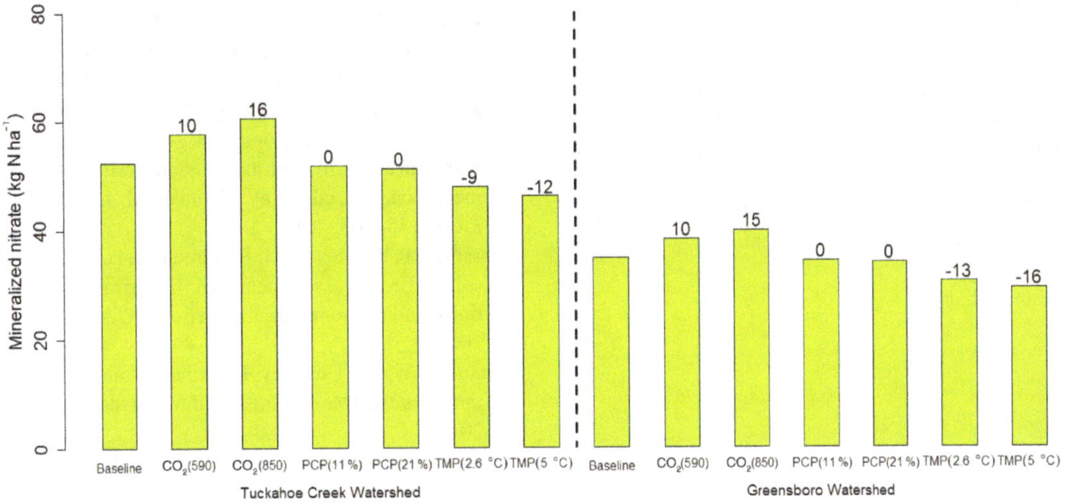

Figure A1. The 14-year average of annual mineralized nitrate during winter seasons (October–March) under the baseline and climate sensitivity scenarios at the watershed scale. Note: the black numerical values above the bar graph indicate the relative changes (%) in hydrologic variables for climate sensitivity scenarios relative to the baseline scenario (relative change (%) = (sensitivity scenarios – baseline) / baseline × 100). PCP and TMP stand for precipitation and temperature, respectively.

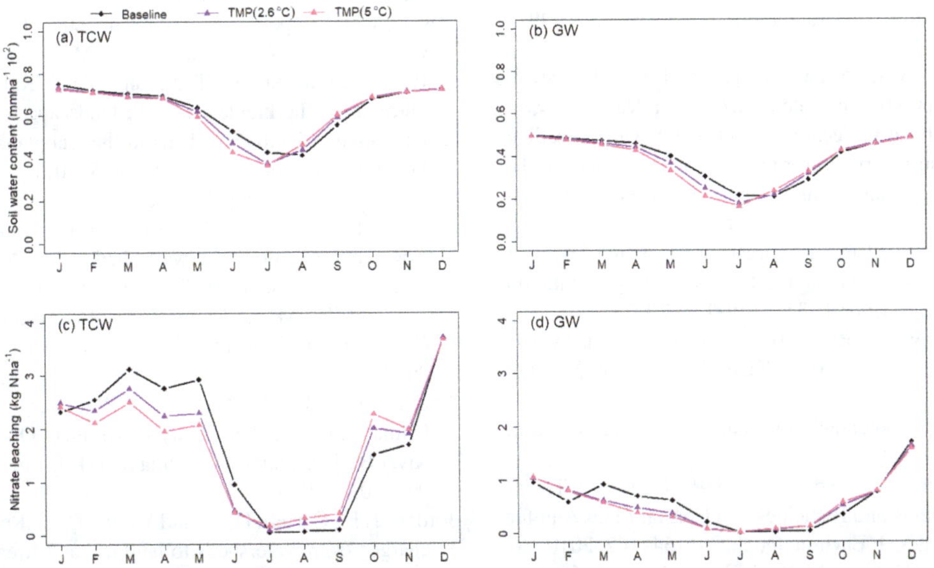

Figure A2. Changes in **(a, b)** soil water content and **(c, d)** nitrate leaching under temperature increase for the Tuckahoe Creek Watershed (TCW) and the Greensboro Watershed (GW). Note: TMP stands for temperature.

Competing interests. The authors declare that they have no conflict of interest.

Disclaimer. The USDA and USGS are equal opportunity providers and employers. Any use of trade, firm, or product names is for descriptive purposes only and does not imply endorsement by the US Government. The findings and conclusions in this article are those of the author(s) and do not necessarily represent the views of the US Fish and Wildlife Service or the USGS.

Special issue statement. This article is part of the special issue "Coupled terrestrial-aquatic approaches to watershed-scale water resource sustainability". It does not belong to a conference.

Acknowledgements. This research was supported by the US Department of Agriculture (USDA) Conservation Effects Assessment Project (CEAP), the National Aeronautics and Space Administration (NASA) Land Cover and Land Use Change (LCLUC) Program (award no: NNX12AG21G), and the US Geological Survey (USGS) Land Change Science Program.

Edited by: Xuesong Zhang

References

Abler, D. G. and Shortle, J. S.: Climate change and agriculture in the Mid-Atlantic Region, Climate Res., 14, 185–194, 2000.

Ainsworth, E. A. and Rogers, A.: The response of photosynthesis and stomatal conductance to rising [CO_2]: mechanisms and environmental interactions, Plant Cell Environ., 30, 258–270, 2007.

Arnold, J. G., Moriasi, D. N., Gassman, P. W., Abbaspour, K. C., White, M. J., Srinivasan, R., Santhi, C., Harmel, R. D., Van Griensven, A., Van Liew, M. W., and Kannan, N.: SWAT: Model use, calibration, and validation, T. ASABE, 55, 1491–1508, 2012.

Brekke, L., Thrasher, B. L., Maurer, E. P., and Pruitt, T.: Downscaled CMIP3 and CMIP5 climate projections: release of downscaled CMIP5 climate projections, comparison with preceding information, and summary of user needs, US Department of the Interior, Bureau of Reclamation, Technical Service Center, Denver, Colorado, 2013.

Chaplot, V.: Water and soil resources response to rising levels of atmospheric CO_2 concentration and to changes in precipitation and air temperature, J. Hydrol., 337, 159–171, 2007.

Chesapeake Bay Program: Bay 101, available at: http://www.chesapeakebay.net/discover/bay101/facts, last access: 31 May 2016.

Chiang, S. L.: A runoff potential rating table for soils, J. Hydrol., 13, 54–62, 1971.

Denver, J. M., Tesoriero, A. J., and Barbaro, J. R.: Trends and Transformation of Nutrients and Pesticides in a Coastal Plain Aquifer System, United States, J. Environ. Qual., 39, 154–167, 2010.

Denver, J. M., Ator, S. W., Lang, M. W., Fisher, T. R., Gustafson, A. B., Fox, R., Clune, J. W., and McCarty, G. W.: Nitrate fate and transport through current and former depressional wetlands in an agricultural landscape, Choptank Watershed, Maryland, United States, J. Soil Water Conser., 69, 1–16, 2014.

Eckhardt, K. and Ulbrich, U.: Potential impacts of climate change on groundwater recharge and streamflow in a central European low mountain range, J. Hydrol., 284, 244–252, 2003.

Ficklin, D. L., Luo, Y., Luedeling, E., and Zhang, M.: Climate change sensitivity assessment of a highly agricultural watershed using SWAT, J. Hydrol., 374, 16–29, 2009.

Ficklin, D. L., Stewart, I. T., and Maurer, E. P.: Climate change impacts on streamflow and subbasin-scale hydrology in the upper Colorado River Basin, PLOS ONE, 8, e71297, https://doi.org/10.1371/journal.pone.0071297, 2013.

Field, C. B., Jackson, R. B., and Mooney, H. A.: Stomatal responses to increased CO_2: implications from the plant to the global scale, Plant Cell Environ., 18, 1214–1225, 1995.

Fisher, T. R., Jordan, T. E., Staver, K. W., Gustafson, A. B., Koskelo, A. I., Fox, R. J., Sutton, A. J., Kana, T., Beckert, K. A., Stone, J. P., McCarty, G., and Lang, M.: The Choptank Basin in transition: intensifying agriculture, slow urbanization, and estuarine eutrophication, in: Coastal Lagoons: critical habitats of environmental change, edited by: Kennish, M. J. and Paerl, H. W., CRC Press, 135–165, 2010.

Gassman, P. W., Reyes, M. R., Green, C. H., and Arnold, J. G.: The soil and water assessment tool: historical development, applications, and future research directions, T. ASABE, 50, 1211–1250, 2007.

Gitau, M. W. and Chaubey, I.: Regionalization of SWAT Model Parameters for Use in Ungauged Watersheds, Water, 2, 849–871, 2010.

Glancey, J., Brown, B., Davis, M., Towle, L., Timmons, J., and Nelson, J.: Comparison of Methods for Estimating Poultry Manure Nutrient Generation in the Chesapeake Bay Watershed, available at: http://www.csgeast.org/2012annualmeeting/documents/Glancey.pdf (last access: 25 September 2014), 2012.

Gombault, C., Madramootoo, C. A., Michaud, A., Beaudin, I., Sottile, M. F., Chikhaoui, M., and Ngwa, F.: Impacts of climate change on nutrient losses from the Pike River watershed of southern Québec, Can. J. Soil Sci., 95, 337–358, 2015.

Hively, W. D., Hapeman, C. J., McConnell, L. L., Fisher, T. R., Rice, C. P., McCarty, G. W., Sadeghi, A. M., Whitall, D. R., Downey, P. M., de Guzmán, G. T. N., and Bialek-Kalinski, K.: Relating nutrient and herbicide fate with landscape features and characteristics of 15 subwatersheds in the Choptank River watershed, Sci. Total Environ., 409, 3866–3878, 2011.

Howarth, R. W., Swaney, D. P., Boyer, E. W., Marino, R., Jaworski, N., and Goodale, C.: The influence of climate on average nitrogen export from large watersheds in the Northeastern United States, Biogeochemistry, 79, 163–186, 2006.

Jha, B. and Jha, M. K.: Rating Curve Estimation of Surface Water Quality Data Using LOADEST, J. Environ. Prot., 4, 849–856, 2013.

Jha, M., Arnold, J. G., Gassman, P. W., Giorgi, F., and Gu, R. R.: Climate Chhange Sensitivity Assessment On Upper Mississippi River Basin Streamflows Using Swat1, J. Am. Water Resour. As., 997–1015, 2006.

Jordan, T. E., Correll, D. L., and Weller, D. E.: Relating nutrient discharges from watersheds to land use and streamflow variability, Water Resour. Res., 33, 2579–2590, 1997.

Kiptala, J. K., Mul, M. L., Mohamed, Y. A., and van der Zaag, P.: Modelling stream flow and quantifying blue water using a modified STREAM model for a heterogeneous, highly utilized and data-scarce river basin in Africa, Hydrol. Earth Syst. Sci., 18, 2287–2303, https://doi.org/10.5194/hess-18-2287-2014, 2014.

Lee, S., Yeo, I. Y., Sadeghi, A. M., McCarty, G. W., and Hively, W. D.: Prediction of climate change impacts on agricultural watersheds and the performance of winter cover crops: Case study of the upper region of the Choptank River Watershed, Proceedings of the ASABE 1st Climate Change Symposium: Adaptation and Mitigation, Chicago, IL, 3–5 May, 2015.

Lee, S., Yeo, I.-Y., Sadeghi, A. M., McCarty, W. M., Hively, W. D., and Lang, M. W.: Impacts of Watershed Characteristics and Crop Rotations on Winter Cover Crop Nitrate Uptake Capacity within Agricultural Watersheds in the Chesapeake Bay Region, PLOS ONE, 11, e0157637, https://doi.org/10.1371/journal.pone.0157637, 2016a.

Lee, C. J., Hirsch, R. M., Schwarz, G. E., Holtschlag, D. J., Preston, S. D., Crawford, C. G., and Vecchia, A. V.: An evaluation of methods for estimating decadal stream loads, J. Hydrol., 542, 185–203, 2016b.

Lee, S., Sadeghi, A. M., Yeo, I.-Y., McCarty, W. M., and Hively, W. D.: Climate, crop rotation, and stream flow data used to run the SWAT model in the Tuckahoe and Greensboro subwatersheds of the Choptank River watersheds, Maryland: US Geological Survey data release, https://doi.org/10.5066/F7DB80RP, 2017.

McCarty, G. W., McConnell, L. L., Hapeman, C. J., Sadeghi, A., Graff, C., Hively, W. D., Lang, M. W., Fisher, T. R., Jordan, T., Rice, C. P., and Codling, E. E.: Water quality and conservation practice effects in the Choptank River watershed, J. Soil Water Conserv., 63, 461–474, 2008.

McCarty, G. W., Hapeman, C. J., Rice, C. P., Hively, W. D., McConnell, L. L., Sadeghi, A. M., Lang, M. W., Whitall, D. R., Bialek, K., and Downey, P.: Metolachlor metabolite (MESA) reveals agricultural nitrate-N fate and transport in Choptank River watershed, Sci. Total Environ., 473, 473–482, 2014.

Mearns, L. O., Hulme, M., Carter, T. R., Leemans, R., Lal, M., Whetton, P., Hay, L., Jones, R. N., Kittel, T., Smith, J., and Wilby, R.: Climate scenario development, chap. 13, in: Climate Change 2001: Working Group I: The Scientific Basis, Cambridge University Press, UK, 2001.

Meinshausen, M., Smith, S. J., Calvin, K., Daniel, J. S., Kainuma, M. L. T., Lamarque, J. F., Matsumoto, K., Montzka, S. A., Raper, S. C. B., Riahi, K., and Thomson, A. G.: The RCP greenhouse gas concentrations and their extensions from 1765 to 2300, Clim. Change, 109, 213–241, 2011.

Moriasi, D. N., Arnold, J. G., Van Liew, M. W., Bingner, R. L., Harmel, R. D., and Veith, T. L.: Model evaluation guidelines for systematic quantification of accuracy in watershed simulations, T. ASABE, 50, 885–900, 2007.

Najjar, R., Patterson, L., and Graham, S.: Climate simulations of major estuarine watersheds in the Mid-Atlantic region of the US, Climatic Change, 95, 139–168, 2009.

Najjar, R. G., Pyke, C. R., Adams, M. B., Breitburg, D., Hershner, C., Kemp, M., Howarth, R., Mulholland, M. R., Paolisso, M., Secor, D., and Sellner, K.: Potential climate-change impacts on the Chesapeake Bay, Estuar. Coast. Shelf S., 86, 1–20, 2010.

Neitsch, S. L., Arnold, J. G., Kiniry, J. R., and Williams, J. R.: Soil and Water Assessment Tool. Theoretical Documentation; Version 2009, Texas Water Resources Institute Technical Report No. 406, Texas A&M University System, College Station, TX, 2011.

Parry, M. L., Rosenzweig, C., Iglesias, A., Livermore, M., and Fischer, G.: Effects of climate change on global food production under SRES emissions and socio-economic scenarios, Global Environ. Chang., 14, 53–67, 2004.

Pervez, M. S. and Henebry, G. M.: Assessing the impacts of climate and land use and land cover change on the freshwater availability in the Brahmaputra River basin, J. Hydrol., 3, 285–311, 2015.

Praskievicz, S.: Impacts Of Projected Climate Changes On Streamflow And Sediment Transport For Three Snowmelt-Dominated Rivers In The Interior Pacific Northwest, River Res. Appl., 32, 4–17, https://doi.org/10.1002/rra.2841, 2014.

Qiu, L., Zheng, F., and Yin, R.: SWAT-based runoff and sediment simulation in a small watershed, the loessial hilly-gullied region of China: capabilities and challenges, Int. J. Sediment Res., 27, 226–234, 2012.

Rogers, C. E. and McCarty, J. P.: Climate change and ecosystems of the Mid-Atlantic Region, Climate Res., 14, 235–244, 2000.

Runkel, R. L., Crawford, C. G., and Cohn, T. A.: Load Estimator (LOADEST): A FORTRAN program for estimating constituent loads in streams and rivers, US Geological Survey Paper, Reston, Virginia, 2004.

Seo, M., Yen, H., Kim, M. K., and Jeong, J.: Transferability of SWAT Models between SWAT2009 and SWAT2012, J. Environ Qual., 43, 869–880, 2014.

Sexton, A. M., Sadeghi, A. M., Zhang, X., Srinivasan, R., and Shirmohammadi, A.: Using NEXRAD and rain gauge precipitation data for hydrologic calibration of SWAT in a northeastern watershed, T. ASABE, 53, 1501–1510, 2010.

Sharifi, A., Lang, M. W., McCarty, G. W., Sadeghi, A. M., Lee, S., Yen, H., Rabenhorst, M. C., Jeong, J., and Yeo, I. Y.: Improving Model Prediction Reliability through Enhanced Representation of Wetland Soil Processes and Constrained Model Auto Calibration – A Paired Watershed Study, J. Hydrol., 541, 1088–1103, 2016.

Shrestha, R. R., Dibike, Y. B., and Prowse, T. D.: Modelling of climate-induced hydrologic changes in the Lake Winnipeg watershed, J. Great Lakes Res., 38, 83–94, 2012.

Singh, A., Imtiyaz, M., Isaac, R. K., and Denis, D. M.: Assessing the performance and uncertainty analysis of the SWAT and RBNN models for simulation of sediment yield in the Nagwa watershed, India, Hydrol. Sci. J., 59, 351–364, 2014.

Suddick, E. C., Whitney, P., Townsend, A. R., and Davidson, E. A.: The role of nitrogen in climate change and the impacts of nitrogen–climate interactions in the United States: foreword to thematic issue, Biogeochemistry, 114, 1–10, 2013.

Tiner, R. W. and Burke, D. G.: Wetlands of Maryland, US Fish and Wildlife Service, Hadly, Massachusetts, 261 pp., 1995.

Uniyal, B., Jha, M. K., and Verma, A. K.: Assessing climate change impact on water balance components of a river basin using SWAT model, Water Resour. Manag., 29, 4767–4785, 2015.

Van Liew, M. W., Feng, S., and Pathak, T. B.: Climate change impacts on streamflow, water quality, and best management practices for the shell and logan creek watersheds in Nebraska, USA, Int. J. Agric. Biol. Eng., 5, 13–34, 2012.

Wang, R., Kalin, L., Kuang, W., and Tian, H.: Individual and combined effects of land use/cover and climate change on Wolf Bay watershed streamflow in southern Alabama, Hydrol. Process., 28, 5530–5546, 2014.

Woznicki, S. A., Nejadhashemi, A. P., and Parsinejad, M.: Climate change and irrigation demand: Uncertainty and adaptation, J. Hydrol., 3, 247–264, 2015.

Wu, Y., Liu, S., and Gallant, A. L.: Predicting impacts of increased CO_2 and climate change on the water cycle and water quality in the semiarid James River Basin of the Midwestern USA, Sci. Total Environ., 430, 150–160, 2012a.

Wu, Y., Liu, S., and Abdul-Aziz, O. I.: Hydrological effects of the increased CO_2 and climate change in the Upper Mississippi River Basin using a modified SWAT, Clim. Change, 110, 977–1003, 2012b.

Yang, Q. and Zhang, X.: Improving SWAT for simulating water and carbon fluxes of forest ecosystems, Sci. Total Environ., 569, 1478–1488, 2016.

Yeo, I.-Y., Lee, S., Sadeghi, A. M., Beeson, P. C., Hively, W. D., McCarty, G. W., and Lang, M. W.: Assessing winter cover crop nutrient uptake efficiency using a water quality simulation model, Hydrol. Earth Syst. Sci., 18, 5239–5253, https://doi.org/10.5194/hess-18-5239-2014, 2014.

Zhang, H., Yuan, W., Dong, W., and Liu, S.: Seasonal patterns of litterfall in forest ecosystem worldwide, Ecol. Complex, 20, 240–247, 2014.

5

Searching for the optimal drought index and timescale combination to detect drought: a case study from the lower Jinsha River basin, China

Javier Fluixá-Sanmartín[1], Deng Pan[2], Luzia Fischer[3], Boris Orlowsky[4], Javier García-Hernández[1], Frédéric Jordan[5], Christoph Haemmig[3], Fangwei Zhang[6], and Jijun Xu[2]

[1]Centre de Recherche sur l'Environnement Alpin (CREALP), 1951 Sion, Switzerland
[2]Changjiang River Scientific Research Institute, Changjiang Water Resources Commission, Wuhan Hubei, 430010, China
[3]GEOTEST AG, 3052 Zollikofen, Switzerland
[4]climate-babel, 8047 Zurich, Switzerland
[5]Hydrique Ingénieurs, 1052 Le Mont-sur-Lausanne, Switzerland
[6]Bureau of Hydrology, Changjiang Water Resources Commission, Wuhan Hubei, 430017, China

Correspondence: Javier Fluixá-Sanmartín (javier.fluixa@crealp.vs.ch)

Abstract. Drought indices based on precipitation are commonly used to identify and characterize droughts. Due to the general complexity of droughts, the comparison of index-identified events with droughts at different levels of the complete system, including soil humidity or river discharges, relies typically on model simulations of the latter, entailing potentially significant uncertainties.

The present study explores the potential of using precipitation-based indices to reproduce observed droughts in the lower part of the Jinsha River basin (JRB), proposing an innovative approach for a catchment-wide drought detection and characterization. Two indicators, namely the Overall Drought Extension (ODE) and the Overall Drought Indicator (ODI), have been defined. These indicators aim at identifying and characterizing drought events on the basin scale, using results from four meteorological drought indices (standardized precipitation index, SPI; rainfall anomaly index, RAI; percent of normal precipitation, PN; deciles, DEC) calculated at different locations of the basin and for different timescales. Collected historical information on drought events is used to contrast results obtained with the indicators.

This method has been successfully applied to the lower Jinsha River basin in China, a region prone to frequent and severe droughts. Historical drought events that occurred from 1960 to 2014 have been compiled and cataloged from different sources, in a challenging process. The analysis of the indicators shows a good agreement with the recorded historical drought events on the basin scale. It has been found that the timescale that best reproduces observed events across all the indices is the 6-month timescale.

1 Introduction

Drought is a natural phenomenon that results from persistent deficiency of precipitations over an extended period of time compared with some long-term average condition (e.g., precipitation), resulting in a water shortage for some activity, group, or environmental sector (Landsberg, 1982). It generally affects larger areas than other hazards and more people than any other natural catastrophe (Keyantash and Dracup, 2002; Wilhite, 2000).

In China, droughts represent the most severe natural threat for socioeconomic development and ecosystems (Mei and Yang, 2014). Drought events occur in the Jinsha River basin (JRB) and surrounding regions with high frequency. They affect a wide range of areas and cause huge losses to the agriculture sector (He et al., 2013). The clustering of severe and sustained droughts in southwest China during the last decade has resulted in tremendous losses, including crop failure, lack of drinking water, ecosystem degradation, health problems and even deaths (Wang et al., 2015).

To reduce and anticipate such drought impacts, a comprehensive characterization of the phenomenon is essential. Effective and accurate analysis of hydrometeorological data is a key input. Drought indices are useful for tracking droughts and providing a quantitative assessment of the severity, location, timing and duration of such events (World Meteorological Organization and Global Water Partnership, 2016), but also for real-time monitoring (Niemeyer, 2008), risk analysis (Hayes et al., 2004) and drought early warning (Kogan, 2000).

Some organizations and agencies already rely on the use of indices in their decision-making processes, thus enhancing proactive drought management policies (Wilhite, 2000). An example is the U.S. Drought Monitor (USDM, 2017), an index-based drought map that policymakers use in discussions of drought and in allocating drought relief. Other platforms such as the European Drought Observatory (Joint Research Centre, 2017); China's Department of Climate Change, National Development and Reform Commission (Department of Climate Change, National Development and Reform Commission, 2017); or the experimental African Flood and Drought Monitor (Land Surface Hydrology Group – Princeton University, 2017) also use this approach for the assessment, diagnosing and forecasting of droughts.

The choice of index should be based on the type of drought (meteorological, agricultural, hydrological or socioeconomical), the climate regime and the regions affected, as well as the available data. It was found that measured meteorological data were limited in the study region and that precipitation was the single most reliable type of exploitable information. The present study thus focuses on the use of meteorological indices based only on precipitation data. The main advantages are their ease of use, the limited data requirements and the capacity for early detection of drought events, while extensive literature and calculation tools are widely accessible (World Meteorological Organization and Global Water Partnership, 2016).

The use of integrated indices such as the Palmer drought severity index (PDSI; Palmer, 1965) or standardized precipitation–evapotranspiration index (SPEI; Vicente-Serrano et al., 2010), relying also on potential evapotranspiration (PET), could improve the scope and quality of this study. However, no reliable information necessary to calculate PET was accessible for the study region. Although approximations may be applied to estimate this variable, for example, by only considering temperature data, some studies (Jeevananda Reddy, 1995; Shaw and Riha, 2011; Staage et al., 2014) showed a high sensitivity of the PET to the chosen equation. A deeper analysis that helps in selecting and applying such methods should be performed prior to the use of these indices. Therefore, it has been decided to base this study on the standardized precipitation index (SPI, McKee et al., 1993, 1995), the rainfall anomaly index (RAI, Van Rooy, 1965), the percent of normal precipitation (PN, Barua et al., 2011) and the deciles index (DEC, Gibbs and Maher, 1967).

To fill the lack of specific drought-related information, most studies assess the performance of drought indices against results from hydrological soil water models (Halwatura et al., 2016; Hao and AghaKouchak, 2013; Trambauer et al., 2014; Vasiliades et al., 2011; Wanders et al., 2010). However, the performance of these types of studies depends on the accuracy of the models. Their limitations and uncertainties represent an important drawback and should be addressed (Mishra and Singh, 2011). An alternative that often requires more time-consuming work is the compilation of historical records of drought events from different sources. Consequently, their duration, the water scarcity levels, and the drought impacts on population and agriculture can be estimated and then integrated into the analysis. This enables one to identify other types of droughts such as socioeconomical droughts that are hard to assess with hydrological models.

Regarding their spatial resolution, the available drought indices may be based on local measurements (Zhou et al., 2012) and index calculations are usually applied to stations or cells of gridded precipitation datasets; overall spatial patterns on catchment or sub-catchment scales are thus hardly captured. As stated above, droughts affect large areas whose limits are often vaguely demarcated. In addition, water resources are part of a more complex interrelated network which links the source to the point of consumption, where isolated rainfall deficiencies do not necessarily imply a shortage of water availability or even a drought event. Some work (Bhalme and Mooley, 1980; Fleig et al., 2011; Mitchell et al., 1979) suggests the use of drought area indices for the study of droughts that considers areal coverage. The use of overall indicators capable of capturing in a single value the effect of the rainfall deficiency at a regional level is thus convenient and will be applied in this study based on the above-mentioned work.

The objective of this study is to capitalize on the collection of drought events that the authors have registered in the lower part of the JRB since 1960 to evaluate and calibrate two indicators capable of identifying drought occurrence and characterizing their intensity on the catchment scale. These indicators are based on commonly used meteorological drought indices for particular timescales.

2 Investigation area and data

The JRB is a sensitive zone in terms of water resources, food security, ecosystem management and human well-being where glacier and climatic variability greatly influence the water regimes and availability. Originating from the southern glacier at Jianggendiru peak, the highest point of the Geladaindong snowy mountain in the middle of the Tanggula Mountains, the JRB, constitutes the upper part of the Yangtze River basin. It is located between 24°28′ and 35°46′ N longitude and between 90°23′ and 104°37′ E latitude in

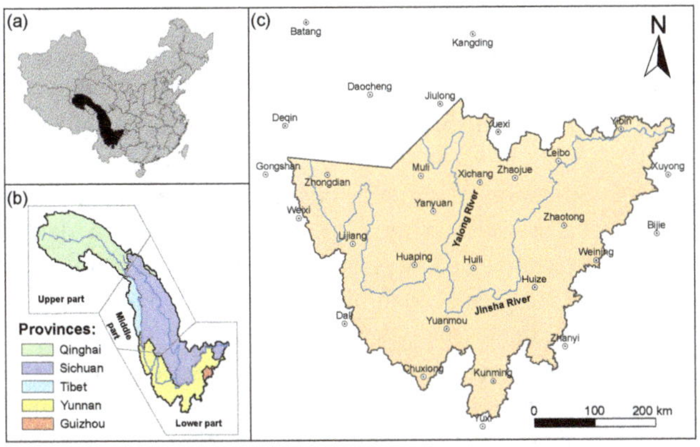

Figure 1. (a) Location of the JRB in China; **(b)** subdivision of the JRB for this study; **(c)** overview of the lower JRB with the location of the 29 meteorological stations.

southwestern China, with a catchment area of 473 200 km^2 (Fig. 1). The total length of the river is 3500 km from Yibin city, with a total fall of 5100 m. This part of the Yangtze River accounts for 55.5 % of its length and 95 % of its total fall.

The lower part of JRB is a hot–dry valley region characterized by a southwest monsoon climate. The hydrologic regime is characterized by a pronounced seasonal cycle with an annual average precipitation of 600–800 mm yr^{-1}. Dry season (November to April) precipitation accounts for 10 to 22 % of the annual precipitation. Evaporation is 10 to 20 times that of precipitation during the dry season, which could be the major reason for the frequent occurrence of winter drought or winter–spring droughts in lower JRB (Mei and Yang, 2014; Yang et al., 2013). However, the limited information available makes unapproachable the assessment of evapotranspiration (ET) in the region for the study period. Thus, given the role that ET can play in the annual moisture budget, results should be addressed carefully.

Droughts occurring in the lower JRB and surrounding areas affect a wide range of areas, causing huge losses in agriculture (Wu et al., 2011): more than 4 million people and 3 million livestock face a drinking water shortage, and more than 1 million hm^2 of cultivated area is susceptible to severe droughts and water shortages, with expected direct economic losses of hundreds of millions of US dollars (Wu, 1999).

Figure 1 shows the division of the JRB in three parts (upper, middle and lower) and the locations of the meteorological stations used. This study focuses on the analysis of drought events in the lower JRB. The precipitation data needed in this study have been obtained from the China Meteorological Data Service Center (CMA) and downloaded from its data sharing service system (CMDC, 2017). A preliminary quality check and correction of datasets (including data gap filling) is performed by CMA before uploading them to the system. The monthly precipitation data of 29 meteorological stations within or around JRB, recorded from 1960 to 2014, have been collected and processed. More than 50 years of continuous data are thus available, except for the Batang and Yuanmou stations where only 46 years are available. The spatial distribution of the stations is assumed to be adequate for the purposes of the study: the stations are distributed relatively evenly both in the zonal and meridional directions, with no zones having a significantly denser presence of stations that could overestimate their importance.

3 Catalog of historical droughts

In order to obtain a good basis for the evaluation of drought indices' performance, historical drought events have been collected since 1960. The information required for the identification and characterization of major droughts in the lower JRB has been compiled from different sources, including scientific literature, inventories (e.g., international disaster database, Chinese inventories), governmental reports and yearbooks, and newspaper and internet articles.

Detailed information is available for the major drought events over the past 20 years. Before 1980, much less information about droughts in the lower JRB is available. Moreover, detailed information prior to 1960 could not be found.

Compiling and harmonizing the information from these different data sources was a challenge. The drought event descriptions in the scientific literature often give an overview about the entire event in a descriptive way without detailed information about the affected area and damage. The literature often provides information about the meteorological conditions and the duration of the event. Available information from government reports and databases, in contrast, generally contain information with a high level of detail for a specific county (e.g., affected areas in km^2) but neither give information about the entire affected area if the event affected several counties nor about prevailing meteorological conditions.

Information at very different levels of detail and various contents were collected. As a first step, all available information was registered in a database. For this purpose, a web-based event registration platform and database (GEOTEST AG, 2017) has been developed to provide a standardized analytical framework with a quantitative description of the drought characteristics. In particular, drought events since 2000 are all mentioned in a different source, which significantly enhances the reliability of their existence and related information. Most of the drought events before 2000 are documented in detail in He (2010), and some of these major dry periods are also mentioned in scientific literature such as He et al. (2016) and Wang et al. (2015). Even if the amount of data and level of detail is lower for these older events, their occurrence and temporal positioning can be assumed to be reliable.

In a second step, the level of detail was harmonized for the most relevant information and summarized in Table 1. This catalog of the most relevant drought characteristics focuses on the affected area, the start date of the drought events, their duration, their spatial and temporal distribution, their severity, and the impacts on the population and agriculture, including damage and financial losses. The time of occurrence and duration is given in seasonal units, the affected area is described on a county scale, and the indicators and impacts are summarized in a descriptive way, as better accuracy was not feasible for all events.

In total, 13 major drought episodes have been registered from 1960 until 2014. However, this dataset is probably not complete, as non-documented events likely have occurred. A clustering of severe and sustained droughts in the JRB has been observed from 2009 to 2014. Another period with high drought activity and severity can be detected between 1980 and 1990. Although the droughts identified from 2009 until 2014 were extremely serious, this was not the worst period in the long term because the drought episodes that occurred around 1940 were of similar intensity and duration (Wang and Chen, 2012). The registered drought events are often strongly correlated with low precipitation, but the analyses also reveal that the registered droughts often occur during periods with temperatures above average.

4 Meteorological drought indices

Four different commonly used meteorological drought indices have been applied in this study: the standardized precipitation index, the rainfall anomaly index, the percent of normal precipitation and the deciles index. Their definition basically rests upon the comparison of precipitation values with the normal value (the definition of "normality" may vary from one index to another), resulting in a single number. This allows characterizing drought conditions and thus facilitating the index's interpretation and use in strategic planning and operational applications (Tigkas et al., 2013).

This comparison must be month or season specific. For instance, for the index calculation of January 2000, the precipitation of this month should be compared to the normal precipitation extracted taking into account only the Januaries from a reference period. The same applies when calculating the index for the time window January–February–March 2000: the precipitation for these 3 months will be compared to the sum of precipitation of all the groups of January-February-March registered in the reference period.

4.1 Standardized precipitation index (SPI)

The widely used standardized precipitation index was formulated by McKee et al. (1993, 1995) to quantify the precipitation deficit from long-term recording and for multiple timescales.

The long-term record of precipitation values is fitted to a probability distribution which is then transformed into a standard normal distribution, of which mean and variance are 0 and 1, respectively (Edwards and McKee, 1997). The datasets are most commonly adjusted to the gamma function (McKee et al., 1993; Sönmez et al., 2005; Tsakiris et al., 2007) although some studies show better adjustments to other functions (Akbari et al., 2015).

A classification of drought conditions based on the SPI values was established by McKee et al. (1993) to define drought intensities and is presented in Table 2. Positive SPI values indicate greater than normal precipitation, and negative values indicate lower than normal precipitation.

As mentioned earlier, the SPI was designed to quantify the precipitation deficit for multiple timescales or moving time windows (World Meteorological Organization, 2012). These timescales reflect the drought impacts on different water resources which are needed by decision makers:

- The 3-month SPI reflects short- and medium-term moisture conditions and provides a seasonal estimation of precipitation.
- The 6-month SPI indicates seasonal to medium-term trends in precipitation and may be very effective in showing the precipitation anomaly over distinct seasons. Information from a 6-month SPI may also be associated with anomalous streamflow and reservoir levels, depending on the region and time of year.
- The 12-month up to 24-month SPI reflects long-term precipitation patterns and is usually tied to streamflow, reservoir levels and even groundwater levels on longer timescales.

4.2 Rainfall anomaly index (RAI)

The rainfall anomaly index was developed by Van Rooy (1965). The RAI indices are computed by comparing the average precipitation over a given time window with the mean of the 10 highest (for positive anomalies) and the 10 lowest (for negative anomalies) precipitation records. Despite its simplicity, this index requires a series of complete data to be calculated.

The RAI values are classified (Van Rooy, 1965) as shown in Table 3. Olukayode Oladipo (1985) found that differences between the RAI and the more complicated indices of the Palmer drought index (Palmer, 1965) and Bhalme and Mooley drought index (Bhalme and Mooley, 1980) were negligible.

4.3 Percent of normal precipitation (PN)

The percent of normal precipitation is one of the simplest measurements of precipitation value for a location. It is calculated by dividing precipitation during a given time window by normal precipitation of that same time window over the reference period (typically considered to be a 30-year average). For PN values over 100 %, the precipitation is higher

Table 1. Catalog of historical drought collected for the lower JRB (DJF: December–January–February; MAM: March–April–May; JJA: June–July–August; SON: September–October–November).

ID	Year	Seasons	Affected area	Reference	Other indicators	Impacts
I	1962–1963	SON, DJF, MAM, JJA	Yunnan, southern part of Sichuan	He (2010)	Precipitation deficit of 50 % from November 1962 to April 1963.	Drinking water shortage for 900 000 people. Impacts on 3700 km^2 of agricultural land.
II	1978–1979	DJF, MAM, JJA	Yunnan	Liu (2012)		Impacts on 7000 km^2 of agricultural land, poor harvest and crop loss.
III	1981–1982	DJF, MAM, JJA	Sichuan, northern Yunnan	He (2010)		Drinking water shortage for 2 or 3 million people and 2 million livestock.
IV	1987	Exact duration unknown	Mainly Yunnan			Impacts on 6000 km^2 of agricultural land, poor harvest and crop loss.
V	1992	MAM, JJA, SON	Yunnan, southern Sichuan	He (2010)	Maximum in precipitation deficit: 50–80 %.	Drinking water shortage for 2 million people and 1 million livestock. Impacts on 9300 km^2 of agricultural land.
VI	1998–1999	DJF, MAM	Mainly Yunnan	He (2010)	Temperatures in Yunnan province 2–3 °C higher than long-term average. Dayao County: 150 days without rain.	8000 km^2 of damaged agricultural area.
VII	2000–2001	DJF, MAM	Sichuan, Yunnan	WCB (2001)	Temperatures in Yunnan province 2–3 °C higher than long-term average. The cities of Dali, Baoshan, Dehong, Chuxiong and Lincang have almost no rainfall during the whole winter.	Drinking water shortage for 3 million people and 2 million livestock. Impacts on 5800 km^2 of agricultural land.
VIII	2005	MAM, JJA	Large parts of Yunnan	Yang et al. (2012), Liu et al. (2007)	High temperatures; in April to early June, the temperature is 1 °C above the same historical period in most parts of Yunnan province. Precipitation deficit of 20–80 % in May–June. 56 days without precipitation.	Drinking water shortage for 6 million people and 4 million livestock. Impacts on 15 200 km^2 of agricultural land and poor harvest.
IX	2009–2010	SON, DJF, MAM	Parts of Yunnan, Sichuan and Guizhou	Yang et al. (2012), Wang et al. (2015)	Precipitation deficit. 119 days without precipitation. Average temperature anomaly of plus 1 °C.	Drinking water shortage for 21 million people and 11 million livestock. Impacts on 43 500 km^2 of agricultural land and poor harvest.
X	2011	MAM, JJA, SON	Large areas in southwest China	Yang et al. (2012), Wang et al. (2015)	Temperatures 0.4–1.1 °C higher than normal. From June to September 2011, persistent high-temperature weather conditions. Precipitation deficit of 20–60 %.	Drinking water shortage for 12 million people and 9 million livestock. Impacts on 19 000 km^2 of agricultural land. Cargo shipping has been suspended.
XI	2011–2012	DJF, MAM	Large areas in southwest China	Wang et al. (2013)	Precipitation deficit.	Drinking water shortage for 2.4 million people and 1.6 million livestock. Impacts on 6500 km^2 of agricultural land.
XII	2012–2013	Oct–Apr	Southwest China	Guha-Sapir et al. (2017), Hu et al. (2015)	From October to April, temperatures are 0.5 °C higher than normal and in February 2.5 °C higher than long-term average. Jan–Feb: precipitation deficit of 45–55 %.	More than 3 million people and about 2 million large livestock had a drinking water shortage with varying degrees. 323 small rivers and 331 small reservoirs dried up. 23 300 km^2 of agricultural area affected (whereof 15 500 km^2 is forest).
XIII	2014	DJF, MAM	Central Yunnan and south Sichuan	Duan et al. (2015)	Spring temperatures 2–4 °C higher than historic values in SW China. Spring precipitation in central Yunnan and south Sichuan province was 50–90 % less than the average of the same period.	Drinking water shortage for 1.6 million people in Yunnan province. 106 rivers and 76 reservoirs dried up. Affected area: 6080 km^2.

than the average precipitation (and vice versa): the higher the PN value, the wetter the considered month is.

The main advantage of this index is its simplicity and transparency, which makes it practical to communicate drought levels to the public (Keyantash and Dracup, 2002). Analyses using PN are very effective when used for a single region and/or a specific season.

Even if no threshold ranges have been widely established in the technical literature for the PN, some studies (Barua et al., 2011; Morid et al., 2006) propose a classification similar to the SPI. For this study, the classification proposed by Barua et al. (2011) has been adopted (Table 4).

Table 2. Classification of drought conditions according to the SPI values.

SPI	Classification
≥ 2.0	Extremely wet
1.5 to 1.99	Very wet
1.0 to 1.49	Moderately wet
−0.99 to 0.99	Near normal
−1.49 to −1.0	Moderately dry
−1.99 to −1.5	Severely dry
≤ −2.0	Extremely dry

Table 3. Classification of the period according to the values of the RAI.

RAI	Classification
≥ 3.00	Extremely wet
2.00 to 2.99	Very wet
1.00 to 1.99	Moderately wet
0.50 to 0.99	Slightly wet
−0.49 to 0.49	Near normal
−0.99 to −0.50	Slightly dry
−1.99 to −1.00	Moderately dry
−2.99 to −2.00	Very dry
≤ −3.00	Extremely dry

4.4 Deciles (DEC)

Another drought-monitoring technique consists in dividing the monthly precipitation data into deciles. This method, developed by Gibbs and Maher (1967), was selected as the meteorological measurement of drought for the Australian Drought Watch System (Lee, 1979; Sivakumar et al., 2010) because it is relatively simple to calculate and requires less data and fewer assumptions than the Palmer drought severity index (Smith et al., 1993). The procedures have also been adopted by the World Meteorological Organization to monitor drought on a worldwide scale (World Meteorological Organization, 1985).

The threshold ranges of deciles used to classify drought conditions are presented in Table 5 (Gibbs and Maher, 1967).

5 Approach for the identification of drought events on the basin scale

An indicator (or indicators) capable of adequately characterizing historical droughts must be able to capture the following characteristics:

- The beginning and the end of the event, which defines its duration.
- The drought intensity, derived from the index value.
- The geographical area affected by the drought.

Table 4. Classification of drought conditions according to the PN values.

PN	Classification
180 % or more of normal rainfall	Extremely wet
161 to 180 % of normal rainfall	Very wet
121 to 160 % of normal rainfall	Moderately wet
81 to 120 % of normal rainfall	Near normal
41 to 80 % of normal rainfall	Moderately dry
21 to 40 % of normal rainfall	Severely dry
20 % or less of normal rainfall	Extremely dry

Table 5. Classification of drought conditions according to the values of the deciles.

DEC	Percent	Classification
Deciles 1–2	Lowest 20 %	Significantly below normal
Deciles 3–4	Next lowest 20 %	Below normal
Deciles 5–6	Middle 20 %	Near normal
Deciles 7–8	Next highest 20 %	Above normal
Deciles 9–10	Highest 20 %	Significantly above normal

The following guidelines specify the approach proposed in this study to characterize drought events on the basin scale based on precipitation data available at each station and how to contrast these results with the cataloged historical events.

First, following the previous definitions (Sect. 4), precipitation data are used to calculate the four above-described meteorological drought indices (SPI, PN, RAI and DEC) for each station and for different timescales (1-, 3-, 6-, 12-, 24- and 48-month timescales) using the 1951–2000 reference period. Then, according to the criteria presented below, these values are used to detect potential drought events at a given station and at a given time.

In order to aggregate results from all stations of the basin, two indicators are proposed in this study: the Overall Drought Extension (ODE) and Overall Drought Indicator (ODI). The results of these indicators will then be contrasted with historical recorded events to define the best combination of the index and timescale used for the definition of the ODE and ODI indicators.

5.1 Use of indices to detect droughts on the station scale

According to McKee et al. (1993), a drought event occurs at the station level any time the SPI is continuously negative and the SPI reaches a value of −1.0 or less, which corresponds to a moderately dry condition (Table 2) or drier. The drought begins when the SPI first falls below zero (mean of the normalized precipitation) and ends with the positive value of SPI following a value of −1.0 or less. The drought magnitude is the positive sum of the SPI for each month during the drought event. The intensity of a drought is defined as the magnitude of this event divided by its duration.

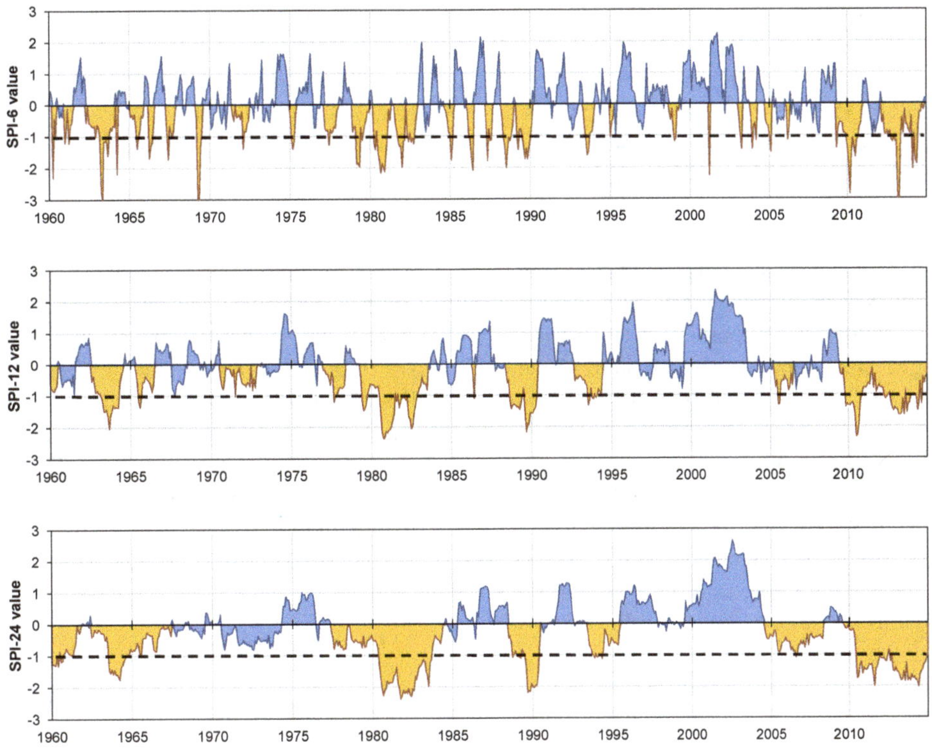

Figure 2. Example of the SPI-6, SPI-12 and SPI-24 series at the Chuxiong station, indicating drought periods in orange (lower threshold in red).

Figure 2 shows an example of the SPI-6, SPI-12 and SPI-24 series calculated at the Chuxiong station. Drought periods are colored in orange and the lower threshold that defines their occurrence in red. The influence of the timescale on the number and duration of detected droughts is clearly apparent. It is worth noting that there are periods in Fig. 2 that identify very short droughts (1 or 2 months long), which is due to the identification criteria based on the index values.

In the present study, the above-mentioned principles used to detect drought events based on the SPI classification (McKee et al., 1993) have been standardized to be applicable to the other three indices (PN, RAI and DEC) as follows:

- A drought event occurs any time the index is continuously below its normal value and reaches the moderately dry condition class.
- The drought is considered to begin when the index first falls below its normal value.
- The drought ends when the index exceeds its normal value.

Table 6 summarizes the thresholds for each index that specify the drought event's start and end criteria, which correspond, respectively, to the limit of the moderately dry class and to the index normal value. Although the "normal value" of DEC would be 50 % (which corresponds to the median of the precipitation records), in this study the drought end criterion suggested for this index is 60 %, which is the limit between near- and above-normal conditions (Jain et al., 2015; Tsakiris et al., 2007).

Table 6. Values of the thresholds defining the start and the end of the drought events for each index.

Index	Start (moderately dry condition)	End (normal value)
SPI	−1	0
RAI	−1	0
PN	80	100
DEC	40 %	60 %

5.2 Identification of drought occurrence

When analyzing directly these meteorological indices, the results only concern each station's surroundings without capturing the patterns of neighboring areas. However, available historical records refer to regional droughts characterized by larger areas that cover several stations.

In order to consider the basin as a whole in the definition of drought occurrence, duration and intensity, the resulting indices must be consistently extended to the entire area and then combined in overall indicators. For that purpose, a regular grid divides the lower JRB into a 400 × 300 cells raster (400 rows and 300 columns) adapted to the study region and the available information, which represents a resolution of 1 cell / 3.2 km^2. The chosen grid resolution must be adapted to the data, the computational requirements and the performance needs of each case. After trimming off the areas sticking out of the basin boundaries, the raster possesses 44 133 cells. Index values have been calculated at each grid cell by applying the inverse distance weighting spatial interpolation from the values available at the stations.

For this study, we followed the approaches taken for the definition of drought area indices (Bhalme and Mooley, 1980; Mitchell et al., 1979) and regional drought area indices (Fleig et al., 2011). It is thus considered that a basin-wide event is ongoing when a substantial part of the basin is under drought conditions. It is therefore necessary to identify the portion of the territory for which the calculated index indicates a drought. An indicator to detect drought occurrence on the basin scale has been set up based on the criteria described above to identify an event considering the index values (Fig. 2).

Based on the interpolation of the index, drought events are detected for each time step and at each grid cell of the described raster. This allows us to define an indicator, named here the Overall Drought Extension, expressed as the percentage of the lower JRB area suffering a drought. It is calculated as the number of cells indicating a drought at a precise date (N_drought) divided by the total number of cells of the raster (in this case, N_TOTAL = 44 133) as shown in Eq. (1):

$$\mathrm{ODE} = \frac{\mathrm{N\_drought}}{\mathrm{N\_TOTAL}} \cdot 100\,\%. \tag{1}$$

The ODE ranges from 0 % (when no drought is occurring at any point of the basin) to 100 % (when the entire basin is suffering an event). It highlights the coverage of a drought, allowing a direct comparison between registered historical information and calculated results. Moreover, it helps define the temporal component of droughts as it states the beginning and the end of an event. However, it does not take into account its intensity.

5.3 Characterization of drought intensities

Regarding the intensity of the droughts, in this study a complementary indicator is applied to integrate the intensities computed at every grid cell. The Overall Drought Indicator is defined as the average index value across the cells under drought conditions at a precise date, as shown in Eq. (2):

$$\mathrm{ODI} = \frac{\sum_{i=1}^{\mathrm{N\_drought}} (\mathrm{Index}_i)}{\mathrm{N\_drought}}. \tag{2}$$

The ODI expresses the average severity in the drought-affected part of the basin. It gives information about the meteorological stress level of the areas being effectively affected by a drought. Moreover, this indicator may help complete the collected historical records which include little information on the magnitude of the events. From indications of Tables 2, 3, 4 and 5, lower values of this indicator denote drier conditions. Undefined values occur when no cells are under drought conditions.

Only cells under drought conditions have been considered to define this indicator. If the ODI had been calculated as an average value for the entire basin (as adopted for instance in Trambauer et al., 2014) higher (or lower) indicator values in a part of the basin may have compensated lower (or higher) indicator values in the rest of the basin, yielding an overall value close to normal precipitation. Therefore, the ODI must always be used together with the ODE: whenever a drought has been detected with the ODE, its overall intensity may be assessed with the corresponding value of the ODI.

5.4 Evaluating indicator-based results with cataloged historical events

In order to support the choice of an index and timescale combination for the definition of the ODE and ODI, an assessment of the quality of the forecasts performed with the different variants is recommended. The hypothesis is that detected drought events (i.e., the forecasts) correspond to the cases when the ODE value exceeds a given threshold, which indicates that a certain area is affected by an event. The temporal coincidence of these forecasts has to be then contrasted with the occurrence of recorded droughts (i.e., the observations). As stated above, Fig. 2 shows that very short index-based events risk being forecasted. In order to avoid an overestimation of droughts, an additional 3-month criterion for beginning and ending forecasted droughts was established: an event will be effectively detected when the ODE value exceeds the threshold for at least 3 consecutive months.

For the matching between forecasts and observations, two monthly series of events were created (one for the events detected according to the ODE values and one for the historical events), where for each month either a "drought" or a "no drought" condition is assigned. Different scores for contrasting this type of dichotomous forecasts (occurrence vs. no occurrence) exist: the Peirce skill score, PSS (Hanssen and Kuipers, 1965; Murphy and Daan, 1985; Peirce, 1884); the Heidke skill score, HSS (Heidke, 1926); the Gilbert skill score, GSS (Schaefer, 1990); or the odds ratio skill score, ORSS (Stephenson, 2000). As recommended by Candogan Yossef et al. (2012), the PSS is used in this study. For its calculation, the miss rate (M) and the false alarm rate (F) are defined in Eqs. (3) and (4), respectively:

$$M = \frac{c}{a+c}, \tag{3}$$

Table 7. Contingency table of the comparison between forecasts and observations.

		Observation	
		Drought	No drought
Forecast	Drought	a (hit)	b (false alarm)
	No drought	c (miss)	d (correct rejection)

$$F = \frac{b}{b+d}, \tag{4}$$

where a, b, c and d represent the number of cases for each possible forecast outcome, respectively (Table 7):

- hit: when one detected drought corresponds with an observed drought;
- false alarm: when a drought appears during a month where no observed event has occurred;
- miss: when, during a month where a drought has been observed, no event has been detected;
- correct rejection: when, during a month where no drought has been observed, no drought is detected.

The miss rate (M) indicates how many of the observed events are not forecasted (related to the Type 1 errors) while the false alarm rate (F) is the proportion of non-occurrences that are incorrectly forecasted (Jolliffe and Stephenson, 2003). The PSS is expressed as shown in Eq. (5):

$$\mathrm{PSS} = 1 - M - F. \tag{5}$$

The PSS ranges from -1 to $+1$: perfect forecasts receive a score of one, random forecasts receive a score of zero and negative values indicate less skill than a random prediction. A suitable combination of the index and timescale will then lead to higher PSS values.

However, high values of the PSS score may be obtained purely by chance, especially when using only a small number of forecasts. Such is the case of the present work, where only 13 independent events have been documented during the 55 years of record keeping. This could lead to overestimating the goodness of a combination of the index and timescale. Therefore, sampling uncertainties must be addressed. A statistical test was applied to check if the calculated PSS values were significantly different from zero, at least at a 95 % confidence level. Assuming independence of the miss and false alarm rates, the standard error in the Peirce skill score is simply the square root of the sum of the squared standard errors in the miss and false alarm rates (Stephenson, 2000), as expressed in Eq. (6):

$$\mathrm{SE_{PSS}} = \sqrt{(\mathrm{SE}_M)^2 + (\mathrm{SE}_F)^2}, \tag{6}$$

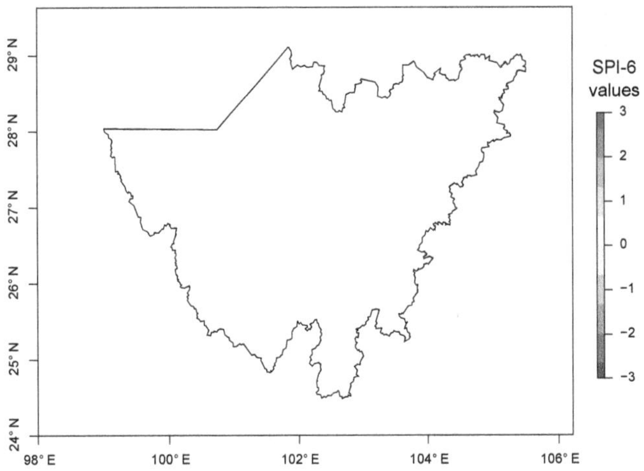

Figure 3. Extrapolated SPI-6 values in October 2012 for the entire lower JRB.

where the standard errors in miss (SE_M) or false alarm (SE_F) rates can be estimated by interpolating the values of the score confidence interval discussed in Agresti and Coull (1998) (cf. Table 8), as suggested by Thornes and Stephenson (2001). If the $\mathrm{PSS} \pm 1.96 \cdot \mathrm{SE_{PSS}}$ interval does not include zero, then the null of a random forecast can be rejected at a 95 % confidence level.

6 Results and discussion

Following the previous approach, the series of the SPI, PN, RAI and DEC indices were calculated for different timescales (1-, 3-, 6-, 12-, 24- and 48-month timescales) within the period of 1960–2014. First computed at the 29 stations, these indices were then extrapolated to the rest of the lower JRB. Figure 3 shows the example of the standardized precipitation index for a 6-month timescale (SPI-6) calculated in October 2012 (corresponding to the drought event XII) and spatially distributed at the lower JRB, where brown colors represent regions under drier conditions.

According to the criteria proposed in Table 6, detected drought events were identified based on the index values. Then, the ODE and ODI indicators were calculated for the lower JRB. An example of the resulting ODE and ODI series is shown in Fig. 4 for the SPI-6 and RAI-6 combinations and in Appendix A for all the timescales and indices analyzed, along with the recorded historical droughts shaded in orange.

The objective is to establish a combination of timescale and index that offers an optimum identification of historical droughts. As stated before, the main criteria used to contrast the performance of the forecasts is that a drought event is supposed to happen when the ODE value exceeds a threshold that is to be defined. The combination finally retained

Table 8. Standard error in estimated miss or false alarm rate calculated using the 95 % score confidence interval as discussed in Agresti and Coull (1998).

Events	Estimated miss or false alarm rate										
	0	0.1	0.2	0.3	0.4	0.5	0.6	0.7	0.8	0.9	1
5	0.111	0.134	0.15	0.16	0.166	0.168	0.166	0.16	0.15	0.134	0.111
10	0.071	0.099	0.116	0.126	0.132	0.134	0.132	0.126	0.116	0.099	0.071
20	0.041	0.07	0.086	0.095	0.101	0.102	0.101	0.095	0.086	0.07	0.041
30	0.029	0.057	0.071	0.08	0.084	0.086	0.084	0.08	0.071	0.057	0.029
40	0.022	0.049	0.062	0.07	0.074	0.076	0.074	0.07	0.062	0.049	0.022
50	0.018	0.043	0.056	0.063	0.067	0.068	0.067	0.063	0.056	0.043	0.018
100	0.009	0.03	0.04	0.045	0.048	0.049	0.048	0.045	0.04	0.03	0.009
500	0.002	0.013	0.018	0.02	0.022	0.022	0.022	0.02	0.018	0.013	0.002
1000	0.001	0.01	0.013	0.014	0.015	0.016	0.015	0.014	0.013	0.01	0.001

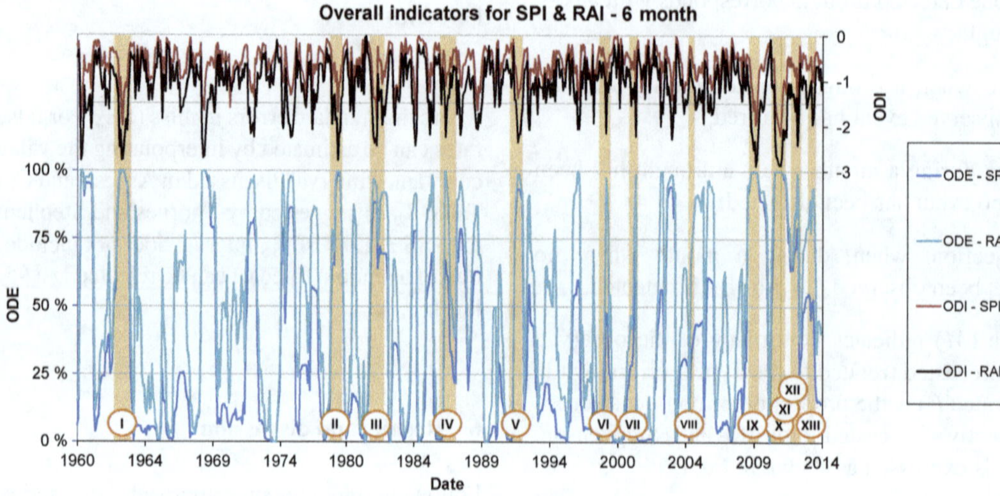

Figure 4. ODE and ODI values using the 6-month timescales of SPI and RAI indices, compared with the 13 detected historical droughts (in orange).

should maximize the number of hits and minimize the misses between the forecasts and the observed events.

The 1-month-scale overall indices show rapid fluctuations that correspond to short periods of precipitation deficiency not captured in the catalog of historical droughts. This is mainly due to punctual, large rainfall events that have an important influence in the indices, which may indicate that the drought had ceased when it is not the case (Barua et al., 2011). The use of this timescale is not recommended for drought monitoring since long drought events are hardly identified. The opposite effect occurs when using the 48-month scale. The inertia of the rainfall shortage tendencies may mask shorter droughts and overestimate their durations. Since most of the episodes last 1 year or less (Table 1), they are hardly detected using the 48-month scale. The droughts which occurred from 2009 to 2014 (droughts IX to XIII) illustrate this phenomenon: even if five different droughts have been cataloged, a unique one is detected using the 48-month scale, according to the ODE time series. Therefore, using the 1- and 48-month scales do not provide any substantial information about the occurrence and duration of the droughts and have been excluded from the performance analysis.

For the rest of the timescales (3-, 6-, 12- and 24-month timescale), the ODE threshold indicating the occurrence of a drought is required for the computing of the PSS that will serve as a support for the selection of the best combination of the index and timescale. Traditionally, cross-validation techniques are used to define optimum thresholds, for when within a training subset the threshold maximizing the PSS is identified and validated in a non-overlapping validation subset. However, the authors consider that the limited number of 13 independent events recorded represents an important limitation to a robust and meaningful cross validation. Instead, another approach was followed. A sensitivity analysis was performed using the same threshold across all of the combinations and exploring the effect of varying it in a reasonable

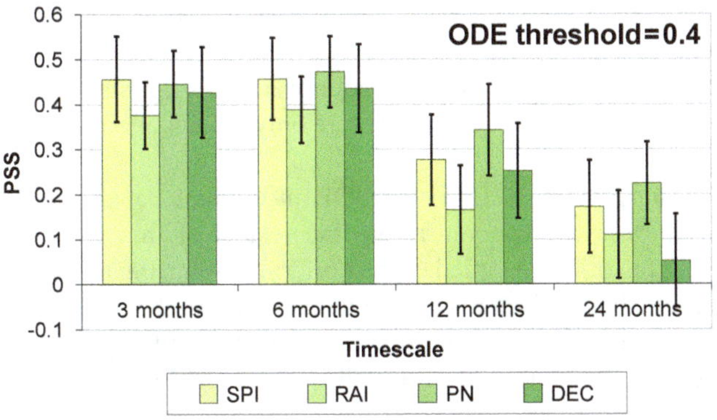

Figure 5. Example of graphic representation of the PSS results for an ODE threshold of 0.4, with the black error bars representing the 95 % confidence interval (±1.96 standard errors) when sampling uncertainties are considered.

range (in this case, from 0.3 to 1 by 0.1 steps). The resulting PSS values are shown in Fig. B1 of Appendix B and an example of PSS for an ODE threshold of 0.4 is presented in Fig. 5. The black bars indicate the statistical error estimates (confidence intervals) at 95 % confidence, due to sampling uncertainties, assessed with the statistical significance test described in Sect. 5.4, which allows indicating whether the score is significantly different from zero.

According to the results, most of the PSS confidence intervals do not include zero, disproving that skill scores could have identified drought events by chance sampling fluctuations. For some of the indices on the 24-month scale (e.g., RAI-24 for an ODE threshold = 0.7), results cannot assert that skill scores are significantly different from zero and thus these combinations should not be considered.

Attending to the PSS values (Fig. B1), results show a consistent tendency across all ODE thresholds of higher PSS at the 3- and the 6-month timescales. Moreover, there is no single index that clearly produces better results. Indeed, based on the PSS values and taking into account their uncertainty, there are no statistically significant differences across the different indices for the 3- and 6-month timescales. This indicates that, for these timescales, all the indices perform similarly well in capturing the events, which is consistent with the fact that they all rely on the same type of data (precipitation). PSS results are independent of the specific threshold and thus they are considered robust. However, it is worth mentioning that, in general, higher PSS values for the 3- and the 6-month timescales are produced using ODE thresholds between 0.4 and 0.6.

Regarding the 6-month ODE series (Figs. A5 and A6 of Appendix A), it is important to highlight some relevant aspects:

- All the observed drought events have their corresponding ODE peaks.
- Although event VIII has an estimated duration of 3 months, ODE and ODI results consistently show a longer drought. The exact period of this drought is not well defined as indicated in the catalog, leaving room for a longer duration of the real episode.
- In general, all the indices are well correlated, identifying most of the recorded droughts.
- Several droughts are consistently detected between event I (1962) and II (1979) even if no drought has been chronicled (false alarms). This may correspond to the above-mentioned scarcity of reliable information on droughts prior to 1980.
- The drought events IX, X, XI, XII and XIII are well captured. As shown in Figs. A5 and A6, the different events during this period (2009–2014) match with the consecutive increases in the ODE values for all the indices (DEC, PN, RAI, SPI).
- However, the 6-month series of ODE suggest some false positive detections: more drought events than the observed are calculated.

Regarding the 3-month ODE series (Figs. A3 and A4), results suggest an overestimation of the number of detected events, as sometimes several detected events combine into one (longer) observed event. The 6-month timescale appears as more appropriate.

In summary, according to the ODE series presented in Appendix A and to the forecast verification carried out with the Peirce skill score (Appendix B), it seems that the best timescale for the identification of droughts is at 6-months. Results show an equally effective performance of the ODE series for all the indices. However, the risk of false positives must be addressed carefully, as the observation record likely misses events, in particular between 1962 and 1979.

Despite the good performance shown by the overall indicator ODE in drought detection, caution is advised. In particular, the choice of meteorological indices as a basis for the calculation of the ODE and ODI can lead to errors when assessing drought occurrence. Temperature variability, not considered here, can play a significant role in the onset of agricultural drought. Meteorological indices may not be fully capable of capturing the impacts on water scarcity and could be complemented with other types of indices, such as agricultural or hydrological. The same approach proposed in this study is recommended using more comprehensive indices in order to better capture the complex drought processes.

The performance assessment of the ODE indicator to detect droughts relies basically on the comparison with the historical events cataloged in this study. The search and compilation of this information from different data sources, often scarce and ambiguous, represents a challenge. Different information sources often provide only partial information for one episode, and for some events the differences in the available information complicate the harmonization of data. As a result, the accuracy of the collected information may impact on the applicability of the developed methodology.

7 Conclusions

This study aims at applying overall drought indicators representing the drought status within the entire lower JRB investigation area. This work represents an attempt at building a tool for drought monitoring and risk management purposes on the basin scale. It is based on established meteorological indices for the identification of droughts and a method for a catchment-wide drought assessment and characterization, which is compared to historical drought events of the lower JRB.

The information used for the identification and characterization of major historic droughts was compiled from different sources. A total of 13 major droughts between 1960 and 2014 were identified in the lower JRB and cataloged using a web-based registration platform, allowing for a comparison of the different events.

Drought indices typically assess local water deficits while available historical records usually refer to regional droughts. To overcome this problem, two drought area indicators, the Overall Drought Extension and the Overall Drought Indicator, have been used to characterize the occurrence and intensity of an event within a specific investigation area. These indicators are based on four common meteorological indices on different timescales: the standardized precipitation index, the rainfall anomaly index, the percent of normal precipitation and the deciles index. By relying exclusively on precipitation, the proposed procedure serves as a basis for further studies in other regions where only precipitation data are available.

The performance of the ODE in drought detection has been assessed by contrasting the results of this indicator with historical recorded events, offering promising results. It seems that the best results are independent of the index used and produced using the 6-month timescale. Although results suggest the same patterns for all ODE thresholds, it has been noticed that the highest PSS values are produced for thresholds between 0.4 and 0.6, which can be defined as a trigger to detect the occurrence of a drought in the lower JRB.

Considering the challenge that the compilation of historical drought information represents and the identified limitations, this is a good method for the monitoring of drought episodes within an entire catchment. The use and contrast of drought indicators on the basin scale with historical collected information represent the main innovative aspects of this study. Since meteorological droughts are the first stage in the progression of subsequent agricultural or hydrological droughts, this methodology could be used to activate a management response for a drought event, which starts at a specific threshold value. Additionally, this methodology can be used to complete lacking information on droughts' duration, geographical extension or intensity.

Appendix A: Series of ODE and ODI indicators

The series of the Overall Drought Extension and the Overall Drought Indicator have been calculated for the 1-, 3-, 6-, 24 and 48-month timescales. Graphic results are presented in Figs. A1 to A12 below.

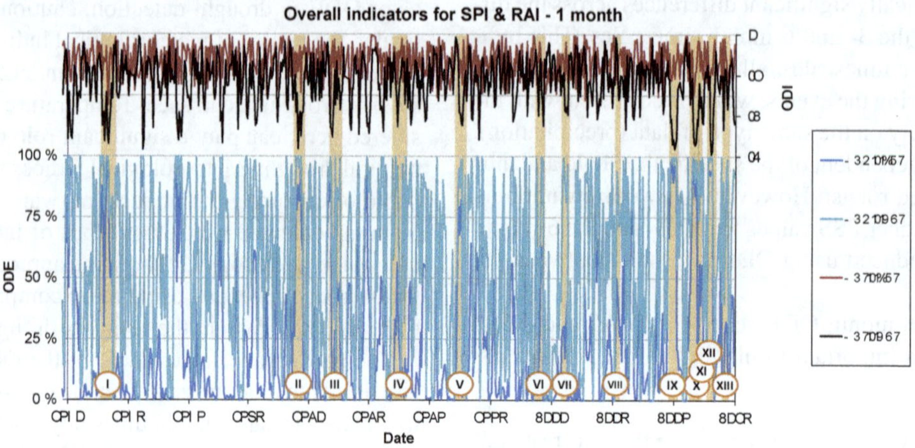

Figure A1. ODE and ODI values using the 1-month timescales of SPI and RAI indices, compared with the 13 detected historical droughts (in orange).

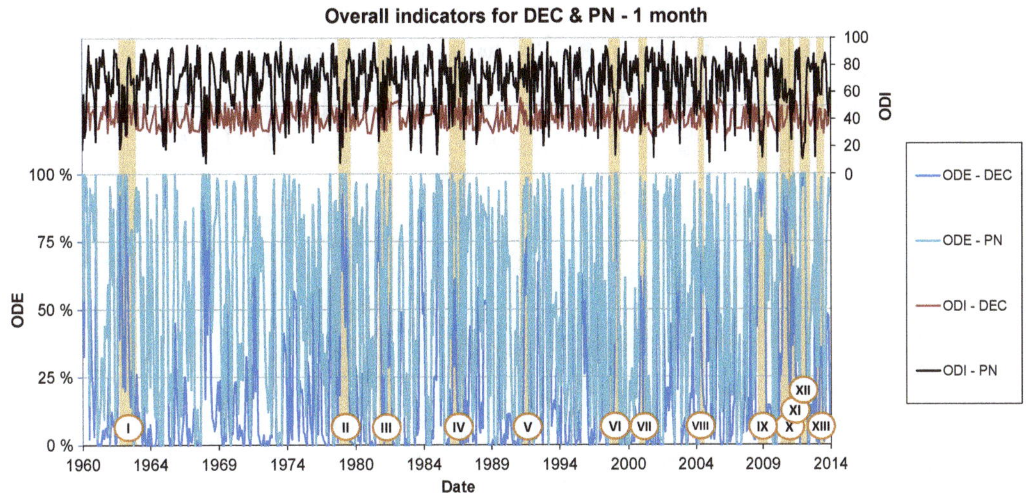

Figure A2. ODE and ODI values using the 1-month timescales of DEC and PN indices, compared with the 13 detected historical droughts (in orange).

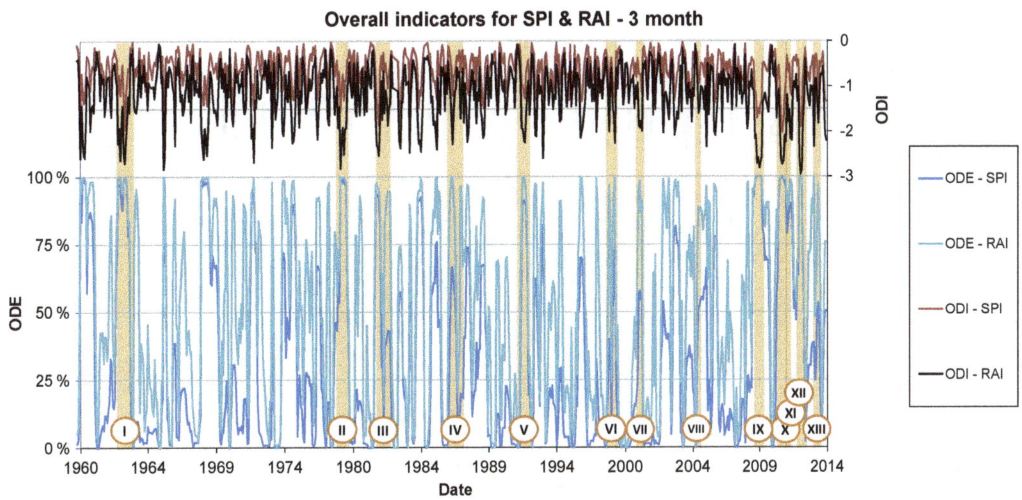

Figure A3. ODE and ODI values using the 3-month timescales of SPI and RAI indices, compared with the 13 detected historical droughts (in orange).

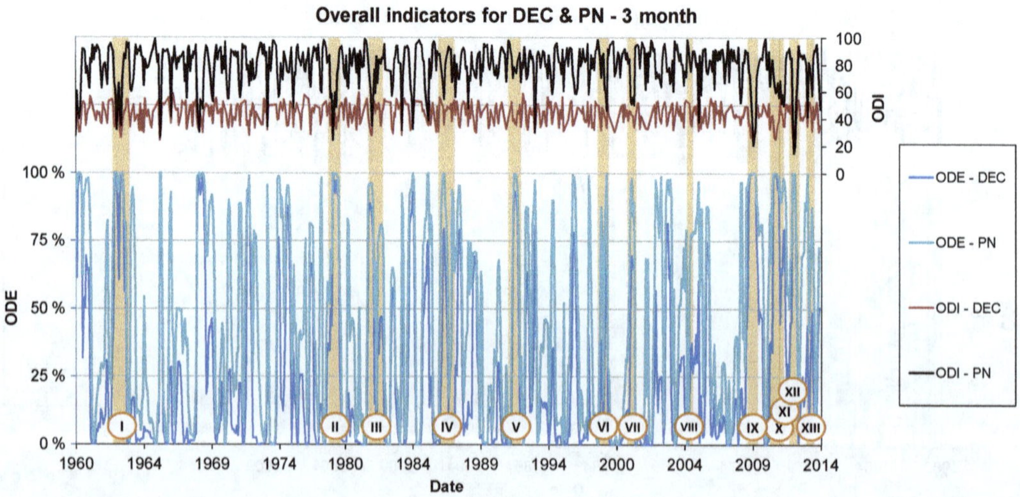

Figure A4. ODE and ODI values using the 3-month timescales of DEC and PN indices, compared with the 13 detected historical droughts (in orange).

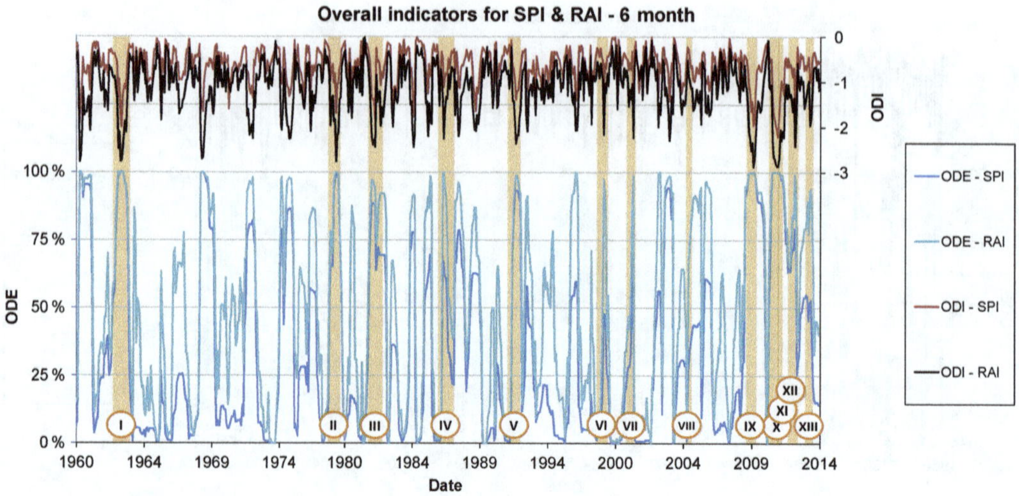

Figure A5. ODE and ODI values using the 6-month timescales of SPI and RAI indices, compared with the 13 detected historical droughts (in orange).

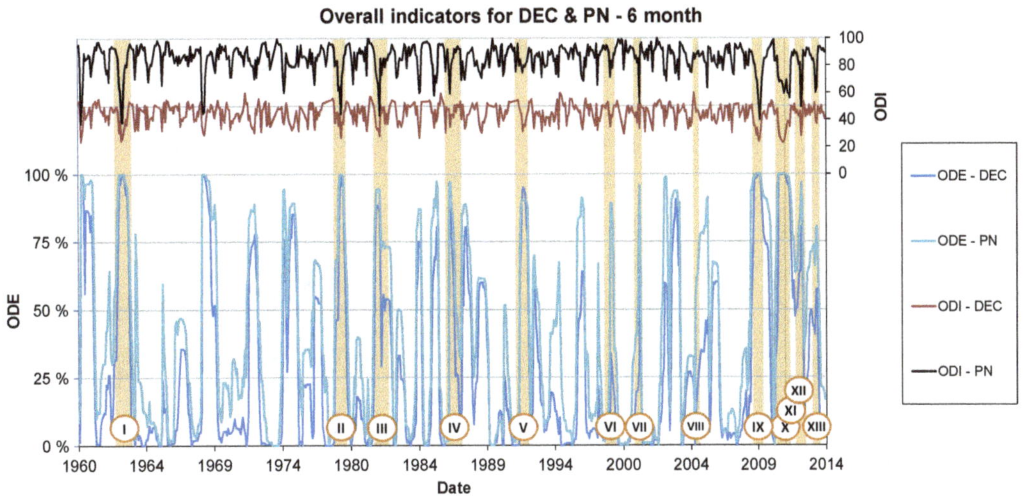

Figure A6. ODE and ODI values using the 6-month timescales of DEC and PN indices, compared with the 13 detected historical droughts (in orange).

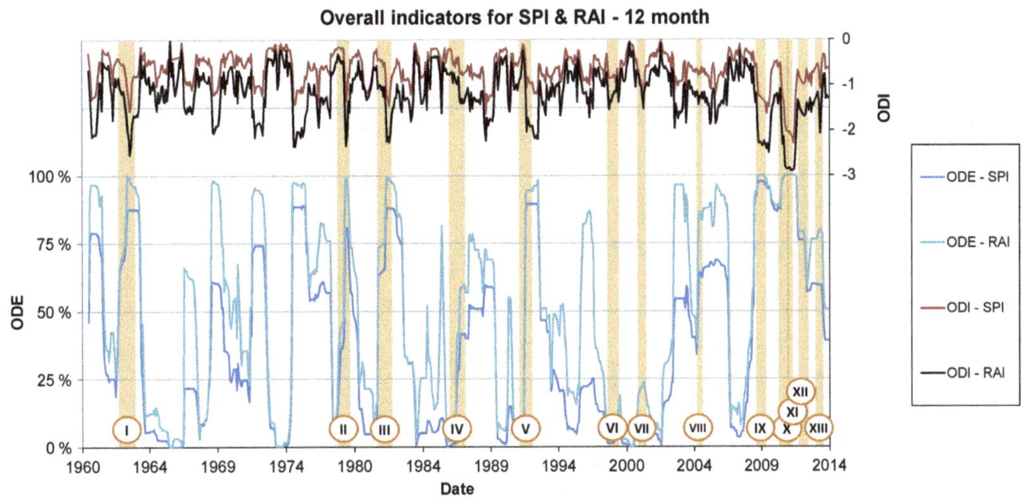

Figure A7. ODE and ODI values using the 12-month timescales of SPI and RAI indices, compared with the 13 detected historical droughts (in orange).

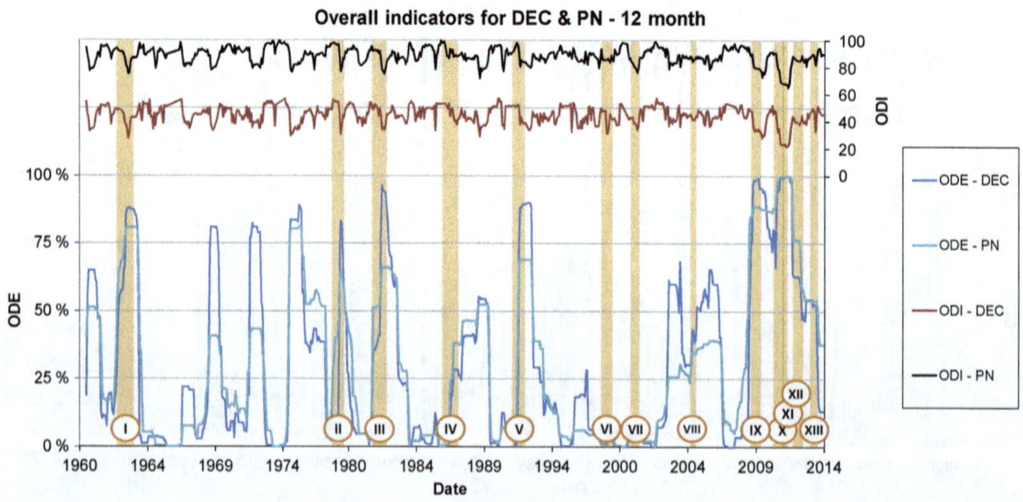

Figure A8. ODE and ODI values using the 12-month timescales of DEC and PN indices, compared with the 13 detected historical droughts (in orange).

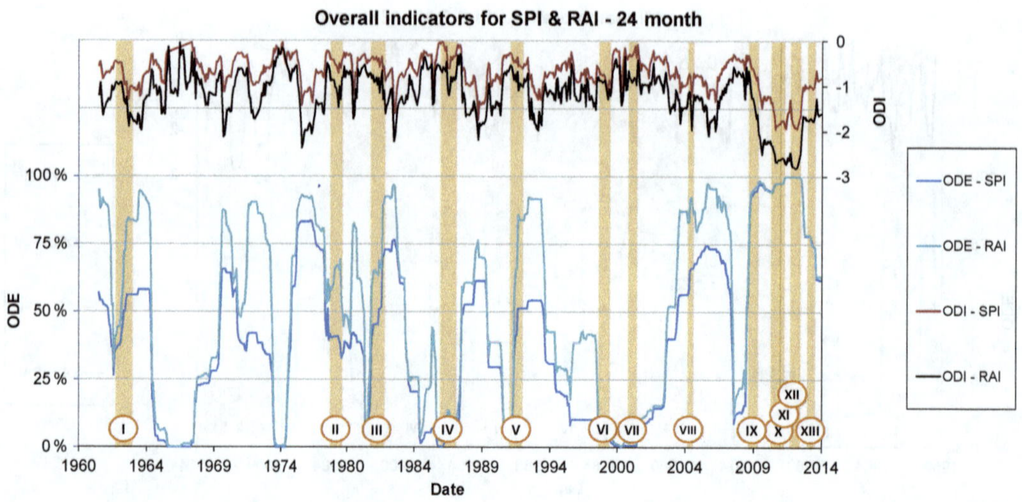

Figure A9. ODE and ODI values using the 24-month timescales of SPI and RAI indices, compared with the 13 detected historical droughts (in orange).

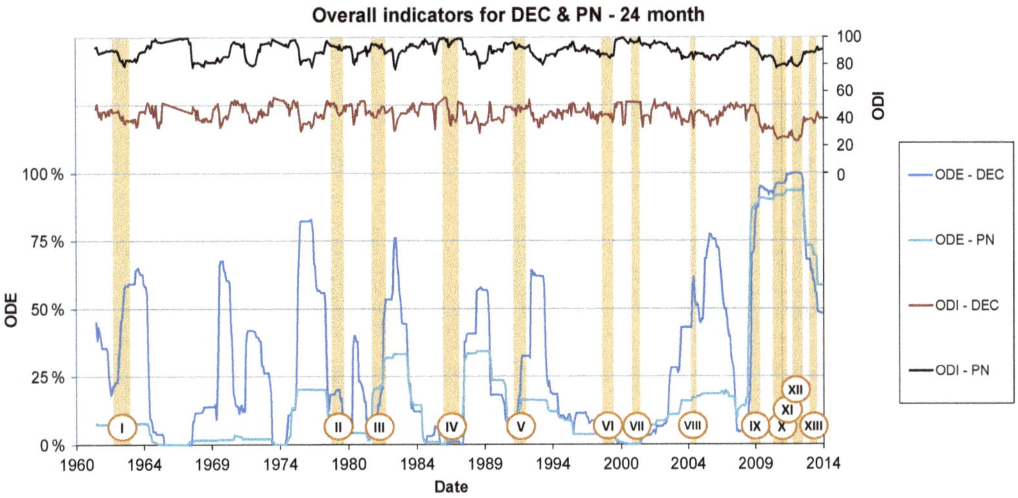

Figure A10. ODE and ODI values using the 24-month timescales of DEC and PN indices, compared with the 13 detected historical droughts (in orange).

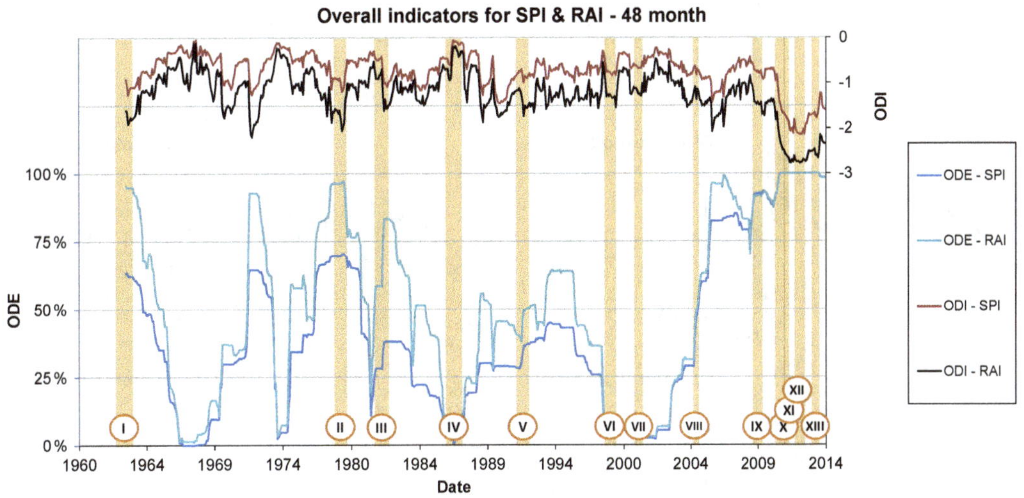

Figure A11. ODE and ODI values using the 48-month timescales of SPI and RAI indices, compared with the 13 detected historical droughts (in orange).

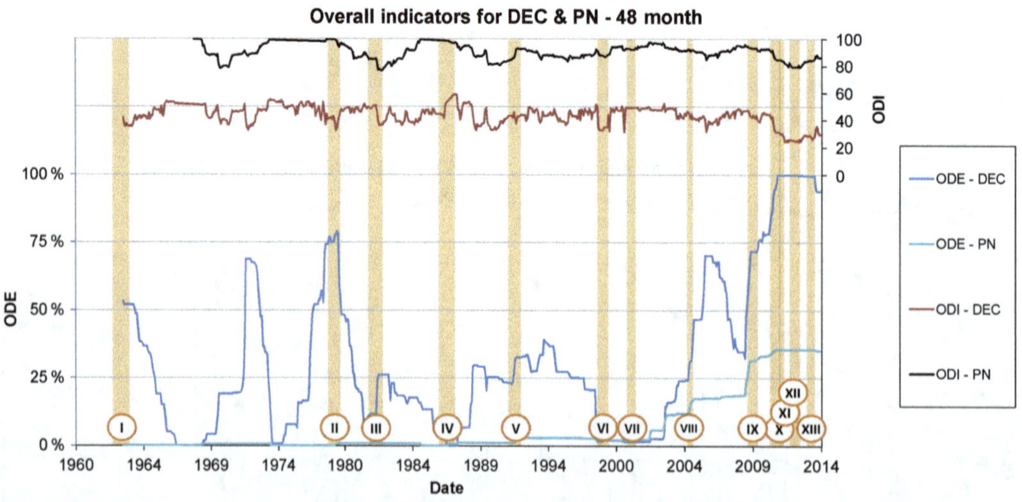

Figure A12. ODE and ODI values using the 48-month timescales of DEC and PN indices, compared with the 13 detected historical droughts (in orange).

Appendix B: PSS results

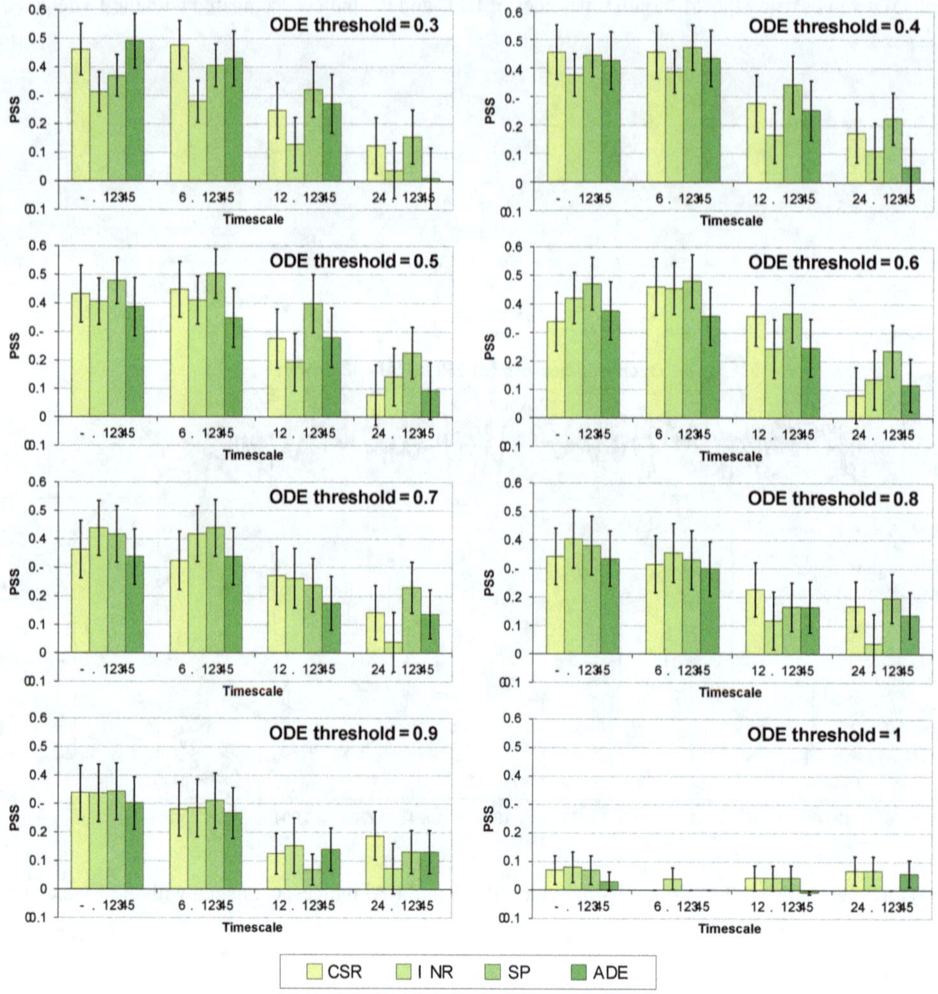

Figure B1. Graphical PSS results for different index and timescale combinations, for a range of ODE thresholds between 0.3 and 1. The black error bars represent the 95 % confidence interval.

Competing interests. The authors declare that they have no conflict of interest.

Acknowledgements. This study is based on the project "Jinsha River Basin (JRB): Integrated Water Resources and Risk Management under a Changing Climate" funded by the Swiss Agency for Development and Cooperation (SDC) and has been supported financially by the International Science and Technology Cooperation Program of China (grant no. 2014DFA71910), the National Natural Science Foundation of China (grant no. 51709009), and the Fundamental Research Funds for Central Public Welfare Research Institutes (grant no. CKSF2016024). The authors would like to acknowledge and thank the Bureau Of Hydrology (BOH) and the Changjiang River Scientific Research Institute (CRSRI) from the Changjiang Water Resources Commission for providing the data and the enterprise Ernst Basler + Partner AG for the coordination and support throughout the entire project. Comments by anonymous reviewers and the journal editor are gratefully acknowledged.

Edited by: Pierre Gentine

References

Agresti, A. and Coull, B. A.: Approximate Is Better than "Exact" for Interval Estimation of Binomial Proportions, Am. Stat., 52, 119–126, https://doi.org/10.2307/2685469, 1998.

Akbari, H., Rakhshandehroo, G. R., Sharifloo, A. H., and Ostadzadeh, E.: Drought Analysis Based on Standardized Precipitation Index (SPI) and Streamflow Drought Index (SDI) in Chenar Rahdar River Basin, Southern Iran, American Society of Civil Engineers, 11–22, 2015.

Barua, S., Ng, A. W. M., and Perera, B. J. C.: Comparative Evaluation of Drought Indexes: Case Study on the Yarra River Catchment in Australia, J. Water Res. Plan. Man., 137, 215–226, https://doi.org/10.1061/(ASCE)WR.1943-5452.0000105, 2011.

Bhalme, H. N. and Mooley, D. A.: Large-Scale Droughts/Floods and Monsoon Circulation, Mon. Weather Rev., 108, 1197–1211, https://doi.org/10.1175/1520-0493(1980)108<1197:LSDAMC>2.0.CO;2, 1980.

Candogan Yossef, N., van Beek, L. P. H., Kwadijk, J. C. J., and Bierkens, M. F. P.: Assessment of the potential forecasting skill of a global hydrological model in reproducing the occurrence of monthly flow extremes, Hydrol. Earth Syst. Sci., 16, 4233–4246, https://doi.org/10.5194/hess-16-4233-2012, 2012.

CMDC: China Meteorological Data Service Center, available at: http://data.cma.cn/en, last access: 20 November 2017.

Department of Climate Change, National Development and Reform Commission: China Climate Change, available at: http://www.ccchina.gov.cn/, last access: 31 August 2017.

Duan, H., Wang, S., and Feng, J.: Drought events and its Influence in 2014 in China, J. Arid Meteorol., 33, 349–360, 2015.

Edwards, D. C. and McKee, T. B.: Characteristics of 20th century drought in the United States at multiple time scales, Climatology Report, Department of Atmospheric Science, Colorado State University, Fort Collins, Colorado, 1997.

Fleig, A. K., Tallaksen, L. M., Hisdal, H., and Hannah, D. M.: Regional hydrological drought in north-western Europe: linking a new Regional Drought Area Index with weather types, Hydrol. Process., 25, 1163–1179, https://doi.org/10.1002/hyp.7644, 2011.

GEOTEST AG: Jinsha River Basin – Natural Hazard Event Registration Platform, available at: http://jinsha.geomaps.ch/Home/, last access: 31 August 2017.

Gibbs, W. J. and Maher, J. V.: Rainfall deciles as drought indicators, Bureau of Meteorology, Melbourne, Australia, 1967.

Guha-Sapir, D., Below, R., and Hoyois, P.: EM-DAT: The CRED/OFDA International Disaster Database, available at: www.emdat.be, last access: 31 August 2017.

Halwatura, D., McIntyre, N., Lechner, A. M., and Arnold, S.: Reliability of meteorological drought indices for predicting soil moisture droughts, Hydrol. Earth Syst. Sci. Discuss., https://doi.org/10.5194/hess-2016-467, 2016.

Hanssen, A. and Kuipers, W. J. A.: On the relationship between the frequency of rain and various meteorological parameters, K. Ned. Meteorol. Inst. Mededelin-Gen En Verh., 102, 2–15, 1965.

Hao, Z. and AghaKouchak, A.: Multivariate Standardized Drought Index: A parametric multi-index model, Adv. Water Resour., 57, 12–18, https://doi.org/10.1016/j.advwatres.2013.03.009, 2013.

Hayes, M. J., Wilhelmi, O. V., and Knutson, C. L.: Reducing Drought Risk: Bridging Theory and Practice, Nat. Hazards Rev., 5, 106–113, https://doi.org/10.1061/(ASCE)1527-6988(2004)5:2(106), 2004.

He, B., Wu, J., Lü, A., Cui, X., Zhou, L., Liu, M., and Zhao, L.: Quantitative assessment and spatial characteristic analysis of agricultural drought risk in China, Nat. Hazards, 66, 155–166, https://doi.org/10.1007/s11069-012-0398-8, 2013.

He, J., Yang, X., Li, Z., Zhang, X., and Tang, Q.: Spatiotemporal Variations of Meteorological Droughts in China During 1961–2014: An Investigation Based on Multi-Threshold Identification, Int. J. Disaster Risk Sc., 7, 63–76, https://doi.org/10.1007/s13753-016-0083-8, 2016.

He, L.: Summary of General Information on Severe Drought Disaster in Yunnan Province Since the Founding of New China, 2010 (in Chinese).

Heidke, P.: Berechnung des Erfolges und der Güte der Windstärkevorhersagen im Sturmwarnungsdienst (Measures of success and goodness of wind force forecasts by the gale-warning service), Geogr. Ann., 8, 301–349, https://doi.org/10.2307/519729, 1926.

Hu, X., Xu, P., Ning, G., and Wang, S.: Causes of Continuous Drought in Southwest China from Autumn of 2012 to Spring of 2013, J. Desert Res., 35, 763–773, 2015.

Jain, V. K., Pandey, R. P., Jain, M. K., and Byun, H.-R.: Comparison of drought indices for appraisal of drought characteristics in the Ken River Basin, Weather Clim. Extrem., 8, 1–11, https://doi.org/10.1016/j.wace.2015.05.002, 2015.

Jeevananda Reddy, S.: Sensitivity of some potential evapotranspiration estimation methods to climate change, Agr. Forest Meteorol., 77, 121–125, https://doi.org/10.1016/0168-1923(95)02239-T, 1995.

Joint Research Centre: European Drought Observatory, available at: http://edo.jrc.ec.europa.eu/edov2/php/index.php?id=1000, last access: 31 August 2017.

Jolliffe, I. T. and Stephenson, D. B. (Eds.): Forecast verification: a practitioner's guide in atmospheric science, Wiley, Chichester, 2003.

Keyantash, J. and Dracup, J. A.: The Quantification of Drought: An Evaluation of Drought Indices, B. Am. Meteorol. Soc., 83, 1167–1180, https://doi.org/10.1175/1520-0477(2002)083<1191:TQODAE>2.3.CO;2, 2002.

Kogan, F.: Contribution of remote sensing to drought early warning, in: Early Warning Systems for Drought Preparedness and Drought Management, edited by: Wilhite, D. A., Sivakumar, M. V. K., and Wood, D. A., Geneva, Switzerland, World Meteorological Organization, 75–87, 2000.

Landsberg, H. E.: Climatic Aspects of Droughts, B. Am. Meteorol. Soc., 63, 593–597, https://doi.org/10.1175/1520-0477(1982)063<0593:CAOD>2.0.CO;2, 1982.

Land Surface Hydrology Group – Princeton University: African Drought Monitor, available at: http://hydrology.princeton.edu/~justin/research/project_global_monitor/overview.html, last access: 31 August 2017.

Lee, D. M.: Australian drought watch system, M.T. Hinchey, Gaborone, Botswana, Botswana Society, 173–187, 1979.

Liu, H.: Study on Historical Drought Disasters and Prevention and Control Measures Solutions in Yunnan Province, Master Dissertation, Northwest Agriculture & Forestry University, China, 2012 (in Chinese).

Liu, Y., Zhao, E.-X., Huang, W., Sun, D., and Ju, J.-H.: Diagnostic Analysis of the Server Drought over Yunnan area in the Early Summer of 2005, J. Trop. Meteorol., 23, 35–40, 2007.

McKee, T. B., Doesken, N. J., and Kliest, J.: The relationship of drought frequency and duration to time scales, American Meterological Society, Anaheim, CA, 179–184, 1993.

McKee, T. B., Doesken, N. J., and Kliest, J.: Drought monitoring with multiple time scales, in: Proceedings of the 9th Conference of Applied Climatology, American Meteorological Society, Boston, MA, Dallas, TX, 233–236, 1995.

Mei, X. and Yang, X.: Drought conditions and management strategies in China, Hanoi, Vietnam, 2014.

Mishra, A. K. and Singh, V. P.: Drought modeling – A review, J. Hydrol., 403, 157–175, https://doi.org/10.1016/j.jhydrol.2011.03.049, 2011.

Mitchell, J. M., Stockton, C. W., and Meko, D. M.: Evidence of a 22-year Rhythm of Drought in the Western United States Related to the Hale Solar Cycle since the 17th Century, in: Solar-Terrestrial Influences on Weather and Climate, edited by: McCormac, B. M. and Seliga, T. A., Springer Netherlands, Dordrecht, 125–143, 1979.

Morid, S., Smakhtin, V., and Moghaddasi, M.: Comparison of seven meteorological indices for drought monitoring in Iran, Int. J. Climatol., 26, 971–985, https://doi.org/10.1002/joc.1264, 2006.

Murphy, A. H. and Daan, H.: Forecast evaluation, in: Probability, Statistics and Decision Making in the Atmospheric Sciences, edited by: Murphy, A. H. and Katz, R. W., Westview Press, Boulder, Colorado, 379–437, 1985.

Niemeyer, S.: New drought indices, in: Options Méditerranéennes?: Série A. Séminaires Méditerranéens, edited by: López-Francos, A., Zaragoza, CIHEAM, 80, 267–274, 2008.

Olukayode Oladipo, E.: A comparative performance analysis of three meteorological drought indices, J. Climatol., 5, 655–664, https://doi.org/10.1002/joc.3370050607, 1985.

Palmer, W. C.: Meteorological drought, Weather Bureau Research Paper, US Department of Commerce, Washington, DC, 1965.

Peirce, C. S.: The numerical measure of the success of predictions, Science, 4, 453–454, https://doi.org/10.1126/science.ns-4.93.453-a, 1884.

Schaefer, J. T.: The Critical Success Index as an Indicator of Warning Skill, Weather Forecast., 5, 570–575, https://doi.org/10.1175/1520-0434(1990)005<0570:TCSIAA>2.0.CO;2, 1990.

Shaw, S. B. and Riha, S. J.: Assessing temperature-based PET equations under a changing climate in temperate, deciduous forests, Hydrol. Process., 25, 1466–1478, https://doi.org/10.1002/hyp.7913, 2011.

Sivakumar, M. V. K., Wilhite, D. A., Svoboda, M. D., Hayes, M., and Motha, R.: Drought risk and meteorological droughts, in: 2011 Global Asessment Report on Disaster Risk Reduction, United Nations International Strategy for Disaster Reduction, Geneva, Switzerland, 2010.

Smith, D. I., Hutchinson, M. F., and McArthur, R. J.: Australian climatic and agricultural drought: Payments and policy, Drought Netw. News, 5, 11–12, 1993.

Sönmez, F. K., Kömüscü, A. ümran, Erkan, A., and Turgu, E.: An Analysis of Spatial and Temporal Dimension of Drought Vulnerability in Turkey Using the Standardized Precipitation Index, Nat. Hazards, 35, 243–264, https://doi.org/10.1007/s11069-004-5704-7, 2005.

Staage, J. H., Tallaksen, L. M., Xu, C.-Y., and Van Lanen, H. A. J.: Standardized precipitation-evapotranspiration index (SPEI): Sensitivity to potential evapotranspiration model and parameters, in: Hydrology in a changing world: environmental and human dimensions?, Proceedings of FRIEND-Water, Montpellier, France, 2014.

Stephenson, D. B.: Use of the "Odds Ratio" for Diagnosing Forecast Skill, Weather Forecast., 15, 221–232, https://doi.org/10.1175/1520-0434(2000)015<0221:UOTORF>2.0.CO;2, 2000.

Thornes, J. E. and Stephenson, D. B.: How to judge the quality and value of weather forecast products, Meteorol. Appl., 8, 307–314, https://doi.org/10.1017/S1350482701003061, 2001.

Tigkas, D., Vangelis, H., and Tsakiris, G.: The Drought Indices Calculator (DrinC), in: Proceedings of 8th International Conference of EWRA Water Resources Management in an Interdisciplinary and Changing Context, Porto, Portugal, 26–29 June 2013.

Trambauer, P., Maskey, S., Werner, M., Pappenberger, F., van Beek, L. P. H., and Uhlenbrook, S.: Identification and simulation of space–time variability of past hydrological drought events in the Limpopo River basin, southern Africa, Hydrol. Earth Syst. Sci., 18, 2925–2942, https://doi.org/10.5194/hess-18-2925-2014, 2014.

Tsakiris, G., Loukas, A., Pangalou, D., Vangelis, H., Tigkas, D., Rossi, G. and Cancelliere, A.: Drought characterization [Part 1. Components of drought planning. 1.3. Methodological component], in: Drought management guidelines technical annex, edited by: Iglesias, A., López-Francos, A., and Moneo, M., Zaragoza, CIHEAM/EC MEDA Water, 58, 85–102, available at: http://om.ciheam.org/om/pdf/b58/00800535.pdf (last access: 31 January 2018), 2007.

USDM: United States Drought Monitor, available at: http://droughtmonitor.unl.edu/Home.aspx, last access: 31 August 2017.

Van Rooy, M. P.: A rainfall anomaly index independent of time and space, Notos, 14, 43–48, 1965.

Vasiliades, L., Loukas, A., and Liberis, N.: A Water Balance Derived Drought Index for Pinios River Basin, Greece, Water Resour. Manag., 25, 1087–1101, https://doi.org/10.1007/s11269-010-9665-1, 2011.

Vicente-Serrano, S. M., Beguería, S., and López-Moreno, J. I.: A Multiscalar Drought Index Sensitive to Global Warming: The Standardized Precipitation Evapotranspiration Index, J. Climate, 23, 1696–1718, https://doi.org/10.1175/2009JCLI2909.1, 2010.

Wanders, N., Lanen, H. A. J., and Loon, A. F.: Indicators for drought characterization on a global scale, Technical Report, 2010.

Wang, L. and Chen, W.: Characteristics of Multi-timescale Variabilities of the Drought over Last 100 Years in Southwest China, Adv. Meteorol. Sci. Technol., 2, 21–26, 2012.

Wang, L., Chen, W., Zhou, W., and Huang, G.: Drought in Southwest China: A Review, Atmos. Ocean. Sci. Lett., 8, 339–344, https://doi.org/10.3878/AOSL20150043, 2015.

Wang, X., Zhang, R., Zhao, L., and Guo, W.: Analysis on Dynamical Diagnosis of Drought in Southwest Region in Winter from 2011 to 2012, Sci.-Tech. Innov. Product., 235, 63–65, 2013.

WCB: Report of Yunnan Province Water Resources Bulletin, Online report, 2001 (in Chinese).

Wilhite, D. A. (Ed.): Chapter 1 Drought as a natural hazard: Concepts and definitions, in: Drought: A Global Assessment, I, 3–18, 2000.

World Meteorological Organization: Climate System Monitoring, Monthly Bulletin, World Meteorological Organization, World Climate Data Programme, Geneva, 1985.

World Meteorological Organization: Standardized precipitation index user guide, 2012.

World Meteorological Organization and Global Water Partnership: Handbook of drought indicators and indices, Integrated Drought Management Programme (IDMP), Integrated Drought Management Tools and Guidelines Series 2, edited by: Svoboda, M. and Fuchs, B. A., Geneva, 2016.

Wu, J., He, B., Lü, A., Zhou, L., Liu, M., and Zhao, L.: Quantitative assessment and spatial characteristics analysis of agricultural drought vulnerability in China, Nat. Hazards, 56, 785–801, https://doi.org/10.1007/s11069-010-9591-9, 2011.

Wu, R. L.: Overview of Flood and Drought Disaster in Yunnan Province, Pearl River, 1, 18–19, 1999.

Yang, H., Song, J., Yan, H., and Li, C.: Cause of the severe drought in Yunnan Province during winter of 2009 to 2010, Clim. Environ. Res., 17, 315–326, 2012.

Yang, S., Dong, X., and Dai, S.: Introduction to Application of River Channel Surveying Technology in Jinsha River, The Yellow River Water Conservancy Press, Zhengzhou, China, 2013.

Zhou, L., Wu, J.-J., Leng, S., Liu, M., Zhang, J., Zhao, L., Diao, C., Zhang, J., Luo, H., Zhang, F., and Shi, Y.: Using a new integrated drought monitoring index to improve drought detection in mid-eastern China, in: IEEE International Geoscience and Remote Sensing Symposium (IGARSS), Munich, Germany, 883–886, 2012.

6

Formulating and testing a method for perturbing precipitation time series to reflect anticipated climatic changes

Hjalte Jomo Danielsen Sørup[1,2], Stylianos Georgiadis[1,3], Ida Bülow Gregersen[4], and Karsten Arnbjerg-Nielsen[1,2]

[1]Technical University of Denmark, Global Decision Support Initiative, Lyngby, Denmark
[2]Technical University of Denmark, Department of Environmental Engineering, Lyngby, Denmark
[3]Technical University of Denmark, Department of Applied Mathematics and Computer Science, Lyngby, Denmark
[4]Ramboll Danmark A/S, Department of Climate Adaptation and Green Infrastructure, Copenhagen, Denmark

Correspondence to: Hjalte Jomo Danielsen Sørup (hjds@env.dtu.dk)

Abstract. Urban water infrastructure has very long planning horizons, and planning is thus very dependent on reliable estimates of the impacts of climate change. Many urban water systems are designed using time series with a high temporal resolution. To assess the impact of climate change on these systems, similarly high-resolution precipitation time series for future climate are necessary. Climate models cannot at their current resolutions provide these time series at the relevant scales. Known methods for stochastic downscaling of climate change to urban hydrological scales have known shortcomings in constructing realistic climate-changed precipitation time series at the sub-hourly scale. In the present study we present a deterministic methodology to perturb historical precipitation time series at the minute scale to reflect non-linear expectations to climate change. The methodology shows good skill in meeting the expectations to climate change in extremes at the event scale when evaluated at different timescales from the minute to the daily scale. The methodology also shows good skill with respect to representing expected changes of seasonal precipitation. The methodology is very robust against the actual magnitude of the expected changes as well as the direction of the changes (increase or decrease), even for situations where the extremes are increasing for seasons that in general should have a decreasing trend in precipitation. The methodology can provide planners with valuable time series representing future climate that can be used as input to urban hydrological models and give better estimates of climate change impacts on these systems.

1 Introduction

Climate change impacts water management worldwide as the water cycle is an essential part of the climate system. The planning horizon for water infrastructure is often very long, making reliable predictions of future climate crucial (Arnbjerg-Nielsen et al., 2015b). In the design of water infrastructure, precipitation data are needed. Especially for urban infrastructure the time resolution of precipitation data needed for design and planning is much finer than what is provided by climate models (Berndtsson and Niemczynowicz, 1988; Schilling, 1991). Hence a lot of effort is put into giving reliable estimates of what the expected change in precipitation will be at these fine scales (Fowler et al., 2007; Kendon et al., 2014; Mayer et al., 2015). Expected changes in precipitation, however, do not translate directly into changes in floods or overflows from structures. To determine these changes, urban hydrological models have to be run, driven by the changed precipitation (Olsson et al., 2009; Willems et al., 2012). By definition, fine-resolution precipitation time series for future climates are not observable, and hence a multitude of statistical approaches have been developed to enable generation of time series with properties that for a large range of metrics have the same characteristics as the expected future precipitation (Willems, 1999; Olsson and Burlando, 2002; Cowpertwait, 2006; Molnar and Burlando, 2008; Burton et al., 2010; Willems et al., 2012; Sørup et al., 2016a).

Expectations to precipitation at event level under climate change are often non-linear with the anticipation that changes in occurrence and size of extreme events will be higher than changes in seasonal or yearly precipitation (Boberg et al., 2010). This is a problem often solved by weather generators or other similar downscaling techniques (Fowler et al., 2007; Burton et al., 2010), but these often have difficulty in presenting realistic time series at the sub-hourly to hourly timescales relevant for urban infrastructure (Segond et al., 2006; Verhoest et al., 2010; Sørup et al., 2016a). Several studies have tested the applicability of Markov models for simulation of high-resolution precipitation series (Srikanthan and McMahon, 1983; Thyregod et al., 1998; Ailliot et al., 2009; Gelati et al., 2010; Sørup et al., 2012). The approach has the advantage that realistic chronology is created in the output. However, for very high resolutions the sensing method of the gauge may have an impact on the signal, giving an upper bound to the temporal resolution of the model, as has been shown for e.g. tipping bucket gauges (Thyregod et al., 1998; Sørup et al., 2012).

In the present study, we develop and demonstrate a novel non-linear methodology that perturbs existing precipitation time series to reflect complex expectations to precipitation in a changed future climate. The method incorporates regional historical knowledge about precipitation through the use of Intensity-Frequency-Duration (IDF) relationships (WMO, 2009) and knowledge about the expected changes in these due to climate change. Thus, the method generates time series for a changed climate which are chronologically identical to the observations used as input, but perturbed to reflect climate change. These series can be used as input for hydraulic or hydrologic models where the climate change effect has to be assessed for all possible rain conditions.

The presented methodology is based on the assumption that precipitation can be scaled according to identified expectations to climate changes. In its simplest form, this assumption is identified as the Delta Change (DC) method (Fowler et al., 2007). The basic assumption is that relative changes in output from climate models might represent expectations to climate change well even though the output itself could be wrongly scaled in absolute values. A more elaborate use of this assumption is provided by Distribution Based Scaling (DBS) presented by Yang et al. (2010). In this approach parameters are derived from regional climate model data to estimate present and future distribution functions for rainfall intensities. The relative change in the distribution parameters is applied to a similar distribution function based on observational data. Thereby, perturbation of rainfall intensities due to climate change relies on the rarity of the individual events and changes markedly from average to extreme events with a high impact on hydrological responses of simulation models (van Roosmalen et al., 2011). Unlike the study by Yang et al. (2010), the expected changes in this study are not calculated directly using the DC method on Regional Climate Model output; they are derived from comprehensive state-of-the-art studies where all available data are used to determine realistic expectations to changes to precipitation due to climate change (e.g. Giorgi, 2006; Kendon et al., 2008; Christensen et al., 2015).

2 Methodology

In urban water management, the relevant time frame to consider is most often that of the rain event (Willems, 1999). The determination of robust IDF relationships for present climate at the relevant timescales is a prerequisite. For developed countries where high-resolution precipitation is generally available, these two prerequisites are very often met (Arnbjerg-Nielsen et al., 2015a), making the methodology generally relevant. The general flow of the methodology is presented in Fig. 1, and how to proceed with each step is presented in the following sections (2.1–2.5).

2.1 Modelling framework

Let us consider a system S that describes precipitation over a time period. The original data are expressed as a time series of precipitation intensity over fixed time steps. This time series alternates between a dry period (no precipitation) and a rainy one. A given event is characterised as dry, extreme or non-extreme with respect to the amount of precipitation during the event.

We denote by E the state space of the system S, with at least three states, i.e. $|E| \geq 3$. Also let D_0, D_1 and D_2 be the non-empty sets of states of dry periods, non-extreme and extreme events, respectively, with $|D_0| = 1$, $|D_1| = d_1 \geq 1$ and $|D_2| = d_2 \geq 1$, i.e. there exists exactly one state for dry periods, $D_0 = D^{\text{dry}}$, but a different number of extreme and non-extreme states (d_1 and d_2, respectively) can be defined for both the non-extreme and extreme events. An extreme event can be further categorised according to the severity of the phenomenon, expressed in terms of the return period of the measured intensity. Non-extreme events can be categorised according to the season in which they appear. Hence, the state space E is partitioned into three disjoint subsets as follows:

$$E = D_0 \cup D_1 \cup D_2, \tag{1}$$

where $D_i \cap D_j = \varnothing, i \neq j, ij \in \{0, 1, 2\}$. We link the non-extreme events to the seasonality of the phenomenon and thus $D_1 = \{D^{\text{winter}}, D^{\text{spring}}, D^{\text{summer}}, D^{\text{autumn}}\}$, that is $d_1 = 4$. D_2 can likewise be partitioned into one or several states appropriate for describing extreme precipitation which may have different return periods or different hydro-climatic origins. In this study, we use a partition based on return periods with $D_2 = \{D^2, D^{10}, D^{100}\}$, referring to states that classify the extremes as either 2-, 10- or 100-year events based on return level.

By definition there is always a dry period between two events, and we assume that there is no dependence between consecutive events. We define the following processes that describe the evolution of a semi-Markov system (Barbu and Limnios, 2008):

Figure 1. Flow diagram showing the general process involved in the presented methodology.

- $J := (J_n)_{n\in N}$ is a Markov chain with state space E, where J_n is the state of the system at the nth event;
- $U := (U_n)_{n\in N}$ is the sequence of jump times between states with state space $\mathbb{N}$ and $U_0 = 0$; and
- $Z := (Z_k)_{k\in N}$ is a discrete-time process with states on E, with Z_k to be the state of the system at a time step k.

The processes J and Z are related through the formula

$$Z_k = J_{N}(k), k \in N, \tag{2}$$

where $N(k)$ is the discrete-time counting process of events in $[1,k] \subset N$, i.e.

$$N(k) := \max\{n \in N : U_n \leq k\}. \tag{3}$$

The corresponding transition matrix of the chain J is very simple. Figure 2b illustrates the evolution of the stochastic system described above.

2.2 Framework for determining state of individual events

There is no unique way to assign a state to an extreme event. In the literature some studies apply hydro-climatic regimes for this classification (Gelati et al., 2010; Svoboda et al., 2016), while others apply event statistics (Madsen et al., 2009; Sørup et al., 2016a). For any given application, the most appropriate classification depends on the data available. In this paper, various methods based on the maximum mean intensities are used to define the event state. For all investigated methods the changes to extremes are evaluated by calculation of IDF curves based on return levels, z_is, at event level for a selection of return periods, I (WMO, 2009). The return period (T) of individual events across all intensities is determined using the median plotting position (Rosbjerg, 1988):

$$T_{\text{median}} = \frac{T_{\text{total}} + 0.4}{\text{rank} - 0.3}, \tag{4}$$

where T_{total} is the length of the time series and rank is the rank number of the individual event.

Using data with observations every minute and a minimum dry weather separation between events of 60 min, the mean maximum intensities over 5, 10, 30, 60, 180, 360 and

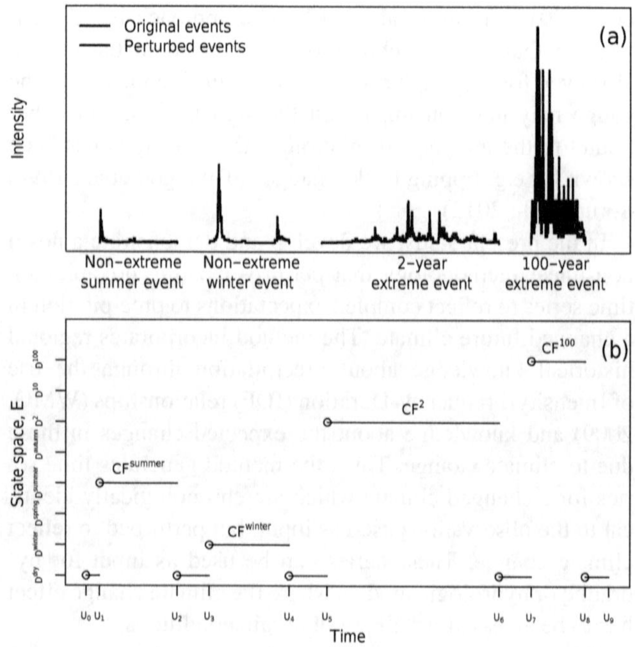

Figure 2. (a) Illustration of the magnitude of perturbation of events for non-extreme summer and winter events as well as 2- and 100-year extreme events, with summer events being perturbed with a factor below 1 and factors for the winter and the extremes being above 1. Factors for extremes are higher than for the winter events, and factors for the very extreme are higher than for the more moderate extreme. **(b)** Illustration of the states associated with the different events if they were to happen in the shown chronology; the dry state, D^{dry}, is present between all wet states.

720 min are calculated for each event. At shorter time frames, e.g. 1 min, the variability of the observed extremes is expected to be very large due to the inherent sampling error (Fankhauser, 1998), and at very long time frames, e.g. 1 day (i.e. 1440 min), the extremes often consist of several events following one another and a different event definition would be necessary to ensure that the real extremes are identified (Madsen et al., 2009). A representative return period for the event is derived based on a mathematical comparison to regional IDF estimates (Fig. 3). This return period is then in turn used to define the state of the event. We test four different selection criteria which define the state of extreme events as either D^2, D^{10} or D^{100}. The selection criteria are listed in Sect. 3.3.

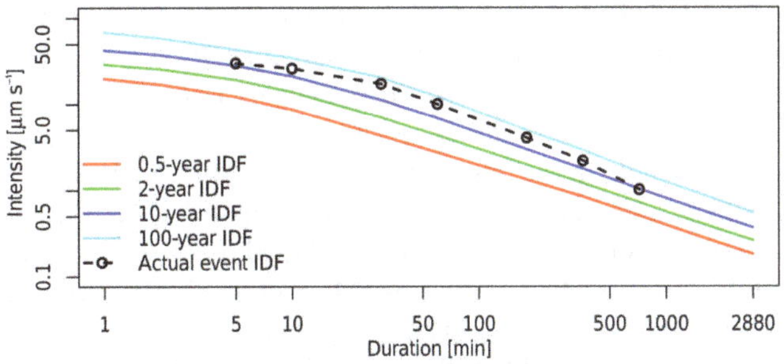

Figure 3. The IDF curve for an extreme event in comparison to the regional IDF curves for 0.5-, 2-, 10- and 100-year return periods, respectively (based on Madsen et al. (2017)).

2.3 Perturbation and change factor

With each event of a time series classified according to a state, the time series can be perturbed using the following methodology linking the time series to the states of the individual events.

Let R_k, $k \in N$, be the precipitation intensity at time step k and $R := (R_k)_{k \in N}$ the corresponding process describing these intensities. The process of perturbed precipitation in each time step k is denoted by $R^* := \left(R_k^*\right)_{k \in N}$.

Similarly to the state space E, we introduce the state space of the change factors, denoted by E_{CF}, $|E_{\mathrm{CF}}| = |E|$. We can then write

$$E_{\mathrm{CF}} = C_0 \cup C_1 \cup C_2, \tag{5}$$

with $|C_0| = 1$, $|C_1| = d_1$ and $|C_2| = d_2$.

We consider the process CF: $= (\mathrm{CF}_n)_{n \in N}$ with state space E_{CF}, where CF_n is the change factor at the nth event. Let $W := (W_k)_{k \in N}$ be the chain, with state space E_{CF}, of change factors in time steps $k \in N$, that is

$$W_k = \mathrm{CF}_{N(k)} \tag{6}$$

with $N(k)$ to be the counting process defined in (Eq. 3). Under the above notation, the original and perturbed sequences of precipitation, R_k and R_k^*, are written as

$$R_k^* = W_k \, R_k. \tag{7}$$

This means that, for a sequence of events, some events will be perturbed more than others, and for extreme cases some might be reduced while others are increased, depending on the local expectations to climate change. Figure 2a shows an example where a non-extreme summer event is perturbed to a lesser volume than the original while a winter non-extreme is increased marginally and both 2- and 100-year extremes are increased considerably more (both in absolute numbers as well as in relative percentages). Figure 2b shows how the state space changes if these four events were to happen chronologically in time with the state jump times marked on the x axis.

2.4 Volume correction based on seasonal dependence of extremes

The extreme part of precipitation is only expected to constitute a smaller fraction of the total precipitation volume on an annual basis (Sørup et al., 2016b), but as extreme precipitation is often associated with a particular season (see e.g. Sørup et al., 2012), the volumetric part of the extremes might be higher for sub-annual considerations. This implies that situations where the expectations for changes to the extremes are very different from the expectations to changes to seasonal precipitation have to be handled through volumetric corrections in order to accommodate the fact that both expectations to changes in extremes and overall seasonal changes are correct. How to do this best will be very much dependent on the local conditions. In our case this is described in Sect. 3.4.

2.5 Evaluation of perturbed time series

The evaluation of the perturbed time series is done against the original time series and against the expected changes.

The average percent-wise difference between the perturbed return levels, $z_{i,j,m}^*$, of the modelled time series, R_k^*, perturbed with the time-dependent change factors, W_k, against the same return levels, $z_{i,j,m}$, of the original time series, R_k, multiplied by the target change factor, CF_e^j, can be defined as

$$\Phi_{i,j,m} = \left(1 - \frac{z_{i,j,m}^*}{z_{i,j,m}\mathrm{CF}_e^j}\right) 100\,\%, \tag{8}$$

across all IDF points, i, all extremity levels and seasonality, j, and all perturbed time series, m. A combined skill score, Φ, across all considered metrics that describe the average deviance from the expectations, can then be defined as

$$\Phi = \sum\nolimits_{i \in I} \sum\nolimits_{j \in J} \sum\nolimits_{m \in M} \frac{\left|1 - \Phi_{i,j,m}\right|}{|I|\,|J|\,|M|}, \tag{9}$$

with $|I|\,|J|\,|M|$ being the product of the total number of IDF points, I, the total number of extreme levels considered plus seasonality, J, and the total number of time series perturbed, M, as a normalization factor.

2.6 Sensitivity analysis

The robustness of the methodology is tested by evaluating its sensitivity to the actual magnitude of the target parameters for both extreme and seasonal changes. Low (L), mean (M) and high (H) scenarios are constructed and paired in all possible combinations to assess both the individual and combined influence of these (Table 1). As this increases the number of scenarios with which the precipitation time series substantially are perturbed, this is not done until after an initial evaluation of the state selection criteria.

Table 1. Tested combinations of extreme and seasonal changes.

Seasonality	Extremes		
	Low expected change	Mean expected change	High expected change
Low expected change	LL	ML	HL
Mean expected change	LM	MM	HM
High expected change	LH	MH	HH

3 Case study: Denmark

To showcase the methodology, it is applied to Danish conditions where the situation is that complex non-linear changes are expected with respect to precipitation in a changed climate.

3.1 Data

3.1.1 Observational data

Precipitation data from the Danish SVK rain gauge network are used in this study (Mikkelsen et al., 1998; Madsen et al., 2002). For this study 10 time series from different parts of Denmark with lengths of approximately 33 years between 1979 and 2012 are used. To distinguish individual events, a dry weather period between individual events of at least 60 min is applied.

3.1.2 IDF curves

For present climate IDF curves are extracted from a regional model for extremes originally developed by Madsen et al. (1998) and updated by Madsen et al. (2009) and Madsen et al. (2017). The IDF curves vary across Denmark, but a single mean regional curve is chosen for this study independently of the location of the gauge considered. Table 2 summarises the IDF values used.

3.1.3 Expectations to climate change

The official recommendations regarding climate change for urban infrastructure in Denmark was determined by Gregersen et al. (2014) on the basis of the ENSEMBLES data set (van der Linden and Mitchell, 2009), with the addition of a few simulations using high-end scenarios. The data set indicates that in general precipitation amounts and intensities will increase and that extremes will increase more than the expected mean increases for Denmark. Furthermore, the results show that it is very likely that increases will be more pronounced for the very rare extremes compared to the more frequent extremes. Table 3 sums up these official expectations for the three return periods that has to be assessed in Danish urban hydrological contexts.

In addition, the Danish Meteorological Institute has published expectations regarding climate change on a seasonal basis (Olesen et al., 2014). The analysis is performed for a range of climate variables and focuses on utilizing the data available in the best possible way to create realistic uncertainty intervals for the expected changes. The estimated change factors for precipitation are based on analysis of the RCP2.6 and RCP8.5 scenarios (Moss et al., 2010), hence a low-end emission scenario and a high-end emission scenario, respectively. Table 4 lists these expectations as well as a simple mean average of the two to represent the mean expected change. To match the change factors for extreme precipitation in Gregersen et al. (2014), which primarily is based on the more average emission A1B scenario (Nakicenovic et al., 2000), simple scaling of the seasonal expectations to a midpoint is applied, as scalability has been shown to be a valid assumption across most scales and most indices (Christensen et al., 2015). The A1B scenario does not lie exactly in the middle between the RCP2.6 and RCP8.5 scenarios, but definitely somewhere between these, and the original estimates from Olesen et al. (2014) are kept as low and high expected changes for the sensitivity analysis.

3.2 Defining states

For Denmark the state space (Eq. 1) is defined with a total of eight states based on the expectations to climate change listed in Tables 3 and 4 with four seasonal states defined for the non-extreme events and three states for the different extreme event levels:

$$E = \{D^{\text{dry}}, D^{\text{winter}}, D^{\text{spring}}, D^{\text{summer}}, D^{\text{autumn}}, D^2, D^{10}, D^{100}\}. \quad (10)$$

Correspondingly the change factors used to perturb the time series are, as a starting point, determined based on the mean expectations listed in Tables 3 and 4.

3.3 Determining state of individual events

For the determination of the state of the individual extreme events four different selection criteria are investigated, with the purpose of defining a representative return period for each event. All points mentioned refer to the return periods of the events intensity points, $\boldsymbol{T} = \{T_5, T_{10}, T_{30}, T_{60}, T_{180}, T_{360}, T_{720}\}$, shown in a situation as depicted in Fig. 3:

Criterion A The maximum return period is used to define the return period of the whole event (based on one point);

$$T_{\text{event}} = T_1^* = \max \boldsymbol{T}. \quad (11)$$

Criterion B The mean of the three largest return periods is used to define the events (based on three points),

Table 2. IDF intensities ($\mu m\, s^{-1}$) for various return periods for Denmark extracted from the model presented by Madsen et al. (2017).

Return period (years)	Duration (min)						
	5	10	30	60	180	360	720
$T = 100$	43.67	34.80	20.63	12.47	5.21	3.11	1.72
$T = 10$	28.62	21.43	11.37	6.95	3.09	1.86	1.09
$T = 2$	19.54	14.08	7.08	4.38	2.04	1.25	0.75
$T = 0.5$	12.40	8.73	4.33	2.75	1.33	0.84	0.51

Table 3. Expected changes in extreme precipitation for Denmark. All values from Table 1 of Gregersen et al. (2014).

Change factor for extreme precipitation (–)	2-year event (CF^2)	10-year event (CF^{10})	100-year event (CF^{100})
Low expected change	1.0	1.0	1.0
Mean expected change	1.2	1.3	1.4
High expected change	1.45	1.7	2.0

$$T_{\text{event}} = \frac{1}{3}\sum\nolimits_{i=1}^{3} T_i^*, \tag{12}$$

where T_2^* and T_3^* are the second and third maxima, respectively, i.e. $T_2^* = \max\{\boldsymbol{T} \setminus T_1^*\}$ and $T_3^* = \max\{\boldsymbol{T} \setminus (T_1^* \cap T_2^*)\}$.

Criterion C The mean of all the return periods is used to define the events (based on all seven points):

$$T_{\text{event}} = \overline{\boldsymbol{T}}. \tag{13}$$

Criterion D A customised step-wise threshold selection criterion is constructed where the event-specific IDF curve is compared to regional IDF levels.

Criterion D is important to test as this allows for construction of a criterion that is closely linked to specific knowledge on the place-specific precipitation dynamics, i.e. for how many duration points on the IDF curve a given return period has to be exceeded for it to be essential for the classification of the event.

Following these selection criteria, four different systems, $S_i, i \in \{A, B, C, D\}$, are constructed and analysed.

Options S_A to S_C are straightforward based on Eqs. (11)–(13), but option S_D is determined specifically for the case study. Table 5 summarises the methodology for option S_D used in this study; specifically it is reflected that for very extreme events, fewer durations have to be extreme for the event as a whole to be considered extreme compared to the more moderate 2-year return level.

3.4 Volume correction based on seasonal dependence of extremes

In previous studies using the SVK data set, it has been shown that

1. the extreme events account for at most 25 % of the total rainwater volume on an annual basis (Sørup et al., 2016b), and
2. the extreme events occur mostly in the summer season (Sørup et al., 2012).

Furthermore, in the summer season the expected seasonal change (−10 %) differs mostly from the expected change in extremes (+20–40 %); see Tables 4 and 3, respectively. Based on this information the seasonal change factor for non-extreme summer events has to be adjusted to reach the overall change factors reported in Table 4. We estimate a partition between non-extreme and extreme events of $\{f_{\text{non-extreme}}, f_{\text{extreme}}\} = \{0.8, 0.2\}$, and the change factor for 2-year events, CF^2, is used to represent the extremes as the largest seasonal volume by far is for the more frequent extremes (Sørup et al., 2016b). In this way the change factor for summer, CF^{summer}, can be adjusted from its value listed in Table 4 (0.9) as

$$CF^{\text{summer}}_{\text{adjusted}} = \frac{CF^{\text{summer}} - CF^2 f_{\text{extreme}}}{f_{\text{non-extreme}}} = \frac{0.9 - 1.2 \cdot 0.2}{0.8} = 0825. \tag{14}$$

In other words the change factors for non-extreme summer events are modified from −10 to −17.5 % in order to compensate for the positive change of +20–40 % to the extremes occurring in the summer period. For the other seasons such an adjustment is not needed.

4 Results

4.1 Evaluation of selection criteria

The 10 time series are perturbed using the four different state selection criteria ($S_A - S_D$) and the evaluation metric is calculated using Eq. (9) with the extreme events having return periods closest to 2, 10 and 100 years (Table 6). Overall, state

Table 4. Expected seasonal changes to precipitation in Denmark based on data from Table 5 of Olesen et al. (2014) and linearly scaled midpoint values.

Change factor for seasonal precipitation (–)	Winter (CF^{winter})	Spring (CF^{spring})	Summer (CF^{summer})	Autumn (CF^{autumn})
Low expected change (RCP2.6)	1.0	1.0	1.0	1.0
Mean expected change	1.1	1.05	0.9	1.05
High expected change (RCP8.5)	1.2	1.1	0.8	1.1

Table 5. Selection criterion S_D for choosing T_{event}s at event level.

A T_{event} is chosen of a	If	Or
2-year event	At least four points from the event have a return period above 0.5 years	At least two points from the event have a return period above 2 years
10-year event	At least three points from the event have a return period above 2 years	At least two points from the event have a return period above 10 years
100-year event	At least three points from the event have a return period above 10 years	At least two points from the event have a return period above 100 years
non-extreme event	None of the above criteria are met	

Table 6. Calculated skill scores, Φ, for the four selection criteria A–D calculated using Eq. (10).

	S_A	S_B	S_C	S_D
Φ	9.3 %	8.5 %	12 %	6.4 %

selection criterion S_D outperforms the other alternatives even though all selection criteria seem reasonable, as all estimated deviances are below 13 % of the expected changes.

In order to study the performance for each state, we construct the skill score variable of Eq. (8) and plot it against the duration for the individual extremes and against months for seasonal precipitation (Fig. 4). Plotted this way 100 % represents a perfect fit, 0 % represents no change and everything positive represents a change in the right direction. For the 2-year return levels both state selection criteria S_B and S_D perform similarly and with a relative change close to 100 %. State selection criterion S_A overestimates the 2-year return level by approximately 10 % on average and state selection criterion S_C underestimates it with a similar magnitude, which still corresponds to a positive change for the events (Fig. 4a). For the 10-year return level, all state selection criteria perform similarly very well (Fig. 4b). When the 100-year return level is evaluated, the reason for criterion S_D's better overall performance becomes clear: it is the only criterion that does not systematically underestimate this return level (Fig. 4c). Even so, all criteria produce results where the direction of change is correct. Given the inherent uncertainty in estimating the actual levels of such events, obtaining close to 85 % of the expected change is considered good. With respect to the seasonal behaviour, all state selection criteria have approximately the same performance at a level close to 100 % (Fig. 4d).

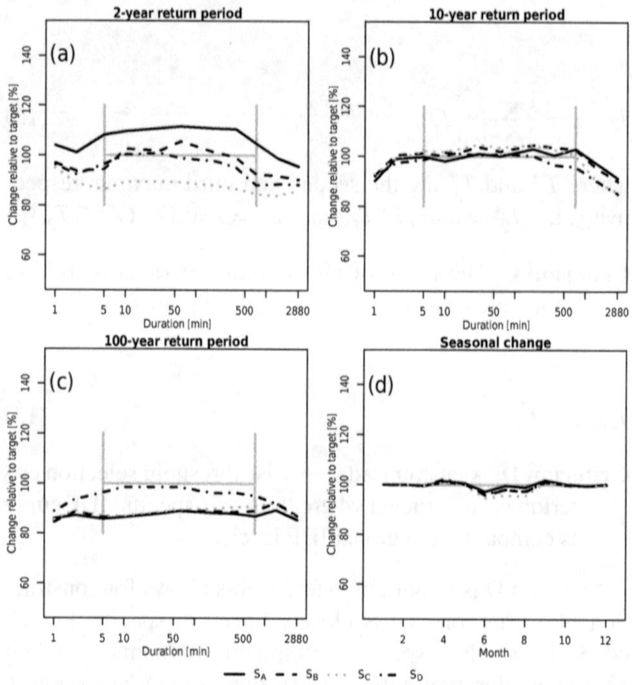

Figure 4. Performance of the different selection criteria, $S_A - S_D$, in producing **(a)** 2-year extremes, **(b)** 10-year extremes, **(c)** 100-year extremes and **(d)** seasonal changes according to the perturbation schemes listed in Tables 3 and 4.

The performance of all the state selection criteria drops when considering durations that are both shorter and longer than the durations used in the state selection methodology

Table 7. Calculated skill scores, Φ, for selection criterion S_D for the nine different sensitivity scenarios listed in Table 1 calculated using Eq. (9).

Φ		Extremes		
		Low	Mean	High
Seasonality	Low	0.0 %	6.0 %	8.6 %
	Mean	1.0 %	6.4 %	8.8 %
	High	1.2 %	6.3 %	8.8 %

(5–720 min). At the minute scale, this is of minor importance, but at 2 days (2880 min) the tendency is very robust across different state selection criteria and extremity levels. This is most likely because these average extreme events are caused by several events with dry periods in between. Hence the individual events are each assessed to be non-extreme and they are adjusted towards lower volumes, even though combined they are rather extreme.

4.2 Sensitivity analysis with selection criterion D

The sensitivity analysis is carried out for the best state selection criterion only, i.e. criterion S_D. The resulting skill scores for the nine individual sensitivity scenarios are listed in Table 7. The highest sensitivity is found when changing between the different extreme precipitation scenarios, with a large increase in the metric when moving from low to mean scenarios and also a notable increase when moving from mean to high scenarios. As such the performance of the methodology drops with the magnitude of the expected changes to extremes, but even for the high extremes the performance is similar to the performance of state selection criteria S_A to S_D in Table 6. The methodology, on the other hand, shows very little sensitivity to the variation in expectations to seasonal changes, not even for the combination where the difference between expectations to seasonal summer precipitation (−20 %) and the extremes become very high.

For all extreme (+45–100 %) indices (Fig. 5a–c), the sensitivity of the expected change in extremes is notable and, especially for the 100-year return level, it is clear that performance drops with increased magnitude of the expected changes to extremes (Fig. 5c), but only to levels comparable to that of the state selection criteria $S_A - S_C$ as shown in Fig. 4. Again, the performance for 2-day events (2880 min) is worse than average, as also seen in Fig. 4. For seasonality (Fig. 5d), the general picture is that the sensitivities of both expectations to seasonality and extremes are of less importance and at a similar level, which in general is a lower level than the one observed for the three extreme indices.

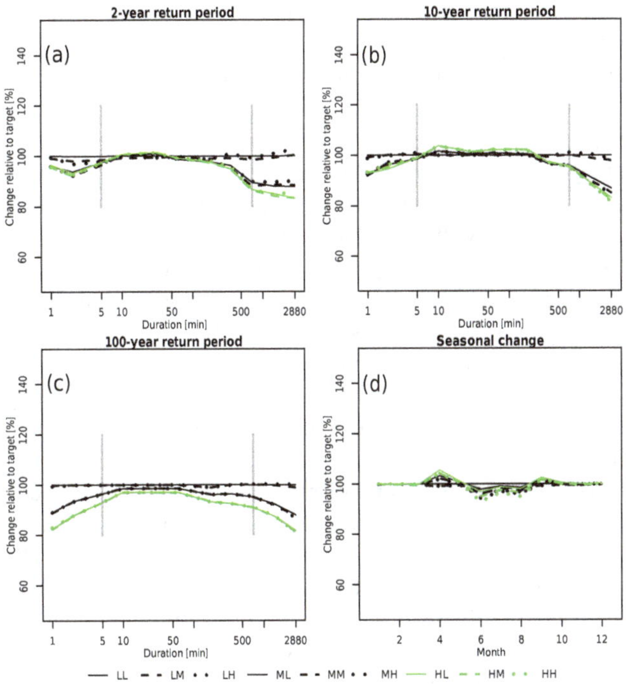

Figure 5. Performance of selection criterion S_D for different parameter values as specified in Table 1 for **(a)** 2-year extremes, **(b)** 10-year extremes, **(c)** 100-year extremes and **(d)** seasonal changes under climate change.

5 Discussion

The proposed framework is very flexible and the separation of dry, non-extreme and extreme weather makes it possible to very effectively perturb time series to reflect different changes in different categories. The presented case study uses eight states to distinguish between different levels of extremes and different seasons and is able to produce time series that satisfactorily represent the expected changes listed in Tables 3 and 4. For other places a different number of states could be relevant and the seasonal partition could be different depending on the local climate and expectation to climate change. The proposed modelling framework fully supports these spatial variations.

Four different state selection criteria over specified event durations are tested in the present study (see Sect. 2.2), as these covered realistic possibilities for the data set used in this study and the focus on urban hydrology. As such, different state selection criteria for different event durations could be relevant in different contexts and could, as illustrated by state selection criterion S_D, be specified as very subjective and case-specific criteria. In this study, the subjective state selection criterion S_D outperforms the other criteria (see Table 6 and Fig. 2), but the superiority is mainly due to its ability to produce the largest changes for the very large, and very uncertain, extreme events. If this part of the evaluation is disregarded, criteria S_B and S_D have a very similar performance, pointing to criterion S_B as being a good onset for investigating data sets where no presumptions exist and no case-specific criterion can be constructed.

All state selection criteria showed a drop in performance for longer duration events than the ones used in the methodology; this is likely due to the used event definition with a minimum of 60 min of dry weather between individual events, which will mean that very long lasting extremes are likely split into several events and therefore not identified as extremes. A different event definition with a longer minimum dry period between events could probably partly solve this, but it would reduce the number of events markedly and increase the chance of small events close to extremes being seen as part of the extreme, with a somewhat false classification as a consequence.

The methodology is relatively sensitive to the magnitude of the perturbation factors (see Sect. 4.2), but the sensitivity is not very dominant and is only at the same size as the sensitivity of the different state selection criteria. Also, the methodology does not address the possibilities of changes to dry spells or changes to the occurrence rate of extremes in general. A future research direction could be to study how the state selection criteria along with the semi-Markov system applied here can be used to generate fully stochastic time series where both the inter-event time and the occurrence probability of the extreme states will be included as criteria that can be changed to meet the expectations to climate change.

6 Conclusions

The proposed methodology is a promising way of creating artificially perturbed precipitation time series, which can represent a changed climate and be used as input in hydrologic and hydraulic models. The methodology perturbs existing time series based on a semi-Markov system where precipitation time series are split into events characterised as dry, extreme or non-extreme. The wet events are divided into different states based on an Intensity-Duration-Frequency relationship based state selection criterion. Of the four tested state selection criteria, the case-specific ones show the best results, but the more general criteria too could be of use when less knowledge about the precipitation regime is available. The sensitivity of the methodology was tested against very different expectations to climate change, both with respect to seasonal changes and changes to extremes, and is generally very robust, also regarding seasons where the general change is negative while the expectation to extremes is positive. The produced time series satisfactorily reproduce changes across all seasons and across all levels of extremes relevant for urban hydrology.

Competing interests. The authors declare that they have no conflict of interest.

Edited by: C. Onof

References

Ailliot, P., Thompson, C., and Thomson, P.: Space-time modelling of precipitation by using a hidden Markov model and censored Gaussian distributions, J. Roy. Stat. Soc. C-App., 58, 405–426, doi:10.1111/j.1467-9876.2008.00654.x, 2009.

Arnbjerg-Nielsen, K., Funder, S. G., and Madsen, H.: Identifying climate analogues for precipitation extremes for Denmark based on RCM simulations from the ENSEMBLES database, Water Sci. Technol., 71, 418–425, doi:10.2166/wst.2015.001, 2015a.

Arnbjerg-Nielsen, K., Leonardsen, L., and Madsen, H.: Evaluating adaptation options for urban flooding based on new high-end emission scenario regional climate model simulations, Clim. Res., 64, 73–84, doi:10.3354/cr01299, 2015b.

Barbu, V. and Limnios, N.: Semi-Markov Chains and Hidden Semi-Markov Models toward Applications: Their Use in Reliability and DNA Analysis, Springer, New York, NY, USA, doi:10.1007/978-0-387-73173-5, 2008.

Berndtsson, R. and Niemczynowicz, J.: Spatial and temporal scales in rainfall analysis: Some aspects and future perspectives, J. Hydrol., 100, 293–313, doi:10.1016/0022-1694(88)90189-8, 1988.

Boberg, F., Berg, P., Thejll, P., Gutowski, W. J., and Christensen, J. H.: Improved confidence in climate change projections of precipitation further evaluated using daily statistics from ENSEMBLES models, Clim. Dynam., 35, 1509–1520, doi:10.1007/s00382-009-0683-8, 2010.

Burton, A., Fowler, H. J., Blenkinsop, S., and Kilsby, C. G.: Downscaling transient climate change using a Neyman-Scott Rectangular Pulses stochastic rainfall model, J. Hydrol., 381, 18–32, doi:10.1016/j.jhydrol.2009.10.031, 2010.

Cowpertwait, P. S. P.: A spatial-temporal point process model of rainfall for the Thames catchment, UK, J. Hydrol., 330, 586–595, doi:10.1016/j.jhydrol.2006.04.043, 2006.

Christensen, O. B., Yang, S., Boberg, F., Maule, C. F., Thejll, P., Olesen, M., Drews, M., Sørup, H. J. D., and Christensen, J. H.:. Scalability of regional climate change in Europe for high-end scenarios, Clim. Res., 64, 25–38, doi:10.3354/cr01286, 2015.

Fankhauser, R.: Influence of systematic errors from tipping bucket rain gauges on recorded rainfall data, Water Sci. Technol., 37, 121–129, doi:10.1016/S0273-1223(98)00324-2, 1998.

Fowler, H. J., Blenkinsop, S., and Tebaldi, C.: Review linking climate change modelling to impacts studies: recent advances in downscaling techniques for hydrological modelling, Int. J. Climatol., 27, 1547–1578, doi:10.1002/joc.1556, 2007.

Gelati, E., Christensen, O. B., Rasmussen, P. F., and Rosbjerg, D.: Downscaling atmospheric patterns to multi-site precipitation amounts in southern Scandinavia, Hydrol. Res., 41, 193–210, doi:10.2166/nh.2010.114, 2010.

Giorgi, F.: Climate change hot-spots, Geophys. Res. Lett., 33, L08707, doi:10.1029/2006GL025734, 2006.

Gregersen, I. B., Madsen, H., Linde, J. J., and Arnbjerg-Nielsen, K.: Opdaterede klimafaktorer og dimensionsgivende regnintensiteter (Updated climate factors and design rain intensities) – Spildevandskomiteen, Skrift nr. 30, The Danish Water and Wastewater Committee under the Danish Engineering Society, Copenhagen, Denmark, https://ida.dk/sites/prod.ida.dk/files/svk_skrift30_0.pdf (last access: 21 December 2016), 2014 (in Danish).

Kendon, E. J., Rowell, D. P., Jones, R. G., and Buonomo, E.: Robustness of future changes in local precipitation extremes, J. Climate, 21, 4280–4297, doi:10.1175/2008JCLI2082.1, 2008.

Kendon, E. J., Roberts, N. M., Fowler, H. J., Roberts, M. J., Chan, S. C., and Senior, C. A.: Heavier summer downpours with climate change revealed by weather forecast resolution model, Nature Climate Change, 4, 570–576, 2014.

Madsen, H., Mikkelsen, P. S., Rosbjerg, D., and Harremoes, P.: Estimation of regional intensity-duration-frequency curves for extreme precipitation, Water Sci. Technol., 37, 29–36, doi:10.1016/s0273-1223(98)00313-8, 1998.

Madsen, H., Mikkelsen, P. S., Rosbjerg, D., and Harremoes, P.: Regional estimation of rainfall intensity-duration-frequency curves using generalized least squares regression of partial duration series statistics. Water Resour. Res., 38, 21-1–21-11, doi:10.1029/2001wr001125, 2002.

Madsen, H., Arnbjerg-Nielsen, K., and Mikkelsen, P. S.: Update of regional intensity-duration-frequency curves in Denmark: Tendency towards increased storm intensities, Atmos. Res., 92, 343–349, 2009.

Madsen, H., Gregersen, I. B., Rosbjerg, D., and Arnbjerg-Nielsen, K.: Regional frequency analysis of short duration rainfall extremes in Denmark from 1979 to 2012, Water Sci. Technol., in review, 2017.

Mayer, S., Maule, C. F., Sobolowski, S., Christensen, O. B., Sørup, H. J. D., Sunyer, M., Arnbjerg-Nielsen, K., and Barstad, I.: Identifying added value in high-resolution climate simulations over Scandinavia, Tellus A, 67, 24941, doi:10.3402/tellusa.v67.24941, 2015.

Mikkelsen, P. S., Madsen, H., Arnbjerg-Nielsen, K., Jørgensen, H. K., Rosbjerg, D., and Harremoës, P.: A rationale for using local and regional point rainfall data for design and analysis of urban storm drainage systems, Water Sci. Technol., 37, 7–14, doi:10.1016/s0273-1223(98)00310-2, 1998.

Molnar, P., and Burlando, P.: Variability in the scale properties of high-resolution precipitation data in the Alpine climate of Switzerland, Water Resour. Res., 44, W10404, doi:10.1029/2007wr006142, 2008.

Moss, R. H., Edmonds, J. A., Hibbard, K. A., Manning, M. R., Rose, S. K., van Vuuren, D. P., Carter, T. R., Emori, S., Kainuma, M., Kram, T., Meehl, G. A., Mitchell, J. F. B., Nakicenovic, N., Riahi, K., Smith, S. J., Stouffer, R. J., Thomson, A. M., Weyant, J. P., and Wilbanks, T. J.: The next generation of scenarios for climate change research and assessment, Nature, 463, 747–756, doi:10.1038/nature08823, 2010.

Nakicenovic, N., Alcamo, J., Davis, J., de Vries, B., Fenhann, J., Gaffin, S., Gregory, K., Grübler, A., Jung, T. Y., Kram, T., Lebre La Rovere, E., Michaelis, L., Mori, S., Morita, T., Pepper, W., Pitcher, H., Price, L., Riahi, K., Roehrl, A., Rogner, H.-H., Sankovski, A., Schlesinger, M., Shukla, P., Smith, S., Swart, R., van Rooijen, S., Victor, N., and Dadi, Z.: Special report on emission scenarios. A special report of Working Group III for the Intergovernmental Panel on Climate Change, Cambridge University Press, New York, 2000.

Olesen, M., Madsen, K. S., Ludwigsen, C. A., Boberg, F., Christensen, T., Cappelen, J., Christensen, O. B., Andersen, K. K., and Christensen, J. H.: Fremtidige klimaforandringer i Danmark (Future climate changes in Denmark), Danmarks Klimacenter rapport nr. 6 2014, Danish Meteorological Institute, Copenhagen, Denmark, https://www.dmi.dk/fileadmin/user_upload/Rapporter/DKC/2014/Klimaforandringer_dmi.pdf (last access: 21 December 2016), 2014 (in Danish).

Olsson, J. and Burlando, P.: Reproduction of temporal scaling by a rectangular pulses rainfall model, Hydrol. Process., 16, 611–630, doi:10.1002/hyp.307, 2002.

Olsson, J., Berggren, K., Olofsson, M., and Viklander, M.: Applying climate model precipitation scenarios for urban hydrological assessment: a case study in Kalmar City, Sweden, Atmos. Res., 92, 364–375, doi:10.1016/j.atmosres.2009.01.015, 2009.

Rosbjerg, D.: Defence of the median plotting position, Progress Report – Institute of Hydrodynamics and Hydraulic Engineering, Technical University of Denmark, 1988.

Schilling, W.: Rainfall data for urban hydrology: what do we need?, Atmos. Res., 27, 5–22, doi:10.1016/0169-8095(91)90003-F, 1991.

Segond, M.-L., Onof, C., and Wheater, H. S.: Spatiat-temporal disaggregation of daily rainfall from a generalized linear model, J. Hydrol., 331, 674–689, doi:10.1016/j.jhydrol.2006.06.019, 2006.

Sørup, H. J. D., Madsen, H., and Arnbjerg-Nielsen, K.: Descriptive and predictive evaluation of high resolution Markov chain precipitation models, Environmetrics, 23, 623–635, doi:10.1002/env.2173, 2012.

Sørup, H. J. D., Christensen, O. B., Arnbjerg-Nielsen, K., and Mikkelsen, P. S.: Downscaling future precipitation extremes to urban hydrology scales using a spatio-temporal Neyman-Scott weather generator, Hydrol. Earth Syst. Sci., 20, 1387–1403, doi:10.5194/hess-20-1387-2016, 2016a.

Sørup, H. J. D., Lerer, S. M., Arnbjerg-Nielsen, K., Mikkelsen, P. S., and Rygaard, M.: Efficiency of stormwater control measures under varying rain conditions: Quantifying the Three Points Approach (3PA), Environ. Sci. Policy, 63, 19–26, doi:10.1016/j.envsci.2016.05.010, 2016b.

Srikanthan, R. and McMahon, T. A.: Sequential generation of short time-interval rainfall data, Nord. Hydrol., 14, 277–306, 1983.

Sunyer, M. A., Madsen, H., Rosbjerg, D., and Arnbjerg-Nielsen, K.: A Bayesian Approach for Uncertainty Quantification of Extreme Precipitation Projections Including Climate Model Interdependency and Nonstationary Bias, J. Climate, 27, 7113–7132, doi:10.1175/JCLI-D-13-00589.1, 2014.

Svoboda, V., Hanel, M., Máca, P., and Kyselý, J.: Projected changes of rainfall event characteristics for the Czech Republic, J. Hydrol. Hydromech., 64, 415–425, doi:10.1515/johh-2016-0036, 2016.

Thyregod, P., Arnbjerg-Nielsen, K., Madsen, H., and Carstensen, N. J.: Modelling the embedded rainfall process using tipping bucket data, Water Sci. Technol., 37, 57–64, doi:10.1016/S0273-1223(98)00316-3, 1998.

van der Linden, P. and Mitchell, J. F.: Ensembles: Climate change and its impacts: Summary of research and results from the ensembles project, Technical Report, Met Office Hadley Centre, Exeter, UK, 2009.

van Roosmalen, L., Sonnenborg, T. O., Jensen, K. H., and Christensen, J. H.: Comparison of Hydrological Simulations of Climate Change Using Perturbation of Observations and Distribution-Based Scaling, Vadose Zone J., 10, 136–150, doi:10.2136/vzj2010.0112, 2011.

Verhoest, N. E. C., Vandenberghe, S., Cabus, P., Onof, C., Meca-

Figueras, T., and Jameleddine, S.: Are stochastic point rainfall models able to preserve extreme flood statistics?, Hydrol. Process., 24, 3439–3445, doi:10.1002/hyp.7867, 2010.

Willems, P.: Stochastic generation of spatial rainfall for urban drainage areas, Water Sci. Technol., 39, 23–30, doi:10.1016/s0273-1223(99)00212-7, 1999.

Willems, P., Arnbjerg-Nielsen, K., Olsson, J., and Nguyen, V.-T.-V.: Climate change impact assessment on urban rainfall extremes and urban drainage: methods and shortcomings, Atmos. Res., 103, 106–118, doi:10.1016/j.atmosres.2011.04.003, 2012.

WMO: Guide to hydrological practices. Volume II: Management of Water Recourses and Application of hydrological practices, WMO report 168, 6th edn., World Meteorological Organization, Geneva, Switzerland, p. 302, 2009.

Yang, W., Andreasson, J., Graham, L. P., Olsson, J., Rosberg, J., and Wetterhall, F.: Distribution-based scaling to improve usability of regional climate model projections for hydrological climate change impacts studies, Hydrol. Res., 40, 211–229, doi:10.2166/nh.2010.004, 2010.

7

Analyzing the future climate change of Upper Blue Nile River basin using statistical downscaling techniques

Dagnenet Fenta Mekonnen[1,2] **and Markus Disse**[1]

[1]Chair of Hydrology and River Basin Management, Faculty of Civil, Geo and Environmental Engineering, Technische Universität München, Arcisstrasse 21, 80333 Munich, Germany

[2]Amhara National Regional State Water, Irrigation and Energy Development Bureau, Bahirdar, Ethiopia

Correspondence: Dagnenet Fenta Mekonnen (dagnfenta@yahoo.com)

Abstract. Climate change is becoming one of the most threatening issues for the world today in terms of its global context and its response to environmental and socioeconomic drivers. However, large uncertainties between different general circulation models (GCMs) and coarse spatial resolutions make it difficult to use the outputs of GCMs directly, especially for sustainable water management at regional scale, which introduces the need for downscaling techniques using a multimodel approach. This study aims (i) to evaluate the comparative performance of two widely used statistical downscaling techniques, namely the Long Ashton Research Station Weather Generator (LARS-WG) and the Statistical Downscaling Model (SDSM), and (ii) to downscale future climate scenarios of precipitation, maximum temperature (T_{max}) and minimum temperature (T_{min}) of the Upper Blue Nile River basin at finer spatial and temporal scales to suit further hydrological impact studies. The calibration and validation result illustrates that both downscaling techniques (LARS-WG and SDSM) have shown comparable and good ability to simulate the current local climate variables. Further quantitative and qualitative comparative performance evaluation was done by equally weighted and varying weights of statistical indexes for precipitation only. The evaluation result showed that SDSM using the canESM2 CMIP5 GCM was able to reproduce more accurate long-term mean monthly precipitation but LARS-WG performed best in capturing the extreme events and distribution of daily precipitation in the whole data range.

Six selected multimodel CMIP3 GCMs, namely HadCM3, GFDL-CM2.1, ECHAM5-OM, CCSM3, MRI-CGCM2.3.2 and CSIRO-MK3 GCMs, were used for downscaling climate scenarios by the LARS-WG model. The result from the ensemble mean of the six GCM showed an increasing trend for precipitation, T_{max} and T_{min}. The relative change in precipitation ranged from 1.0 to 14.4 % while the change for mean annual T_{max} may increase from 0.4 to 4.3 °C and the change for mean annual T_{min} may increase from 0.3 to 4.1 °C. The individual result of the HadCM3 GCM has a good agreement with the ensemble mean result. HadCM3 from CMIP3 using A2a and B2a scenarios and canESM2 from CMIP5 GCMs under RCP2.6, RCP4.5 and RCP8.5 scenarios were downscaled by SDSM. The result from the two GCMs under five different scenarios agrees with the increasing direction of three climate variables (precipitation, T_{max} and T_{min}). The relative change of the downscaled mean annual precipitation ranges from 2.1 to 43.8 % while the change for mean annual T_{max} and T_{min} may increase in the range from 0.4 to 2.9 °C and from 0.3 to 1.6 °C respectively.

1 Introduction

The impacts of climate change on the hydrological cycle in general and on water resources in particular are of high significance due to the fact that all natural and socioeconomic systems critically depend on water. The direct impact of climate change can be variation and changing pattern of water resources availability and hydrological extreme events such as floods and droughts, with many indirect effects on agriculture, food and energy production and overall water infrastructure (Ebrahim et al., 2013). The impact may be worse on transboundary rivers like the Upper Blue Nile River where competition for water is becoming high from different economic, political and social interests of the riparian countries and when runoff variability of upstream countries can greatly affect the downstream countries (Kim, 2008; Semenov and Barrow, 1997).

According to IPCC (2007), between 75 and 250 million people are projected to be exposed to increased water stress due to climate change in Africa by 2020. The increasing water demand of upstream countries in the Nile Basin coupled with climate change impacts can affect the availability of water resources for downstream countries and in the basin, which could result in resource conflicts and regional insecurities. Moreover, climate variability (i.e., the way climate fluctuates yearly and seasonally above or below a long-term average value, caused by changes in forcing factors such as variation in seasonal extent of the intertropical convergence zone like El Niño and La Niña events) is already imposing a significant challenge to Ethiopia by affecting food security, water and energy supply, poverty reduction, and sustainable socioeconomic development efforts. To mitigate these challenges, the Ethiopian government therefore carried out a series of studies on Upper Blue Nile River basin (UBNRB), which has been identified as an economic "growth corridor", focused on identifying irrigation and hydropower potential of the basin (BCEOM, 1998; USBR, 1964; WAPCOS, 1990). As a result, large-scale irrigation and hydropower projects including the Grand Ethiopian Renaissance Dam (GERD), the largest hydroelectric power plant in Africa, have been identified and are being constructed as mitigation measures for the impacts of climate change. However, most studies have put less emphasis on climate change and its impact on the hydrology of the basin, and hence identifying local impacts of climate change at basin level is quite important. This is especially important in the UBNRB for the sustainability of large-scale water resource development projects, for proper water resource management leading to regional security and for finding possible mitigation measures to avoid catastrophic consequences.

To this end, several individual studies have been done to study the impacts of climate change on the water resources of Upper Blue Nile River basin. Taye et al. (2011) reviewed some of the research outputs and concluded that clear discrepancies were observed, particularly on the projection of precipitation. For instance, as the results obtained from Bewket and Conway (2007), Conway (2000) and Gebremicael et al. (2013) showed, there is no significant trend observed in the amount of seasonal and annual rainfall, while Mengistu et al. (2014) reported statistically nonsignificant increasing trends in annual and seasonal rainfall. For the future projection, expected changes in precipitation amount are unclear. For instance, Kim (2008) used the outputs of six GCMs for the projection of future precipitations and temperature, and the result suggested that the changes in mean annual precipitation from the six GCMs range from −11 to 44 % with a change of 11 % from the weighted average scenario in the 2050s. However, the changes in mean annual temperature range from 1.4 to 2.6 °C with a change of 2.3 °C from the weighted average scenario. Likewise, Yates and Strzepek (1998a) used three GCMs and the result revealed that the changes in precipitation range from −5 to 30 % and the changes in temperature range from 2.2 to 3.5 °C. Yates and Strzepek (1998b) also used six GCMs and the result showed in the range from −9 to 55 % for precipitation while temperature increased from 2.2 to 3.7 °C. Another study done by Elshamy et al. (2009), used 17 GCMs and the result showed that changes in total annual precipitation range between −15 to 14 % but the ensemble mean of all models showed almost no change in the annual total rainfall, while all models predict the temperature to increase between 2 and 5 °C. Gebre and Ludwig (2015) used five bias-corrected 50 km × 50 km spatial resolution GCMs for RCP4.5 and RCP8.5 scenarios to downscale the future climate change of four watersheds (Gilgel Abay, Gumara, Ribb and Megech) located in Tana subbasin for the time period of 2030s and 2050s (the time periods of 2011–2040, 2041–2070 and 2071–2100 are referred to throughout as the 2030s, 2050s and 2080s, respectively). The result suggested that the selected five GCMs disagree on the direction of future prediction of precipitation, but multimodal average monthly and seasonal precipitation may generally increase over the watersheds.

For the historical context, the discrepancies could be due to the period and length of data analyzed and the failure to consider stations which can represent the spatial variability of the basin and also errors induced from observed data. For the future context, besides the abovementioned reasons, discrepancies could be due to the differences in GCMs and scenarios used for downscaling, the downscaling techniques applied (can be dynamical and statistical), selection of representative predictors, the period of analysis, and spatial and temporal resolution of observed and predictor data sets.

To address uncertainty in projected climate changes, the IPCC (2014) thus recommends using a large ensemble of climate change scenarios produced from various combinations of atmospheric ocean general circulation models (AOGCMs) and forcing scenarios. However, it can become prohibitively time consuming to assess the climate change using many climate change scenarios and many statistical downscaling models simultaneously. As a result, researchers typically assess climate change and its impacts under only one or a few climate change scenarios, selected arbitrarily with no justification, for instance those that used only A1B and A2 scenarios. Yet there is no any hard rule to select an appropriate subset of climate change scenarios among the wide range of possibilities (Casajus et al., 2016).

GCMs perform reasonably well at larger spatial scales but poorly at finer spatial and temporal scales, especially precipitation, which is of interest in hydrological impact analysis (Goly et al., 2014). Hence, the processes of downscaling that ensure that the scale discrepancy between the coarse-scale GCMs and the required local-scale climate variables for hy-

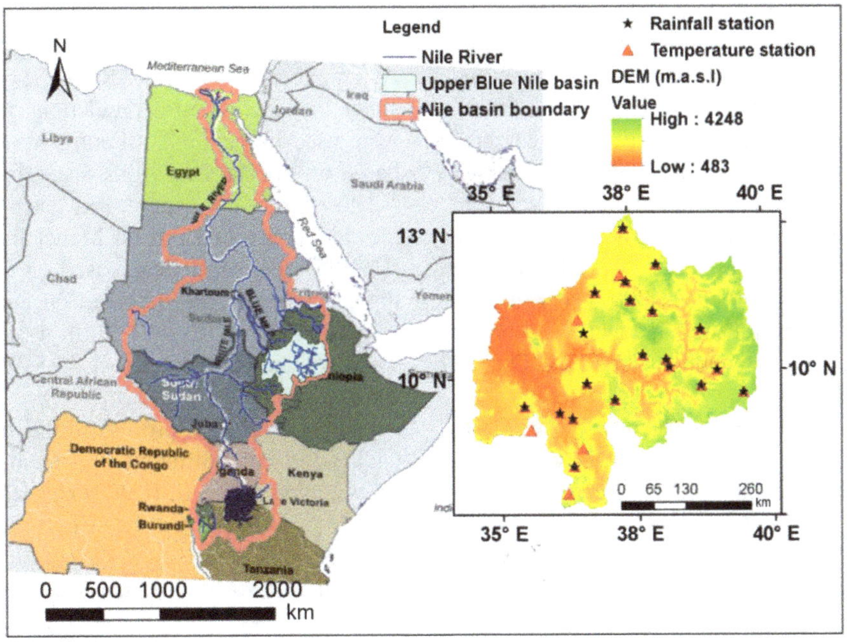

Figure 1. Location map of the study area.

drological models will be narrowed down should be investigated for their contribution, which has often been overlooked in previous studies on climate change analysis in the UBNRB. Many researchers have tried to compare the comparative skill of downscaling methods in different study areas (Dibike and Coulibaly, 2005; Ebrahim et al., 2013; Fiseha et al., 2012; Goodarzi et al., 2015; Hashmi et al., 2011; Khan et al., 2006; Qian et al., 2004; Wilby et al., 2004; Wilby and Wigley, 1997; Xu, 1999). However, no single model has been found to perform well over all the regions and timescales. Thus, evaluations of different models is critical to understanding the applicability of the existing models.

Apart from the GCMs and downscaling techniques, most of the previous studies (e.g., Beyene et al. (2010), Elshamy et al. (2009) and Kim (2008)) used CRU, NFS and other gridded data sets constructed based on the interpolation of a few stations in Ethiopia, which are relatively less accurate compared with the station-based data (Worqlul et al., 2014). Therefore, the objective of this study is (i) to evaluate the comparative performance of two widely used statistical downscaling techniques, namely the Long Ashton Research Station Weather Generator (LARS-WG) and the Statistical Downscaling Model (SDSM) over the UBNRB, and (ii) to downscale future climate scenarios of precipitation, maximum temperature (T_{max}) and minimum temperature (T_{min}) at acceptable spatial and temporal resolution, which can be used directly for further hydrological impact studies. This can be achieved through applying a multimodel approach, to minimize the uncertainty of GCMs and incorporate acceptable number of weather stations which have long time series and reliable observed climate data to minimize the errors coming from the less accurate gridded data sets.

Generally, downscaling methods are classified into dynamic and statistical downscaling (Fowler et al., 2007; Wilby et al., 2002). Dynamic downscaling nests higher-resolution regional climate models (RCMs) into coarse-resolution GCMs to produce complete set of meteorological variables which are consistent with each other. The outputs from this method are still not at the scale that the hydrological model requires. Statistical downscaling overcomes this challenge; moreover it is computationally undemanding, simple to apply and provides the possibility of uncertainty analysis (Dibike et al., 2005; Semenov et al., 1997; Wilby et al., 2002). Extensive details on the strength and weakness of the two methods can be found in Wilby et al. (2007, 1997). Among the different possibilities, two well-recognized statistical downscaling tools, a regression-based SDSM (Wilby et al., 2002) and a stochastic weather generator, LARS-WG (Semenov et al., 1997, 2002) were chosen for this study. They have been tested in various regions (e.g., Chen et al., 2013; Dibike et al., 2005; Dile et al., 2013; Elshamy et al., 2009; Fiseha et al., 2012; Hashmi et al., 2011; Hassan et al., 2014; Maurer and Hidalgo, 2008; Yimer et al., 2009) under different climatic conditions of the world.

2 Description of study area

The Upper Blue Nile River basin extends from 7°45′ to 13° N and 34°30′ and 37°45′ E; see Fig. 1. It is one of the most important major basins of Ethiopia because it contributes to

45 % of the countries surface water resources, 20 % of the population, 17 % of the landmass, 40 % of the nation's agricultural product and a large portion of the hydropower and irrigation potential of the country (Elshamy et al., 2009). The whole UBNRB has an area coverage of 199 812 km^2 (BCEOM, 1998). For this study, Rahad, Gelegu and Dinder subcatchments that do not flow through the main river stem to the Republic of Sudan are excluded. Hence, the basin area coverage is 176 000 km^2 which is about 15 % of total area of 1.12 million km^2 (Awulachew et al., 2007) of Ethiopia. The elevation ranges between 489 m a.s.l. downstream on the western side to 4261 m a.s.l. upstream at Mount Ras Dashen in the northeastern part.

The Upper Blue Nile River itself has an average annual runoff of about 49 billion m^3. In addition, the Dinder, Galegu and Rahad rivers have a combined annual runoff of about 5 billion m^3. The rivers of the UBNRB contribute on average about 62 % of the Nile total at Aswan. Together with the contributions of the Baro-Akobo and Tekeze rivers, Ethiopia accounts for 86 % of runoff at Aswan (BCEOM, 1998). The climate of Ethiopia is mainly controlled by the seasonal migration of the intertropical convergence zone, following the position of the sun relative to the earth and the associated atmospheric circulation. It is also highly influenced by the complex topography. The whole UBNRB has long-term mean annual rainfall, minimum temperature and maximum temperature of 1452 mm yr^{-1}, 11.4 and 24.7 °C respectively, as calculated by this study from 15 rainfall and 26 temperature gauging stations from the period 1984–2011. The mean seasonal rainfall based on the above data showed that about 238, 1065 and 148 mm occurred in Belg (October–January), Kiremit (June–September) and Bega (February–May) respectively, in which about 74 % of rainfall is concentrated between June and September (Kiremit season).

3 Data sets

3.1 Local data sets

The historical precipitation, maximum temperature and minimum temperature data for the study area were obtained from the Ethiopian Meteorological Agency (EMA), which were analyzed and checked for further quality control. A considerable length of time series data were missed in almost all available stations, and hence 15 rainfall and 25 temperature stations which have long time series and relatively short time missing records were selected. Filling missed or gap records was the first task for further meteorological data analysis. This task was done using the well-known methodology of the inverse distance weighting method. To check the quality of the data, the double mass curve (DMC) analysis was used. DMC analysis is a cross correlation between the accumulated totals of the gauge in question against the corresponding totals for a representative group of nearby gauges.

3.2 Large-scale data sets

A new version of the LARS-WG5.5 was applied for this study that incorporates predictions from 15 GCMs which were used in the IPCC's Fourth Assessment Report (AR4) based on Special Emissions Scenarios SRES B1, A1B and A2 for three time windows as listed in Table 1. However, the fifth phase of Coupled Model Intercomparison Project (CMIP5) climate models, based on the new radiative forcing scenarios (Representative Concentration Pathway, RCP) which were used for IPCC Fifth Assessment Report (AR5), were not incorporated into LARS-WG at the time of the study.

As it is difficult to process all the incorporated 15 CMIP3 GCMs and large differences in predictions of climate variables among the GCMs are expected, the performance of GCMs in simulating the current climate variables of the study area (UBNRB) should be evaluated, and the best-performing GCMs were selected. The MAGICC/SCEGEN computer program tool was used for the performance evaluation of the 15 GCMs found in the LARS WG5.5 database, as it is a standard method for selecting models on the basis of their ability to accurately represent current climate, either for a particular region or for the globe. In this study, we used a semiquantitative skill score that rewards relatively good models and penalizes relatively bad models as suggested by the Wigley (2008) user manual. The statistics used for model selection are pattern correlation (R^2), Root mean square error (RMSE), bias (B), and a bias-corrected RMSE (RMSE-corr). The analysis was done separately for precipitation and temperature and finally an average score value was taken for model selection. The six best-performing GCMs have been selected for this study, namely HadCM3, GFDL-CM2.1, ECHAM5-OM, CCSM3, MRI-CGCM2.3.2 and CSIRO-MK3 in the order of their performance to construct future precipitation, maximum temperature and minimum temperature in the UBNRB for the time periods of the 2030s, 2050s and 2080s under A1B, A2 and B1 scenarios; see Table 1.

Moreover, atmospheric large-scale predictor variables used for representing the present condition were obtained from the National Centre for Environmental Prediction (NCEP) reanalysis data set. CanESM2, a second-generation Canadian earth system model (ESM) developed by Canadian Centre for Climate Modelling and Analysis (CCCma) of Environment Canada that represents CMIP5 and HadCM3 outputs from the Hadley Centre, United Kingdom (UK) representing CMIP3 were used in SDSM for the construction of daily local meteorological variables corresponding to their future climate scenario.

The reason for selecting these two GCMs was that they are models that made daily predictor variables freely available to be directly fed into the SDSM, covering the study area with a better resolution. Additionally, HadCM3 is the most-used GCM in previous studies such as Dibike et al. (2005), Dile

Table 1. Selected global climate models from IPCC AR4 incorporated into LARS-WG.

Research center	Country	GCM	Model acronym	Grid resolution	Emission scenarios	Time periods
Common Wealth Scientific and Industrial Research Organization	Australia	CSIRO-MK3	CSMK3	1.9° × 1.9°	A1B, B1	B, T1, T2, T3
Max-Plank Institute for Meteorology	Germany	ECHAM5-OM	MPEH5	1.9° × 1.9°	A1B, A2, B1	B, T1, T2, T3
National Institute for Environmental Studies	Japan	MRI-CGCM2.3.	MIHR	2.8° × 2.8°	A1B, B1	B, T1, T2, T3
UK Meteorological Office	UK	HadCM3	HADCM3	2.5° × 3.75°	A1B, A2, B1	B, T1, T2, T3
Geophysical Fluid Dynamics Lab	USA	GFDL-CM2.1	GFCM21	2° × 2.5°	A1B, A2, B1	B, T1, T2, T3
National Centre for Atmospheric Research	USA	CCSM3	NCCCS	1.4° × 1.4°	A1B, B1	B, T1, T2, T3

B: baseline; T1: 2011–2030; T2: 2046–2065; T3: 2081–2100

Table 2. Name and description of all NCEP predictors on HadCM3 and canESM2 grid.

Variables	Descriptions	Variables	Descriptions
temp	Mean temperature at 2 m	s500+	Specific humidity at 500 hPa height
mslp	Mean sea level pressure	s850+	Specific humidity at 850 hPa height
p500	500 hPa geopotential height	[b]_f	Geostrophic air flow velocity
p850	850 hPa geopotential height	[b]_z	Vorticity
rhum[a]	Near-surface relative humidity	[b]_u	Zonal velocity component
r500[a]	Relative humidity at 500 hPa	[b]_v	Meridional velocity component
r850[a]	Relative humidity at 850 hPa	[b]zh	Divergence
shum	Near-surface specific humidity	[b]thas	Wind direction
Prec+	Total precipitation		

[a] Refers to predictors only found from HadCM3. [b] Refers to different atmospheric levels: the surface (p_), 850 hPa height (p8) and 500 hPa height (p5). + Refers predictors only for canESM2.

et al. (2013), Hassan et al. (2014) and Yimer et al. (2009), and HadCM3 ranked first in performance evaluation done by MAGICC/SCEGEN computer program tools and its downscaled results match with the ensemble mean of the six GCMs used in the LARS-WG model. Furthermore, they can represent two different scenario generations describing the amount of greenhouse gases (GHGs) in the atmosphere in the future. HadCM3 GCM used emission scenarios of A2 (separated world scenario), in which the CO_2 concentration was projected to be 414, 545 and 754 ppm, and B2 (the world of technological inequalities), where the CO_2 concentration was expected to be 406, 486 and 581 ppm at the time periods of the 2020s, 2050s and 2080s respectively (Semenov and Stratonovitch, 2010) that were used in the CMIP3 for the IPCC's AR4 (IPCC, 2007). The canESM2 GCM represents the latest plausible radiative forcing scenarios, a wide range which includes a very low forcing level (RCP2.6), where radiative forcing peaks at approximately $3\,W\,m^{-2}$, peaks approximately 490 ppm CO_2 equivalent before 2100, and then declines to $2.6\,W\,m^{-2}$; two medium stabilization scenarios were used for the IPCC's AR5, RCP6 and RCP 4.5, in which radiative forcing is stabilized at $6\,W\,m^{-2}$ (approximately 850 ppm CO_2 equivalent) and $4.5\,W\,m^{-2}$ (approximately 650 ppm CO_2 equivalent) after 2100 respectively, and one very high baseline emission scenario (RCP8.5) was used for which radiative forcing reaches $> 8.5\,W\,m^{-2}$ (1370 ppm CO_2 equivalent) by 2100 and continues to rise for some time (IPCC, 2014).

The NCEP data set was normalized over the complete 1961–1990 period data, and interpolated to the same grid as HadCM3 (2.5° latitude × 3.75° longitude) and canESM2 (2.8125° latitude × 2.8125° longitude) from its horizontal resolution of (2.5° latitude × 2.5° longitude), to represent the current climate conditions. NCEP reanalysis data were normalized and interpolated as follows (Hassan et al., 2014):

$$u_n = \frac{(u_t - u_a)}{\sigma u}, \tag{1}$$

in which u_n is the normalized atmospheric variable at time t, u_t is the original data at time t, u_a is the multiyear average during the period, and σu is the standard deviation.

The canESM2 outputs for three different climate scenarios are RCP 2.6, RCP 4.5 and RCP 8.5 for the period 2006–2100, while the outputs of HadCM3 for A2a (medium–high) and B2a (medium–low) emission scenarios for the period 1961–2099 were obtained on a grid-by-grid-box basis for the study area from the Environment Canada website http://ccds-dscc.ec.gc.ca/index.php?page=dst-sdi (the "a" in A2a and B2a refers the ensemble member in the HadCM3 A2 and B2 experiments). The archive of canESM2 and HadCM3 GCM output contains 26 daily predictor variables, each listed in Table 2.

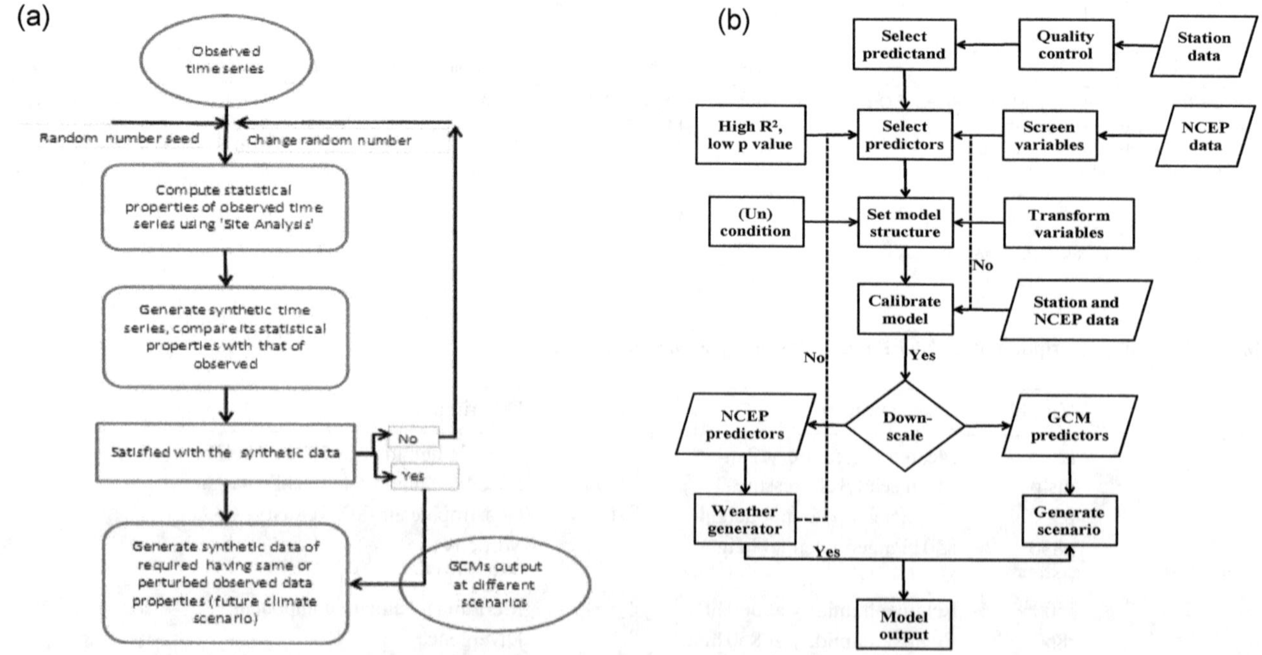

Figure 2. Schematic diagram of **(a)** LARS WG analysis and **(b)** SDSM analysis. Source: Wilby et al. (2002).

4 Methodology

4.1 Description of LARS-WG model

LARS-WG is a stochastic weather generator which can be used for the simulation of weather data at a single station under both current and future climate conditions. These data are in the form of daily time series for a group of climate variables, namely precipitation, maximum temperature and minimum temperature, and solar radiation (Chen et al., 2013; Semenov et al., 1997). LARS-WG uses a semiempirical distribution (SED) that is defined as the cumulative probability distribution function (CDF) to approximate probability distributions of dry and wet series, daily precipitation, minimum temperatures and maximum temperatures.

$$\text{EPM} = \{a_0, a_i, h_i, i = 0, \ldots, 23\} \tag{2}$$

EPM is a histogram of the distribution of 23 different intervals (a_{i-1}, a_i) where $a_{i-1} < a_i$ (Semenov et al., 2002), which offers a more accurate representation of the observed distribution compared with the 10 used in the previous version. By perturbing parameters of distributions for a site with the predicted changes of climate derived from global or regional climate models, a daily climate scenario for this site could be generated and used in conjunction with a process-based impact model for assessment of impacts. In general, the process of generating synthetic weather data can be categorized into three distinct steps: model calibration, model validation and scenario generation as represented in Fig. 2a, which are briefly described by (Semenov et al., 2002) as follows.

The inputs to LARS-WG are the series of daily observed data (precipitation, minimum temperature and maximum temperature) of the base period (1984–2011) and site information (latitude, longitude and altitude). After the input data preparation and quality control, the observed daily weather data at a given site were used to determine a set of parameters for probability distributions of weather variables. These parameters are used to generate a synthetic weather time series of arbitrary length by randomly selecting values from the appropriate distributions, with the same statistical characteristics as the original observed data but differing on a day-to-day basis. The LARS-WG distinguishes wet days from dry days based on whether the precipitation is greater than zero. The occurrence of precipitation is modeled by alternating wet and dry series approximated by semiempirical probability distributions. The statistical characteristics of the observed and synthetic weather data during calibration of the model are analyzed to determine if there are any statistically significant differences using a chi-square goodness-of-fit test (Kolmogorov–Smirnov, KS) and the means and standard deviation using t and F test respectively. This can be done by changing the parameters of LARS-WG (number of years and seed number).

To generate climate scenarios at a site for a certain future period with a selected emission scenario, the LARS-WG baseline parameters, which are calculated from observed weather for a baseline period (1984–2011), are adjusted by the Δ-changes for the future period and the emissions predicted by a GCM for each climatic variable for the grid covering the site. In this study, the local-scale climate scenarios based on the SRES A2, A1B and B1 scenarios simulated by the selected six GCMs are generated for the time periods of 2011–2030, 2046–2065 and 2080–2099 to predict the future change of precipitation and temperature in UBNRB.

Δ-changes were calculated as relative changes for precipitation and absolute changes for minimum and maximum temperatures (Eqs. 3 and 4) respectively. No adjustments for distributions of dry and wet series and temperature variability were made, because this would require daily output from the GCMs which is not readily available from the LARS-WG data set (Semenov et al., 2010).

$$\Delta T_i = (\overline{T}_{\mathrm{GCM,FUT},i} - \overline{T}_{\mathrm{synt,Base},i}) \tag{3}$$

$$\Delta P_i = \left(\frac{\overline{P}_{\mathrm{GCM,FUT},i}}{\overline{P}_{\mathrm{synt,Base},i}}\right) \tag{4}$$

In the above equations, ΔT_i and ΔP_i are climate change scenarios of the temperature and precipitation, respectively, for long-term averages for each month ($1 \le i \le 12$); $\overline{T}_{\mathrm{GCM,FUT},i}$ and $\overline{P}_{\mathrm{GCM,FUT},i}$ are the long-term average temperature and precipitation respectively, simulated by the GCM in the future periods per month for three time periods, and $\overline{T}_{\mathrm{Synth,Base},i} and \overline{P}_{\mathrm{Synth,Base},i}$ are the long-term average temperature and precipitation respectively, simulated by the model in the period similar to observation period (in this study 1984–2011) for each month. For obtaining time series of future climate scenarios, climate change scenarios are added to the observed values by employing the change factor (CF) method (Eqs. 5 and 6) (in this study 1984–2011).

$$T = T_{\mathrm{obs}} + \Delta T \tag{5}$$

$$P = P_{\mathrm{obs}} + \Delta P \tag{6}$$

T and P are time series of the future climate scenarios of temperature and precipitation (2011–2100), and T_{obs} and P_{obs} are observed temperature and precipitation. So, in LARS-WG downscaling, unlike in SDSM, large-scale atmospheric variables are not directly used in the model, but rather are based on the relative mean monthly changes between current and future periods predicted by a GCM. Local station climate variables are adjusted proportionately to represent climate change (Dibike et al., 2005).

4.2 Description of SDSM

The SDSM is best described as a hybrid of the stochastic weather generator and regression based on the family of transfer function methods, due to the fact that a multiple linear regression model is developed between a few selected large-scale predictor variables (Table 2) and local-scale predictands such as temperature and precipitation in order to condition local-scale weather parameters from large-scale circulation patterns. The stochastic component of SDSM enables the generation of multiple simulations with slightly different time series attributes, but the same overall statistical properties (Wilby et al., 2002). It requires two types of daily data: the first type corresponds to local predictands of interest (e.g., temperature, precipitation) and the second type corresponds to the data of large-scale predictors (NCEP and GCM) of a grid box closest to the station.

The SDSM model categorizes the task of downscaling into a series of discrete processes such as quality control and data transformation, screening of predictor variables, model calibration, and weather and scenario generation as shown in Fig. 2b. Detail procedures and steps can be found in Wilby et al. (2002) for further reading. Screening potentially useful predictor–predictand relationships for model calibration is one of the most challenging but very crucial stages in the development of any statistical downscaling model. It is because of the fact that the selection of appropriate predictor variables largely determines the success of SDSM and also the character of the downscaled climate scenario (Wilby et al., 2007). After routine screening procedures, the predictor variables that provide physically sensible meaning in terms of their high explained variance, correlation coefficient (r) and the magnitude of their probability (p value) were selected.

The model calibration process in SDSM was used to construct downscaled data based on multiple regression equations given daily weather data (predictand) and the selected predictor variables at each station. The model was structured as a monthly model for both daily precipitation and temperature using the same set of the selected NCEP predictors for the calibration period. Hence, 12 regression equations were developed for 12 months. Bias correction and variance inflation factor were adjusted until the model replicated the observed data. Model validation was carried out by testing the model using an independent data set. To compare the observed and simulated data, SDSM has provided a summary statistics function that summarizes the result of both the observed and simulated data. Time series of station data and large-scale predictor variables (NCEP reanalysis data) were divided into two groups: for the periods 1984–1995 (1984–2000) and 1996–2001 (2001–2005) for model calibration and validation of HadCM3 (canESM2) GCMs.

The scenario generator operation produces ensembles of synthetic daily weather series given observed daily atmospheric predictor variables supplied by a GCM for either current or future climates (Wilby et al., 2002). The scenario generation produced 20 ensemble members of synthetic weather data for 139 years (1961–2099) from HadCM3 A2a and B2a scenarios and for 95 years (2006–2100) from canESM2 for RCP2.6, 4.5 and 8.5 scenarios, and the mean of the ensemble members was calculated and used for further climate change analysis. The generated scenario was divided into three time windows of 30 years of data: 2011–2040, 2041–2070 and 2071–2100.

4.3 Downscaling model performance evaluation criteria

A number of statistical tests were carried out to compare the skills of the two downscaling models categorized into two main classes. First, quantitative statistical tests using metrics, such as mean absolute error (MAE), root mean square error and bias. These metrics are by far the most widely used and accepted of the many possible numerical metrics (Amirabadizadeh et al., 2016; Bennett et al., 2013) to evaluate the comparative performance of the models to simulate the current climate variable of precipitation on the basis of long-term monthly averages defined by using Eqs. (7)–(9). In this study correlation and correlation-based measures such as coefficient of determination (R^2) and coefficient of efficiency (Nash–Sutcliffe efficiency, NSE) are not included due to the fact that these measures are oversensitive to extreme values and are insensitive to additive and proportional differences between model simulations and observations (Legates and McCabe, 1999). Evaluation was done in two steps as suggested by Goly et al. (2014): (i) equally weighing the metrics and (ii) varying the weights of metrics. For the case of equally weighted the following steps were applied. (a) Comparison of the values of the performance metrics among the models and ranking (obtaining individual model rankings for each performance metrics) at station level. Here, score 1 will be given to the model that has smaller metrics value and score 3 to the one having larger value and 2 for the model having the value in between. (b) Summing up the score pertaining to each model across all the stations. (c) Once the final scores are obtained for each evaluation metric, the models are ranked again based on the totals by summing up the metrics score value for each models.

$$\text{MAE} = \frac{\sum_{i=1}^{n} |X_i - Y_i|}{n} \tag{7}$$

$$\text{RMSE} = \sqrt{\frac{1}{n}\sum_{i=1}^{n} (X_i - Y_i)^2} \tag{8}$$

$$\text{Bias} = \frac{\sum_{i=1}^{i=n} X_i}{n} - \frac{\sum_{i=1}^{i=n} Y_i}{n} \tag{9}$$

In the above equations X_i and Y_i are the ith observation and simulated data by the model, respectively, μx and μy are the average of all data of X_i and Y_i in the study population, and n is the number of all samples to be tested.

Additionally, the varying weights technique was applied to the performance metrics as given in Eq. (10) to rank the models according to their skills. To avoid the discrepancy in weighing the performance measures because of differences in the order of their magnitudes, each performance measure is normalized (divided by the maximum value) and then the cumulative weighted performance measure for each downscaling model is calculated (Goly et al., 2014). The weights of metrics are arranged in such a way that more emphasis is given to MAE and RMSE, followed by bias (0.5, 0.35 and 0.15 respectively).

$$W_i = W_{\text{MAE}}\frac{\text{MAE}_i}{\text{MAE}_{\max}} + W_{\text{RMSE}}\frac{\text{RMSE}_i}{\text{RMSE}_{\max}} + W_{\text{Bias}}\frac{\text{Bias}_i}{\text{Bias}_{\max}}, \tag{10}$$

where the index i refers to a downscaling model, W_i refers to overall performance measure, and $0 \le W_i \le 1$.

Secondly, qualitative tests, comparing the skill of models in regard to capturing the distribution of the observed data to the whole range and in capturing the extreme precipitation events. For this purpose, statistical metrics such as IRF, ABC, 99p, 95p, 1daymax, R1, R10, R20 and SDII and graphical representations of box–whisker plots and KS cumulative distribution test were applied. KS is used to compare the probability distribution function (PDF) of the observations to the PDF of the simulated precipitation (Simard and L'Ecuyer, 2011). These plots provide a convenient visual summary of several statistical properties of the data set as they vary over time. A scoring technique is applied to compare the accuracy of the models. In this scoring technique, the bias of an evaluation metric for each station is used: score 1 will be given to the model that has smaller bias, score 3 to the one with a larger bias and 2 for the model with a value in between. Afterwards, evaluation was carried out using an equally weighted method only due to the assumption that the metrics have equal weights, as discussed above for model ranking. For the Kolmogorov–Smirnov cumulative distribution test, the observed and the simulated precipitation data from each model were compared using a p value at the significance level of 5 % for each station. The computed p value is lower than the significance level $\alpha = 0.05$, which indicates the simulated fail to follow the same distribution as the observed. Furthermore, the F test and t test are applied to test the equality of monthly variances of precipitation and equality of monthly mean respectively.

IRF and ABC are recommended by Campozano et al. (2016), while 95p, 99p, 1day max, R1, R10, R20 and SDII are recommended by Expert on Climate Change Detection and Indices (ETCCDI). The interquartile relative fraction (IRF): to evaluate the modeled variability representation relative to the observed is defined by Eq. (11):

$$\text{IRF} = \frac{Q_3^{\text{m}} - Q_1^{\text{m}}}{Q_3^{\text{o}} - Q_1^{\text{o}}}, \tag{11}$$

where Q_3^{m} and Q_3^{o} are the 75th modeled and observed percentile, and Q_1^{m} and Q_1^{o} are the 25th modeled and observed percentile respectively. A value of IRF > 1 represents overestimation of the variability, IRF = 1 is a perfect representation of the variability, and IRF < 1 is an underestimation of the variability. To evaluate the bias of the 25th, 50th and 75th

Table 3. Calibration results of the average statistical tests comparing the observed data from 26 stations with synthetic data generated through LARS-WG. The numbers in the table show the average numbers of tests gave a p value of less than 5 % significance level.

Tests	KS test		t test	F test	KS test	t test	KS test	t test
Parameters	WDseries	RainD	RMM	RMV	T_{min}D	T_{min}M	T_{max}D	T_{max}M
Average	0.04	0.00	0.27	2.08	0	0.12	0	0.12
Total	8	12	12	12	12	12	12	12
% failed	0.48	0.00	2.24	17.31	0	1	0	1

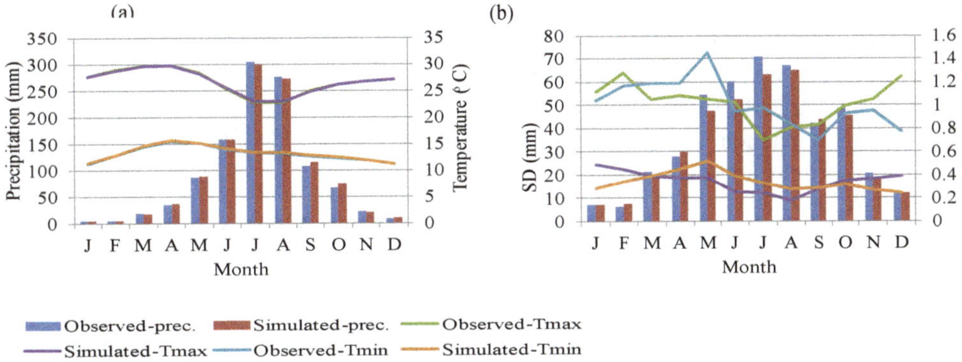

Figure 3. Observed and simulated **(a)** mean monthly precipitation, T_{max} and T_{min}; **(b)** standard deviation of precipitation, T_{max} and T_{min} using LARS-WG.

percentiles, the absolute cumulative bias (ACB) is defined as Eq. (12):

$$\begin{aligned}\mathrm{ACB} = \mathrm{abs}(Q_1^m - Q_1^o) + \mathrm{abs}(Q_2^m - Q_2^o) \\ + \mathrm{abs}(Q_3^m - Q_3^o),\end{aligned} \quad (12)$$

where Q_3^m and Q_3^o are the 75th modeled and observed percentile, Q_2^m and Q_2^o are the 50th modeled and observed percentile Q_1^m, and Q_1^o are the 25th modeled and observed percentile respectively. A value of ACB = 0 is a perfect representation of the modeled and observed distributions, while under- or overestimation indicates a divergence of ACB from zero to positive values. The terms 95p and 99p denote the 95th and 99th percentiles of daily precipitation amount respectively; 1 daymax is the highest 1-day precipitation amount; R1, R10 and R20 are number of precipitation days ($\geq$ 1 mm), heavy precipitation days ($\geq$ 10 mm) and extreme heavy precipitation days $\geq$ 20 mm respectively; and SDII is the simple daily intensity index calculated as the ratio of total precipitation to the number of wet days ($\geq$ 1 mm).

5 Results and analysis

5.1 Calibration and validation of LARS-WG

To verify the performance of LARS-WG, in addition to the graphic comparison, some statistical tests were performed. The KS test is performed to test equality of the seasonal distributions of wet and dry series (WDSeries), distributions of daily rainfall (RainD), and distributions of daily maximum ($T_{max}D$) and minimum ($T_{min}D$) temperature. The F test is performed to test equality of monthly variances of precipitation (RMV) while the t test is performed to verify equality of monthly mean rainfall (RMM), monthly mean of daily maximum temperature ($T_{max}M$), and monthly mean of daily minimum temperature ($T_{min}M$). All of the tests calculate a p value, which is used to accept or reject the hypotheses that the two sets of data (observed and generated) could have come from the same distribution at the 5 % significance level. Therefore, the average number of p values less than 5 % recorded from 26 stations and the percentage that failed from the total of 8 seasons or 12 months have been presented in Table 3. The result revealed that LARS-WG performs very well for all parameters except RMM and RMV. However, an average of 2.2 and 17.3 % of the months of a year obtained a p value < 5 % for the monthly mean and variance of precipitation respectively. From these numbers, it can be noted that the model is less capable of simulating the monthly variances than the other parameters.

For illustrative purposes, a graphical representation of monthly mean and standard deviation of the simulated and observed precipitation, T_{max} and T_{min}, was constructed (see Fig. 3) for the randomly chosen Gondar station as it has been difficult to present the result of all stations. It can be seen from the result that observed and simulated monthly mean precipitation, T_{max} and T_{min}, match very well. However, as it is known to be difficult to simulate the standard deviations in most statistical downscaling studies, the performance of the standard deviation is less accurate compared to the mean (Fig. 3b).

5.2 Screening variable, model calibration and validation of SDSM

Initially, offline correlation analysis was performed using SPSS software between predictands and NCEP reanalysis predictors to identify an optimal lag and physically sensible predictors for climate variables of precipitation, T_{max} and T_{min}. Analysis of the offline correlation revealed that an optimal lag or time shift does not improve the correlation of predictands and predictors for this particular study. Average partial correlation of observed precipitation with predictors as shown in Fig. 5 indicates all stations followed the same correlation pattern (both in magnitude and direction) that illustrates all stations can have identical physically sensible predictors, with a few exceptions. Furthermore, there are a number of predictors that have correlation coefficient values in the range of 20 to 45 % for precipitation across all stations. This range is considered to be acceptable when dealing with precipitation downscaling (Wilby et al., 2002) because of its complexity and high spatial and temporal variability to downscale.

The predictor variables identified for each downscaling GCM and for the corresponding local climate variables showed that different large-scale atmospheric variables control different local variables. For instance, the set of temp, mslp, s500, s850, p8_v, p500, shum comprises the most potential or meaningful predictors for temperature, and the set of s500, s850, p8_u, p_z, pzh, p500 performs best for predicting precipitation of the study area, which is consistent with the result of offline correlation analysis. After carefully screening predictor variables, model calibration and validation was carried out. The graphical comparison between the observed and generated rainfall, T_{max} and T_{min}, was run to enhance the confidence of the model performance, as shown in Figs. 6 and 7 for Gondar station only. Examination of Fig. 6 shows that the calibrated model reproduces the monthly mean precipitation and mean standard deviation of daily T_{max} and T_{min} values quite well. However, the model is less accurate in reproducing variance of observed precipitation. As Wilby et al. (2004) point out, downscaling models are often regarded as less able to model the variance of the observed precipitation with great accuracy.

The statistical performance metrics of MAE and RMSE values for the monthly precipitation modeled from canESM2 range from 3.5 to 14.8 mm and 4.9 to 22.4 mm, which shows that canESM2 performs better than HadCM3, with the MAE and RMSE values ranging from 6.2 to 48.6 mm and 7.6 to 73.4 mm respectively. The result of statistical analysis revealed that the model is much better in simulating T_{max} and T_{min} than precipitation, because the high dynamical properties of precipitation make it difficult to simulate. After accomplishing a satisfactory calibration, the multiple regression model is validated using an independent set of data outside the period for which the model is calibrated. The validation result revealed that the model is successfully validated but with less accuracy compared to calibration for both GCMs as shown in Fig. 7. In general, the result analysis of performance measure and graphical representation of observed and simulated scenarios, both for calibration and validation, revealed that the model performs very well in simulating the climate variables.

5.3 Downscaling with LARS-WG

Since the performance of LARS-WG during calibration and validation was very good, downscaling of the climate scenario can be done from six selected multimodel CMIP3 GCMs under three scenarios (A1B, B1 and A2) for three time periods. After downscaling the future climate scenarios at all stations from the selected six GCMs, the projected precipitation analysis for the areal UBNRB was calculated from the point rainfall stations using the Thiessen polygon method. The result analysis (Fig. 4a) revealed that GCMs disagree on the direction of precipitation change: two GCMs (CSMK3 and GFCM21) showed decreasing trends, and a majority, or four, GCMs (NCCSM, Hadcm3, MPEH5 and MIHR) showed increasing trends from the reference period in all three time periods. By the 2030s, the relative change in mean annual precipitation is projected in the range between −2.3 and 6.5 % for A1B, −2.3 and 7.8 % for B1, and −3.7 and 6.4 % for A2 emission scenarios. In the 2050s, the relative changes in precipitation range between −8 and 22.7 % for A1B, −2.7 and 22 % for B1, and −7.4 and 8.7 % for A2 scenarios. In the time of 2080s, the relative changes in precipitation projected may vary between −7.5 and 29.9 % for A1B, −5.3 and 13.7 % for B1, and −5.9 and 43.8 % for A2 emission scenarios. The multimodel average result showed that in the future precipitation may generally increase over the basin in the range of 1 to 14.4 %, which is in line with the result from the HadCM3 GCM (0.8 to 16.6 %).

In a different way from precipitation, the projections of mean annual T_{max} and T_{min} have showed coherent increasing trends from the six GCMs under all scenarios in all three future time periods (Fig. 4b). The result calculated from the ensemble mean showed that mean annual T_{max} may increase up to 0.5, 1.8 and 3.6 °C by 2030s, 2050s and 2080s respectively under the A2 scenario, which is in line with the results from both GFCM21 and HadCM3 GCMs. Likewise, the UBNRB may experience an increase in mean annual T_{min} up to 0.6, 1.8 and 3.6 °C by the 2030s, 2050s and 2080s respectively from the multimodel average.

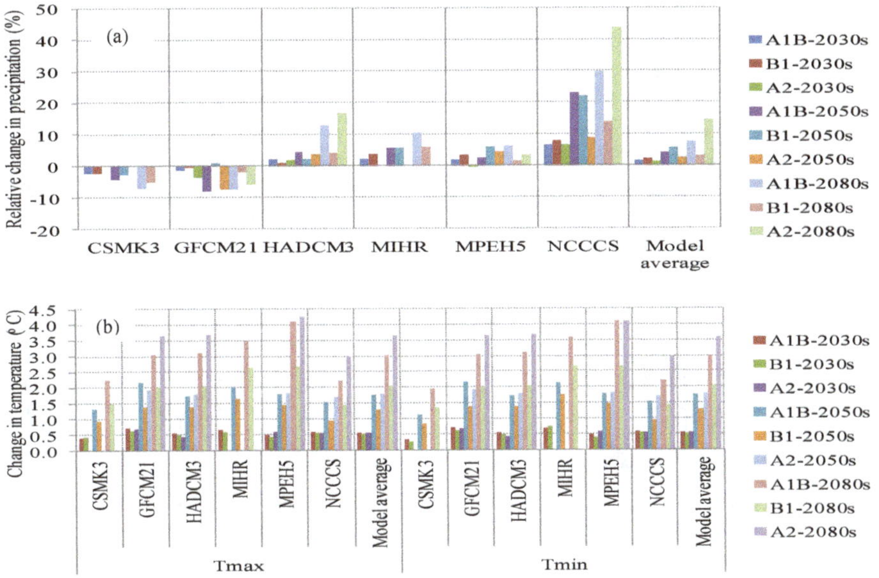

Figure 4. **(a)** Relative change mean annual precipitation and **(b)** change in T_{max} and T_{min} modeled from six GCMs for three time periods of UBNRB compared from the reference period of 1984–2011 by using LARS-WG.

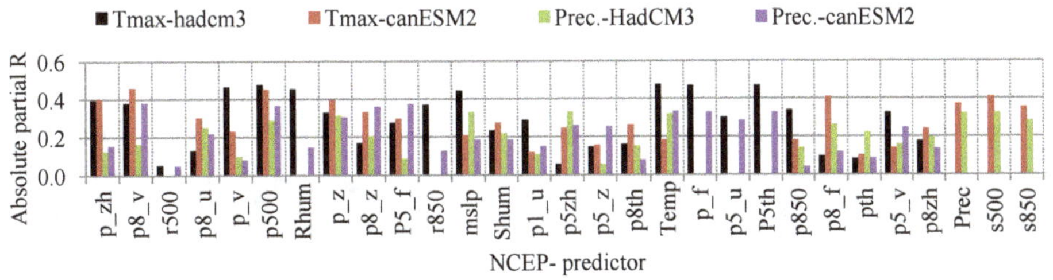

Figure 5. Average partial correlation coefficient values of all stations for precipitation and T_{max} with NCEP predictors.

5.4 Downscaling with SDSM

Here, as it is difficult to process all the selected six CMIP3 GCM3 using SDSM, we choose the HadCM3 GCM as the best due to the fact that the downscaling result of HadCM3 using LARS-WG fits with the downscaling result of the ensemble mean model. Also, canESM2 from the CMIP5 GCMs was selected to test the improvements of CMIP5 over CMIP3. Results of downscaling future climate scenario of areal UBNRB using SDSM calculated from all stations using Thiessen polygon methods are summarized in Fig. 8. The overall analysis of the result indicates a general increase in mean annual precipitation for three time windows (2030s, 2050s and 2080s) for all five scenarios (A2a and B2a for HadCM3 and RCP2.6, RCP4.5 and RCP8.5 for canESM2) in the range of 2.1 to 43.8 %. The maximum (minimum) relative change of mean annual precipitation is projected to be 43.8 % (6.2 %), 29.5 % (3.5 %) and 19 % (2.1 %) in the 2080s, 2050s and 2030s under the RCP8.5 scenario of canESM2 (B2a) scenario of HadCM3. In general, the RCP8.5 scenario of canESM2GCM resulted in pronounced increases in all three time periods, whereas scenario B2a of the HadCM3 GCM reported minimum change over the study area.

Regarding temperature, the downscaling result of T_{max} and T_{min} showed an increasing trend consistently in all months and seasons in three time periods under all scenarios with mean annual value ranging from 0.5 to 2.6 °C and 0.3 to 1.6 °C under scenario RCP8.5 and B2a respectively. The RCP 8.5 scenario reported maximum change while B2a scenario reported minimum change both for T_{max} and T_{min} in all three time periods compared to other scenarios. The analysis of the downscaling result illustrates maximum temperature may become much hotter compared to minimum temperature in all scenarios and time periods in the future across the UBNRB.

5.5 Comparative performance evaluation of LARS-WG and SDSM models

Chen et al. (2013) argued that though major sources of uncertainty are linked to GCMs and emission scenarios, uncertainty related to the choice of downscaling methods give

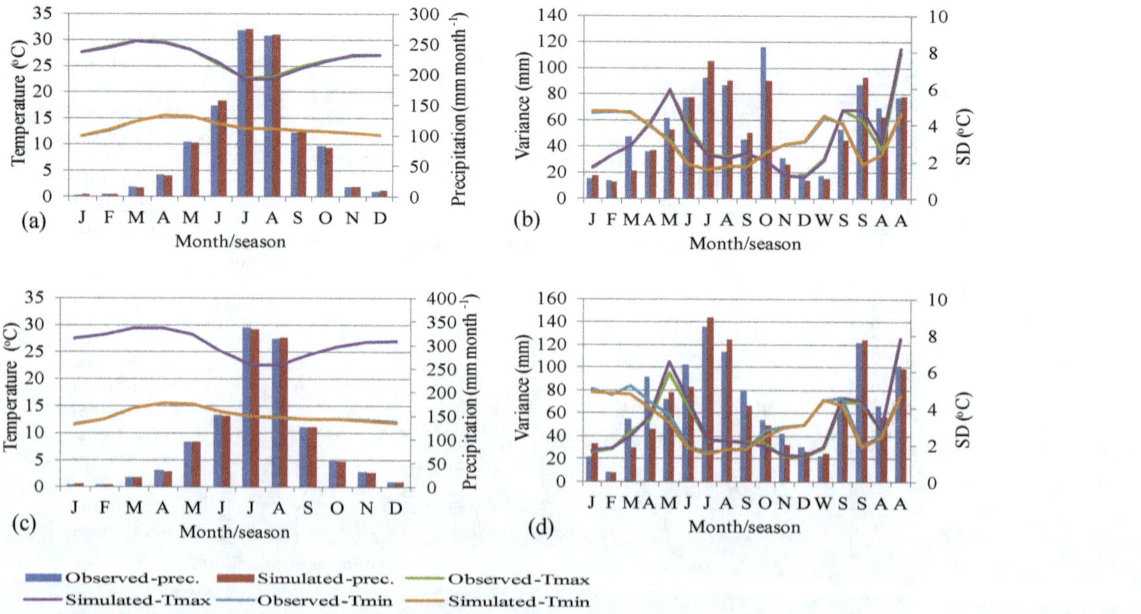

Figure 6. Calibration of observed and simulated of precipitation, maximum temperature and minimum temperature for the Gondar station using SDSM from canESM2 and HadCM3 from **(a, b)** to **(c, d)**.

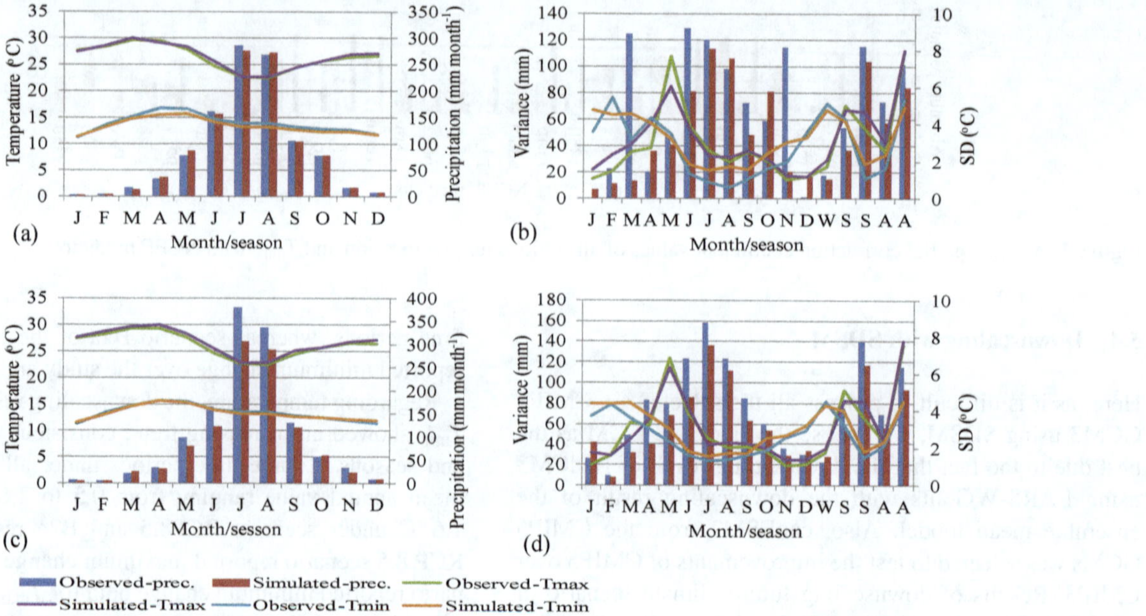

Figure 7. Validation of observed and simulated of precipitation, maximum temperature and minimum temperature for Gondar station using SDSM from canESM2 and HadCM3 from**(a, b)** to **(c, d)** respectively.

less attention to climate change analysis. Therefore, in this study, comparative performance evaluation of the downscaling methods has been given due emphasis and carried out in a number of statistical and graphical tests both quantitatively and qualitatively. The model skill was evaluated and ranked at each site for each metric as shown in Table 4 for metrics of RMSE. The overall rank obtained by summing up the score of each model for each metric is presented in Tables 5 and 6, for quantitative and qualitative measures respectively.

The result revealed that SDSM/canESM2 narrowly performed best in simulating the long-term average values in both equally weighted and varying weights of the quantitative metrics. However, LARS-WG performed best in qualitative measures in reproducing the distribution and extreme

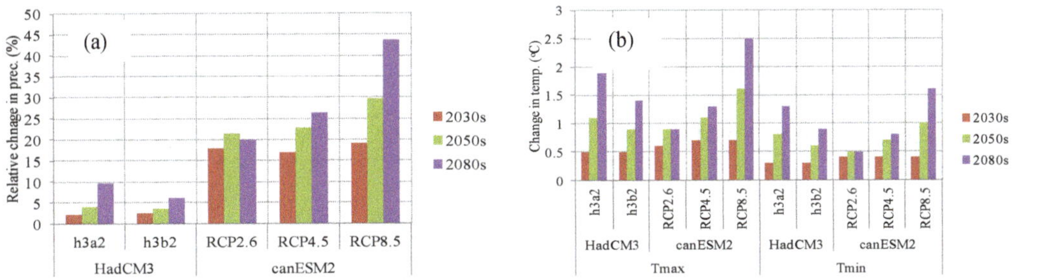

Figure 8. **(a)** Relative change of mean annual precipitation, and **(b)** change of mean annual $T_{\max}$ and $T_{\min}$ for three time periods compared to the baseline period of UBNRB using SDSM for HadCM3 and canESM2 GCMs under different scenarios.

Table 4. Performance measure and ranking of models during the baseline period (1984–2011) for evaluation metric RMSE.

	RMSE value			Equally weighted score			Varying weights score		
Stations	SDSM/canESM2	SDSM/HadCM3	LARS-WG	SDSM/canESM2	SDSM/HadCM3	LARS-WG	SDSM/canESM2	SDSM/HadCM3	LARS-WG
Abaysheleko	7.4	15.7	18.9	1	2	3	0.14	0.29	0.35
Alemketema	19.4	7.6	10.5	3	1	2	0.35	0.14	0.19
Anger	11.1	13.1	10.0	2	3	1	0.29	0.35	0.27
Angerguten	8.2	16.1	9.8	1	3	2	0.18	0.35	0.21
Bahirdar	8.5	21.7	11.5	1	3	2	0.14	0.35	0.19
Bedele	6.4	46.1	14.7	1	3	2	0.05	0.35	0.11
Dangila	13.2	53.8	9.0	2	3	1	0.09	0.35	0.06
Dedesa	8.2	18.0	13.8	1	3	2	0.16	0.35	0.27
Dmarkos	5.0	19.1	12.5	1	3	2	0.09	0.35	0.23
Dtabor	22.4	39.4	10.7	2	3	1	0.20	0.35	0.10
Fitche	17.8	11.2	10.8	3	2	1	0.35	0.22	0.21
Gimijabet	14.5	32.4	11.4	2	3	1	0.16	0.35	0.12
Gondar	5.0	18.2	3.6	2	3	1	0.10	0.35	0.07
Nedjo	8.4	15.4	11.7	1	3	2	0.19	0.35	0.27
Shambu	8.6	15.7	10.7	1	3	2	0.19	0.35	0.24
Overall score				24	41	25	2.67	4.85	2.88

events of daily precipitation. For instance, absolute bias for the 95th percentile of daily precipitation (95p) ranges from 4.35 to 12.4 mm for SDSM/canESM2, from 3.2 to 12.2 mm for SDSM/HadCM3 and from 0.07 to 3.7 mm for LARS-WG; for the mean of daily precipitation amount (SDII), absolute bias ranges from 1.3 to 6.3 mm for SDSM/canESM2, from 2.1 to 5.6 mm for SDSM/HadCM3 and from 0.01 to 3 mm for LARS-WG.

Furthermore, as the Kolmogorov–Smirnov test from Table 7 shows, LARS-WG captures the distribution of the observed precipitation of 93.3 % from all stations while SDSM captures only 20 % of the 15 stations equally both from canESM2 and HadCM3 GCMs at 5 % significance level. The t test result revealed that 86.7 % of the simulated precipitation by LARS-WG and SDSM/HadCM3 models are capturing their perspective mean values from all stations while the SDSM and hadCM3 models capture only 66.7 %. The F test showed 93.3 % of the simulated and the observed precipitation are normally distributed around their respective variance value in all three models. In general, the comparative performance test revealed that the LARS-WG model performed best in qualitative measures while SDSM/canESM2 is best in quantitative measures in UBNRB. In addition, Figs. 9 and 10 confirmed graphically the ability of the LARS-WG model in capturing the distribution and extreme events of the precipitation in representative stations (randomly chosen) by a whisker box plot and Kolmogorov–Smirnov test respectively.

For future simulation, the HadCM3 GCM A2 scenario was used in common for two (LARS-WG and SDSM) downscaling methods to test whether the downscaling methods may affect the GCM result under the same forcing scenario. The results obtained from the two downscaling models were found reasonably comparable and both approaches showed increasing trends for precipitation, $T_{\max}$ and $T_{\min}$. However, the magnitude of the downscaled climate data from the two methods as presented in Fig. 11 indicates that LARS-WG overpredicts precipitation and temperature compared to SDSM. The relative change of mean annual precipitation using LARS-WG is about 16.1 % and an average increase in mean annual $T_{\max}$ and $T_{\min}$ is about 3.7 and 3.6 °C respectively in the 2080s. SDSM predicts the relative change of mean annual precipitation of only about 9.7 % and an average increase in $T_{\max}$ and $T_{\min}$ of about 2 and 1.3 °C respectively in the same period. The differences in the future predictions are the result of the differences in the basic concepts behind the two downscaling techniques. The SDSM uses large-scale predictor variables from GCM outputs which can be considered as more reliable for climate change prediction using

Table 5. Statistical downscaling models ranking during the baseline period (1984–2011) for quantitative measures. The numbers in the table show the total ranking scores summed up from 15 stations.

	Equally weighted overall score			Varying weights overall score			
Evaluation metrics	SDSM/canESM2	SDSM/HadCM3	LARS-WG	Weight	SDSM/canESM2	SDSM/HadCM3	LARS-WG
RMSE	24	41	25	0.35	2.67	4.85	2.88
MAE	24	41	25	0.50	3.99	7.47	4.64
Bias	23	39	28	0.15	1.29	1.83	0.70
Total	71	121	78	1.0	7.94	14.15	8.22
Rank	1	3	2		1	3	2

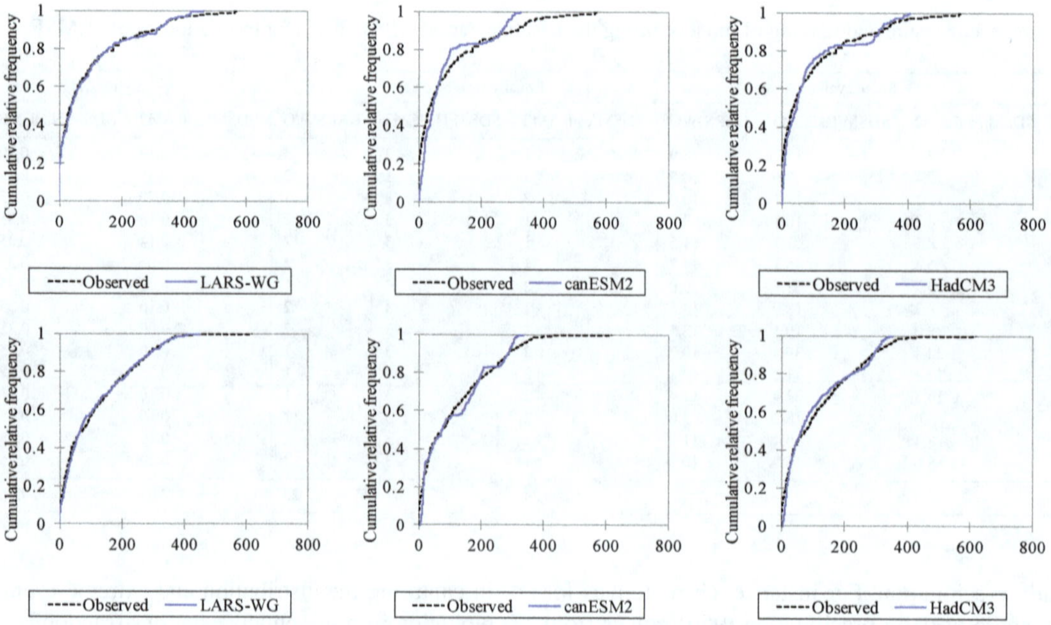

Figure 9. Kolmogorov–Smirnov test to compare the skill of the models for the observed precipitation distribution (upper three Alemketema stations, lower three Debre Markos stations).

multiple linear regression. However, the LARS WG uses the relative change factors (RCFs) derived from the direct GCM output of only those variables which directly correspond to the predictands. Hence, because of the well-known fact that GCMs are not very reliable in simulating precipitation, the error induced from the GCM output for precipitation will propagate the error of downscaling that makes the performance of LARS-WG to downscale precipitation needs more caution (Dibike et al., 2005).

6 Discussions and conclusions

The uncertainty related to climate change analysis can be due to climate models and downscaling methods among many other factors. In this study, we employed a multimodel approach to see that the uncertainties came from different GCMs. In total, 21 systematically selected future climate scenarios were produced for each time period, which we might think representative to understand fully and to project plausibly the future climate change in the study area and to retain information about the full variability of GCMs. Moreover, we applied two widely used statistical downscaling methods, namely the regression downscaling technique (SDSM) and the stochastic weather generation method (LARS WG) for this particular study.

The performance of the three models (HadCM3/SDSM, canESM2/SDSM and LARS-WG) were tested for the baseline period of 1984–2011 in representing the current situation, particularly for precipitation, as it is the most difficult climate variable to model. The result suggested that SDSM using canESM2 GCM captures the long-term monthly average very well at most of the stations and it ranked first from others. This could be attributed to the increasing performance of GCMs from time to time (i.e., CMIP5 GCMs performs

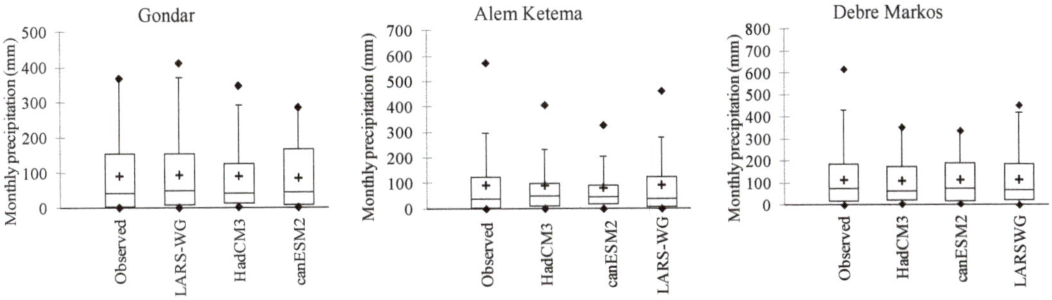

Figure 10. Box plot showing the model performance for three stations on a monthly basis. Box boundaries indicate the 25th and 75th percentiles, the line within the box marks the median, whiskers below and above the box indicate the 10th and 90th percentiles, and dots indicate the extremes.

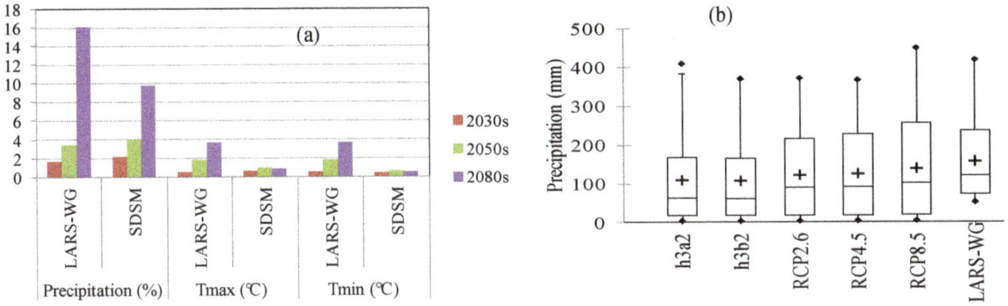

Figure 11. Comparison of climate change scenarios **(a)** downscaled using LARS-WG and SDSM from HadCM3 GCM for the a2 scenario and **(b)** downscaled from different scenarios (LARS-WG using hadCM3 a2 scenario).

Table 6. Ranking of statistical downscaling models during the baseline period (1984–2011) for qualitative measures (distribution and extreme events of daily precipitation). The numbers in the table show the total ranking scores obtained from 15 stations.

Evaluation metrics	SDSM/ canESM2	SDSM/ HadCM3	LARS-WG
95p	42	33	15
99p	41	34	15
1 day max	39	36	15
SDII	36	38	16
R20	42	33	16
R10	37	34	19
R1	40	35	15
1-IRF	32	29	29
ACB	33	33	24
Total score	342	305	164
Rank	3	2	1

better than CMIP3 GCMs) due to the fact that modeling was based on the new set of radiative forcing scenarios that replaced SRES emission scenarios, constructed for IPCC AR5 where the impacts of land use and land cover change on the environment and climate are explicitly included. However, LARS-WG performed best in qualitative measures in capturing the distribution and extreme events of the daily precipitation than SDSM. The better performance of LARS-WG in capturing the distribution and extreme events of the daily precipitation may be associated with the use of 23 interval histograms for the construction of semiempirical distribution, which offers a more accurate representation of the observed distribution compared with the 10 intervals used in the previous version (Semenov et al., 2010). The poor performance of SDSM would indicate the difficulty in finding climate variables from the NCEP data that could explain the variability of daily precipitation well. Therefore, LARS-WG would be more preferred in areas of the UBNRB where there is high climatic variability to correctly simulate the distribution and extreme events of the precipitation, which is crucial for a realistic assessment of flood events and agricultural production.

The downscaling result reported from the six GCMs used in LARS-WG showed large intermodel differences: two GCMs reported that precipitation may decrease while four GCMs reported that precipitation may increase in the future. The large intermodel differences in the GCMs showed the uncertainties of GCMs associated with their differences of resolution and assumptions of physical atmospheric processes to represent local-scale climate variables, which are typical characteristics for Africa and because of low convergence in climate model projections in the area of UBNRB

Table 7. Kolmogorov–Smirnov, t and F tests during the baseline period (1984–2011) for qualitative measures.

	Kolmogorov–Smirnov test			t test			F test		
Station	HadCM3	canESM2	LARS-WG	HadCM3	canESM2	LARS-WG	HadCM3	canESM2	LARS-WG
Total stations	15	15	15	15	15	15	15	15	15
Passed ($p > 5$ %)*	3	3	14	14	14	14	10	1013	13
% passed	20	20	93.3	93.3	93.3	93.3	66.7	86.7	86.7

* Number of stations with p value > 5 % (pass to simulate the distribution of precipitation).

(Gebre and Ludwig, 2015). These results further reinforce multimodel strategies for conducting climate change studies. The multimodel average result showed that in the future precipitation may generally increase over the basin in the range of 1 to 14.4 % which is in line with the result from HadCM3 GCM (0.8 to 16.6 %); this indicates that HadCM3 from CMIP3 GCMs has a better representation of local-scale climate variables in the study area, consistent with the previous study result by Kim and Kaluarachchi (2009) and Dile et al. (2013) in the same study area.

LARS-WG produces synthetic climate data of any length with the same characteristics as the input record, and it simulates weather separately for a single site. Therefore, the resulting weather series for different sites are independent of each other, which can cause loss of a very strong spatial correlation that exists in real weather data during simulation. However, a few stochastic models have been developed to produce weather series simultaneously at multiple sites, preserving the spatial correlation, mainly for daily precipitation, such as space–time models, nonhomogeneous hidden Markov models and nonparametric models that typically use a K nearest-neighbor (K-NN) procedure (King et al., 2015). They are complicated in both calibration and implementation and are unable to adequately reproduce the observed correlations (Khalili et al., 2007). In this study, the simple Pearson's correlation coefficient (R^2) value was checked in two stations before and after simulation of the observed data to test the capability of LARS-WG in preserving the spatial correlation of stations. The result revealed that the spatial correlation of the stations distorted or decreased from the original is insignificant.

In conclusion, a multimodel average from LARS-WG and individual model result from SDSM showed a general increasing trend for all three climatic variables (precipitation, T_{max} and T_{min}) in all three time periods. The positive change in precipitation in future can be a good opportunity for the farmers who are engaged in rainfed agriculture to maximize their agricultural production and to change their livelihoods. However, this information cannot be a guarantee for irrigation farming because precipitation is not the only factor affecting the flow of the river, which is the main source for irrigation. Evapotranspiration, dynamics of land use land cover, proper water resource management and other climatic factors which are not yet assessed by this study can influence the flow of the river directly and indirectly. Furthermore, the result from this study revealed that maximum positive precipitation change may occur in Autumn (September–November), when most agricultural crops mature and harvesting begins, while minimum precipitation change may occur during summer (June–August), when about 80 % of the annual rainfall occurs; this climate variability can be a potential threat to the farmers, who have limited ability to cope with the negative impacts of climate variability and overall ongoing economic development efforts in the basin.

In general, this study has shown that climate change will likely occur that may affect the water resources and hydrology of the UBNRB. On the basis of the results obtained in this study, both SDSM and LARS-WG models can be adopted with reasonable confidence as downscaling tools to undertake climate change impact assessment studies for the future. However, LARS-WG is more suitable for extreme precipitation impact assessment studies, such as those dealing with floods and droughts. Moreover, the paper provides substantial information that the choice of downscaling methods has a contribution in the uncertainty of future climate prediction. The authors would also like to suggest further assessment of the study area using a large ensemble of CMIP5 GCMs. Further relative performance of downscaling techniques for other climatic variables such as T_{max}, T_{min}, dry spell length, wet spell length, inter-annual and seasonal cycle of precipitation using additional PDF-based metrics such as the Brier score and the skill score might counteract the limitations of this paper.

Competing interests. The authors declare that they have no conflict of interest.

Acknowledgements. We are grateful to the Ethiopian Meteorological Agency (EMA) for providing us the meteorological data for free. The first author received financial support for traveling money from the DAAD water–food–energy NeXus project. The authors are indebted to the two anonymous reviewers and the editor, Dimitri Solomatine, for their critique and constructive suggestions and comments on earlier versions of this paper, which were helpful in the improvement of the manuscript.

Edited by: Dimitri Solomatine

References

Amirabadizadeh, M., Ghazali, A. H., Huang, Y. F., and Wayayok, A.: Downscaling daily precipitation and temperatures over the Langat River Basin in Malaysia: a comparison of two statistical downscaling approaches, International Journal of Water Resources and Environmental Engineering, 8, 120–136, 2016.

Awulachew, S. B., Yilma, A. D., Loulseged, M., Loiskandl, W., Ayana, M., and Alamirew, T.: Water Resources and Irrigation Development in Ethiopia, Working Paper 123, International Water Management Institute, Colombo, Sri Lanka, 78 pp., 2007.

BCEOM: Abbay River Basin Integrated Development Master Plan, section II, volume V – Water Resources Development, part 1 – Irrigation and Drainage, Ministry of Water Resources, Addis Ababa, Ethiopia, 1998.

Bennett, N. D., Croke, B. F., Guariso, G., Guillaume, J. H., Hamilton, S. H., Jakeman, A. J., Marsili-Libelli, S., Newham, L. T., Norton, J. P., and Perrin, C.: Characterising performance of environmental models, Environ. Modell. Softw., 40, 1–20, 2013.

Bewket, W. and Conway, D.: A note on the temporal and spatial variability of rainfall in the drought-prone Amhara region of Ethiopia, Int. J. Climatol., 27, 1467–1477, 2007.

Beyene, T., Lettenmaier, D. P., and Kabat, P.: Hydrologic impacts of climate change on the Nile River Basin: implications of the 2007 IPCC scenarios, Climatic Change, 100, 433–461, 2010.

Campozano, L., Tenelanda, D., Sanchez, E., Samaniego, E., and Feyen, J.: Comparison of statistical downscaling methods for monthly total precipitation: case study for the Paute River Basin in southern Ecuador, Adv. Meteorol., 2016, 6526341, https://doi.org/10.1155/2016/6526341, 2016.

Casajus, N., Périé, C., Logan, T., Lambert, M.-C., de Blois, S., and Berteaux, D.: An objective approach to select climate scenarios when projecting species distribution under climate change, PloS one, 11, e0152495, https://doi.org/10.1371/journal.pone.0152495, 2016.

Chen, H., Guo, J., Zhang, Z., and Xu, C.-Y.: Prediction of temperature and precipitation in Sudan and South Sudan by using LARS-WG in future, Theor. Appl. Climatol., 113, 363–375, 2013.

Conway, D.: The climate and hydrology of the Upper Blue Nile River, Geogr. J., 166, 49–62, 2000.

Dibike, Y. B. and Coulibaly, P.: Hydrologic impact of climate change in the Saguenay watershed: comparison of downscaling methods and hydrologic models, J. Hydrol., 307, 145–163, 2005.

Dile, Y. T., Berndtsson, R., and Setegn, S. G.: Hydrological response to climate change for Gilgel Abay River, in the Lake Tana Basin–Upper Blue Nile Basin of Ethiopia, PloS one, 8, e79296, https://doi.org/10.1371/journal.pone.0079296, 2013.

Ebrahim, G. Y., Jonoski, A., van Griensven, A., and Di Baldassarre, G.: Downscaling technique uncertainty in assessing hydrological impact of climate change in the Upper Beles River Basin, Ethiopia, Hydrol. Res., 44, 377–398, 2013.

Elshamy, M. E., Seierstad, I. A., and Sorteberg, A.: Impacts of climate change on Blue Nile flows using bias-corrected GCM scenarios, Hydrol. Earth Syst. Sci., 13, 551–565, https://doi.org/10.5194/hess-13-551-2009, 2009.

Fiseha, B., Melesse, A., Romano, E., Volpi, E., and Fiori, A.: Statistical downscaling of precipitation and temperature for the Upper Tiber Basin in Central Italy, International Journal of Water Sciences, 1, https://doi.org/10.5772/52890, 2012.

Fowler, H., Blenkinsop, S., and Tebaldi, C.: Linking climate change modelling to impacts studies: recent advances in downscaling techniques for hydrological modelling, Int. J. Climatol., 27, 1547–1578, 2007.

Gebre, S. L. and Ludwig, F.: Hydrological response to climate change of the Upper Blue Nile River Basin: based on IPCC Fifth Assessment Report (AR5), Journal of Climatology & Weather Forecasting, 3, 121, https://doi.org/10.4172/2332-2594.1000121, 2015.

Gebremicael, T., Mohamed, Y., Betrie, G., van der Zaag, P., and Teferi, E.: Trend analysis of runoff and sediment fluxes in the Upper Blue Nile Basin: a combined analysis of statistical tests, physically-based models and landuse maps, J. Hydrol., 482, 57–68, 2013.

Goly, A., Teegavarapu, R. S., and Mondal, A.: Development and evaluation of statistical downscaling models for monthly precipitation, Earth Interact., 18, 1–28, 2014.

Goodarzi, E., Dastorani, M., and Talebi, A.: Evaluation of the change-factor and LARS-WG methods of downscaling for simulation of climatic variables in the future (Case study: Herat Azam Watershed, Yazd-Iran), Ecopersia, 3, 833–846, 2015.

Hashmi, M. Z., Shamseldin, A. Y., and Melville, B. W.: Comparison of SDSM and LARS-WG for simulation and downscaling of extreme precipitation events in a watershed, Stoch. Env. Res. Risk A., 25, 475–484, 2011.

Hassan, Z., Shamsudin, S., and Harun, S.: Application of SDSM and LARS-WG for simulating and downscaling of rainfall and temperature, Theor. Appl. Climatol., 116, 243–257, 2014.

IPCC: Climate Change 2007. Synthesis Report. Contribution of Working Groups I, II and III to the Fourth Assessment Report of the Intergovernmental Panel on Climate Change, edited by: Core Writing Team, Pachauri, R. K., and Reisinger, A., 104 pp., IPCC, Geneva, Switzerland, 2007.

IPCC: Climate Change 2014: Synthesis Report. Contribution of Working Groups I, II and III to the Fifth Assessment Report of the Intergovernmental Panel on Climate Change, edited by: Core Writing Team, Pachauri, R. K., and Meyer, L. A., 151 pp., IPCC, Geneva, Switzerland, 2014.

Khalili, M., Leconte, R., and Brissette, F.: Stochastic multisite generation of daily precipitation data using spatial autocorrelation, J. Hydrometeorol., 8, 396–412, 2007.

Khan, M. S., Coulibaly, P., and Dibike, Y.: Uncertainty analysis of statistical downscaling methods, J. Hydrol., 319, 357–382, 2006.

Kim, U. and Kaluarachchi, J. J.: Climate change impacts on water resources in the Upper Blue Nile River Basin, Ethiopia, Wiley Online Library, 45, 1361–1378, https://doi.org/10.1111/j.1752-1688.2009.00369.x, 2009.

Kim, U., Kaluarachchi, J. J., and Smakhtin, V. U.: Climate change impacts on hydrology and water resources of the Upper Blue Nile River Basin, Ethiopia, IWMI Research Report 126, International Water Management Institute, Colombo, Sri Lanka, 27 pp., 2008.

King, L. M., McLeod, A. I., and Simonovic, S. P.: Improved weather generator algorithm for multisite simulation of precipitation and temperature, J. Am. Water Resour. As., 51, 1305–1320, 2015.

Legates, D. R. and McCabe, G. J.: Evaluating the use of "goodness-of-fit" measures in hydrologic and hydroclimatic model validation, Water Resour. Res., 35, 233–241, 1999.

Maurer, E. P. and Hidalgo, H. G.: Utility of daily vs. monthly large-scale climate data: an intercomparison of two statistical downscaling methods, Hydrol. Earth Syst. Sci., 12, 551–563, https://doi.org/10.5194/hess-12-551-2008, 2008.

Mengistu, D., Bewket, W., and Lal, R.: Recent spatiotemporal temperature and rainfall variability and trends over the Upper Blue Nile River Basin, Ethiopia, Int. J. Climatol., 34, 2278–2292, 2014.

Qian, B., Hayhoe, H., and Gameda, S.: Evaluation of the stochastic weather generators LARS-WG and AAFC-WG for climate change impact studies, Clim. Res., 29, 3–21, 2004.

Semenov, M. A. and Barrow, E. M.: Use of a stochastic weather generator in the development of climate change scenarios, Climatic Change, 35, 397–414, 1997.

Semenov, M. A. and Barrow, E. M.: LARS-WG: a stochastic weather generator for use in climate impact studies, Version 3.0, User Manual, Hertfordshire, UK, 2002.

Semenov, M. A. and Stratonovitch, P.: Use of multi-model ensembles from global climate models for assessment of climate change impacts, (Open Access for articles 4 years old and older), Clim. Res., 41, 1–14, https://doi.org/10.3354/cr00836, 2010.

Simard, R. and L'Ecuyer, P.: Computing the two-sided Kolmogorov–Smirnov distribution, J. Stat. Softw., 39, 1–18, 2011.

Taye, M. T., Ntegeka, V., Ogiramoi, N. P., and Willems, P.: Assessment of climate change impact on hydrological extremes in two source regions of the Nile River Basin, Hydrol. Earth Syst. Sci., 15, 209–222, https://doi.org/10.5194/hess-15-209-2011, 2011.

USBR: Land and Water Resources of the Blue Nile Basin, Ethiopia, Main Report, United States Bureau of Reclamation, Washington, DC, 1964.

WAPCOS: Preliminary Water Resources Development Master Plan for Ethiopia, Prepared for EVDSA, Addis Ababa, Final report, 1990.

Wigley, T. M.: MAGICC/SCENGEN 5.3: User manual (version 2), NCAR, Boulder, CO, 80, 2008.

Wilby, R., Charles, S., Zorita, E., Timbal, B., Whetton, P., and Mearns, L.: Guidelines for Use of Climate Scenarios Developed from Statistical Downscaling Methods, available from the DDC of IPCC TGCIA 27, 2004.

Wilby, R., Dawson, C., and Barrow, E.: SDSM User Manual – a Decision Support Tool for the Assessment of Regional Climate Change Impacts, London, UK, 2007.

Wilby, R. L. and Wigley, T.: Downscaling general circulation model output: a review of methods and limitations, Prog. Phys. Geogr., 21, 530–548, 1997.

Wilby, R. L., Dawson, C. W., and Barrow, E. M.: SDSM – a decision support tool for the assessment of regional climate change impacts, Environ. Modell. Softw., 17, 145–157, 2002.

Worqlul, A. W., Maathuis, B., Adem, A. A., Demissie, S. S., Langan, S., and Steenhuis, T. S.: Comparison of rainfall estimations by TRMM 3B42, MPEG and CFSR with ground-observed data for the Lake Tana basin in Ethiopia, Hydrol. Earth Syst. Sci., 18, 4871–4881, https://doi.org/10.5194/hess-18-4871-2014, 2014.

Xu, C.-Y.: From GCMs to river flow: a review of downscaling methods and hydrologic modelling approaches, Prog. Phys. Geogr., 23, 229–249, 1999.

Yates, D. N. and Strzepek, K. M.: An assessment of integrated climate change impacts on the agricultural economy of Egypt, Climatic Change, 38, 261–287, 1998a.

Yates, D. N. and Strzepek, K. M.: Modeling the Nile Basin under climatic change, J. Hydrol. Eng., 3, 98–108, 1998b.

Yimer, G., Jonoski, A., and Van Griensven, A.: Hydrological response of a catchment to climate change in the Upper Beles River Basin, Upper Blue Nile, Ethiopia, Nile Basin Water Engineering Scientific Magazine, 2, 49–59, 2009.

8

Climate-driven disturbances in the San Juan River sub-basin of the Colorado River

Katrina E. Bennett[1], **Theodore J. Bohn**[2,3], **Kurt Solander**[1], **Nathan G. McDowell**[1,a], **Chonggang Xu**[1], **Enrique Vivoni**[3,4], **and Richard S. Middleton**[1]

[1]Earth and Environmental Sciences, Los Alamos National Lab, Los Alamos, NM 87545, USA
[2]Julie Ann Wrigley Global Institute of Sustainability, Arizona State University, Tempe, AZ 85287, USA
[3]School of Earth and Space Exploration, Arizona State University, Tempe, AZ 85287, USA
[4]School of Sustainable Engineering and the Built Environment, Arizona State University, Tempe, AZ 85287, USA
[a]current address: Pacific Northwest National Laboratory, Richland, WA 99354, USA

Correspondence: Katrina E. Bennett (kbennett@lanl.gov)

Abstract. Accelerated climate change and associated forest disturbances in the southwestern USA are anticipated to have substantial impacts on regional water resources. Few studies have quantified the impact of both climate change and land cover disturbances on water balances on the basin scale, and none on the regional scale. In this work, we evaluate the impacts of forest disturbances and climate change on a headwater basin to the Colorado River, the San Juan River watershed, using a robustly calibrated (Nash–Sutcliffe efficiency 0.76) hydrologic model run with updated formulations that improve estimates of evapotranspiration for semi-arid regions. Our results show that future disturbances will have a substantial impact on streamflow with implications for water resource management. Our findings are in contradiction with conventional thinking that forest disturbances reduce evapotranspiration and increase streamflow. In this study, annual average regional streamflow under the coupled climate–disturbance scenarios is at least 6–11 % lower than those scenarios accounting for climate change alone; for forested zones of the San Juan River basin, streamflow is 15–21 % lower. The monthly signals of altered streamflow point to an emergent streamflow pattern related to changes in forests of the disturbed systems. Exacerbated reductions of mean and low flows under disturbance scenarios indicate a high risk of low water availability for forested headwater systems of the Colorado River basin. These findings also indicate that explicit representation of land cover disturbances is required in modeling efforts that consider the impact of climate change on water resources.

1 Introduction

Widespread forest disturbances are projected to increase with climate change (McDowell et al., 2016; Allen et al., 2010; Van Mantgem et al., 2009) and this will have major implications for ecosystem services (Anderegg et al., 2013). These ecosystem services impact the provision of food, water, and energy and therefore necessitate a more robust understanding of (a) how the landscape will respond to the associated shifts in timing and frequency of streamflow and (b) the dominant processes driving these changes. However, the impacts of coupled disturbances (e.g., climate-induced pest outbreaks, fires, drought) in forest ecosystems on water resources remain understudied, despite its importance for natural resource management and energy production in affected basins around the globe. For instance, climate-induced changes in forests will feedback to the climate system by altering the albedo and reducing the carbon sink, which is anticipated to further transform ecosystems in either positive or negative ways (Dale et al., 2001). This is particularly salient for regions of the USA such as the Colorado River basin (CRB), where forest cover is anticipated to be significantly impacted by a higher incidence of wildfire, drought, and pest infestations (Williams et al., 2013).

To date, predictions of future streamflow in forested river basins under future changes in climate and land cover have exhibited wide disagreement as to the strength and even the direction of change. Water yields and peak streamflows in North American river basins are anticipated to either increase, decrease, or show no response to changing forest cover (see Table 2 in Adams et al., 2012; Schnorbus et al., 2010; Guardiola-Claramonte et al., 2011; Somor, 2010; McDowell et al., 2018). Causes of the reported changes have been related to topography (Schnorbus et al., 2010) and climate variability (Allen et al., 2015). Other reported causes are secondary impacts occurring as a result of forest mortality that enhance processes such as radiation (Royer et al., 2011; Varhola et al., 2010), changes in albedo (Winkler et al., 2010), evapotranspiration (Zou et al., 2010; Kang et al., 2006), groundwater availability (Bearup et al., 2014, 2016), and soil moisture states (Dale et al., 2001). Snow has been reported to play an important role in changes in streamflow (Solander et al., 2017; Bennett et al., 2015) through increased snow accumulation and snowmelt (Bewley et al., 2010; Boon, 2007), snow cover duration (Boon, 2009), reduced interception, canopy sublimation, and evapotranspiration in disturbed forests (Livneh et al., 2015b). Temperature changes were noted to play a role in decreasing streamflow in one of eight catchments examined in Somor (2010). Some research points to the fact that we have a limited understanding on why or how streamflow changes in the future under land cover shifts (Bonan, 2008; Buma and Livneh, 2015, 2017).

Disturbances inducing forest loss are poorly represented or entirely absent from earth system models (ESMs) and eco-hydrologic models (Brovkin et al., 2013a; Scheller and Mladenoff, 2007; McDowell et al., 2011). In the Coupled Model Intercomparison Project phase 5 (CMIP5), several ESMs contain dynamic vegetation models (Collins et al., 2011; Watanabe et al., 2011; Brovkin et al., 2013b). ESMs in CMIP5 model fractional plant functional types (PFTs) and include feedbacks to the climate and land surface driven by atmospheric simulations. However, in general, ESMs contain simplifications of the explicit scenario of evolving landscape ecology and do not include a full suite of disturbances (e.g., pests, drought, wildfire), which are extremely difficult to simulate due to the computational expense associated with the coupling between disturbances and the disparate timescales involved. Therefore, current ESMs are limited in their assessment of the feedbacks driven by the disturbances and the threat of complete system crash, e.g., the loss of large tracts of forests and/or strongly declining water resources (McDowell et al., 2018).

Climate impact studies from the southwestern USA highlight the strong influence of changing temperature and precipitation on forest distributions (Dale et al., 2001), forest mortality (Allen et al., 2015), and streamflow (McCabe and Wolock, 2007; Nash and Gleick, 1991). Specifically, the strong interaction between forests and water (Anderson et al., 1976) means that forest disturbances will have a large impact on water volume and the timing of streamflow. Streamflow decreases were previously reported for the upper CRB at Lee's Ferry using the Variable Infiltration Capacity (VIC) hydrologic model (Liang et al., 1994, 1996) forced with future climate change scenarios alone and reported decreased future projected streamflow (Christensen and Lettenmaier, 2002). A separate study reported that the CRB streamflow is highly sensitive to precipitation and temperature shifts, with large reductions in streamflow estimated for small increases in precipitation and temperature (Nash and Gleick, 1991) resulting from enhanced rates of evapotranspiration.

The strong link between forests and water, coupled with the emerging threat to future forest viability and integrity, provides a major impetus to study the combined effects of climate and land cover change on streamflow in the CRB. To this end, we apply a hydrologic model that incorporates projected climate changes as well as forest and land cover changes based on recent research. Unlike many other studies, we have included the impact of regrowth of shrubs into our modeling approach since this is the most likely outcome in the CRB (Rother et al., 2015). We determine the direction and shift in streamflow under different scenarios of climate and land cover changes, and we also identify the physical alterations occurring within watersheds across scales and forest compositions to reveal the driving mechanisms behind the streamflow changes. Knowledge of the physical mechanisms and dominant triggers for streamflow alterations is critical because it will allow decision makers to make more-informed assessments of near- and long-term water operations in a water-constrained world with a changing climate (Zou et al., 2010).

This paper is arranged as follows: Sect. 2 details information on forest disturbances and the impact of climate change on streamflow, Sect. 3 outlines our study site and methodology, and Sects. 4, 5, and 6 provide results, discussion, and conclusions, respectively.

2 Climate-driven forest disturbances and climate change impacts on streamflow

As fire, pests, and drought modify forests in response to climate change, a number of important energy and water fluxes become altered (Adams et al., 2012). Land covers with a low canopy profile and a small crown cover, e.g., shrubs or bare ground, partition water and energy in different manners compared to forests that have a large crown cover. For example, shrubs retain snow pack later into the melt season thus snow is able to reach the ground and accumulate (Boon, 2007, 2009). This is largely due to reduced canopy cover which results in reduced interception and sublimation or transpiration (Pomeroy et al., 1998; Bearup et al., 2014) from this above-ground storage reservoir, as well as higher solar radiation and wind speeds through the open areas (Harpold et al., 2014). Snow also melts quicker in response to increased shortwave radiation and turbulent heat transfers linked to negative longwave radiative fluxes (Burles and Boon, 2011). The responses have also been linked to decreased albedo due to higher litter and/or darker soils associated with dying trees (Bewley et al., 2010; Winkler et al., 2010). Other effects of tree mortality on the water balance include changes to soil moisture states, changes in groundwater recharge (Bearup et al., 2016), and potential feedbacks to the atmosphere (Bonan, 2008).

Variations in the type and magnitude of forest disturbances can also strengthen or dampen some of these effects (Buma and Livneh, 2015). For example, forest wildfires change soil properties and litter depths, which in turn alter infiltration and runoff processes. In addition, mountain-pine-beetle infestations change canopy conditions but do not change soil properties or moisture states. Drought may affect the forest canopy and the ability of soils to infiltrate water as trees die and soils desiccate (Adams et al., 2012). Effects may also be impacted by the type of post-disturbance regrowth and extent of disturbance that occurs, such as increases or a difference in shrubs, grasses, forbs coverage, or differences in the composition of tree species that replace the forests (Buma and Livneh, 2017). For instance, water-yield declines have been associated with changes in species compositions of forests in the southern Appalachian Mountains (Caldwell et al., 2016). A recent study indicates that soil water residence time is a key factor in water availability post-disturbance owing to limited (or enhanced) evapotranspiration processes (Buma and Livneh, 2017).

Selection of the appropriate timescale is an important aspect to consider in vegetation disturbance studies. For example, research focused on forest disturbance over a single summer season, such as post-fire hydrology in New Mexico shows that surface runoff and recharge both rise following fire (Atchley et al., 2018). However, these studies do not measure the year-long change in water balance, which is critical for snow-driven systems. Examination of water partitioning in disturbed forests within time frames of less than 5 years may not adequately resolve effects such as forest succession on the hydrologic regime of study basins (Brown et al., 2005), which can result in a significant overestimation of the impacts of changing forests on water yields over the long term (Pugh and Gordon, 2013). Indeed, most studies that encompass forest disturbance monitoring of greater than 5 years point to increased evapotranspiration from understory regrowth and an associated decline or mitigating effect of the forest cover removal (Brown et al., 2014; Biederman et al., 2014; Guardiola-Claramonte et al., 2011). Other research has shown that vegetation management, such as cutting or thinning of forests in the face of climate change, could be used to ameliorate the impacts of reduced streamflow in the CRB, noting that these effects would only last for a period of approximately 10 years (Zou et al., 2010).

Finally, spatial heterogeneity also plays an important role in terms of changing hydro-ecology in the wake of disturbances. Studies that focus on plot-scale results, where the disturbed forest is the primary cover type, illustrate different responses compared to locations where impacted trees are only a component of the overall land cover, including grasses, shrubs, and non-impacted tree species (Biederman et al., 2014; Pugh and Gordon, 2013). Recent studies examining climate change and extreme wildfire on runoff erosion found that peak streamflow sediment yield will increase with climate change and fire severity due to the lack of spatial heterogeneity in land cover types (Gould et al., 2016). A study by Penn et al. (2016) compared hillslope to watershed-scale responses and found a muted effect on the watershed scale in a headwater basin of the Colorado Rocky Mountains.

3 Methods

3.1 Study site

To understand the impact of forest disturbances on streamflow under climate changes, on different temporal scales and spatial settings, we implemented the VIC model for the San Juan River basin, a sub-basin of the CRB, to simulate streamflow responses to future changes in temperature, precipitation, and land cover. The San Juan is a major headwater basin of the CRB, accounting for 15 % of streamflow and 22 % of the area of the upper CRB. Spanning four states – the Four Corners – the San Juan watershed is also critical for thermoelectric and hydropower generation, substantial oil and gas development, and extensive irrigated agriculture. Temperature ranges from −2 to 23 °C from January to July, respectively, while average annual precipitation is ~ 666 mm. The San Juan basin captures the diversity present across the CRB. For instance, high-elevation (> 4000 m) Colorado mountain ranges and large snowmelt-driven rivers comprise the upper San Juan basin. The lower San Juan basin, located in New Mexico and Arizona, is flat, semi-arid, and representative of the lower CRB, with intermittent streams that drain into the main tributary of the San Juan during the summer when they are charged by summer monsoonal rains. The San Juan River eventually drains into the CRB, just below the town of Bluff, Utah (Fig. 1).

3.2 Hydrologic model

For this work, we used the Variable Infiltration Capacity (VIC) model version 4.2 (Bohn and Vivoni, 2016; Liang et al., 1994) at a 1/16° (6 km) spatial resolution. In each grid cell, VIC simulates vertical energy and water dynamics at a 1 h time step for a mosaic of land cover tiles underlain by a 3-layer soil column. Sub-grid heterogeneity in infiltration is represented by a statistical distribution (the variable infiltration capacity curve). Surface runoff is generated via saturation excess, while sub-surface runoff is characterized by the non-linear baseflow curve of Franchini and Pacciani (1991). VIC version 4.2 includes fractional canopy coverage derived

Table 1. Calibration, validation, and entire period of record monthly statistics for simulated to naturalized streamflow for the San Juan at Archuleta, NM, and San Juan at Bluff, UT.

	Objective function	San Juan at Archuleta, NM	San Juan at Bluff, UT
Calibration (2006–2010)	volume bias	0.91	0.90
	Nash–Sutcliffe efficiency	0.78	0.76
	Nash–Sutcliffe efficiency log	0.77	0.75
Validation (2001–2005)	volume bias	0.93	0.93
	Nash–Sutcliffe efficiency	0.83	0.60
	Nash–Sutcliffe efficiency log	0.77	0.43
Entire period (2001–2010)	volume bias	0.90	0.91
	Nash–Sutcliffe efficiency	0.81	0.67
	Nash–Sutcliffe efficiency log	0.78	0.58

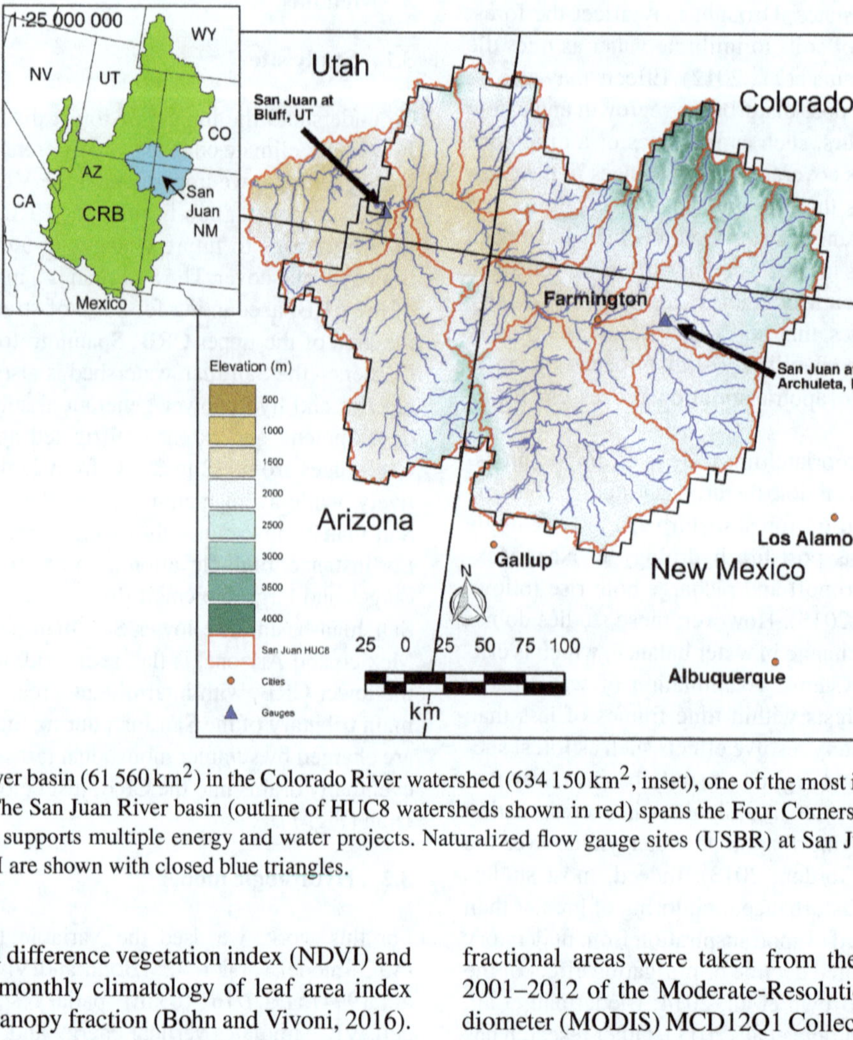

Figure 1. San Juan River basin (61 560 km^2) in the Colorado River watershed (634 150 km^2, inset), one of the most important water resources in the western USA. The San Juan River basin (outline of HUC8 watersheds shown in red) spans the Four Corners region (CO, UT, AZ and NM) of the USA, and supports multiple energy and water projects. Naturalized flow gauge sites (USBR) at San Juan at Bluff, UT and San Juan at Archuleta, NM are shown with closed blue triangles.

from the normalized difference vegetation index (NDVI) and a spatially varying monthly climatology of leaf area index (LAI), albedo, and canopy fraction (Bohn and Vivoni, 2016). Historical climate data used to run VIC (daily precipitation, minimum and maximum temperature, and wind speed) were obtained from existing gridded data sets for the USA (Livneh et al., 2015a). These daily fields were disaggregated to hourly intervals within the VIC model via algorithms as described in Bohn et al. (2013), which also estimated hourly short- and long-wave radiation and humidity. Land cover fractional areas were taken from the average of the years 2001–2012 of the Moderate-Resolution Imaging Spectroradiometer (MODIS) MCD12Q1 Collection 5 Plant Functional Type (PFT) product of Friedl et al. (2010), using the International Geosphere-Biosphere Program (IGBP) classification. Repeating climatological seasonal cycles of vegetation parameters (LAI, canopy fraction, and albedo) were derived from the MODIS Collection 5 MOD15A2, MCD43A3, and MOD13A1 products (Myneni et al., 2002; Schaaf et al., 2002; Huete et al., 2002) over the period 2000–2012, and aggregated spatially over the MCD12Q1 land cover classes within each 1/16° grid cell.

Soil physical properties (e.g., bulk density, saturated hydraulic conductivity, quartz content) were derived from global data sets such as the United Nations Food and Agriculture Organization (FAO) Digital Soil Map of the World (FAO, 1998). Vegetation structural parameters, such as stomatal and canopy resistances, were taken from Ducoudré et al. (1993). Several other parameters were calibrated empirically: D_2 and D_3 (the thicknesses of the 2nd and 3rd soil layers); b_{infilt} (the infiltration capacity curve shape parameter); D_{s}, W_{s}, and $D_{\mathrm{s_{max}}}$ (non-linear baseflow parameters); and α_{snow} (the albedo of newly fallen snow). The model was calibrated using an automatic calibration tool (Yapo et al., 1998) to correct streamflow biases against the USA Bureau of Reclamation (USBR) naturalized gauged monthly streamflow (2006–2010) for the San Juan River basin at Bluff, UT (Fig. 1; Table 1). Our calibration achieved a Nash–Sutcliffe efficiency of 0.76 over the calibration period and 0.60 over the validation period for the San Juan at Bluff, UT, USBR monthly naturalized flow data (Table 1).

3.3 Climate and vegetation change scenarios

Our study focused on three vegetation projections and four climate projections for a total of 12 different scenarios (Fig. 2). The four climate projections employed to drive VIC were based on ESM simulations from CMIP5 (Taylor et al., 2012) climate data including daily temperature, precipitation, and wind speed, downscaled using the multivariate adaptive constructed analogue (MACA) approach (Fig. 2a and b; Abatzoglou and Brown, 2012) and again disaggregated to hourly intervals via the algorithms described in Bohn et al. (2013). We selected the four ESMs from CMIP5 because they implemented dynamic vegetation processes: HadGEM2-ES (Collins et al., 2011; Cox, 2001), MIROC-ESM (SIEB-DGVM; Watanabe et al., 2011; Sato et al., 2007), MPI-ESM-LR (JSBACH; Giorgetta et al., 2013; Reick et al., 2013), and IPSL-CM5B-LR (ORCHIDEE; Krinner et al., 2005). We used the representative concentration pathway (RCP) 8.5 emissions scenario, which stipulates strongly increasing emissions by 2100 (Van Vuuren et al., 2011) and corroborates current emissions on par with RCP 8.5 (Le Quéré et al., 2015).

The three vegetation projections used in this study are: (1) climate-only, which assumes static land cover (i.e., vegetation types do not change); (2) dynamic, which uses the dynamic vegetation changes present within the four CMIP5 ESMs (Fig. 2c); and (3) disturbed, which uses disturbance projections based on empirical statistical estimates of forest mortality in the southwest USA (Fig. 2d; McDowell et al., 2016). Vegetation classes were aggregated to six dominant cover types from 16 classes in the IGBP vegetation classification and from 9–13 classes in the ESMs (Table 2). Vegetation changes observed in the dynamic and disturbed scenarios were applied to the historical MODIS vegetation to alter forest coverage for future runs using a simple delta-change approach. For both the dynamic and disturbed scenarios, historical forest cover fractions were reduced and concordantly replaced by shrubs in increments of ~ 10 years (2006–2010, 2010–2020, and so forth); we then ran the projections in 10 year segments, with each segment having a new (constant) forest fraction and starting with the state from the previous time period. Forest cover was reduced by approximately 90 % by 2100 for the disturbed scenario (McDowell and Allen, 2015; McDowell et al., 2016; Fig. 2c, d). Table 3 contains forest and shrub vegetation fractions for each scenario, and LAI, canopy fraction, and albedo for forest and shrubs used in all scenarios from the average of grid cells with greater than 50 % forest cover in the San Juan River basin.

We ran the 12 different scenarios for 1950–2099. We analyzed daily, seasonal, and annual streamflow as well as monthly statistics of temperature, precipitation, evapotranspiration, and snow water equivalent (SWE) to understand changes in the water balance in the San Juan River basin. In addition, we investigated the aridity effect upon water availability under forest disturbances using a one-cell analysis. Some studies have suggested an aridity effect, whereby basins with less than 500 mm annual precipitation will see streamflow decrease and vice-versa (Guardiola-Claramonte et al., 2011; Adams et al., 2012), although this finding is not supported in all work (Brown et al., 2010; Caldwell et al., 2016). For the single cell, we considered climate change by adding +3 °C (warm) and +6 °C (hot) to the temperature time series and changing precipitation by 20 or −20 %. We then changed vegetation characteristics in the single cell to simulate changes in LAI or fractional vegetation spacing (canopy spacing).

4 Results

4.1 Changing climate and land cover

Temperature and precipitation changes are consistent with previous modeling efforts for the region (Gangopadhyay et al., 2011). The four ESMs projected consistent increases in the annual average temperature (4.3–7.1 °C, mean change of 5.7 °C) but variable changes in the annual average precipitation (both increases and decreases, −6.6–8.2 %, mean change of 1.7 %) for the San Juan River basin by comparing the last 30 years of this century to the last 30 years of the previous century (not shown). For both temperature and precipitation, changes are most dramatic and variable after the 2050s, concurrent with increasing greenhouse gas emissions. These four ESMs represent the range of warm/hot and wet/dry changes for the San Juan River basin by the 2080s in CMIP5 (Taylor et al., 2012; Brekke et al., 2013). Seasonal variations in temperature and precipitation change indicate

Table 2. IGBP classification names and codes from MODIS, our new classification, and remapped values (RM). Values for each ESM follows MIROC-ESM, MPI-ESM-LR, HadGEM2-ES, and IPSL-CM5A-LR. For each ESM, the original code and the remapped (RM) values are given.

IGBP classification	Code	Our class	RM	MIROC-ESM	Code	RM	MPI-ESM-LR	Code	RM	HadGEM2-ES	Code	RM	IPSL-CM5B-LR	Code	RM
Evergreen needleleaf forest	1	Evergreen forests	1	Tropical forests	1	1	Glacier	1	nc	Broadleaf trees	1	2	Bare soil	1	6
Evergreen broadleaf forest	2	Evergreen forests	1	Temperate/boreal evergreen	2/4	1	Tropical broadleaf evergreen	2	1	Needleleaf trees	2	1	Tropical broad-leaved evergreen	2	1
Deciduous needleleaf forest	3	Deciduous forests	2	Temperate/boreal deciduous	3/5	2	Tropical broadleaf deciduous	3	2	C_3/C_4 grass	3/4	3	Tropical broad-leaved raingreen	3	2
Deciduous broadleaf forest	4	Deciduous forests	2	C_3/C_4 grass	6/7	3	Extra-tropical evergreen	4	1	Shrubs	5	4	Temperate/boreal needleleaf evergreen	4/7	1
Mixed forest	5	Evergreen forests	1	Crop	8	5	Extra-tropical deciduous	5	2	Urban/inland water/bare/ice	6-9	6	Temperate broad-leaved evergreen	5	1
Closed shrublands	6	Shrubs	4	Pasture	9	3	Raingreen shrubs	6	4				Boreal needleleaf summergreen	9	2
Open shrublands	7	Shrubs	4	Bare ground/residual	10/11	6	Deciduous shrubs	6	4				Temperate/boreal broad-leaved summergreen	6/8	2
Woody savannas (40–60 % tree cover)	8	Grass	3				C_3/C_4 grass	8/9	3				C_3/C_4 grass	10/11	3
Savannas (10–40 % tree cover)	9	Grass	3				C_3/C_4 pasture	10/11	3				C_3/C_4 agriculture	12/13	5
Grasslands	10	Grass	3				C_3/C_4 crops	12/13	5						
Permanent wetlands (assuming like grass)	11	Grass	3												
Croplands (corn)	12	Crops	5												
Urban and built-up (assuming sparse grass)	13	Urban/water/bare	6												
Cropland/natural vegetation mosaic	14	Crops	5												
Snow and ice (assuming sparse grass)	15	Urban/water/bare	6												
Barren or sparsely vegetated (assuming sparse grass)	16	Urban/water/bare	6												

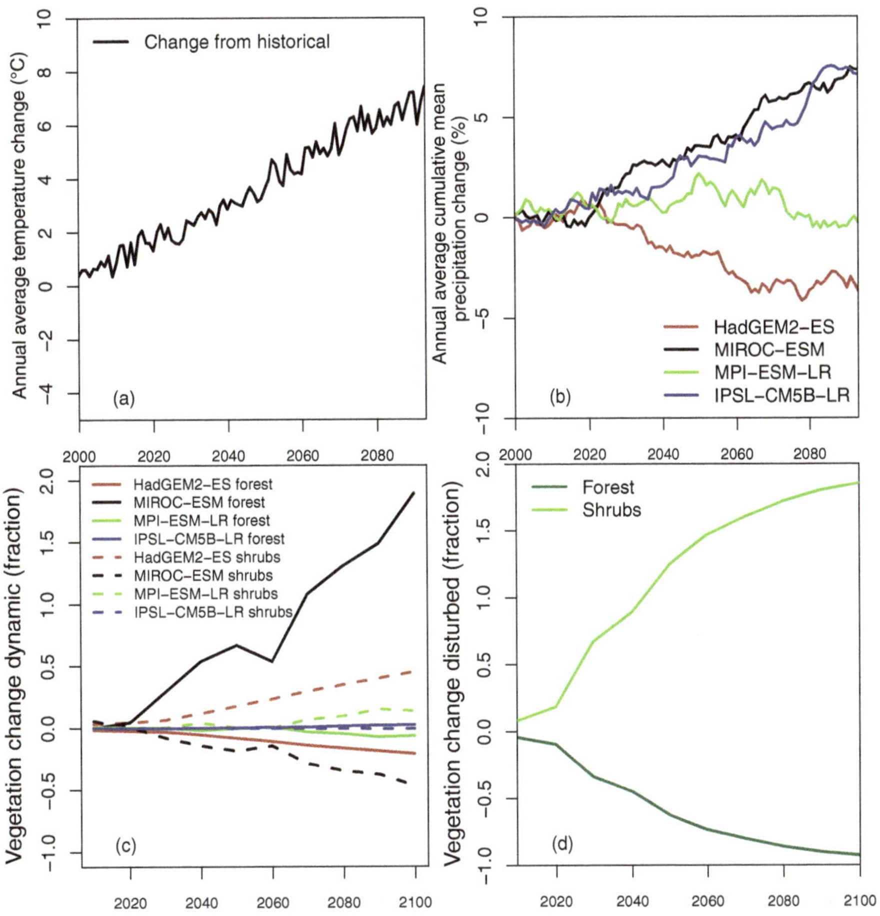

Figure 2. Temperature (°C, **a**), precipitation (%, **b**), and land cover (**c, d**) changes to the 2080s (2070–2099) averaged over the ESMs and spatially over the San Juan River basin (61 560 km^2 or 1570 grid cells). CMIP5 dynamic forest disturbances for the four ESMs as a difference from the historical climatology (1970–1999) illustrate a more moderate change with HadGEM2-ES projecting the largest decline and MIROC-ESM projecting an increase in forest cover (**c**). The disturbed vegetation scenario based on McDowell et al. (2016) (**d**) illustrates strongly declining forest covers (> 50 % forest loss) and increasing shrub covers.

important regional process shifts. For instance, summer and winter differences show that the summer is warming slightly more than the winter (5.8 °C compared to 5.6 °C). Annual average fall and winter precipitation is projected to increase, while spring and summer precipitation is projected to decrease slightly (−1 %) with a large range in variability across the basin. We note that the signals of change for both temperature and precipitation differ from the results offered by the U.S. Department of the Interior Bureau of Reclamation data sets at 1/8th of a degree and downscaled using a slightly different technique (Bureau of Reclamation, 2011).

Our study identifies a key challenge in representing land cover: land cover change is represented differently by each of the four ESMs due to the variable representations of land use and the application of different dynamic models within each ESM (Arora, 2002). The dynamic trajectory of change is variable depending on the ESM considered (Fig. 2c). The MIROC-ESM model projects increasing forest cover, while the HadGEM2-ES model projects the most amount of change in terms of forest loss. Both IPSL-CM5B-LR and MPI-ESM-LR show only a small amount of change in terms of land cover shifts by the 2080s. The disturbed scenario reflects regional changes expected in the southwestern USA (McDowell et al., 2016) and therefore projects a more severe, and likely realistic, scenario in terms of projected forest cover change in the San Juan River basin (Fig. 2d).

4.2 Changing streamflow and water balances

Annually, simulated streamflow in the San Juan River basin under the climate-only scenario exhibits differences of −15 to 45 %, while the disturbed scenario indicates a shift of −21 to 34 % for the 2080s as compared with historical streamflow, dependent on the ESM (Fig. 3). The dynamic scenario changes the streamflow by −16 to 50 %, but is generally very similar to the climate-only scenario. An exception to this is the MIROC-ESM dynamic scenario, which projects an in-

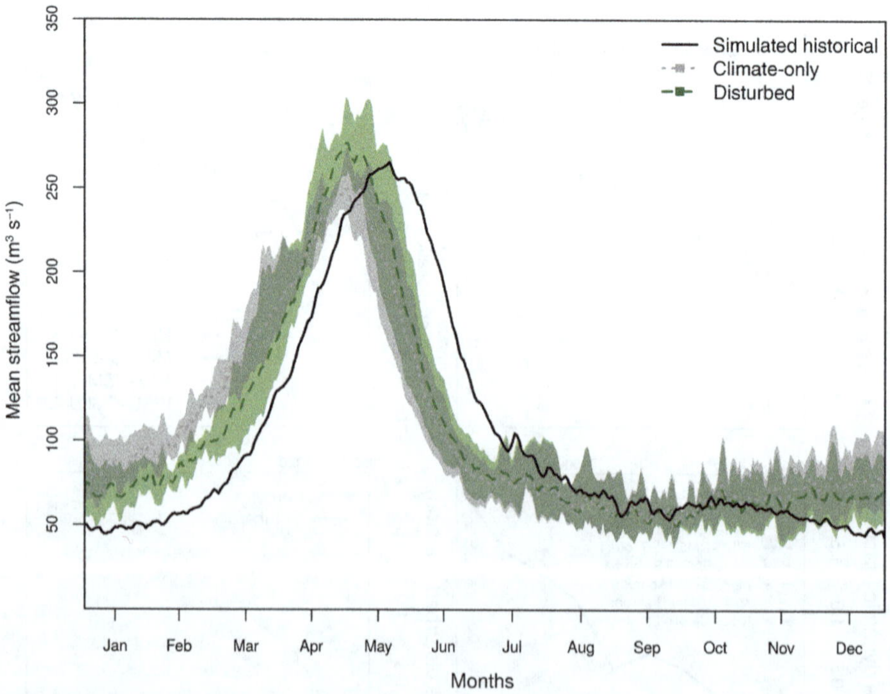

Figure 3. Monthly average future streamflows ($m^3\,s^{-1}$, 2070–2099) compared to historical (1970–1999) for the San Juan River basin at Bluff, UT. The range of responses for each ESM is represented by the semi-transparent envelope around the lines. The simulated historical streamflow is shown in black (solid line, no envelope), the climate-only scenarios are shown in grey (dashed dark grey line, grey envelope), and the disturbed scenarios are shown in green (dashed dark green line, green envelope).

crease in streamflow during winter and a decrease in summer peak flow in response to increasing forest cover and thus decreasing shrubs represented in this model (Fig. 2c). Due to the lack of a large distinction in the vegetation changes under the dynamic scenarios and resultant similarities between the climate-only and dynamic scenarios, we focus on the differences between the climate-only and the disturbed scenarios for the remainder of this paper.

Seasonal streamflow in the climate-only scenario versus historical simulated streamflow illustrates a shift in the timing of peak flow and increased winter streamflow in the San Juan River basin (Fig. 3). Peak streamflow occurs approximately 1 month earlier, owing to earlier snowmelt due to a temperature increase. This shift means that winter flows are higher and also indicates more mid-winter warming events. Under disturbed forest cover conditions, seasonal streamflow shows a different hydrograph that represents a shift in timing of winter streamflow and a change in the magnitude of both low (December–January) and high (April–May) streamflow compared to the climate-only scenario (green envelope and line, Fig. 3). The VIC simulations driven by disturbed scenarios project a lower late-fall and winter streamflow, with a delay in spring ice melt and subsequent increase in the pulse of peak streamflow during April–May. Recessional streamflow (May–July) is also slightly higher in the disturbed scenario than the climate-only scenario, resulting in greater water availability in summer (Fig. 3).

The mechanisms causing these differences in streamflow response to climate change and forest disturbance are illustrated in Fig. 4 for forest-dominant (more than 50 %) regions. In the climate-only scenario, the streamflow response is dominated by the impact of temperature on snow pack. Warmer winter temperatures and reduced March snowfall (Fig. 4a) lead to a reduction in SWE, snowmelt, and sublimation from both the pack and the canopy (Fig. 4b–e). Warmer temperatures, earlier snowmelt, and greater April rainfall subsequently lead to increases in soil evaporation (Fig. 4f) and transpiration (Fig. 4g) in the spring. Warmer temperatures also lead to increased canopy evaporation in late summer (Fig. 4h). However, the higher rates of soil evaporation and transpiration in the spring deplete the middle layer soil moisture (Fig. 4i), which diminishes their rates in the summer (Fig. 4f, g). In the fall, higher rainfall and warmer temperatures lead to greater soil evaporation but only minimal replenishment of soil moisture. Both bottom layer soil moisture (Fig. 4j) and total (surface and subsurface) runoff (Fig. 4l) exhibit earlier and smaller peaks in the spring, reflecting the earlier melting of the reduced snow pack and lower levels in summer due to greater evapotranspiration in the spring.

Two factors in the disturbed scenario further impact the snow pack, partially compensating for the effects of climate change. First, replacement of forest with shrubs leads to an increase (relative to the climate-only scenario) in the on-the-ground snow pack accumulation (Fig. 4b), a prominent

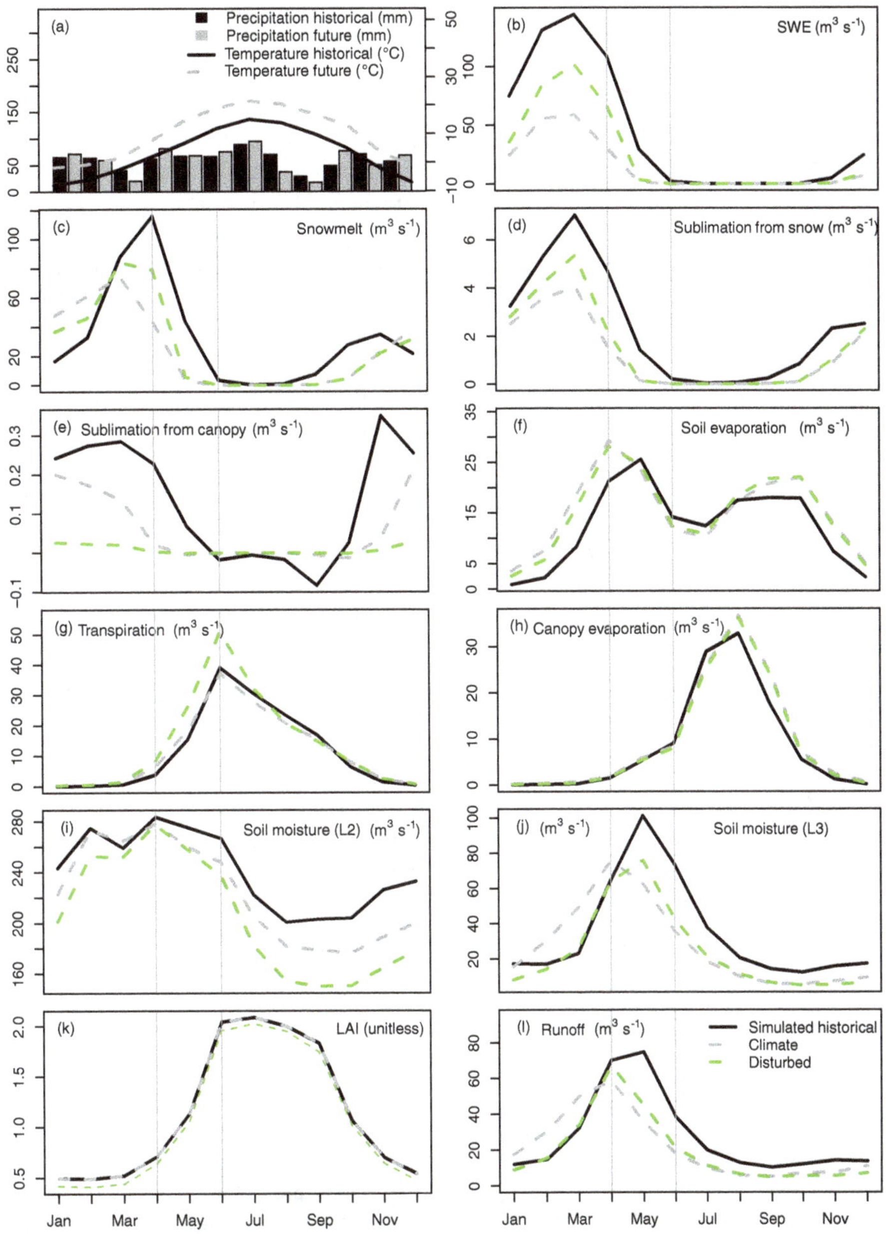

Figure 4. Monthly water balance panels for forested (greater than 50 %) regions in the San Juan River basin. Panels include precipitation (mm) and temperature (°C) for historical (1970–1999, black) and future (2070–2099, grey, **a**), snow water equivalent (SWE, $m^3\,s^{-1}$, **b**), snowmelt ($m^3\,s^{-1}$, **c**), sublimation from snow ($m^3\,s^{-1}$, **d**), sublimation from canopy ($m^3\,s^{-1}$, **e**), soil evaporation ($m^3\,s^{-1}$, **f**), transpiration ($m^3\,s^{-1}$, **g**), canopy evaporation ($m^3\,s^{-1}$, **h**), soil moisture from second soil later (L2, $m^3\,s^{-1}$, **i**), soil moisture from the third soil layer (L3, $m^3\,s^{-1}$, **j**), leaf area index (LAI, unitless, **k**), and runoff ($m^3\,s^{-1}$, **l**) for simulated historical (black line), climate-only (grey line), and disturbed scenarios (green line). The months of April and June are indicated with vertical grey lines.

feature observed in disturbed forests across North America (Boon, 2007; Zou et al., 2010; Biederman et al., 2015; Brown et al., 2014; Harpold et al., 2014). Shrubs in our VIC simulations have no canopy thus they have no mechanism to intercept snow. This disturbance-driven increase in on-the-ground snow pack partially compensates for the climate-induced streamflow increase during this time of the year. In addition, the larger on-the-ground snow pack and smaller canopy snow pack are accompanied by proportionally higher and lower rates of sublimation from the ground and canopy snow packs, respectively, relative to climate-only (Fig. 4d, e). Second, the increase in snow on the ground caused by the higher shrub coverage leads to higher rates of snowmelt (approximately 30 %, Fig. 4c), releasing water during the late spring and early summer (April–June; Fig. 4l). The larger snowmelt flux leads to substantial increases in transpiration in spring and early summer, not only relative to the climate-only scenario but also relative to historical conditions (Fig. 4g). A similar increase in soil evaporation (Fig. 4f) does not occur due to the upper and middle soil layers already being saturated in the spring in the climate-only case (Fig. 4i). The high shrub transpiration rate in the disturbed scenario delays the larger snowmelt flux in reaching the bottom soil layer (Fig. 4j), leading to a delayed peak in runoff (Fig. 4j, l). Meanwhile, LAI values are similar between shrubs and forests (Fig. 4k), indicating that the water and energy balance differences are mainly due to snow process shifts between those two land cover types.

Differences in streamflow among disturbed and climate-only scenario results are most notable at fine scales (100 to $3000\,km^2$, 72 grid cells in the San Juan) where the forest cover dominates (more than 50 %) the land cover (Fig. 5). As basin size increases (smaller circles in Fig. 5), and forest cover becomes sparser with respect to other types of land cover (e.g., mixed forest, deciduous, shrubs and grass covers), the differences between the climate-only scenario and the disturbed scenario begin to decrease, which is corroborated by several other studies (Zhang et al., 2014; Anderegg et al., 2015). The difference between the two scenarios for the densest forest and smallest portion of the basin is approximately 20 %, as observed for the larger circles in Fig. 5. This dynamic occurs even under changing precipitation and temperature projections that would otherwise cause increasing or decreasing streamflow (observed in Fig. 5 color ramps that change from red to blue for precipitation and different symbols for different temperature ranges). Even as the ESMs project increasing precipitation and temperature, we see the variability in responses through the cascade of scale and land cover variability. This finding is consistent with other studies that observed that variability (forest cover composition and topography) in the area and size of forest–shrub conversion can buffer responses of streamflow or evapotranspiration shifts from climate change (Winkler et al., 2014; Harpold et al., 2014; Caldwell et al., 2016).

The dynamic between temperature, precipitation, and vegetation changes is examined in more detail by using a simple single-cell analysis where climate and vegetation is altered in a sensitivity framework. Figure 6 illustrates the single-cell changes in terms of runoff from the grid cell to investigate the aridity effect. We can see that forests and shrubs act similarly in hot/warm and dry environments, while the differences are more pronounced between forests and shrubs in wetter environments (Fig. 6a). For the vegetation changes, we see that changing forest structure (e.g., fraction of the canopy that may occur as a result of pest outbreaks) results in small shifts for forests but very large shifts for shrubs (Fig. 6b). For example, shrubs with 50 % canopy spacing produce more water than the forests for the majority of runoff conditions (below the 3rd quantile, Fig. 6b). On the other hand, under high runoff conditions (above the 3rd quantile, Fig. 6b) shrubs act similarly to regularly spaced shrubs. These changes are, gen-

Table 3. Average values for cells with forest cover greater than 50 %. Climate-only forest and shrub fractional values (top left), dynamic forest percentages for each climate model and all decades, and disturbed forests and shrubs for each decade. Below, leaf area index (LAI, unitless), albedo (unitless), and canopy spacing (fraction) for all months for forest, with shrub values in brackets.

climate-only forest	0.70										
climate-only shrubs	0.08	Decade 10	Decade 20	Decade 30	Decade 40	Decade 50	Decade 60	Decade 70	Decade 80	Decade 90	Decade 100
dynamic forest (HAD)		0.69	0.69	0.68	0.67	0.65	0.63	0.61	0.59	0.57	0.56
dynamic forest (MIR)		0.68	0.70	0.76	0.80	0.82	0.80	0.86	0.88	0.89	0.92
dynamic forest (MPI)		0.71	0.70	0.70	0.69	0.70	0.71	0.68	0.67	0.66	0.66
dynamic forest (IPSL)		0.70	0.70	0.70	0.70	0.70	0.71	0.71	0.71	0.71	0.71
dynamic forest (HAD)		0.09	0.10	0.10	0.12	0.14	0.16	0.18	0.19	0.21	0.23
dynamic forest (MIR)		0.11	0.09	0.07	0.05	0.05	0.06	0.04	0.03	0.03	0.02
dynamic forest (MPI)		0.08	0.08	0.08	0.09	0.08	0.08	0.10	0.11	0.13	0.12
dynamic forest (IPSL)		0.08	0.08	0.08	0.08	0.08	0.08	0.08	0.08	0.08	0.08
disturbed forest		0.67	0.64	0.47	0.39	0.26	0.19	0.14	0.10	0.07	0.05
disturbed shrubs		0.11	0.15	0.32	0.39	0.52	0.60	0.64	0.68	0.71	0.73

HAD = HadGEM2-ES, MIR = MIROC-ESM, MPI = MPI-ESM-LR, IPSL = IPSL-CM5-LR

Months	Jan	Feb	Mar	Apr	May	Jun	Jul	Aug	Sep	Oct	Nov	Dec
LAI (unitless)	0.55 (0.43)	0.54 (0.42)	0.57 (0.45)	0.76 (0.66)	1.20 (1.09)	2.15 (2.01)	2.17 (2.06)	2.07 (1.98)	1.92 (1.78)	1.12 (1.04)	0.75 (0.67)	0.59 (0.48)
Albedo (unitless)	0.21 (0.28)	0.22 (0.28)	0.20 (0.24)	0.15 (0.17)	0.11 (0.12)	0.10 (0.11)	0.11 (0.12)	0.11 (0.11)	0.10 (0.11)	0.10 (0.11)	0.12 (0.14)	0.12 (0.14)
Canopy spacing (fraction)	0.10 (0.05)	0.08 (0.04)	0.09 (0.05)	0.17 (0.13)	0.39 (0.33)	0.68 (0.61)	0.75 (0.69)	0.76 (0.70)	0.69 (0.63)	0.47 (0.41)	0.33 (0.25)	0.15 (0.09)

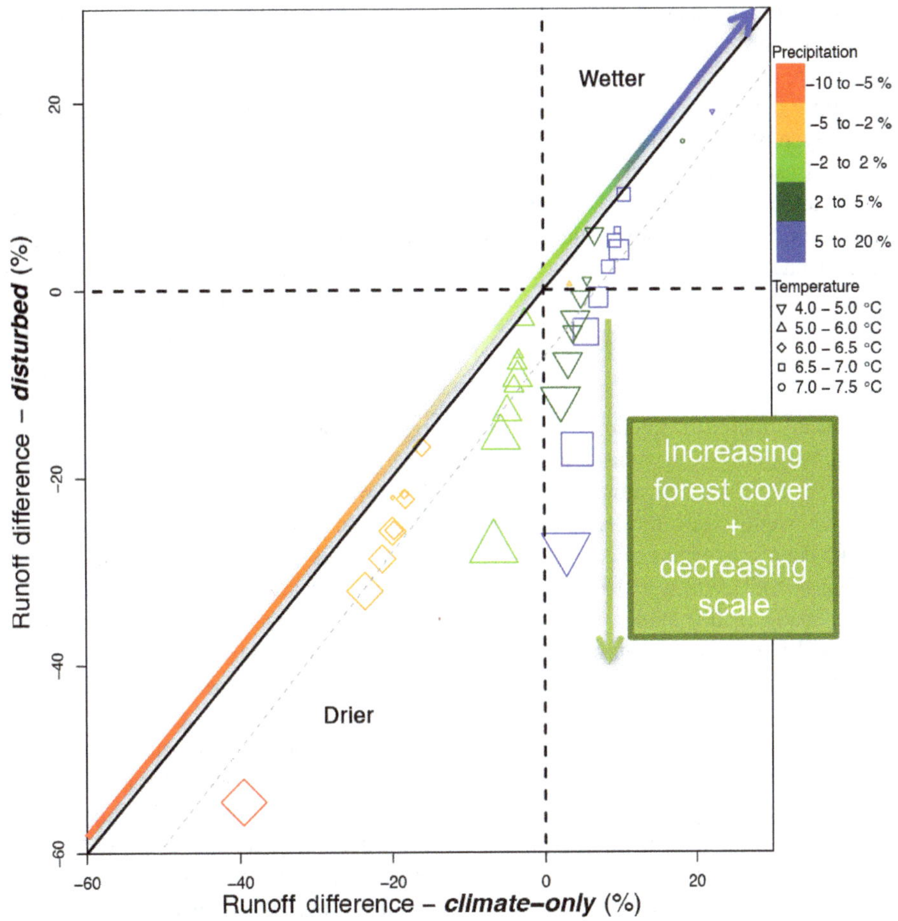

Figure 5. Runoff differences for the climate-only scenario (x axis) plotted against disturbed scenario (y axis). Increasing sizes indicate results from the entire San Juan River basin (smallest), to all forests (next largest size), and then forests with greater than 10, 50, 70, and 90 % coverage (largest). All changes are shown as projected differences from the historical (in % change).

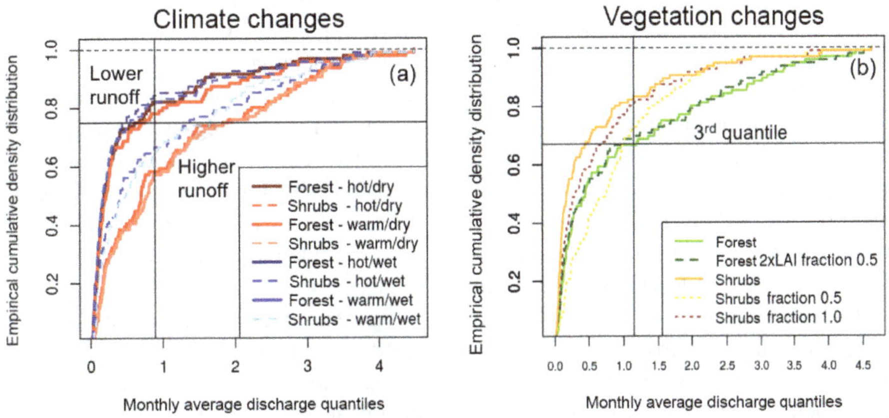

Figure 6. Cumulative density distributions are used here to illustrate the variations in **(a)** climate and **(b)** vegetation parameterizations for a single cell in the San Juan River basin. Climate alterations are shown for a hot/wet, hot/dry, warm/wet and warm/dry scenarios for both forest and shrubs. For the vegetation parameterization, forests, forests with 2 times leaf area index (LAI), shrubs, and shrubs with fractions of 0.5 and 1.0 are shown. A shrub fraction of 1 indicates that there are no spaces between the plants. The third quantile in the average of the distributions is indicated in each plot with intersecting lines.

erally speaking, on the same order as changes within the climate. Therefore, we hypothesize that larger changes in hydrology under disturbances are more likely to occur not from the forest disturbance itself but the secondary effects such as regrowth of shrubs, the type of regrowth, and the pattern of that regrowth on the landscape. This is why timescale is such an important consideration in studies of this nature, as regrowth patterns are key to understanding how water is partitioned across a disturbed landscape.

5 Discussion

Climate change in the CRB is anticipated to cause large impacts to water resource sustainability (Christensen and Lettenmaier, 2007; Rasmussen et al., 2014; Dawadi and Ahmad, 2012). However, to our knowledge, few modeling studies have considered the impacts of climate change coupled with changes in vegetation (Buma and Livneh, 2015; Carroll et al., 2017; Pribulick et al., 2016). In this study, we incorporated changes from the CMIP5 dynamic vegetation models and from an estimate of forest mortality (McDowell et al., 2016) to consider the impacts of both climate and vegetation changes on the water balance in a headwater system of the CRB, the San Juan River basin. We found that failing to consider climate change coupled with vegetation disturbances could result in a $\sim 10\,\%$ over-estimation of the annual water availability for this basin. For a river system such as the CRB that is already gravely stressed, $10\,\%$ less water in the system may lead to significant water management challenges. Furthermore, considering seasonality in flows, the changes we illustrated in our monthly and seasonal scenarios indicate that less water will runoff during spring with more water arriving at peak melt. This could lead to water shortages and flooding, and lead to planning issues for short-term water delivery.

In this work, we considered not only the impacts of changes to streamflow but also the reasons why streamflow is changing. As in other studies, we found that at the size of the basin and the land cover variability can obscure the signal of change (Biederman et al., 2015; Penn et al., 2016). When we consider forested regions only, we are able to understand how and why streamflow is projected to change under the disturbed conditions. The main mechanism that is shifting streamflow is the manner in which shrubs impact the water balance during the cold and warm seasons. Snow pack is retained further into the melt season and when snow starts to melt, it melts more quickly and results in higher peak flows. These peak flows, however, occur during a time that the shrubs are using the water, resulting in large transpiration losses. We found differences in canopy evaporation, soil evaporation, and sublimation, but these differences are quite small overall when comparing the overall volume of water held back in the snowpack (SWE) and leaving via shrub transpiration. Overall, the differences in the mechanisms and delivery of the water changed, and this results in a $20\,\%$ reduction in the amount of water that is available for runoff through the year compared to the historical streamflow conditions for the forested regions of the San Juan River basin.

We also found that climate and disturbance have opposing influences on snow pack. In the early season, the forest disturbance partially compensates for streamflow impacts caused by warmer temperatures (Fig. 3). We see that the shift towards an earlier snowmelt is compensated in part by the fact that shrubs held on to snow for longer into the snowmelt season, releasing snow at a later date. Overall, this effect results in less water in the system and peak flows that occur at the same point in the year, but are higher in magnitude which results in lower soil moisture, such that the effect of having a late melting pack may not be beneficial for water resources. However, the snow on the ground could have important consequences for early season energy balances.

The findings presented herein represent a deviation from the more broadly accepted viewpoint that forest disturbances will lead to reduced evapotranspiration and increased streamflow. Previous findings tend to be based on studies carried out over short time periods (first 1–2 year responses, < 5 year studies), paired-basin analyses in watersheds disturbed by clear cuts, and where climate variability may have obscured results (Brown et al., 2005). Studies pointing towards increased streamflow also broadly found evapotranspiration from the canopy decreased, leading to an increase in runoff. However, work based on observations across scales and encompassing multiple disturbances indicates that the regrowth potential for understory, such as shrubs used in this study, is high and that the regrowth is a major controlling factor for water availability and direction of change for evapotranspiration and runoff (Caldwell et al., 2016; Biederman et al., 2014, 2015; Brown et al., 2014; Pribulick et al., 2016). Moreover, ecologists project that global forest covers are expected to decline and be replaced with species and understory compositions that are more water intensive. It is therefore paramount to treat regrowth correctly within models to align these two currently disparate principles and account accurately for changing streamflow. This is a fundamental issue because ESMs rely upon the research principles developed on plot-scale and watershed-scale observational studies and modeling work.

We also investigated the aridity effect upon water availability under forest disturbances. By comparing vegetation changes to climate shifts in a single cell, we highlighted the impacts of changing temperature and precipitation versus the impacts of changing forest cover properties. We found that the greatest differences in our results for forests and shrubs under climate change occurred under wet conditions, and vice versa for dry conditions. This may be due to the large component of water that is leaving via evaporation in arid environments. Additionally, we found that an important controlling factor for water availability in shrub environments was the canopy spacing between the shrubs that lead to changes in water balances and water partitioning within the environment. Thus, we believe that not only climate (i.e., aridity) but also vegetation characteristics (i.e., canopy spacing) may play a fundamental role in the impact of disturbances on water availability.

Our study did not incorporate fine-scale processes such as lateral flow, as these processes are not possible to implement using the VIC modeling approach. Suggestions from studies examining vegetation structure and patchiness illustrate that interconnected hillslope and riparian vegetation, such as shrublands, can receive water from wet upland catchment areas (Thompson et al., 2011). A recent study using the ParFlow–CLM model, which incorporates lateral flow processes, found results that are very similar to our findings (Pribulick et al., 2016). For example, Pribulick et al. (2016) show non-linear declines in streamflow in response to vegetation and temperature changes, with enhanced responses in snowmelt-driven transects and where percent vegetation change was largest. Other research suggests that when water availability declines, plants adapt and become more efficient at using water (Troch et al., 2009), which is not incorporated in the approaches undertaken in this study or Pribulick et al. (2016). Additionally, VIC model parameters that impact changes in water balances have inherent uncertainty under climate change (Bennett et al., 2017, 2012), and some of these parameters, such as canopy overstory, are represented in a binary fashion, which is not necessarily indicative of the forest and shrubland mixtures observed during forest recovery. This is a clear example of improvements that could be made to the modeling approach presented herein that should be investigated in future work.

Our study also points to the challenges associated with incorporating spatially variable estimates of changing vegetation patterns. Although CMIP5 contains information from dynamic vegetation models, we find a large disparity between the results from CMIP5 (dynamic) and the current estimates from cutting-edge research on forest cover mortality (disturbed); these methods are a scenario-style approach that do not necessary reflect future changing land cover conditions. The importance of accurately representing the effects that are observed on the catchment scale and in offline models, such as VIC, cannot be understated. ESMs require the capability to incorporate these changes in a meaningful way that can be validated using our current understanding of changes to forests in the southwestern USA and globally. This will enable the science community to accurately capture the range of responses and the impacts of changing climate and changing land cover on water resources. ESMs such as the U.S. Department of Energy's E3SM model have started down this path with the incorporation and testing of the FATES dynamic vegetation model (Fisher et al., 2015).

The differences in future streamflow projections in comparison to historical conditions we observed in this study are notable, and suggest that streamflow will decline by a minimum of ~ 10 % for headwater systems of the CRB. However, in the San Juan River basin forest cover accounts for only 17 % of the area, and only 4 % of these cells have greater than 80 % forest cover. Impacts may be much larger for adjacent systems, such as the Gunnison, CO. Future work will investigate the impacts of changing forest cover using sensitivity analysis to understand the cumulative effect of changing climate and vegetation cover across the entire CRB.

6 Conclusion

The implications of declining streamflow due to forest disturbances in the San Juan River basin are significant for the CRB. Given that the San Juan basin is one of the least forested portions of the CRB and assuming similar rates of forest decline will occur nearby (Van Mantgem et al., 2009), other regions of the CRB are likely to experience streamflow declines of much higher severity due to changes in forest cover. Failure to adequately understand the direction and magnitude of changes could have catastrophic consequences for those who rely on this resource. The CRB – roughly 11 % area of the continental USA – directly supports water supply for more than 30 million people, accounts for approximately 15 % of the USA's crops and livestock (Cohen et al., 2013), and 53 GW of power generation capacity (Buono and Eckstein, 2014; Cohen et al., 2013). Pressing work includes improvements in the spatial representation of changing forest covers across the CRB, further investigations into drivers of change including an understanding of the role of aridity and variability, a stronger link between ecological dynamics and hydrological knowledge that is translated into models, and upward propagation of these findings into global climate models and ESMs.

Data availability. Downscaled CMIP5 climate model projections may be downloaded via the MACA web portal: https://climate.northwestknowledge.net/MACA/ (Abatzoglou, 2018). VIC model may be downloaded via GitHub: https://github.com/UW-Hydro/VIC (Computational Hydrology Group, 2018). CMIP5 dynamic vegetation changes may be downloaded via the ESGF portal: https://esgf-node.llnl.gov/projects/esgf-llnl/. Historical VIC forcing data may be obtained from ftp://gdo-dcp.ucllnl.org/pub/dcp/archive/OBS/livneh2014.1_16deg/ (Livneh, 2018). Naturalized streamflow data for the Colorado River basin may be obtained from USBR: https://www.usbr.gov/lc/region/g4000/NaturalFlow/current.html (U.S. Bureau of Reclamation, 2018). Other model parameter files and McDowell et al. (2016) vegetation disturbances may be obtained by contacting the authors.

Competing interests. The authors declare that they have no conflict of interest.

Special issue statement. This article is part of the special issue "Observations and modeling of land surface water and energy exchanges across scales: special issue in Honor of Eric F. Wood". It is a result of the Symposium in Honor of Eric F. Wood: Observations and Modeling across Scales, Princeton, New Jersey, USA, 2–3 June 2016.

Acknowledgements. We acknowledge the support of John Abatzaglou and Katherine Hegewisch for providing us with additional models not in the original MACA data set. We acknowledge the World Climate Research Programme's Working Group on Coupled Modeling, which is responsible for CMIP, and we thank the climate modeling groups for producing and making available their model output. For CMIP, the U.S. Department of Energy's Program for Climate Model Diagnosis and Intercomparison provides coordinating support and led development of software infrastructure in partnership with the Global Organization for Earth System Science Portals. Katrina E. Bennett, Kurt Solander, Richard S. Middleton, Nathan G. McDowell, and Chonggang Xu acknowledge the Los Alamos National Lab's LDRD program for supporting this work. Nathan G. McDowell further acknowledges support of Pacific Northwest National Laboratories LDRD program. Theodore J. Bohn was supported by grant 1216037 from the U.S. National Science Foundation (NSF) Science, Engineering and Education for Sustainability (SEES) Post-Doctoral Fellowship program and NSF grant 1462086, DMUU: Decision Center for a Desert City III: Transformational Solutions for Urban Water Sustainability Transitions in the Colorado River Basin.

Edited by: Reed Maxwell

References

Abatzoglou, J. T. and Brown, T. J.: A comparison of statistical downscaling methods suited for wildfire applications, Int. J. Climatol., 32, 772–780, 2012.

Abatzoglou, J.: Multivariate Adaptive Constructed Analogs (MACA) dataset, available at: https://climate.northwestknowledge.net/MACA/, last access: 24 January 2018.

Adams, H. D., Luce, C. H., Breshears, D. D., Allen, C. D., Weiler, M., Hale, V. C., Smith, A., and Huxman, T. E.: Ecohydrological consequences of drought- and infestation-triggered tree die-off: insights and hypotheses, Ecohydrology, 5, 145–159, 2012.

Allen, C. D., Macalady, A. K., Chenchouni, H., Bachelet, D., McDowell, N., Vennetier, M., Kitzberger, T., Rigling, A., Breshears, D. D., and Hogg, E. T.: A global overview of drought and heat-induced tree mortality reveals emerging climate change risks for forests, Forest Ecol. Manage., 259, 660–684, 2010.

Allen, C. D., Breshears, D. D., and McDowell, N. G.: On underestimation of global vulnerability to tree mortality and forest die-off from hotter drought in the Anthropocene, Ecosphere, 6, 1–55, 2015.

Anderegg, W. R. L., Kane, J. M., and Anderegg, L. D. L.: Consequences of widespread tree mortality triggered by drought and temperature stress, Nature Clim. Change, 3, 30–36, 2013.

Anderegg, W. R. L., Hicke, J. A., Fisher, R. A., Allen, C. D., Aukema, J., Bentz, B., Hood, S., Lichstein, J. W., Macalady, A. K., McDowell, N., Pan, Y., Raffa, K., Sala, A., Shaw, J. D., Stephenson, N. L., Tague, C., and Zeppel, M.: Tree mortality from drought, insects, and their interactions in a changing climate, New Phytologist, 208, 674–683, https://doi.org/10.1111/nph.13477, 2015.

Anderson, H. W., Hoover, M. D., and Reinhart, K. G.: Forests and water: effects of forest management on floods, sedimentation, and water supply. General Technical Report PSW-018, USDA Forest Service Berkeley CA, 115, 1976.

Arora, V.: Modeling vegetation as a dynamic component in soil-vegetation-atmosphere transfer schemes and hydrological models, Rev. Geophys., 40, 3-1–3-26, 2002.

Atchley, A. L., Kinoshita, A. M., Lopez, S., Trader, L., and Middleton, R. S.: Simulating surface and subsurface water balance changes due to fire severity at the hillslope scale, J. Hydrol., in review, 2018.

Bearup, L. A., Maxwell, R. M., Clow, D. W., and McCray, J. E.: Hydrological effects of forest transpiration loss in bark beetle-impacted watersheds, Nature Clim. Change, 4, 481–486, https://doi.org/10.1038/nclimate2198, 2014.

Bearup, L. A., Maxwell, R. M., and McCray, J. E.: Hillslope response to insect-induced land-cover change: an integrated model of end-member mixing, Ecohydrology, 9, 195–203, https://doi.org/10.1002/eco.1729, 2016.

Bennett, K. E., Werner, A. T., and Schnorbus, M.: Uncertainties in hydrologic and climate change impact analyses in headwater basins of British Columbia, J. Climate, 25, 5711–5730, https://doi.org/10.1175/JCLI-D-11-00417.1, 2012.

Bennett, K. E., Cannon, A. J., and Hinzman, L.: Historical trends and extremes in boreal Alaska river basins, J. Hydrol., 527, 590–607, 2015.

Bennett, K. E., Urrego Blanco, J., Jonko, A., Bohn, T. J., Atchley, A. L., Urban, N. M., and Middleton, R. S.: Global sensitivity of simulated water balance indicators under future climate change in the Colorado Basin, Water Resour. Res., 54, published online, https://doi.org/10.1002/2017WR020471, 2017.

Bewley, D., Alila, Y., and Varhola, A.: Variability of snow water equivalent and snow energetics across a large catchment subject to Mountain Pine Beetle infestation and rapid salvage logging, J. Hydrol., 388, 464–479, 2010.

Biederman, J., Harpold, A., Gochis, D., Ewers, B., Reed, D., Papuga, S., and Brooks, P.: Increased evaporation following widespread tree mortality limits streamflow response, Water Resour. Res., 50, 5395–5409, 2014.

Biederman, J. A., Somor, A. J., Harpold, A. A., Gutmann, E. D., Breshears, D. D., Troch, P. A., Gochis, D. J., Scott, R. L., Meddens, A. J., and Brooks, P. D.: Recent tree die-off has little effect on streamflow in contrast to expected increases from historical studies, Water Resour. Res., 51, 9775–9789, 2015.

Bohn, T. J. and Vivoni, E. R.: Process-based characterization of evapotranspiration sources over the North American monsoon region, Water Resour. Res., 52, 358–384, https://doi.org/10.1002/2015WR017934, 2016.

Bohn, T. J., Livneh, B., Oyler, J. W., Running, S. W., Nijssen, B., and Lettenmaier, D. P.: Global evaluation of MTCLIM and related algorithms for forcing of ecological and hydrological models, Agric. Forest Meteorol., 176, 38–49, 2013.

Bonan, G. B.: Forests and Climate Change: Forcings, Feedbacks, and the Climate Benefits of Forests, Science, 320, 1444–1449, https://doi.org/10.1126/science.1155121, 2008.

Boon, S.: Snow accumulation and ablation in a beetle-killed pine stand in Northern Interior British Columbia, J. Ecosyst. Manage., 8, 1–13, 2007.

Boon, S.: Snow ablation energy balance in a dead forest stand, Hydrol. Process., 23, 2600–2610, 2009.

Brekke, L., Thrasher, B. L., Maurer, E. P., and Pruitt, T.: Downscaled CMIP3 and CMIP5 Climate and Hydrology Projections: Release of Downscaled CMIP5 Climate Projections, Comparison with preceding Information, and Summary of User Needs, U.S. Department of the Interior, Bureau of Reclamation, Technical Services Center, Denver, Colorado, 47, 2013.

Brovkin, V., Boysen, L., Arora, V., Boisier, J., Cadule, P., Chini, L., Claussen, M., Friedlingstein, P., Gayler, V., and Van Den Hurk, B.: Effect of anthropogenic land-use and land-cover changes on climate and land carbon storage in CMIP5 projections for the twenty-first century, J. Climate, 26, 6859–6881, 2013a.

Brovkin, V., Boysen, L., Raddatz, T., Gayler, V., Loew, A., and Claussen, M.: Evaluation of vegetation cover and land-surface albedo in MPI-ESM CMIP5 simulations, J. Adv. Model. Earth Syst., 5, 48–57, https://doi.org/10.1029/2012MS000169, 2013b.

Brown, A. E., Zhang, L., McMahon, T. A., Western, A. W., and Vertessy, R. A.: A review of paired catchment studies for determining changes in water yield resulting from alterations in vegetation, J. Hydrol., 310, 28–61, 2005.

Brown, M. G., Black, T. A., Nesic, Z., Foord, V. N., Spittlehouse, D. L., Fredeen, A. L., Bowler, R., Grant, N. J., Burton, P. J., and Trofymow, J.: Evapotranspiration and canopy characteristics of two lodgepole pine stands following mountain pine beetle attack, Hydrol. Process., 28, 3326–3340, 2014.

Brown, R., Derksen, C., and Wang, L.: A multi-data set analysis of variability and change in Arctic spring snow cover extent, 1967–2008, J. Geophys. Res.-Atmos., 115, D16111, https://doi.org/10.1029/2010jd013975, 2010.

Buma, B. and Livneh, B.: Potential Effects of Forest Disturbances and Management on Water Resources in a Warmer Climate, Forest Sci., 61, 895–903, 2015.

Buma, B. and Livneh, B.: Key landscape and biotic indicators of watersheds sensitivity to forest disturbance identified using remote sensing and historical hydrography data, Environ. Res. Lett., 12, 074028, 2017.

Buono, R. M. and Eckstein, G.: Minute 319: a cooperative approach to Mexico–US hydro-relations on the Colorado River, Water Int., 39, 263–276, 2014.

Bureau of Reclamation: West-Wide Climate Risk Assessments: Bias-Corrected and Spatially Downscaled Surface Water Projections. Technical Memorandum No. 86-68210-2011-01, U.S. Department of the Interior, Bureau of Reclamation, Technical Services Center, Denver, Colorado, 138, 2011.

Burles, K. and Boon, S.: Snowmelt energy balance in a burned forest plot, Crowsnest Pass, Alberta, Canada, Hydrol. Process., 25, 3012–3029, https://doi.org/10.1002/hyp.8067, 2011.

Caldwell, P. V., Miniat, C. F., Elliott, K. J., Swank, W. T., Brantley, S. T., and Laseter, S. H.: Declining water yield from forested mountain watersheds in response to climate change and forest mesophication, Glob. Change Biol., 2016.

Carroll, R. W. H., Huntington, J. L., Snyder, K. A., Niswonger, R. G., Morton, C., and Stringham, T. K.: Evaluating mountain meadow groundwater response to Pinyon-Juniper and temperature in a great basin watershed, Ecohydrology, 10, e1792, https://doi.org/10.1002/eco.1792, 2017.

Christensen, N. S. and Lettenmaier, D. P.: Effects of climate change on the hydrology and water resource of the Colorado River basin, Department of Civil and Environmental Engineering, University of Washington, Seattle, WA, 2002.

Christensen, N. S. and Lettenmaier, D. P.: A multimodel ensemble approach to assessment of climate change impacts on the hydrology and water resources of the Colorado River Basin, Hydrol. Earth Syst. Sci., 11, 1417–1434, https://doi.org/10.5194/hess-11-1417-2007, 2007.

Cohen, M., Christian-Smith, J., and Berggren, J.: Water to supply the land: Irrigated agriculture in the Colorado River basin, Pacific Institute, Oakland, CA, 2013.

Collins, W. J., Bellouin, N., Doutriaux-Boucher, M., Gedney, N., Halloran, P., Hinton, T., Hughes, J., Jones, C. D., Joshi, M., Liddicoat, S., Martin, G., O'Connor, F., Rae, J., Senior, C., Sitch, S., Totterdell, I., Wiltshire, A., and Woodward, S.: Development and evaluation of an Earth-System model – HadGEM2, Geosci. Model Dev., 4, 1051–1075, https://doi.org/10.5194/gmd-4-1051-2011, 2011.

Computational Hydrology Group, Department of Civil and Environmental Engineering, University of Washington: The Variable Infiltration Capacity Model, available at: https://github.com/UW-Hydro/VIC, last access: 24 January 2018.

Cox, P. M.: Description of the TRIFFID dynamic global vegetation model, Technical Note 24, Hadley Centre, United Kingdom Meteorological Office, Bracknell, UK, 2001.

Dale, V. H., Joyce, L. A., McNulty, S., Neilson, R. P., Ayres, M. P., Flannigan, M. D., Hanson, P. J., Irland, L. C., Lugo, A. E., Peterson, C. J., Simberloff, D., Swanson, F. J., Stocks, B. J., and Wotton, B. M.: Climate change and forest disturbances, BioScience, 51, 723–734, 2001.

Dawadi, S. and Ahmad, S.: Changing climatic conditions in the Colorado River Basin: Implications for water resources management, J. Hydrol., 430–431, 127–141, https://doi.org/10.1016/j.jhydrol.2012.02.010, 2012.

Ducoudré, N. I., Laval, K., and Perrier, A.: SECHIBA, a new set of parameterizations of the hydrologic exchanges at the land-atmosphere interface within the LMD atmospheric general circulation model, J. Climate, 6, 248–273, 1993.

FAO: United Nations Food and Agriculture Organization. Digital Soil Map of the World and Derived Soil Properties, Land Water, Digital Media Ser., 1 (CD-ROM), Rome, Italy, 1998.

Fisher, R. A., Muszala, S., Verteinstein, M., Lawrence, P., Xu, C., McDowell, N. G., Knox, R. G., Koven, C., Holm, J., Rogers, B. M., Spessa, A., Lawrence, D., and Bonan, G.: Taking off the training wheels: the properties of a dynamic vegetation model without climate envelopes, CLM4.5(ED), Geosci. Model Dev., 8, 3593–3619, https://doi.org/10.5194/gmd-8-3593-2015, 2015.

Franchini, M. and Pacciani, M.: Comparative analysis of several conceptual rainfall-runoff models, J. Hydrol., 122, 161–219, 1991.

Friedl, M. A., Sulla-Menashe, D., Tan, B., Schneider, A., Ramankutty, N., Sibley, A., and Huang, X.: MODIS Collection 5 global land cover: Algorithm refinements and characterization of new datasets, Remote Sens. Environ., 114, 168–182, 2010.

Gangopadhyay, S., Pruitt, T., Brekke, L., and Raff, D.: Hydrologic projections for the western United States, Eos, Transactions American Geophysical Union, 92, 441–442, 2011.

Giorgetta, M. A., Jungclaus, J., Reick, C. H., Legutke, S., Bader, J., Böttinger, M., Brovkin, V., Crueger, T., Esch, M., and Fieg, K.: Climate and carbon cycle changes from 1850 to 2100 in MPI-ESM simulations for the Coupled Model Intercomparison Project phase 5, J. Adv. Model. Earth Syst., 5, 572–597, 2013.

Gould, G. K., Liu, M., Barber, M. E., Cherkauer, K. A., Robichaud, P. R., and Adam, J. C.: The effects of climate change and extreme wildfire events on runoff erosion over a mountain watershed, J. Hydrol., 536, 74–91, 2016.

Guardiola-Claramonte, M., Troch, P. A., Breshears, D. D., Huxman, T. E., Switanek, M. B., Durcik, M., and Cobb, N. S.: Decreased streamflow in semi-arid basins following drought-induced tree die-off: a counter-intuitive and indirect climate impact on hydrology, J. Hydrol., 406, 225–233, 2011.

Harpold, A. A., Biederman, J. A., Condon, K., Merino, M., Korgaonkar, Y., Nan, T., Sloat, L. L., Ross, M., and Brooks, P. D.: Changes in snow accumulation and ablation following the Las Conchas Forest Fire, New Mexico, USA, Ecohydrology, 7, 440–452, https://doi.org/10.1002/eco.1363, 2014.

Huete, A., Didan, K., Miura, T., Rodriguez, E. P., Gao, X., and Ferreira, L. G.: Overview of the radiometric and biophysical performance of the MODIS vegetation indices, Remote Sens. Environ., 83, 195–213, 2002.

Kang, S., Kimball, J. S., and Running, S. W.: Simulating effects of fire disturbance and climate change on boreal forest productivity and evapotranspiration, Sci. Total Environ., 362, 85–102, https://doi.org/10.1016/j.scitotenv.2005.11.014, 2006.

Krinner, G., Viovy, N., de Noblet-Ducoudré, N., Ogée, J., Polcher, J., Friedlingstein, P., Ciais, P., Sitch, S., and Prentice, I. C.: A dynamic global vegetation model for studies of the coupled atmosphere-biosphere system, Global Biogeochem. Cycles, 19, GB1015, https://doi.org/10.1029/2003GB002199, 2005.

Le Quéré, C., Moriarty, R., Andrew, R. M., Canadell, J. G., Sitch, S., Korsbakken, J. I., Friedlingstein, P., Peters, G. P., Andres, R. J., Boden, T. A., Houghton, R. A., House, J. I., Keeling, R. F., Tans, P., Arneth, A., Bakker, D. C. E., Barbero, L., Bopp, L., Chang, J., Chevallier, F., Chini, L. P., Ciais, P., Fader, M., Feely, R. A., Gkritzalis, T., Harris, I., Hauck, J., Ilyina, T., Jain, A. K., Kato, E., Kitidis, V., Klein Goldewijk, K., Koven, C., Landschützer, P., Lauvset, S. K., Lefèvre, N., Lenton, A., Lima, I. D., Metzl, N., Millero, F., Munro, D. R., Murata, A., Nabel, J. E. M. S., Nakaoka, S., Nojiri, Y., O'Brien, K., Olsen, A., Ono, T., Pérez, F. F., Pfeil, B., Pierrot, D., Poulter, B., Rehder, G., Rödenbeck, C., Saito, S., Schuster, U., Schwinger, J., Séférian, R., Steinhoff, T., Stocker, B. D., Sutton, A. J., Takahashi, T., Tilbrook, B., van der Laan-Luijkx, I. T., van der Werf, G. R., van Heuven, S., Vandemark, D., Viovy, N., Wiltshire, A., Zaehle, S., and Zeng, N.: Global Carbon Budget 2015, Earth Syst. Sci. Data, 7, 349–396, https://doi.org/10.5194/essd-7-349-2015, 2015.

Liang, X., Lettenmaier, D. P., Wood, E. F., and Burges, S. J.: A simple hydrologically based model of land surface water and energy fluxes for general circulation models, J. Geophys. Res., 99, 14415–14428, 1994.

Liang, X., Wood, E. F., and Lettenmaier, D. P.: Surface soil moisture parameterization of the VIC-2L model: Evaluation and modification, Global Planet. Change, 13, 195–206, 1996.

Livneh, B., Bohn, T. J., Pierce, D. W., Munoz-Arriola, F., Nijssen, B., Vose, R., Cayan, D. R., and Brekke, L.: A spatially comprehensive, hydrometeorological data set for Mexico, the U.S., and Southern Canada 1950–2013, Scientific Data, 2, 150042, https://doi.org/10.1038/sdata.2015.42, 2015a.

Livneh, B., Deems, J. S., Buma, B., Barsugli, J. J., Schneider, D., Molotch, N. P., Wolter, K., and Wessman, C. A.: Catchment response to bark beetle outbreak and dust-on-snow in the Colorado Rocky Mountains, J. Hydrol., 523, 196–210, https://doi.org/10.1016/j.jhydrol.2015.01.039, 2015b.

Livneh, B.: Daily gridded Hhdrometeorological data set for Mexico, the conterminous U.S., and southern Canada, ftp://gdo-dcp.ucllnl.org/pub/dcp/archive/OBS/livneh2014.1_16deg/, last access: 24 January 2018.

McCabe, G. J. and Wolock, D. M.: Warming may create substantial water supply shortages in the Colorado River basin, Geophys. Res. Lett., 34, L22708, https://doi.org/10.1029/2007GL031764, 2007.

McDowell, N. G. and Allen, C. D.: Darcy's law predicts widespread forest mortality under climate warming, Nature Clim. Change, 5, 669–672, 2015.

McDowell, N. G., Beerling, D. J., Breshears, D. D., Fisher, R. A., Raffa, K. F., and Stitt, M.: The interdependence of mechanisms underlying climate-driven vegetation mortality, Trends Ecol. Evol., 26, 523–532, 2011.

McDowell, N. G., Williams, A. P., Xu, C., Pockman, W. T., Dickman, L. T., Sevanto, S., Pangle, R., Limousin, J., Plaut, J., Mackay, D. S., Ogee, J., Domec, J. C., Allen, C. D., Fisher, R. A., Jiang, X., Muss, J. D., Breshears, D. D., Rauscher, S. A., and Koven, C.: Multi-scale predictions of massive conifer mortality due to chronic temperature rise, Nature Clim. Change, 6, 295–300, https://doi.org/10.1038/nclimate2873, 2016.

McDowell, N. G., Michaletz, S., Bennett, K. E., Solander, K., Xu, C., Maxwell, R. M., Allen, C. D., and Middleton, R. S.: Predicting Chronic Climate-Driven Disturbances and Their Mitigation, Trends Ecol. Evol., 33, 15–27, https://doi.org/10.1016/j.tree.2017.10.002, 2018.

Myneni, R. B., Hoffman, S., Knyazikhin, Y., Privette, J. L., Glassy, J., Tian, Y., Wang, Y., Song, X., Zhang, Y., Smith, G. R., Lotsch, A., Friedl, M., Morisette, J. T., Votava, P., Nemani, R. R., and Running, S. W.: Global products of vegetation leaf area and fraction absorbed PAR from year one of MODIS data, Remote Sens. Environ., 83, 214–231, https://doi.org/10.1016/S0034-4257(02)00074-3, 2002.

Nash, L. L. and Gleick, P. H.: Sensitivity of streamflow in the Colorado basin to climatic changes, J. Hydrol., 125, 221–241, 1991.

Penn, C. A., Bearup, L. A., Maxwell, R. M., and Clow, D. W.: Numerical experiments to explain multiscale hydrological responses to mountain pine beetle tree mortality in a headwater watershed, Water Resour. Res., 52, 3143–3161, https://doi.org/10.1002/2015WR018300, 2016.

Pomeroy, J. W., Parviainen, J., Hedstrom, N., and Gray, D. M.: Coupled modelling of forest snow interception and sublimation, Hydrol. Process., 12, 2317–2337, https://doi.org/10.1002/(SICI)1099-1085(199812)12:15<2317::AID-HYP799>3.0.CO;2-X, 1998.

Pribulick, C. E., Foster, L. M., Bearup, L. A., Navarre-Sitchler, A. K., Williams, K. H., Carroll, R. W., and Maxwell, R. M.: Contrasting the hydrologic response due to land cover and climate change in a mountain headwaters system, Ecohydrology, 9, 1431–1438, 2016.

Pugh, E. and Gordon, E.: A conceptual model of water yield effects from beetle-induced tree death in snow-dominated lodgepole pine forests, Hydrol. Process., 27, 2048–2060, https://doi.org/10.1002/hyp.9312, 2013.

Rasmussen, R., Ikeda, K., Liu, C., Gochis, D., Clark, M., Dai, A., Gutmann, E., Dudhia, J., Chen, F., and Barlage, M.: Climate change impacts on the water balance of the Colorado Headwaters: High-resolution regional climate model simulations, J. Hydrometeor., 15, 1091–1116, 2014.

Reick, C., Raddatz, T., Brovkin, V., and Gayler, V.: Representation of natural and anthropogenic land cover change in MPI-ESM, J. Adv. Model. Earth Syst., 5, 459–482, 2013.

Rother, M. T., Veblen, T. T., and Furman, L. G.: A field experiment informs expected patterns of conifer regeneration after disturbance under changing climate conditions, Can. J. Forest Res., 45, 1607–1616, 2015.

Royer, P. D., Cobb, N. S., Clifford, M. J., Huang, C. Y., Breshears, D. D., Adams, H. D., and Villegas, J. C.: Extreme climatic event-triggered overstorey vegetation loss increases understorey solar input regionally: primary and secondary ecological implications, J. Ecol., 99, 714–723, 2011.

Sato, H., Itoh, A., and Kohyama, T.: SEIB–DGVM: A new Dynamic Global Vegetation Model using a spatially explicit individual-based approach, Ecol. Model., 200, 279–307, https://doi.org/10.1016/j.ecolmodel.2006.09.006, 2007.

Schaaf, C. B., Gao, F., Strahler, A. H., Lucht, W., Li, X., Tsang, T., Strugnell, N. C., Zhang, X., Jin, Y., and Muller, J.-P.: First operational BRDF, albedo nadir reflectance products from MODIS, Remote Sens. Environ., 83, 135–148, 2002.

Scheller, R. M. and Mladenoff, D. J.: An ecological classification of forest landscape simulation models: tools and strategies for understanding broad-scale forested ecosystems, Landscape Ecol., 22, 491–505, 2007.

Schnorbus, M., Bennett, K. E., and Werner, A. T.: Quantifying the Water Resources Impacts of Mountain Pine Beetle and Associated Salvage Harvest Operations Across a Range of Watershed Scales: Hydrologic Modelling of the Fraser River Basin. Information Report BC-X-423, Canadian Forest Service, Pacific Forestry Centre, Victoria, 64, 2010.

Solander, K. C., Bennett, K. E., and Middleton, R. S.: Historical shifts in streamflow extremes in the Colorado River Basin, J. Hydrol.-Regional Studies, 12, 363–377, https://doi.org/10.1016/j.ejrh.2017.05.004, 2017.

Somor, A. J.: Quantifying streamflow change following bark beetle outbreak in multiple central Colorado catchments, The University of Arizona, 2010.

Taylor, K. E., Stouffer, R. J., and Meehl, G. A.: An overview of CMIP5 and the experiment design, B. Am. Meteorol. Soc., 93, 485–498, 2012.

Thompson, S. E., Harman, C. J., Troch, P. A., Brooks, P. D., and Sivapalan, M.: Spatial scale dependence of ecohydrologically mediated water balance partitioning: A synthesis framework for catchment ecohydrology, Water Resour. Res., 47, W00J03, https://doi.org/10.1029/2010WR009998, 2011.

Troch, P. A., Martinez, G. F., Pauwels, V. R. N., Durcik, M., Sivapalan, M., Harman, C., Brooks, P. D., Gupta, H., and Huxman, T.: Climate and vegetation water use efficiency at catchment scales, Hydrol. Process., 23, 2409–2414, https://doi.org/10.1002/hyp.7358, 2009.

U.S. Bureau of Reclamation: Colorado River Basin Natural Flow, 1906–2015, available at: https://www.usbr.gov/lc/region/g4000/NaturalFlow/current.html, last access: 24 January 2018.

Van Mantgem, P. J., Stephenson, N. L., Byrne, J. C., Daniels, L. D., Franklin, J. F., Fulé, P. Z., Harmon, M. E., Larson, A. J., Smith, J. M., and Taylor, A. H.: Widespread increase of tree mortality rates in the western United States, Science, 323, 521–524, 2009.

Van Vuuren, D. P., Edmonds, J., Kainuma, M., Riahi, K., Thomson, A., Hibbard, K., Hurtt, G. C., Kram, T., Krey, V., and Lamarque, J.-F.: The representative concentration pathways: an overview, Clim. Change, 109, 5–31, 2011.

Varhola, A., Coops, N. C., Bater, C. W., Teti, P., Boon, S., and Weiler, M.: The influence of ground-and lidar-derived forest structure metrics on snow accumulation and ablation in disturbed forests, Can. J. Forest Res., 40, 812–821, 2010.

Watanabe, S., Hajima, T., Sudo, K., Nagashima, T., Takemura, T., Okajima, H., Nozawa, T., Kawase, H., Abe, M., Yokohata, T., Ise, T., Sato, H., Kato, E., Takata, K., Emori, S., and Kawamiya, M.: MIROC-ESM 2010: model description and basic results of CMIP5-20c3m experiments, Geosci. Model Dev., 4, 845–872, https://doi.org/10.5194/gmd-4-845-2011, 2011.

Williams, A. P., Allen, C. D., Macalady, A. K., Griffin, D., Woodhouse, C. A., Meko, D. M., Swetnam, T. W., Rauscher, S. A., Seager, R., and Grissino-Mayer, H. D.: Temperature as a potent driver of regional forest drought stress and tree mortality, Nature Clim. Change, 3, 292–297, 2013.

Winkler, R., Boon, S., Zimonick, B., and Baleshta, K.: Assessing the effects of post-pine beetle forest litter on snow albedo, Hydrol. Process., 24, 803–812, 2010.

Winkler, R., Boon, S., Zimonick, B., and Spittlehouse, D.: Snow accumulation and ablation response to changes in forest structure and snow surface albedo after attack by mountain pine beetle, Hydrol. Process., 28, 197–209, 2014.

Yapo, P. O., Gupta, H. V., and Sorooshian, S.: Multi-objective global optimization for hydrologic models, J. Hydrol., 204, 83–97, 1998.

Zhang, L., Podlasly, C., Ren, Y., Feger, K.-H., Wang, Y., and Schwärzel, K.: Separating the effects of changes in land management and climatic conditions on long-term streamflow trends analyzed for a small catchment in the Loess Plateau region, NW China, Hydrol. Process., 28, 1284–1293, https://doi.org/10.1002/hyp.9663, 2014.

Zou, C. B., Ffolliott, P. F., and Wine, M.: Streamflow responses to vegetation manipulations along a gradient of precipitation in the Colorado River Basin, Forest Ecol. Manage., 259, 1268–1276, 2010.

9

Multimodel assessment of climate change-induced hydrologic impacts for a Mediterranean catchment

Enrica Perra[1,2], Monica Piras[1,3], Roberto Deidda[1,3], Claudio Paniconi[2], Giuseppe Mascaro[3,4], Enrique R. Vivoni[4], Pierluigi Cau[5], Pier Andrea Marras[5], Ralf Ludwig[6], and Swen Meyer[6,7]

[1]Dipartimento di Ingegneria Civile, Ambientale ed Architettura, Università degli Studi di Cagliari, Cagliari, Italy
[2]Centre Eau Terre Environnement, Institut National de la Recherche Scientifique, Quebec City, Canada
[3]Consorzio Interuniversitario nazionale per la Fisica dell'Atmosfere e dell'Idrosfere, Tolentino, Italy
[4]School of Sustainable Engineering and the Built Environment and School of Earth and Space Exploration, Arizona State University, Tempe, AZ, USA
[5]Centro di Ricerca, Sviluppo e Studi Superiori in Sardegna, Pula, Cagliari, Italy
[6]Physical Geography and Environmental Modeling, Department of Geography, Ludwig-Maximilians-Universität, Munich, Germany
[7]Leibniz-Institut für Gemüse- und Zierpflanzenbau Großbeeren/Erfurt e.V., Grossbeeren, Germany

Correspondence: Enrica Perra (enrica.perra@unica.it)

Abstract. This work addresses the impact of climate change on the hydrology of a catchment in the Mediterranean, a region that is highly susceptible to variations in rainfall and other components of the water budget. The assessment is based on a comparison of responses obtained from five hydrologic models implemented for the Rio Mannu catchment in southern Sardinia (Italy). The examined models – CATchment HYdrology (CATHY), Soil and Water Assessment Tool (SWAT), TOPographic Kinematic APproximation and Integration (TOPKAPI), TIN-based Real time Integrated Basin Simulator (tRIBS), and WAter balance SImulation Model (WASIM) – are all distributed hydrologic models but differ greatly in their representation of terrain features and physical processes and in their numerical complexity. After calibration and validation, the models were forced with bias-corrected, downscaled outputs of four combinations of global and regional climate models in a reference (1971–2000) and future (2041–2070) period under a single emission scenario. Climate forcing variations and the structure of the hydrologic models influence the different components of the catchment response. Three water availability response variables – discharge, soil water content, and actual evapotranspiration – are analyzed. Simulation results from all five hydrologic models show for the future period decreasing mean annual streamflow and soil water content at 1 m depth. Actual evapotranspiration in the future will diminish according to four of the five models due to drier soil conditions. Despite their significant differences, the five hydrologic models responded similarly to the reduced precipitation and increased temperatures predicted by the climate models, and lend strong support to a future scenario of increased water shortages for this region of the Mediterranean basin. The multimodel framework adopted for this study allows estimation of the agreement between the five hydrologic models and between the four climate models. Pairwise comparison of the climate and hydrologic models is shown for the reference and future periods using a recently proposed metric that scales the Pearson correlation coefficient with a factor that accounts for systematic differences between datasets. The results from this analysis reflect the key structural differences between the hydrologic models, such as a representation of both vertical and lateral subsurface flow (CATHY, TOPKAPI, and tRIBS) and a detailed treatment of vegetation processes (SWAT and WASIM).

1 Introduction

Climate studies agree on the prediction that the Mediterranean area will be particularly affected by changes under global warming (IPCC, 2014). This region, in fact, has been singled out as one of the hotspots in future climate change predictions (Giorgi, 2006), due to higher susceptibility to more frequent and more intense extreme events. In addition, observations during the last decades indicate that mean and extreme temperatures have increased in several Mediterranean regions (Xoplaki et al., 2003; Del Río et al., 2011; El Kenawy et al., 2011; Acero et al., 2014) and that precipitation has diminished, especially in the warm season (Giorgi and Lionello, 2008; Sousa et al., 2011; Vicente-Serrano and Cuadrat-Prats, 2007).

Climate change impact assessment at the catchment scale is usually conducted through a procedure that involves the following steps (e.g., Xu et al., 2005): (i) selection of global climate models (GCMs) and regional climate models (RCMs) for future climate predictions; (ii) correction of the discrepancies between simulated and observed climatological features; (iii) application of downscaling techniques to increase the coarse scale of climate model outputs to the finer resolutions required by hydrologic models; and (iv) use of downscaled outputs as forcing for the calibrated hydrologic models to simulate the basin hydrologic response (Sulis et al., 2011, 2012; Piras et al., 2014; Hawkins et al., 2015; Majone et al., 2016; Meyer et al., 2016). Each of these steps is affected by uncertainties (Xu and Singh, 2004), including the choice of emission scenarios and climate forcings (Giorgi and Mearns, 2002; Tebaldi et al., 2005; Pechlivanidis et al., 2017), the selection of downscaling techniques (Wood et al., 2004; Im et al., 2010) and hydrologic model (Clark et al., 2008; Jiang et al., 2007; Dams et al., 2015), and the availability of observed data required for calibration and validation of both downscaling techniques and hydrologic models. Hawkins and Sutton (2009) estimate that by the end of the century, the emission scenarios will represent the dominant source of uncertainty in climate projections.

One approach to dealing with uncertainties is to use multiple climate and hydrologic models (Bosshard et al., 2013; Cornelissen et al., 2013; Gädeke et al., 2014; Najafi et al., 2011). For example, Bae et al. (2011) compared in a Korean basin three semi-distributed hydrologic models forced with outputs from 13 GCMs and three greenhouse gas emission scenarios. Their results show that the hydrologic models can produce major differences in runoff change considering the same climate change scenarios, in particular during the dry season. Bastola et al. (2011) examined the role of hydrologic model uncertainties (parameter and structural uncertainty) using four conceptual hydrologic models and six climate change scenarios within the generalized likelihood uncertainty estimation (GLUE) and Bayesian model averaging (BMA) methods. The results for the four Irish catchments considered showed a tendency of increasing flow in winter and decreasing flow in summer. Thompson et al. (2013) demonstrated for the Mekong River in Southeast Asia that GCM-related uncertainty in climate change projections is generally larger than that related to the use of three hydrologic models, which simulate the same direction of change in mean discharge. However, hydrologic model-related uncertainty is not negligible and in some cases is of a similar magnitude to GCM-related uncertainty. Vansteenkiste et al. (2014) used an ensemble of hydrologic models, from lumped conceptual to distributed physically based, to assess the impact of climate change on the Grote Nete basin (Belgium). The uncertainty in the hydrologic impact results was evaluated by the relative change in runoff volumes and peak- and low-flow extremes from historical and future climate conditions. Large differences in model predictions were found, especially under low-flow conditions. Using an ANOVA (ANalysis Of VAriance) approach, Bosshard et al. (2013) assessed the uncertainties induced by climate models, bias-correction methods, and hydrological models using the output of eight RCMs for the Upper Rhine. The results indicate that some of the uncertainties are not attributable to individual modeling chain components, but rather they depend on the interactions between these components and that overall the greatest contribution to uncertainty derives from the climate models. Maurer et al. (2010) investigated the effect of hydrologic model structure by comparing a lumped and distributed model driven by 22 climate model outputs for three California watersheds. The projected percent changes in monthly discharge did not significantly differ between the two models, except for extreme flows and during summer months.

In this study, we characterize the agreement between both climate and hydrologic predictions in the Rio Mannu catchment, a small Mediterranean basin located in a semiarid region in Sardinia (Italy). For this aim, we use an ensemble of climate and hydrologic models, including four combinations of GCMs and RCMs and a set of five hydrologic models of varying structural complexity, from conceptual to physically based. This is the first study wherein a wide range of distributed hydrologic models forced with outputs of different climate models is applied to a Mediterranean catchment to assess the impact of climate change. Moreover, unlike many previous studies, we focus on a set of variables characterizing and affecting the water balance at the catchment scale, including precipitation, air temperature, discharge, soil water content in the first meter, and actual evapotranspiration. The results are discussed in the context of the process representations for each model and within a rigorous analysis of agreement framework. For the latter a new metric, proposed by Duveiller et al. (2016), will be used to compare model results for reference and future periods using correlation and bias coefficients.

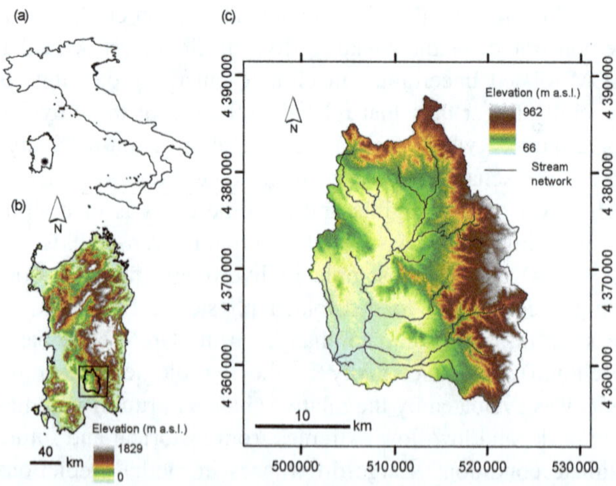

Figure 1. Location of the Rio Mannu basin within Italy **(a)** and Sardinia **(b)**. Boundaries in WGS84 UTM zone 32N coordinates, elevation, and stream network of the basin **(c)**.

2 Study area

The study site is the Rio Mannu di San Sperate at the Monastir basin, located in southern Sardinia, Italy (Fig. 1). The Rio Mannu basin was one of the seven study sites of a European-funded climate change research project (Ludwig et al., 2010). Amongst the reasons for selecting the Rio Mannu site for this project is the presence of an agricultural research station within its boundaries, where extensive field characterization studies could be undertaken, and the vulnerability of this region to climatic extremes (e.g., several prolonged drought periods over the past decades). Over the course of the European project, several field and modeling activities of relevance to this study were undertaken (Cassiani et al., 2012; Marras et al., 2014; Filion et al., 2016; Meyer et al., 2016).

The Rio Mannu catchment drains an area of 473 km^2 and is characterized by a gently rolling topography, with an elevation range from 66 to 962 m a.s.l. and a mean slope of 17 %. The mean annual precipitation is 600 mm and the mean temperature ranges from 9 °C in January to 25 °C in July–August. The climate is typically Mediterranean, with about 90 % of the annual rainfall falling from October to April. The discharge regime is characterized by low flows (less than 1 $m^3\,s^{-1}$) for most of the year (Mascaro et al., 2013a). Precipitation, temperature, and discharge within and around the Rio Mannu catchment have been collected at a daily timescale, albeit intermittently and not always coincidentally, since 1925. As shown in Fig. 2, soil texture in the Rio Mannu catchment is dominated by three classes, including clay loam–clay (37 %), sandy loam–loam (32 %), and sandy loam–sandy clay loam (20 %). Agriculture ($\sim$ 48 %) and sparse vegetation ($\sim$ 26 %) are the dominant land use classes. A more detailed description of the basin land surface properties can be found in Mascaro et al. (2013b).

3 Methods

An impact assessment framework was developed during the precursor European project wherein the four best-performing GCM–RCM combinations from the ENSEMBLES project (van der Linden and Mitchell, 2009) were selected for each study site. The daily GCM–RCM outputs at 25 km resolution for a reference (1971–2000) and future (2041–2070) period were bias-corrected and statistically downscaled. For the Rio Mannu site, the downscaled data were then used to force five hydrologic models for the reference and future periods. The hydrologic models were independently calibrated and validated against observed data, with each modeling group using the type of data most suitable to that model, such as field-scale soil moisture, evapotranspiration patterns, and discharge. More details on the model calibration are given after the model descriptions below.

3.1 Hydrologic models

The five hydrologic models examined in this study are CATchment HYdrology (CATHY), the Soil and Water Assessment Tool (SWAT), TOPographic Kinematic APproximation and Integration (TOPKAPI), the TIN-based Real time Integrated Basin Simulator (tRIBS), and the WAter balance SImulation Model (WASIM). The models differ greatly in their representation of terrain features and physical processes and in their numerical complexity, but they are all able to account for the spatial variability of meteorological inputs and land surface properties, albeit at different levels of detail. Table 1 summarizes the characteristics of each hydrologic model, highlighting the main differences between them. For more detail, the reader is referred to the references provided below in the description of each model.

CATHY is a physically based numerical model that resolves in a detailed manner the interaction between subsurface and surface water (Camporese et al., 2010). The surface module is based on the resolution of a one-dimensional diffusion wave approximation of the Saint-Venant equation for overland and channel routing (Orlandini and Rosso, 1996). The subsurface module solves the three-dimensional Richards equation that describes flow in variably saturated porous media (Paniconi and Wood, 1993). The surface grid, catchment boundaries, and rill and channel flow paths are delineated via topographic analysis of digital elevation maps. Model inputs consist of spatially variable or homogeneous meteorological data and surface properties for each zone and layer of the basin. CATHY outputs include time series of actual fluxes and discharge and at any location in the stream network and spatial maps of several hydrological variables (e.g., pressure, saturation, ponding) at specified times. The CATHY model has been used in many exploratory studies, benchmarking exercises, and real catchment applications, including the assessment of climate change impacts (e.g., Gau-

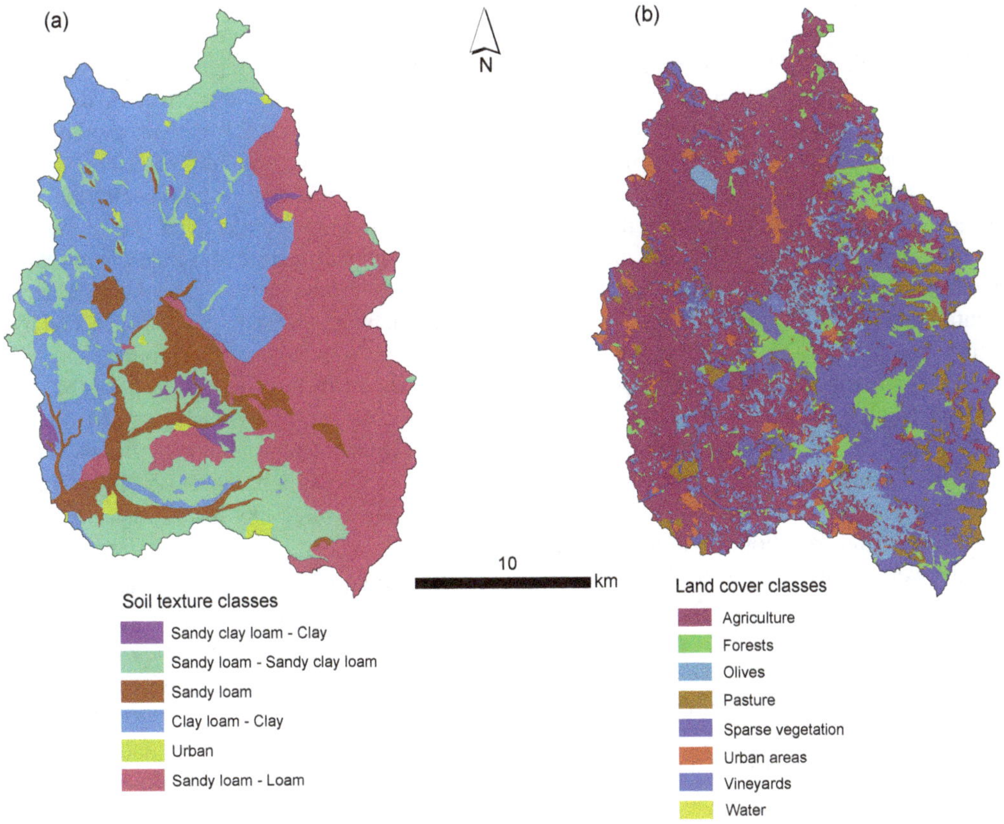

Figure 2. (a) Soil texture and **(b)** land cover maps used for the Rio Mannu basin.

Table 1. Comparison of the structure and characteristics of the five hydrologic models.

Model	Discretization scheme	Infiltration/subsurface flow	Surface flow	Topographic representation
CATHY	Finite element	Richards' equation	Diffusive wave	Regular grid
SWAT	Subwatershed	Tipping bucket	Soil Conservation Service (SCS)	Homogeneous hydrologic units
TOPKAPI	Finite difference	Kinematic wave	Kinematic wave	Regular grid
tRIBS	Finite difference control volume	Modified Green–Ampt	Kinematic wave	Triangulated irregular network
WASIM	Finite difference	Richards' equation	Kinematic wave	Regular grid

thier et al., 2009; Sulis et al., 2011; Gatel et al., 2016; Kollet et al., 2017; Scudeler et al., 2017).

SWAT is a conceptual, semi-distributed model that allows the evaluation of climate and land use impacts on water resources, sediments, and agriculture through a physical representation of hydrologic processes, soil temperature, plant growth, nutrients, pesticides, and land use (Arnold et al., 1998). In SWAT, a watershed is divided into multiple subwatersheds, which are then further divided into hydrologic response units (HRUs) that consist of homogeneous land use, management, topographic, and soil characteristics. The HRUs are represented as a percentage of the subwatershed area and need not be contiguous or spatially identified. A daily time step is adopted in the simulation of hydrologic processes. Surface runoff is estimated using the Soil Conservation Service (SCS) curve number procedure, and the movement of soil moisture vertically within the soil profile is simulated using a one-dimensional tipping bucket approach. Model inputs consist of meteorological data and surface and vegetation properties. SWAT outputs include time series of discharge at any location in the stream network and actual evapotranspiration and soil water content integrated over the basin. Its applications range from engineering/practical aims to research studies (e.g., Arnold et al., 1999; Cau et al., 2005; Mausbach and Dedrick, 2004; Volk et al., 2007).

TOPKAPI is a physically based distributed rainfall–runoff model that combines basin topography with the kinematic approach (Ciarapica and Todini, 2002). The model consists of five modules that simulate the main hydrologic processes, including subsurface flow, overland flow, channel flow, evapotranspiration, and snowmelt. These can be simulated at an hourly time step. Four nonlinear reservoir differential equations solved using a two-dimensional finite difference method are used to describe subsurface, overland, and channel flow. The model uses a regular grid to represent the terrain and is computationally efficient and thus suitable to be applied for real-time flood forecasting. Model inputs consist of meteorological data and spatial maps of surface properties (e.g., soil texture and land cover maps). TOPKAPI outputs include time series of discharge at any location in the stream network and actual evapotranspiration and soil water content integrated over the basin. TOPKAPI has been successfully implemented as a research and operational hydrologic model in several catchments worldwide (e.g., Liu and Todini, 2002; Bartholmes and Todini, 2005; Liu et al., 2005; Martina et al., 2006).

tRIBS is a physically based spatially distributed model that reproduces a range of hydrologic processes (Ivanov et al., 2004), including canopy interception and transpiration, evaporation from bare and vegetated soils, infiltration and soil moisture redistribution, shallow subsurface transport, and overland and channel flows (Mascaro et al., 2013b). Terrain features are represented via triangulated irregular networks (TINs). In each Voronoi polygon derived from TINs the coupled energy and water balances are computed, while the infiltration scheme is based on the resolution of the two-dimensional modified Green–Ampt model. A kinematic wave routing model is used to simulate transport of water in the channel network. Model inputs include spatial maps of surface properties (e.g., soil texture and land cover maps). tRIBS outputs include time series of discharge at any location in the stream network and spatial maps of several hydrological variables (e.g., actual evapotranspiration, soil water content at different depths) at specified times or integrated over the simulation period (Piras, 2014). The model has been applied across a large range of scales in the areas of hydrometeorology, climate change, and ecohydrology (e.g., Liuzzo et al., 2010; Mascaro et al., 2010, 2015; Mahmood and Vivoni, 2014). Recently, Piras (2014) and Piras et al. (2014) applied tRIBS in the Rio Mannu catchment to evaluate the hydrologic impact of climate change.

WASIM is a physically based and fully distributed hydrologic model (Schulla, 2015) originally developed to evaluate the influence of climate change on water balance and runoff regime in pre-alpine and alpine river catchments (Schulla, 1997). WASIM runs in a grid-based structure and represents vertical fluxes in the unsaturated zone by the one-dimensional Richards equation, which is solved with a finite difference scheme. Discharge routing is performed by a kinematic wave approach. After the translation of the wave for all channels, a single linear storage is applied to the routed discharge considering the effect of diffusion and retention (Schulla and Jasper, 2001). Sub-modules are available for various hydrologic variables such as interception, discharge, runoff, snowmelt, and evapotranspiration. Model inputs consist of meteorological data and spatial maps of surface properties (e.g., soil texture and land cover maps). WASIM outputs include time series of discharge at any location in the stream network and spatial maps of several hydrological variables (e.g., actual evapotranspiration and soil water content) at specified times or integrated over the simulation period. WASIM has been previously applied for hydrologic issues such as impact analysis for river basins and hydrologic forecasting (e.g., Cornelissen et al., 2013; Jasper et al., 2002; Kunstmann et al., 2006; Meyer et al., 2016).

The calibration procedures varied from one model to the next, owing to the significant structural differences between the models, and to the fact that each modeling group worked independently of the others, within different project frameworks and timelines. tRIBS was calibrated manually against discharge observations at the outlet for the year 1930, and validated for the 1931–1932 period (Mascaro et al., 2013b). The calibration focused on its two most sensitive parameters, saturated hydraulic conductivity at the surface and the decay parameter that models the variation of conductivity with soil depth. Model performance was assessed based on a comparison between observed and simulated discharge and flood duration curves, quantified using the Nash–Sutcliffe index. For WASIM, Meyer et al. (2016) conducted a soil sampling campaign on the Rio Mannu catchment in 2010–2011. They tested the performance of different regionalization methods for soil texture against soil moisture field measurements, and derived a new soil texture map based on the best of these methods. WASIM was then run with two different soil configurations – the new map and the available regional soil map – and the sensitivity to model outputs was examined. The model was validated with spatially distributed evapotranspiration rates using the triangle method (Jiang and Islam, 1999), and performances were quantified using the coefficient of determination. The SWAT model parameterization was based on a regional-scale calibration of the model's soil parameters against discharge observations (Cau et al., 2005). Model performances were quantified using the correlation coefficient computed from the simulated and observed discharge for all basins in Sardinia with at least 10 years of continuous streamflow data. For CATHY and TOPKAPI, the same dataset and parameter settings as the tRIBS model were used for common parameters, in order to investigate parameter transferability between these three models (Perra et al., 2018).

Table 2. List of global and regional climate models used in this work. The four GCM–RCM combinations used are ECH-RCA, ECH-REM, ECH-RMO, and HCH-RCA.

	Climatological center and model	Abbreviation
GCM	Hadley Centre for Climate Prediction, Met Office, UK, HadCM3 model	HCH
	Max Planck Institute for Meteorology, Germany, ECHAM5/MPI model	ECH
RCM	Swedish Meteorological and Hydrological Institute (SMHI), Sweden RCA model	RCA
	Max Planck Institute for Meteorology, Hamburg, Germany, REMO model	REM
	Koninklijk Nederlands Meteorologisch Instituut (KNMI), Netherlands, RACMO2 model	RMO

3.2 Climate models, bias correction, and statistical downscaling

Deidda et al. (2013) analyzed the open-access outputs of 14 GCM–RCM combinations from the ENSEMBLES project to identify those exhibiting the best performance in terms of representing the intra-annual variability of precipitation and temperature in the present climate for the seven study sites of the precursor European project. For each study site, the selected set of climate model data was validated using the E-OBS dataset, a high-quality pan-European gridded observational dataset of daily precipitation and temperature (Haylock et al., 2008). The models (and their acronyms: ECH-RCA, ECH-REM, ECH-RMO, and HCH-RCA) selected for the Rio Mannu site are listed in Table 2. For these models, outputs were extracted for a reference (1971–2000) and future (2041–2070) period under the A1B emission scenario (Nakićeović et al., 2000), which was considered one of the most realistic and provided the most complete dataset within the ENSEMBLES models. A large-scale bias correction was applied to precipitation and temperature fields using the daily translation method (Wood et al., 2004; Maurer and Hildago, 2008) with the E-OBS dataset as a reference. In addition, downscaling techniques were applied to disaggregate precipitation and temperature from the coarse resolution of the climate models (~ 25 km, 24 h) to finer resolutions (5 km, 1 h) suitable for hydrologic modeling. For precipitation, the multifractal downscaling model of Deidda et al. (1999) and Deidda (2000) was utilized, while temperature was interpolated in space through lapse rate corrections as in Liston and Elder (2006). More details on the bias correction and downscaling techniques are provided in Piras et al. (2014). For the tRIBS, CATHY, and TOPKAPI models, temperature grids were used to derive hourly grids of potential evapotranspiration according to the method described in Mascaro et al. (2013b).

3.3 Metrics to compare climate and hydrologic models

To compare the outputs of (i) the four climate models and (ii) the five hydrologic models forced by the four climate models in the reference and future periods, we first derived the climatological monthly means. Next, we quantified the difference between each pair of climate or hydrologic models by using the Pearson correlation coefficient r and the bias coefficient α, proposed by Duveiller et al. (2016), defined as

$$r = \frac{\sum_{i=1}^{n} \left(X_i - \overline{X}\right)\left(Y_i - \overline{Y}\right)}{n\sigma_X\sigma_Y} \tag{1}$$

$$\alpha = \begin{cases} \dfrac{2}{\left[\frac{\sigma_X}{\sigma_Y} + \frac{\sigma_Y}{\sigma_X} + \frac{(\overline{X}-\overline{Y})^2}{\sigma_X\sigma_Y}\right]}, & r > 0 \\ 0, & r \le 0 \end{cases} \tag{2}$$

where X_i and Y_i are the 30-year mean monthly values of a given response variable simulated by a pair of models, $\overline{X}$ and $\overline{Y}$ are their means, σ_X and σ_Y are their standard deviations, and $n = 12$ is the number of months per year.

The Pearson coefficient, which can range between -1 and 1, is a widely used measure of the degree of linear dependence between two datasets, but it does not give any indication of how similar they are in magnitude. In contrast, the bias coefficient, ranging from 0 (full bias, no agreement) to 1 (no bias, perfect agreement), evaluates possible additive or multiplicative biases between the model outputs. These two indices were recently used in a hydrologic model intercomparison study (Kollet et al., 2017) to evaluate the agreement between seven integrated surface–subsurface models for a series of benchmark test cases. Here, the two indices r and α were computed for all pairs of both climate and hydrologic representative variables. The results are presented in matrix pictures where each element represents the index value for a single model pair, thus allowing easy comparison of each combination of model pairs with all the others. In Fig. 3, an example of a matrix picture between two models A and B is shown: the circles represent correlation r and the squares bias α, with the color and size of the markers proportional to the value of the metric. Four possible levels of model agreement are reported: high, medium, low, and no agreement.

4 Results and discussion

In this section the main meteorological forcing, precipitation, and temperature projected by the climate models are first presented and analyzed in terms of variations between the future

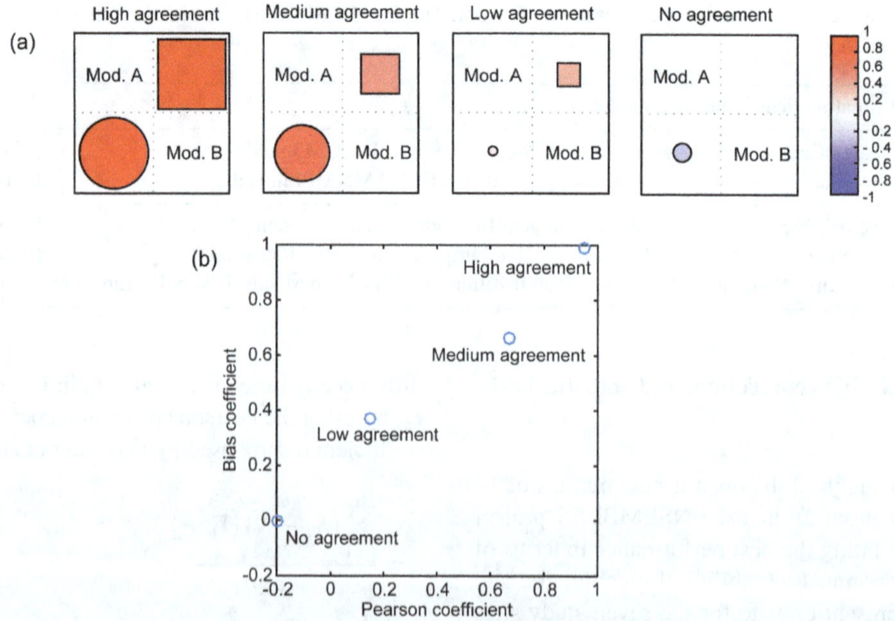

Figure 3. **(a)** Example of a matrix picture between two models A and B. Correlation values ($-1 \leq r \leq 1$) are represented with circles below the diagonal, while bias ($0 \leq \alpha \leq 1$) is plotted as squares above the diagonal. The size and color of symbols are proportional to the coefficient values. White (blank) matrix entries correspond to $r = 0$ or $\alpha = 0$. **(b)** Scatterplot corresponding to the four cases shown in panel **(a)**.

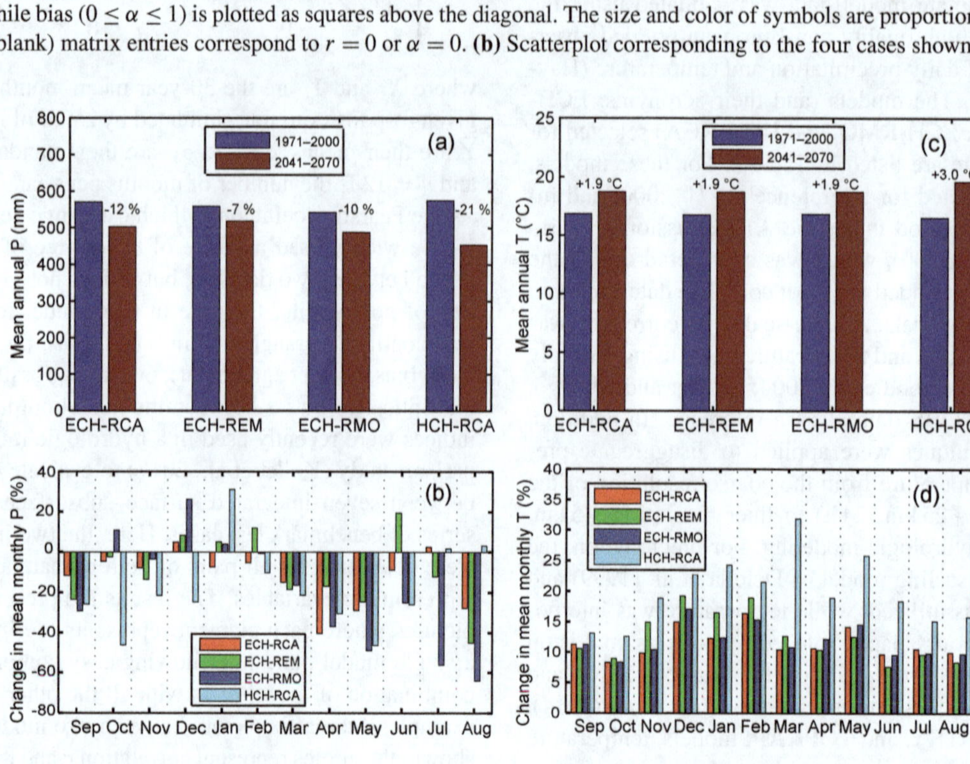

Figure 4. Mean annual precipitation P **(a)** and temperature T **(c)** predicted by the four climate models for the reference (1971–2000), blue bars, and future (2041–2070), red bars, periods. Relative change in mean monthly precipitation P **(b)** and temperature T **(d)** between the reference and future periods for the four climate models.

and reference periods, in order to establish the expected climate change trends for the Rio Mannu catchment. The level of agreement between climate models is then evaluated for the reference and future periods using Pearson correlation values and Duveiller biases. Subsequently, the impact of projected climate change is investigated through application of the five hydrologic models. Water availability and fluxes in terms of discharge, soil water content, and actual evapotranspiration are analyzed for trends and inter-model agreement.

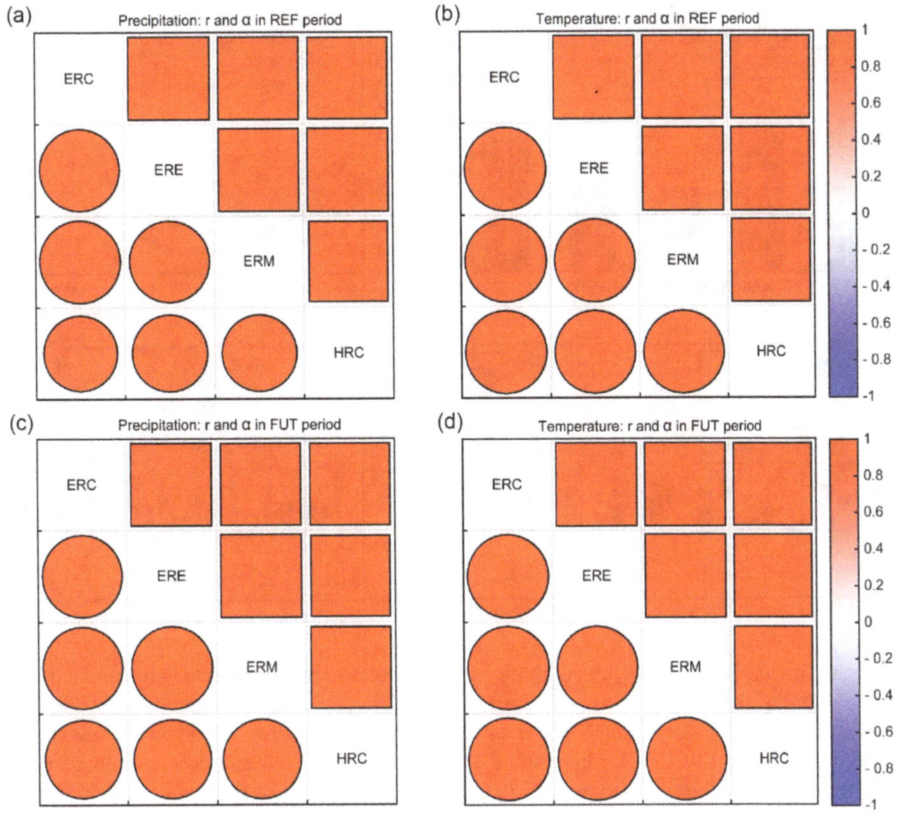

Figure 5. Results of the analysis of agreement for mean monthly precipitation **(a, c)** and temperature **(b, d)** among the four climate models (ERC: ECH-RCA, ERE: ECH-REM, ERM: ECH-RMO, HRC: HCH-RCA) for the reference (REF, 1971–2000, **a, b**) and future (FUT, 2041–2070, **c, d**) periods.

4.1 Climate models: projected changes and agreement analysis

Climate model outputs were bias-corrected and downscaled to provide more reliable inputs to the hydrologic models. Specifically, for each climate model, the climatological means of precipitation (P) and temperature (T) averaged over the catchment were computed at annual and monthly scales. Figure 4 compares results for the reference and future periods. All models predict a decrease in mean annual P, with percent changes ranging from $-7\,\%$ to $-21\,\%$, and an increase in T from 1.9 to 3 °C. All models predict negative changes in P for all months except winter (December–February), where the models simulated an increase in P, and also June for ECH-REM and October for ECH-RMO. T is projected to rise in all months for all models, with the RCMs forced by ECH predicting comparable magnitudes in change, and HCH-RCA simulating the largest increment.

To quantify the agreement of the monthly climatologies of P and T predicted by the models, the correlation coefficient, r, and the bias, α, are plotted in Fig. 5. The left and right panels show the results for, respectively, P and T for the reference (Fig. 5a, b) and future (Fig. 5c, d) periods. In each panel, circles represent r and squares α, while the color and size of the markers are proportional to the metric value. The actual values of the Pearson and bias coefficients are reported in Table S1 in the Supplement. The metrics indicate a general high level of agreement of the climatologies simulated by all models, with r and α for each pair of models always larger than 0.9 for both variables and in both periods. Comparing the same climate model for the reference and future periods, the values of r and α (last row and last column, respectively, of the bottom panels) are also high for both variables: for P and the HCH-RCA model, which is the model that slightly differs from the others, both Pearson and bias coefficients are close to 1 ($r = 0.918$ and $\alpha = 0.912$). As a result, the agreement of seasonal cycles is high, especially in the case of temperature, suggesting that the uncertainty due to climate models can be considered low, although a small bias is found when comparing the three climate models forced by ECH with HCH-RCA, as expected since it is recognized that GCMs exert the major influence on the projected climate change (Graham et al., 2007; Kay et al., 2009).

4.2 Hydrologic impact

A summary of the annual and monthly climatologies of basin-averaged potential evapotranspiration (ETP), runoff

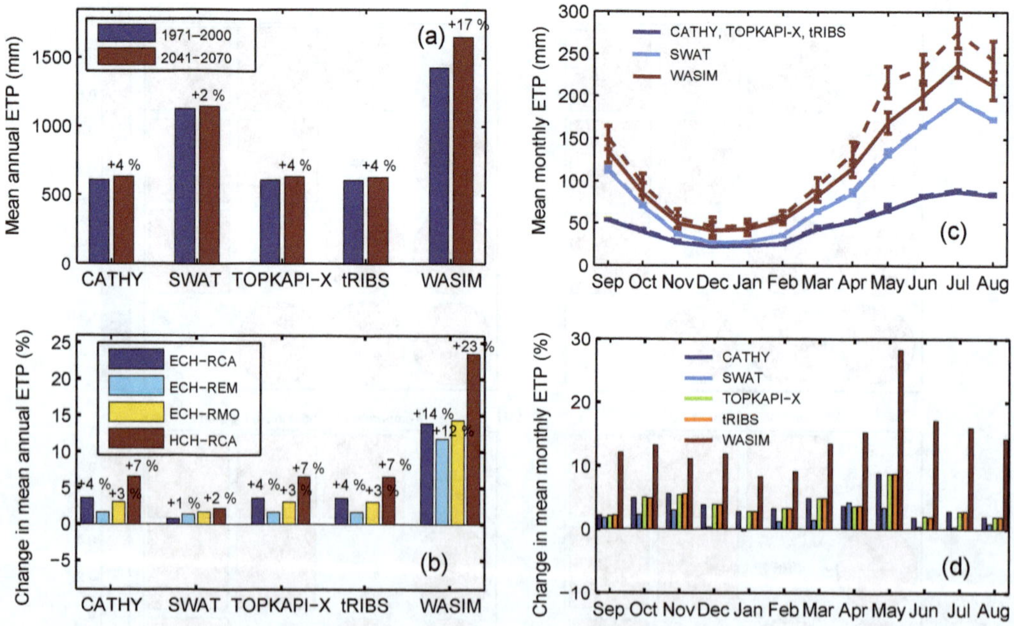

Figure 6. Potential evapotranspiration (ETP) for each hydrologic model forced by the four selected climate models for the reference (REF, 1971–2000) and future (FUT, 2041–2070) periods. **(a)** Mean annual ETP during REF and FUT periods, obtained as an average among the four climate models. **(b)** Relative change in mean annual ETP between REF and FUT periods forced by the four climate models. **(c)** Seasonal distribution of mean monthly ETP during REF (solid line) and FUT (dotted line) periods and corresponding standard deviations (vertical bars), obtained as an average among the four climate models. **(d)** Seasonal distribution of relative change in mean monthly ETP between REF and FUT periods, obtained as an average among the four climate models.

(Q), soil water content (SWC), and actual evapotranspiration (ET_a) simulated by the five hydrologic models, forced by four climate models, is reported in Figs. 6–9. Each figure shows the annual simulated variable in each period, including the percent change from reference to future (panel a); the relative change in mean annual values between reference and future periods forced by the four climate models (panel b); the seasonal distribution of mean monthly values of each variable during reference and future periods and corresponding standard deviations (panel c); and finally the seasonal distribution of relative change in mean monthly values between reference and future periods (panel d).

ETP is predicted to rise by all models on an annual basis (Fig. 6a), mostly due to the projected increment of T. The values of annual and monthly (Fig. 6c) ETP differ among the hydrologic models, due to the different computation methods adopted. For SWAT and WASIM, ETP was computed at a daily timescale by internal routines based on the Hargreaves (Hargreaves et al., 1994; Hargreaves and Allen, 2003) and Penman–Monteith (Penman, 1948; Monteith, 1965) formulas, respectively, producing an annual mean of about 1100 mm for SWAT and 1400 mm for WASIM. For CATHY, TOPKAPI, and tRIBS, a common reliable diurnal cycle for ETP was derived at an hourly timescale using an approach based on the Penman–Monteith and Hargreaves formulas, detailed in Mascaro el al. (2013b), producing an annual mean of about 650 mm, which is consistent with previous estimates for this region (Pulina et al., 1986). From Fig. 6c we can also observe the slight increase in ETP predicted by all hydrologic models in the future period, except for the WASIM model and especially during the summer and spring months. Furthermore, note that the highest increase in ETP is predicted with all hydrologic models under HCH-RCA forcing (Fig. 6b), as expected since this GCM–RCM combination also projects the highest increase in temperature, as already discussed. Among the hydrologic models, WASIM is the one that predicts the higher increase. We can observe also from Fig. 6d that relative changes in potential evapotranspiration are predicted to increase much more during summer and spring.

Results in terms of Q are analyzed in Fig. 7: it is apparent that all models predict decreasing values in the future. Figure 7a, reporting mean annual Q obtained for each hydrologic model in the reference and future periods obtained as an average among the four climate models, shows a reduction that ranges from −12 % according to SWAT to −69 % according to CATHY. Figure 7b reports the relative change between future and reference periods computed for each climate model configuration: we can observe that the reduction varies within the same hydrologic model considering different climate forcing. The largest decrease is always given by configurations forced with HCH-RCA, ranging from −23 %

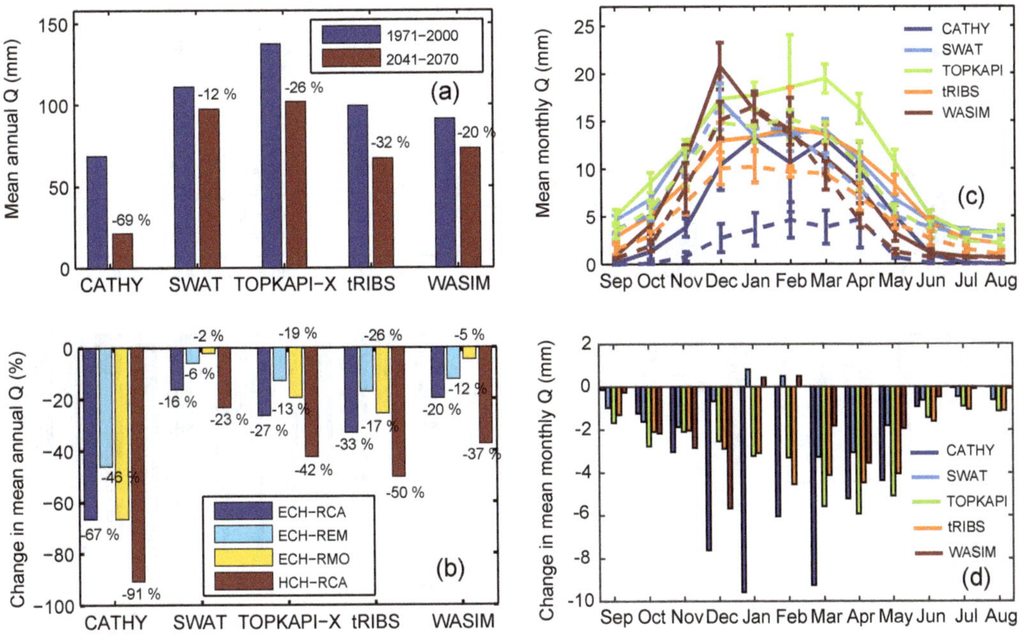

Figure 7. As Fig. 6, but for discharge (Q) simulated by each hydrologic model forced by the four selected climate models for the reference (REF, 1971–2000) and future (FUT, 2041–2070) periods.

Table 3. Change (%) between future and reference periods of mean annual discharge (Q), soil water content (SWC), and actual evapotranspiration (ET_a) for the five hydrologic models (CAT: CATHY, SWA: SWAT, TOP: TOPKAPI, TRI: tRIBS, WAS: WASIM), expressed as mean (μ) and standard deviation (σ) calculated for each and then all hydrologic models (HMs) considering all climate models (ERC: ECH-RCA, ERE: ECH-REM, ERM: ECH-RMO, HRC: HCH-RCA), and for each and then all climate models (CMs) considering all hydrologic models.

		Overall tendencies considering all climate models						Overall tendencies considering all hydrologic models				
		CAT (%)	SWA (%)	TOP (%)	TRI (%)	WAS (%)	ALL HM (%)	ERC (%)	ERE (%)	ERM (%)	HRC (%)	ALL CM (%)
Q	μ	−68	−12	−25	−32	−19	−31	−32	−19	−24	−49	−31
	σ	18	10	13	14	14	14	20	16	26	25	22
SWC	μ	−10	−8	−13	−5	−10	−9	−8	−6	−6	−16	−9
	σ	5	5	6	3	4	5	2	2	3	5	3
ET_a	μ	10	−12	−8	−2	−11	−5	−3	−3	−5	−7	−5
	σ	4	5	3	1	5	3	7	6	9	14	9

for the SWAT model to −91 % for the CATHY model, followed by the ECH-RCA climate model, for which the reduction varies from −16 % for SWAT to −67 % for CATHY. A summary in terms of change (%) between reference and future periods for mean annual Q, simulated by the five hydrologic models and the four climate models, is provided in Table 3. Figure 7c refers to mean monthly Q, showing the mean seasonality in reference (solid line) and future (dotted line) periods with bars indicating the standard deviations within each model. Figure 7d details the monthly variations during the two periods according to the five hydrologic models: the seasonality is quite similar among them even if some differences hold also in this case. The five hydrologic models predict diminished mean monthly Q in the future period throughout the year with the exception of January and February, when SWAT and WASIM simulate a slight increase.

Figure 8 shows mean values and changes in SWC in the first meter depth of soil. All simulations predict a decreasing trend of SWC, but again we can notice some differences among the hydrologic models. For instance, Fig. 8a clearly shows that CATHY presents the highest soil humidity (35 %) and WASIM the lowest (17 %), while the highest and lowest decrements in the future are observed, respectively, for TOPKAPI (−13 %) and tRIBS (−5 %). Figure 8b details the rel-

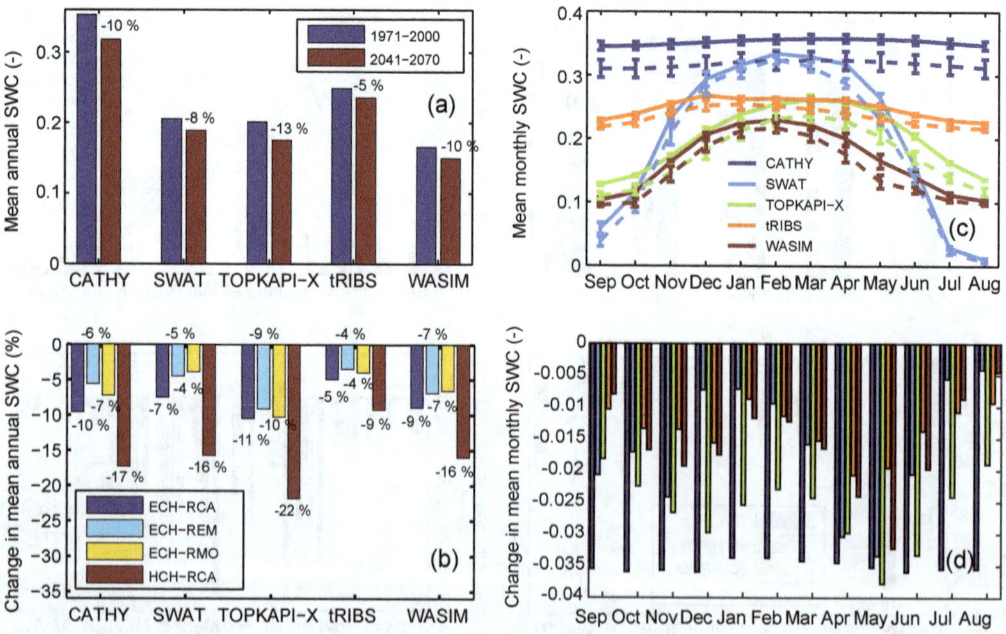

Figure 8. As Fig. 6, but for soil water content (SWC) simulated by each hydrologic model forced by the four selected climate models for the reference (REF, 1971–2000) and future (FUT, 2041–2070) periods. For panel **(d)** colors for each model are the same as panel **(c)**.

ative change between future and reference periods for each configuration of hydrologic and climate models. As for Q, each hydrologic model simulates the maximum SWC reduction under the HCH-RCA configuration. The reduction with this climate forcing, ranging from $-9\,\%$ for tRIBS to $-22\,\%$ for TOPKAPI, can in fact be double with respect to the one obtained with the other climate models. A summary in terms of change (%) between reference and future periods of mean annual SWC, simulated by the five hydrologic models and the four climate models, is provided in Table 3. The mean monthly seasonal distribution of SWC reported in Fig. 8c is quite different among the five hydrologic models. SWAT presents the highest variations from winter/spring to summer month values, while the annual range is more limited in the CATHY and tRIBS simulations. The mean monthly relative changes between reference and future periods represented in Fig. 8d are always negative. CATHY simulates a quite constant diminution (of about 0.035) throughout the year, which is always larger than the other models. The changes exceed 0.03 in May for TOPKAPI and SWAT, which instead predict the lowest reduction in winter months. These reductions during spring months can be related to the higher vegetation activity combined with moderate values of temperature and potential evapotranspiration.

The differences among the hydrologic models in representing soil–vegetation–atmosphere transfers are reflected also in simulations of ET_a processes (Fig. 9). The mean annual values reported in Fig. 9a are predicted to decrease in the future by four models (tRIBS presents the lowest reduction, $-2\,\%$ on average, SWAT the highest, $-12\,\%$ on average), with CATHY being the only one projecting a slight increase. The reason for this is that CATHY simulates the highest soil water content in the first meter depth, as can be appreciated in Fig. 8a and c, and it is the model that simulates the minimum discharge (Fig. 7a and c); thus, it retains more water within the soil zone available for evaporation. Figure 9b shows that the highest variations (both positive and negative) in actual evapotranspiration are again reached in simulations forced by the HCH-RCA model, ranging from $-17\,\%$ for SWAT and WASIM and $+16\,\%$ for CATHY. A summary in terms of change (%) between reference and future periods of mean annual ET_a, simulated by the five hydrologic models and the four climate models, is provided in Table 3. Mean monthly ET_a reported in Fig. 9c presents different patterns, with the hydrologic models divided into two groups: CATHY, TOPKAPI, and tRIBS reach the highest values during summer months, when temperature and potential evapotranspiration are higher; SWAT and WASIM anticipate the seasonal peak in spring when moderate temperatures coincide with vegetation activity, a component that these two models incorporate more fully than the other three models. The different patterns may also be due to limited water availability during the summer: as can be seen from Fig. 8c, SWAT and WASIM simulate quite low soil water content during July and August. Figure 9d, reporting relative changes in monthly values, shows that the increase predicted by CATHY is highest in spring; in fact relative changes in monthly temperatures (Fig. 4d) and potential evapotranspiration (Fig. 6d) during spring are predicted to increase much more than during summer months. The other four models predict instead a future

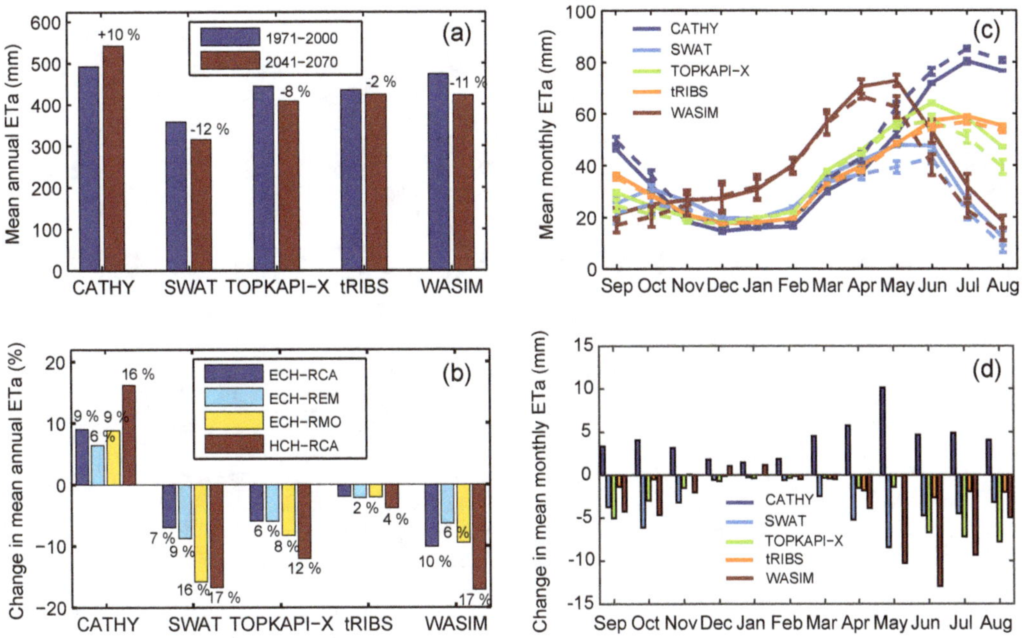

Figure 9. As Fig. 6, but for actual evapotranspiration (ET_a) simulated by each hydrologic model forced by the four selected climate models for the reference (REF, 1971–2000) and future (FUT, 2041–2070) periods.

diminution more pronounced in summer, exceeding -10 mm for the WASIM model. This can be related to the fact that these models in the future simulate during summer months lower soil water content with respect to the reference period (Fig. 8d). In winter months the future evapotranspiration reduces negligibly, or even increases slightly in the case of WASIM.

4.3 Hydrologic model agreement analysis

The agreement among the hydrologic models forced with the different climate configurations is evaluated using the Pearson correlation and Duveiller bias coefficients in Figs. 10–12. The actual values of the Pearson and bias coefficients are reported in Tables S2–S4. Figure 10 shows the results for Q, and each panel summarizes the agreement among hydrologic models forced by a specific GCM–RCM configuration for the reference (Fig. 10a) and future (Fig. 10b) periods. Following the same graphical representation of Fig. 3, Pearson coefficients are displayed as circles in the lowest-left part of each panel, while the bias coefficients are represented with squares in the upper-right part. In both cases the size of symbols is proportional to the magnitude of the corresponding coefficient. We can notice that during the reference period the agreement in terms of both indices between any pair of hydrologic models is high, with the HCH-RCA performance slightly better than for the other climate forcing configurations. The values of the Pearson coefficient r, comparing the hydrologic models in pairs, range from 0.70 (CATHY-WASIM under ECH-RMO forcing) to 0.99 (TOPKAPI-tRIBS under ECH-RMO forcing) for the reference period and from 0.67 (CATHY-WASIM and CATHY-SWAT under ECH-REM forcing) to 0.99 (SWAT-TOPKAPI under HCH-RCA forcing) for the future period. Looking at the bias coefficient we can observe a general agreement ($\alpha > 0.7$) regardless of the climate forcing among the hydrologic models, except for CATHY, which shows in the future period the highest differences with the other models, as expected from Fig. 5, with the lowest agreement occurring for the HCH-RCA configuration. CATHY generates the lowest Q in the future, and this result is reflected in the values of the bias parameter α, which exceeds the value of 0.7 only when CATHY is compared under ECH-REM with tRIBS and WASIM, which use the same equations as CATHY to represent subsurface and surface dynamics, albeit simplified in some way (e.g., lower dimensionality). Bias parameter values range from 0.53 (CATHY-TOPKAPI forced by ECH-RMO) to 0.99 (SWAT-tRIBS forced by ECH-RMO) for the reference period and from 0.09 (CATHY-SWAT under HCH-RCA forcing) to 0.99 (SWAT-TOPKAPI under ECH-RMO forcing) for the future period.

Figure 11 shows a similar comparison for SWC. The agreement among the hydrologic models generally diminishes with respect to the discharge intercomparison. Again the CATHY model presents the lowest correlation with the others, followed by tRIBS. CATHY and tRIBS are in fact the two models that show limited variations of SWC from winter/spring to summer month values with respect to the others, as shown in Fig. 8b. The values of the Pearson coefficient r range from 0.65 (CATHY-tRIBS in the HCH-RCA

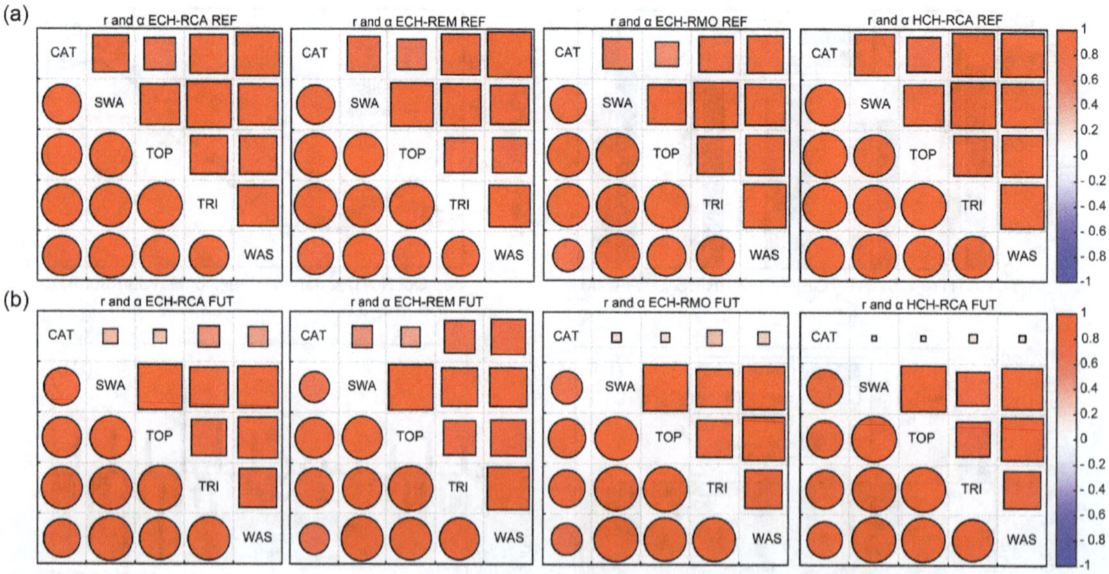

Figure 10. As Fig. 5, but for mean monthly discharge agreement between the five hydrologic models (CAT: CATHY, SWA: SWAT, TOP: TOPKAPI, TRI: tRIBS, WAS: WASIM). Each panel displays the agreement under a specific GCM–RCM forcing, for the reference (REF, **a**) and future (FUT, **b**) periods.

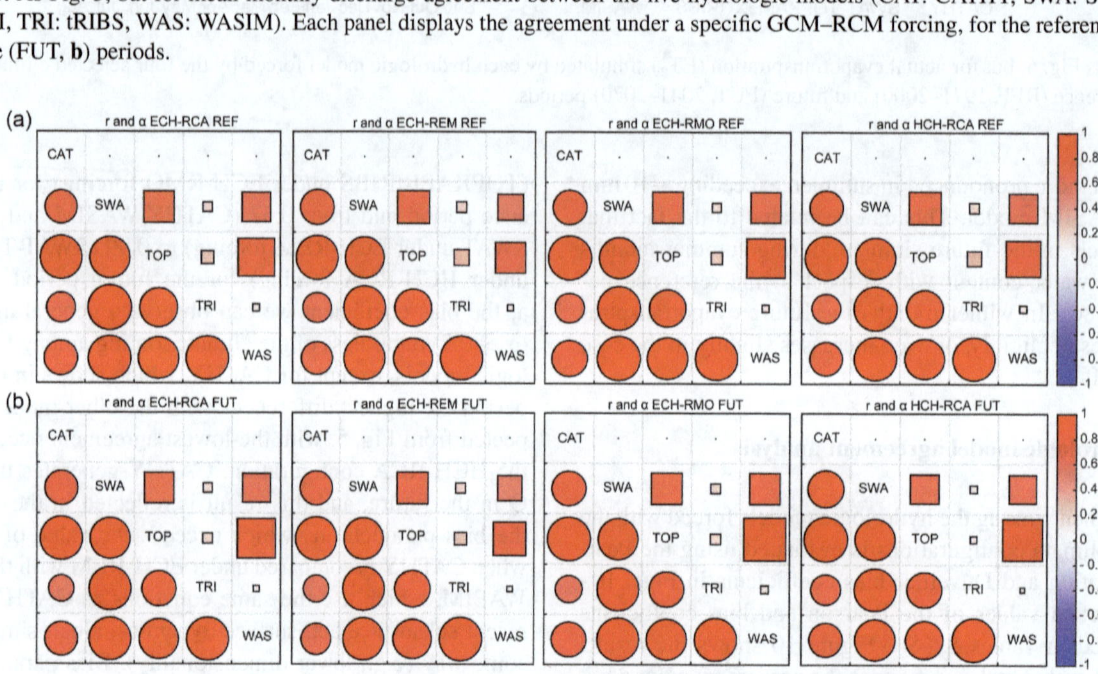

Figure 11. As Fig. 10, but for mean monthly soil water content agreement between the five hydrologic models (CAT: CATHY, SWA: SWAT, TOP: TOPKAPI, TRI: tRIBS, WAS: WASIM) for the reference (REF, **a**) and future (FUT, **b**) periods.

configuration) to 0.98 (SWAT-tRIBS forced by ECH-RCA) for the reference period and from 0.57 (CATHY-tRIBS in ECH-RCA simulations) to 0.97 (CATHY-TOPKAPI forced by HCH-RCA) for the future period. The value of the bias coefficient α is near zero when the models are compared with CATHY and also quite low (about 0.2) when compared with tRIBS. The values of α range from 0.01 (CATHY versus tRIBS and WASIM in both the ECH-RCA and ECH-REM configurations) to 0.86 (TOPKAPI versus WASIM forced by ECH-RMO) for the reference period and from 0.01 (CATHY-WASIM in ECH-REM simulations) to 0.94 (TOPKAPI versus WASIM forced by ECH-RMO) for the future period.

The analysis of agreement presents the lowest Pearson correlation values in the case of ET_a (Fig. 12). The values of r range from −0.21 (CATHY-WASIM in the ECH-REM configuration) to 0.98 (CATHY-tRIBS for all climate model configurations) for the reference period and from −0.29 (CATHY-WASIM in ECH-REM simulations) to 0.99 (CATHY-tRIBS for all configurations) for the future period. Despite the high values of the Pearson coefficient between

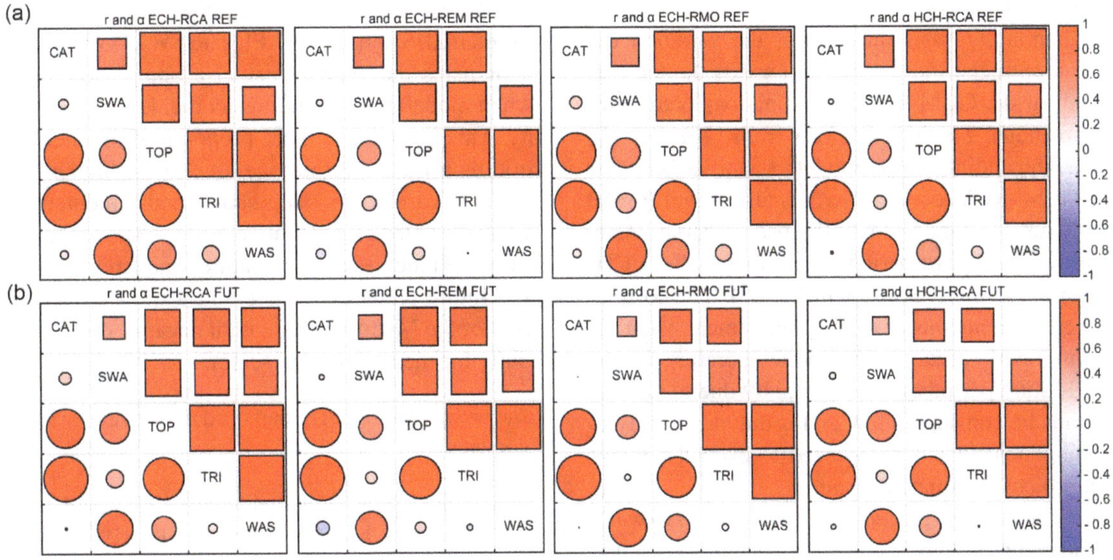

Figure 12. As Fig. 10, but for mean monthly actual evapotranspiration agreement between the five hydrologic models (CAT: CATHY, SWA: SWAT, TOP: TOPKAPI, TRI: tRIBS, WAS: WASIM) for the reference (REF, **a**) and future (FUT, **b**) periods.

CATHY and tRIBS for both the reference and future periods regardless of the GCM–RCM forcing, the Duveiller index displays a worsening ($\alpha \approx 0.6$–0.8). Considering overall results, the values of the bias coefficient α range from 0 (CATHY and tRIBS versus WASIM in the ECH-REM configuration) to 0.99 (TOPKAPI versus tRIBS for all configurations) for the reference period and from 0 (CATHY-WASIM in ECH-RMO simulations) to 0.99 (TOPKAPI versus tRIBS for all configurations) for the future period. From this figure it can be noted that, notwithstanding the differences, Pearson and bias indices for the reference period are similar for CATHY, tRIBS, and TOPKAPI (which are forced with the same ETP values). Furthermore, these models reach the highest values of ET_a during the summer months (Fig. 9c), when the temperature is highest. For the future period this agreement is maintained, with a strong correlation between tRIBS and TOPKAPI. Referring to the SWAT–WASIM pair, they anticipate the peak of ET_a in spring when moderate temperatures coincide with vegetation activity (Fig. 9c). This can be seen for the reference period and less for the future one, when α is slightly lower.

The differences that emerge from the analysis of agreement are consistent with the key structural differences between the hydrologic models. CATHY, for instance, has the most detailed subsurface representation of the five models (fully three-dimensional Richards equation; soil and aquifer zones), and as such will tend to retain more water in subsurface storage, making some of this water available for subsequent evaporation. An additional factor contributing to greater subsurface storage in CATHY is that all the lateral and bottom boundaries of the simulation domain are considered impermeable. In the agreement metrics CATHY tends to align most with TOPKAPI and tRIBS, which, although with a more simplified representation, also account for both vertical and lateral subsurface flow, unlike SWAT and WASIM, which resolve flow only in the vertical direction. These latter two models, on the other hand, show strong agreement, to the exclusion of the other models, for some of the evapotranspiration responses, consistent with the fact that both these models include a quite detailed representation of vegetation processes.

5 Conclusions

Five hydrologic models forced with the outputs of four combinations of global and regional climate models were compared to evaluate climate change consequences on the response of a medium-sized Mediterranean basin, the Rio Mannu catchment. In order to evaluate the agreement between model pairs, a new metric based on Pearson correlation and Duveiller bias coefficients has been used. The hydrologic models, independently calibrated and validated, were applied in cascade with climate models for a reference (1971–2000) and a future (2041–2070) period. Temporal series of different response variables simulated by the hydrologic models were used to evaluate the impacts on the basin of the predicted climate change in terms of water resource availability.

In a first step, climate model outputs, suitably bias-corrected and downscaled, were analyzed for the reference and future periods by comparing mean monthly and annual values of precipitation and temperature and by examining the agreement metrics. All of the GCM–RCM combinations agree that in the future period there will be decreasing mean annual precipitation (average differences of about −12 %), whereas on a monthly basis the sign of the variation depends on the month and the model. As regards the temperature trend, all of the GCM–RCM combinations predict increasing mean annual T values that vary from 11 % (1.9 °C) to 19 % (3 °C) depending on the model. A similar behavior for the four GCM–RCM combinations is also found for the mean monthly temperature trend, with positive variations in every season for the future period, from about 7 % (ECH-REM in June) to 30 % (HCH-RCA in March). The correlation and bias coefficients show favorable agreement when analyzing mean monthly precipitation and temperature for the reference and future periods. The uncertainty due to climate models can thus be considered low and is due principally to the GCM component that is recognized to exert the major influence on projected climate change.

In a second step, hydrologic model outputs related to water availability (namely discharge, soil water content, and actual evapotranspiration) were analyzed. Simulation results show decreasing mean annual runoff and a reduction of the soil water content at 1 m depth for the future period (average decreases of 31 % and 9 %, respectively). Actual mean annual evapotranspiration in the future will diminish according to four of the five hydrologic models due to drier soil conditions (average decrease of 8 %), while it will rise (by 10 %) in the prediction of the CATHY model, which retains the highest water content in its soil profile. For all response variables the biggest decrease is always predicted with the HCH-RCA model. Analyzing hydrologic model outputs at the monthly scale, we can observe variations not perceptible at the annual scale. Discharge for instance is predicted to decrease in the future period in all months except for January and February.

In terms of model agreement, for the reference period we can observe a good concordance between each pair of hydrologic models, while more significant differences emerge for the future period. The model that most differs from the others is CATHY, which generates the lowest discharge in the future, and this result is reflected in the values of the bias parameter. The five hydrologic models confirm the reduction of soil water content throughout the year, and the magnitude of variation depends on the hydrologic model considered. Again the CATHY model yields the lowest correlation with the other models, followed by tRIBS. Both models, in fact, show limited variation of soil water content from winter/spring to summer months with respect to the others, and this as well is reflected in the bias value. Actual evapotranspiration could rise in the future period according to the CATHY model and, during January and February, also according to WASIM, which instead predicts the strongest reductions in summer months. As regards the analysis of agreement for actual evapotranspiration, Pearson and bias indices are similar for CATHY, tRIBS, and TOPKAPI (which are forced with the same values of potential evapotranspiration). Moreover, these models reach the highest values of actual evapotranspiration during summer months. For the future period this agreement is maintained, with a strong correlation between tRIBS and TOPKAPI. The SWAT–WASIM model pair anticipates the peak of actual evapotranspiration in spring when moderate temperatures coincide with vegetation activity. This behavior is more pronounced for the reference period than for the future one, due to the higher bias. Overall the five hydrologic models show good agreement, responding similarly to the climate model predictions of reduced precipitation and increased temperatures and lending strong support to a future scenario of increased water shortages for this region of the Mediterranean, with negative consequences especially for the agricultural sector.

Competing interests. The authors declare that they have no conflict of interest.

Acknowledgements. This study was begun within the CLIMB project (Climate Induced Changes on the Hydrology of Mediterranean Basins, http://www.climb-fp7.eu), funded by the European Commission 7th Framework Programme. Financial support was also provided by Sardinia Region L.R. 7/2007 projects "Valutazione degli impatti sul comportamento idrologico dei bacini idrografici e sulle produzioni agricole conseguenti alle condizioni di cambiamento climatico" (funding call 2008) and "Impatti antropogenici e climatici sul ciclo idrologico a scala di bacino e di versante" (funding call 2013). The authors wish to thank Gabriele Coccia for his help in implementing the TOPKAPI model. The first author gratefully acknowledges the Sardinia Regional Government for financial support during her PhD (P.O.R. Sardegna F.S.E. Operational Programme of the Autonomous Region of Sardinia, European Social Fund 2007–2013 – Axis IV Human Resources, Objective l.3, Line of Activity l.3.1).

Edited by: Alberto Guadagnini

References

Acero, F. J., García, J. A., Cruz Gallego, M., Parey, S., and Dacunha-Castelle, D.: Trends in summer extreme temperatures over the Iberian Peninsula using nonurban station data, J. Geophys. Res.-Atmos., 119, 39–53, https://doi.org/10.1002/2013JD020590, 1002.

Arnold, J. G., Srinivasan, R., Muttiah, R. S., and Williams, J. R.: Large area hydrologic modeling and assessment, part I: model development, J. Am. Water Resour. Assoc., 34, 73–89, https://doi.org/10.1111/j.1752-1688.1998.tb05961.x, 1998.

Arnold, J. G., Srinivasan, R., Muttiah, R. S., and Allen, P. M.: Continental scale simulation of the hydrologic balance, J. Am. Water Resour. Assoc., 35, 1037–1051, https://doi.org/10.1111/j.1752-1688.1999.tb04192.x, 1999.

Bae, D.-H., Jung, I.-W., and Lettenmaier, D. P.: Hydrologic uncertainties in climate change from IPCC AR4 GCM simulations of the Chungju Basin, Korea, J. Hydrol., 401, 90–105, https://doi.org/10.1016/J.JHYDROL.2011.02.012, 2011.

Bartholmes, J. and Todini, E.: Coupling meteorological and hydrological models for flood forecasting, Hydrol. Earth Syst. Sci., 9, 333–346, https://doi.org/10.5194/hess-9-333-2005, 2005.

Bastola, S., Murphy, C., and Sweeney, J.: The role of hydrological modelling uncertainties in climate change impact assessments of Irish river catchments, Adv. Water Resour., 34, 562–576, https://doi.org/10.1016/j.advwatres.2011.01.008, 2011.

Bosshard, T., Carambia, M., Goergen, K., Kotlarski, S., Krahe, P., Zappa, M., and Schär, C.: Quantifying uncertainty sources in an ensemble of hydrological climate-impact projections, Water Resour. Res., 49, 1523–1536, https://doi.org/10.1029/2011WR011533, 2013.

Camporese, M., Paniconi, C., Putti, M., and Orlandini, S.: Surface-subsurface flow modeling with path-based runoff routing, boundary condition-based coupling, and assimilation of multisource observation data, Water Resour. Res., 46, W02512, https://doi.org/10.1029/2008WR007536, 2010.

Cassiani, G., Ursino, N., Deiana, R., Vignoli, G., Boaga, J., Rossi, M., Perri, M. T., Blaschek, M., Duttmann, R., Meyer, S., Ludwig, R., Soddu, A., Dietrich, P., and Werban, U.: Noninvasive Monitoring of Soil Static Characteristics and Dynamic States: A Case Study Highlighting Vegetation Effects on Agricultural Land, Vadose Zone J., 11, https://doi.org/10.2136/vzj2011.0195, 2012.

Cau, P., Cadeddu, A., Gallo, C., Lecca, G., and Marrocu, M.: Estimating the water balance of the Sardinian island using the SWAT model, L'Acqua, 5, 29–38, 2005.

Ciarapica, L. and Todini, E.: TOPKAPI: a model for the representation of the rainfall-runoff process at different scales, Hydrol. Process., 16, 207–229, https://doi.org/10.1002/hyp.342, 2002.

Clark, M. P., Slater, A. G., Rupp, D. E., Woods, R. A., Vrugt, J. A., Gupta, H. V., Wagener, T., and Hay, L. E.: Framework for Understanding Structural Errors (FUSE): A modular framework to diagnose differences between hydrological models, Water Resour. Res., 44, W00B02, https://doi.org/10.1029/2007WR006735, 2008.

Cornelissen, T., Diekkrüger, B., and Giertz, S.: A comparison of hydrological models for assessing the impact of land use and climate change on discharge in a tropical catchment, J. Hydrol., 498, 221–236, https://doi.org/10.1016/J.JHYDROL.2013.06.016, 2013.

Dams, J., Nossent, J., Senbeta, T. B., Willems, P., and Batelaan, O.: Multi-model approach to assess the impact of climate change on runoff, J. Hydrol., 529, 1601–1616, https://doi.org/10.1016/J.JHYDROL.2015.08.023, 2015.

Deidda, R.: Rainfall downscaling in a space-time multifractal framework, Water Resour. Res., 36, 1779–1794, https://doi.org/10.1029/2000WR900038, 2000.

Deidda, R., Benzi, R., and Siccardi, F.: Multifractal modeling of anomalous scaling laws in rainfall, Water Resour. Res., 35, 1853–1867, https://doi.org/10.1029/1999WR900036, 1999.

Deidda, R., Marrocu, M., Caroletti, G., Pusceddu, G., Langousis, A., Lucarini, V., Puliga, M., and Speranza, A.: Regional climate models' performance in representing precipitation and temperature over selected Mediterranean areas, Hydrol. Earth Syst. Sci., 17, 5041–5059, https://doi.org/10.5194/hess-17-5041-2013, 2013.

del Río, S., Herrero, L., Fraile, R., and Penas, A.: Spatial distribution of recent rainfall trends in Spain (1961–2006), Int. J. Climatol., 31, 656–667, https://doi.org/10.1002/joc.2111, 2011.

Duveiller, G., Fasbender, D., and Meroni, M.: Revisiting the concept of a symmetric index of agreement for continuous datasets, Sci. Rep.-UK, 6, 19401, https://doi.org/10.1038/srep19401, 2016.

El Kenawy, A., López-Moreno, J. I., and Vicente-Serrano, S. M.: Recent trends in daily temperature extremes over northeastern Spain (1960–2006), Nat. Hazards Earth Syst. Sci., 11, 2583–2603, https://doi.org/10.5194/nhess-11-2583-2011, 2011.

Filion, R., Bernier, M., Paniconi, C., Chokmani, K., Melis, M., Soddu, A., Talazac, M., and Lafortune, F.-X.: Remote sensing for mapping soil moisture and drainage potential in semi-arid regions: Applications to the Campidano plain of Sardinia, Italy, Sci. Total Environ., 543, 862–876, https://doi.org/10.1016/j.scitotenv.2015.07.068, 2016.

Gädeke, A., Hölzel, H., Koch, H., Pohle, I., and Grünewald, U.: Analysis of uncertainties in the hydrological response of a model-based climate change impact assessment in a subcatchment of the Spree River, Germany, Hydrol. Proc., 28, 3978–3998, https://doi.org/10.1002/hyp.9933, 2014.

Gatel, L., Lauvernet, C., Carluer, N., and Paniconi, C.: Effect of surface and subsurface heterogeneity on the hydrological response of a grassed buffer zone, J. Hydrol., 542, 637–647, https://doi.org/10.1016/J.JHYDROL.2016.09.038, 2016.

Gauthier, M. J., Camporese, M., Rivard, C., Paniconi, C., and Larocque, M.: A modeling study of heterogeneity and surface water-groundwater interactions in the Thomas Brook catchment, Annapolis Valley (Nova Scotia, Canada), Hydrol. Earth Syst. Sci., 13, 1583–1596, https://doi.org/10.5194/hess-13-1583-2009, 2009.

Giorgi, F.: Climate change hot-spots, Geophys. Res. Lett., 33, L08707, https://doi.org/10.1029/2006GL025734, 2006.

Giorgi, F. and Lionello, P.: Climate change projections for the Mediterranean region, Global Planet. Change, 63, 90–104, https://doi.org/10.1016/j.gloplacha.2007.09.005, 2008.

Giorgi, F. and Mearns, L. O.: Calculation of average, uncertainty range, and reliability of regional climate changes from AOGCM simulations via the "Reliability Ensemble Averaging" (REA) method, J. Clim., 15, 1141–1158, https://doi.org/10.1175/1520-0442(2002)015<1141:COAURA>2.0.CO;2, 2002.

Graham, L. P., Hagemann, S., Jaun, S., and Beniston, M.: On interpreting hydrological change from regional climate models, Clim. Change, 81, 97–122, https://doi.org/10.1007/s10584-006-9217-0, 2007.

Hargreaves, G. H.: Defining and Using Reference Evapotranspiration, J. Irrig. Drain. Eng., 120, 1132–1139, https://doi.org/10.1061/(ASCE)0733-9437(1994)120:6(1132), 1994.

Hargreaves, G. H. and Allen, R. G.: History and Evaluation of Hargreaves Evapotranspiration Equation, J. Irrig. Drain. Eng., 129, 53–63, https://doi.org/10.1061/(ASCE)0733-9437(2003)129:1(53), 2003.

Hawkins, E. and Sutton, R.: The potential to narrow uncertainty in regional climate predictions, B. Am. Meteorol. Soc., 90, 1095–1107, https://doi.org/10.1175/2009BAMS2607.1, 2009.

Hawkins, G. A., Vivoni, E. R., Robles-Morua, A., Mascaro, G., Rivera, E., and Dominguez, F.: A climate change projection for summer hydrologic conditions in a semiarid watershed of central Arizona, J. Arid Environ., 118, 9–20, https://doi.org/10.1016/j.jaridenv.2015.02.022, 2015.

Haylock, M. R., Hofstra, N., Klein Tank, A. M. G., Klok, E. J., Jones, P. D., and New, M.: A European daily high-resolution gridded data set of surface temperature and precipitation for 1950–2006, J. Geophys. Res., 113, D20119, https://doi.org/10.1029/2008JD010201, 2008.

Im, E.-S., Jung, I.-W., Chang, H., Bae, D.-H., and Kwon, W.-T.: Hydroclimatological response to dynamically downscaled climate change simulations for Korean basins, Clim. Change, 100, 485–508, https://doi.org/10.1007/s10584-009-9691-2, 2010.

IPCC: Climate Change: Impacts, Adaptation, and Vulnerability. Part A: Global and Sectoral Aspects, Contribution of Working Group II to the Fifth Assessment Report of the Intergovernmental Panel on Climate Change, edited by: Field, C. B., Barros, V. R., Dokken, D. J., Mach, K. J., Mastrandrea, M. D., Bilir, T. E., Chatterjee, M., Ebi, K. L., Estrada, Y. O., Genova, R. C., Girma, B., Kissel, E. S., Levy, A. N., MacCracken, S., Mastrandrea, P. R., and White L. L., Cambridge University Press, Cambridge, UK, New York, NY, USA, 1132 pp., 2014.

Ivanov, V. Y., Vivoni, E. R., Bras, R. L., and Entekhabi, D.: Catchment hydrologic response with a fully distributed triangulated irregular network model, Water Resour. Res., 40, W11102, https://doi.org/10.1029/2004WR003218, 2004.

Jasper, K., Gurtz, J., and Lang, H.: Advanced flood forecasting in Alpine watersheds by coupling meteorological observations and forecasts with a distributed hydrological model, J. Hydrol., 267, 40–52, https://doi.org/10.1016/S0022-1694(02)00138-5, 2002.

Jiang, L. and Islam, S.: A methodology for estimation of surface evapotranspiration over large areas using remote sensing observations, Geophys. Res. Lett., 26, 2773–2776, https://doi.org/10.1029/1999GL006049, 1999.

Jiang, T., Chen, Y. D., Xu, C., Chen, X., Chen, X., and Singh, V. P.: Comparison of hydrological impacts of climate change simulated by six hydrological models in the Dongjiang Basin, South China, J. Hydrol., 336, 316–333, https://doi.org/10.1016/J.JHYDROL.2007.01.010, 2007.

Kay, A. L., Davies, H. N., Bell, V. A., and Jones, R. G.: Comparison of uncertainty sources for climate change impacts: flood frequency in England, Clim. Change, 92, 41–63, https://doi.org/10.1007/s10584-008-9471-4, 2009.

Kollet, S., Sulis, M., Maxwell, R. M., Paniconi, C., Putti, M., Bertoldi, G., Coon, E. T., Cordano, E., Endrizzi, S., Kikinzon, E., Mouche, E., Mügler, C., Park, Y.-J., Refsgaard, J. C., Stisen, S., and Sudicky, E.: The integrated hydrologic model intercomparison project, IH-MIP2: A second set of benchmark results to diagnose integrated hydrology and feedbacks, Water Resour. Res., 53, 867–890, https://doi.org/10.1002/2016WR019191, 2017.

Kunstmann, H., Krause, J., and Mayr, S.: Inverse distributed hydrological modelling of Alpine catchments, Hydrol. Earth Syst. Sci., 10, 395–412, https://doi.org/10.5194/hess-10-395-2006, 2006.

Liston, G. E. and Elder, K.: A Meteorological Distribution System for High-Resolution Terrestrial Modeling (MicroMet), J. Hydrometeorol., 7, 217–234, https://doi.org/10.1175/JHM486.1, 2006.

Liu, Z. and Todini, E.: Towards a comprehensive physically-based rainfall-runoff model, Hydrol. Earth Syst. Sci., 6, 859–881, https://doi.org/10.5194/hess-6-859-2002, 2002.

Liu, Z., Martina, M. L. V., and Todini, E.: Flood forecasting using a fully distributed model: application of the TOPKAPI model to the Upper Xixian Catchment, Hydrol. Earth Syst. Sci., 9, 347–364, https://doi.org/10.5194/hess-9-347-2005, 2005.

Liuzzo, L., Noto, L. V., Vivoni, E. R., and La Loggia, G.: Basin-Scale Water Resources Assessment in Oklahoma under Synthetic Climate Change Scenarios Using a Fully Distributed Hydrologic Model, J. Hydrol. Eng., 15, 107–122, https://doi.org/10.1061/(ASCE)HE.1943-5584.0000166, 2010.

Ludwig, R., Soddu, A., Duttmann, R., Baghdadi, N., Benabdallah, S., Deidda, R., Marrocu, M., Strunz, G., Wendland, F., Engin, G., Paniconi, C., Prettenthaler, F., Lajeunesse, I., Afifi, S., Cassiani, G., Bellin, A., Mabrouk, B., Bach, H., and Ammerl, T.: Climate-induced changes on the hydrology of Mediterranean basins – A research concept to reduce uncertainty and quantify risk, Fresenius Environ. Bull., 19, 2379–2384, 2010.

Mahmood, T. H. and Vivoni, E. R.: Forest ecohydrological response to bimodal precipitation during contrasting winter to summer transitions, Ecohydrology, 7, 998–1013, https://doi.org/10.1002/eco.1424, 2014.

Majone, B., Villa, F., Deidda, R., and Bellin, A.: Impact of climate change and water use policies on hydropower potential in the south-eastern Alpine region., Sci. Total Environ., 543, 965–80, https://doi.org/10.1016/j.scitotenv.2015.05.009, 2016.

Marras, P., Muroni, D., Manca, S., Soru, C., Pusceddu, G., Marrocu, M., and Cau, P.: The SWAT model and a web-based information system to assess the water balance of Sardinia (Italy), SWAT International Conference, Pernambuco, Brazil, 28 July–1 August 2014.

Martina, M. L. V., Todini, E., and Libralon, A.: A Bayesian decision approach to rainfall thresholds based flood warning, Hydrol. Earth Syst. Sci., 10, 413–426, https://doi.org/10.5194/hess-10-413-2006, 2006.

Mascaro, G., Vivoni, E. R., and Deidda, R.: Implications of ensemble quantitative precipitation forecast errors on distributed streamflow forecasting, J. Hydrometeorol., 11, 69–86, https://doi.org/10.1175/2009JHM1144.1, 2010.

Mascaro, G., Deidda, R., and Hellies, M.: On the nature of rainfall intermittency as revealed by different metrics and sampling approaches, Hydrol. Earth Syst. Sci., 17, 355–369, https://doi.org/10.5194/hess-17-355-2013, 2013a.

Mascaro, G., Piras, M., Deidda, R., and Vivoni, E. R.: Distributed hydrologic modeling of a sparsely monitored basin in Sardinia, Italy, through hydrometeorological downscaling, Hydrol. Earth Syst. Sci., 17, 4143–4158, https://doi.org/10.5194/hess-17-4143-2013, 2013b.

Mascaro, G., Vivoni, E. R., and Méndez-Barroso, L. A.: Hyperresolution hydrologic modeling in a regional watershed and its interpretation using empirical orthogonal functions, Adv. Water Resour., 83, 190–206, https://doi.org/10.1016/j.advwatres.2015.05.023, 2015.

Maurer, E. P. and Hidalgo, H. G.: Utility of daily vs. monthly large-scale climate data: an intercomparison of two statistical downscaling methods, Hydrol. Earth Syst. Sci., 12, 551–563, https://doi.org/10.5194/hess-12-551-2008, 2008.

Maurer, E. P., Brekke, L. D., and Pruitt, T.: Contrasting Lumped and Distributed Hydrology Models for Estimating Climate Change Impacts on California Watersheds1, JAWRA J. Am. Water Resour. Assoc., 46, 1024–1035, https://doi.org/10.1111/j.1752-1688.2010.00473.x, 2010.

Mausbach, M. J. and Dedrick, A. R.: The length we go: Measuring environmental benefits of conservation practices, J. Soil Water Conserv., 59, 96–103, 2004.

Meyer, S., Blaschek, M., and Duttmann, R.: Improved hydrological model parametrization for climate change impact assessment under data scarcity – The potential of field monitoring techniques and geostatistics, Sci. Total Environ., 543, 906–923, https://doi.org/10.1016/J.SCITOTENV.2015.07.116, 2016.

Monteith, J. L.: Evaporation and environment, Symp. Soc. Exp. Biol., 19, 205–34, 1965.

Najafi, M. R., Moradkhani, H., and Jung, I. W.: Assessing the uncertainties of hydrologic model selection in climate change impact studies, Hydrol. Process., 25, 2814–2826, https://doi.org/10.1002/hyp.8043, 2011.

Nakićeović, N., Alcamo, J., Davis, G., de Vries, H. J. M., Fenhann, J., Gaffin, S., Gregory, K., Grubler, A., Jung, T. Y., Kram, T., La Rovere, E. L., Michaelis, L., Mori, S., Morita, T., Papper, W., Pitcher, H., Price, L., Riahi, K., Roehrl, A., Rogner, H.-H., Sankovski, A., Schlesinger, M., Shukla, P., Smith, S., Swart, R., van Rooijen, S., Victor, N., and Dadi, Z.: Emissions Scenarios, A Special Report of Working Group III of the Intergovernmental Panel on Climate Change, Cambridge University Press, Cambridge, 599 pp., 2000.

Orlandini, S. and Rosso, R.: Diffusion Wave Modeling of Distributed Catchment Dynamics, J. Hydrol. Eng., 1, 103–113, https://doi.org/10.1061/(ASCE)1084-0699(1996)1:3(103), 1996.

Paniconi, C. and Wood, E. F.: A detailed model for simulation of catchment scale subsurface hydrologic processes, Water Resour. Res., 29, 1601–1620, https://doi.org/10.1029/92WR02333, 1993.

Pechlivanidis, I. G., Arheimer, B., Donnelly, C., Hundecha, Y., Huang, S., Aich, V., Samaniego, L., Eisner, S., and Shi, P.: Analysis of hydrological extremes at different hydro-climatic regimes under present and future conditions, Clim. Change, 141, 467–481, https://doi.org/10.1007/s10584-016-1723-0, 2017.

Penman, H. L.: Natural evaporation from open water, hare soil and grass, P. Roy. Soc. Lond. A Mat., 193, 120–145, https://doi.org/10.1098/RSPA.1948.0037, 1948.

Perra, E., Piras, M., Deidda, R., Mascaro, G., and Paniconi, C.: Investigating parameter transferability across models and events for a semiarid Mediterranean catchment, Hydrol. Process., submitted, 2018.

Piras, M.: Evaluation of climate change impacts on the hydrologic response of a sparsely-monitored basin in Sardinia, Italy, through distributed hydrologic simulations and hydrometeorological downscaling, PhD Thesis, University of Cagliari, Italy, 2014.

Piras, M., Mascaro, G., Deidda, R., and Vivoni, E. R.: Quantification of hydrologic impacts of climate change in a Mediterranean basin in Sardinia, Italy, through high-resolution simulations, Hydrol. Earth Syst. Sci., 18, 5201–5217, https://doi.org/10.5194/hess-18-5201-2014, 2014.

Pulina, M. A.: L'evapotraspirazione potenziale in Sardegna in funzione dello studio del regime idrico dei suoli, Studi Sassaresi, Annali della Facoltà di Agraria dell'Università di Sassari, 1986.

Schulla, J.: Hydrologische Modellierung von Flussgebieten zur Abschatzung der Folgen von Klimaanderungen, Zurcher Geographische Schriften, ETH Zurich, Heft 69, 187 pp., 1997.

Schulla, J.: Model description WaSiM (Water balance Simulation Model), Hydrology Software Consulting, Regensdorferstrasse, Zurich, available at: http://www.wasim.ch/downloads/doku/wasim/wasim_2012_ed2_en.pdf (last access: 18 July 2018), 2015.

Schulla, J., and Jasper, K.: Model Description WASIM-ETH (Water Balance Simulation Model ETH), ETH-Zurich, Zurich, available at: http://www.wasim.ch/downloads/doku/wasim/wasim_2007_en.pdf (last access: 18 July 2018), 2001.

Scudeler, C., Paniconi, C., Pasetto, D., and Putti, M.: Examination of the seepage face boundary condition in subsurface and coupled surface/subsurface hydrological models, Water Resour. Res., 53, 1799–1819, https://doi.org/10.1002/2016WR019277, 2017.

Sousa, P. M., Trigo, R. M., Aizpurua, P., Nieto, R., Gimeno, L., and Garcia-Herrera, R.: Trends and extremes of drought indices throughout the 20th century in the Mediterranean, Nat. Hazards Earth Syst. Sci., 11, 33–51, https://doi.org/10.5194/nhess-11-33-2011, 2011.

Sulis, M., Paniconi, C., Rivard, C., Harvey, R., and Chaumont, D.: Assessment of climate change impacts at the catchment scale with a detailed hydrological model of surface-subsurface interactions and comparison with a land surface model, Water Resour. Res., 47, W01513, https://doi.org/10.1029/2010WR009167, 2011.

Sulis, M., Paniconi, C., Marrocu, M., Huard, D., and Chaumont, D.: Hydrologic response to multimodel climate output using a physically based model of groundwater/surface water interactions, Water Resour. Res., 48, W12510, https://doi.org/10.1029/2012WR012304, 2012.

Tebaldi, C., Smith, R. L., Nychka, D., and Mearns, L. O.: Quantifying uncertainty in projections of regional climate change: A Bayesian approach to the analysis of multimodel ensembles, J. Climate, 18, 1524–1540, https://doi.org/10.1175/JCLI3363.1, 2005.

Thompson, J. R., Green, A. J., Kingston, D. G., and Gosling, S. N.: Assessment of uncertainty in river flow projections for the Mekong River using multiple GCMs and hydrological models, J. Hydrol., 486, 1–30, https://doi.org/10.1016/j.jhydrol.2013.01.029, 2013.

van der Linden, P. and Mitchell, J. F. B.: ENSEMBLES: Climate change and its impacts: Summary of research and results from the ENSEMBLES project, Met Office Hadley Centre, Exeter, UK, 2009.

Vansteenkiste, T., Tavakoli, M., Ntegeka, V., De Smedt, F., Batelaan, O., Pereira, F., and Willems, P.: Intercomparison of hydrological model structures and calibration approaches in climate scenario impact projections, J. Hydrol., 519, 743–755, https://doi.org/10.1016/J.JHYDROL.2014.07.062, 2014.

Vicente-Serrano, S. M. and Cuadrat-Prats, J. M.: Trends in drought intensity and variability in the middle Ebro valley (NE of the Iberian peninsula) during the second half of the twentieth century, Theor. Appl. Climatol., 88, 247–258, https://doi.org/10.1007/s00704-006-0236-6, 2007.

Volk, M., Hirschfeld, J., Schmidt, G., Bohn, C., Dehnhardt, A., Liersch, S., and Lymburner, L.: A SDSS-based Ecological-economic Modelling Approach for Integrated River Basin Management on Different Scale Levels – The Project FLUMAGIS, Water Resour. Manag., 21, 2049–2061, https://doi.org/10.1007/s11269-007-9158-z, 2007.

Wood, A. W., Leung, L. R., Sridhar, V., and Lettenmaier, D. P.: Hydrologic Implications of Dynamical and Statistical Approaches to Downscaling Climate Model Outputs, Clim. Change, 62, 189–216, https://doi.org/10.1023/B:CLIM.0000013685.99609.9e, 2004.

Xoplaki, E., González-Rouco, J. F., Luterbacher, J., and Wanner, H.: Mediterranean summer air temperature variability and its connection to the large-scale atmospheric circulation and SSTs, Clim. Dynam., 20, 723–739, https://doi.org/10.1007/s00382-003-0304-x, 2003.

Xu, C., Widén, E., and Halldin, S.: Modelling hydrological consequences of climate change – Progress and challenges, Adv. Atmos. Sci., 22, 789–797, https://doi.org/10.1007/BF02918679, 2005.

Xu, C.-Y. and Singh, V. P.: Review on Regional Water Resources Assessment Models under Stationary and Changing Climate, Water Resour. Manag., 18, 591–612, https://doi.org/10.1007/s11269-004-9130-0, 2004.

10

Waning habitats due to climate change: the effects of changes in streamflow and temperature at the rear edge of the distribution of a cold-water fish

José María Santiago[1], Rafael Muñoz-Mas[2], Joaquín Solana-Gutiérrez[1], Diego García de Jalón[1], Carlos Alonso[1], Francisco Martínez-Capel[2], Javier Pórtoles[3], Robert Monjo[3], and Jaime Ribalaygua[3]

[1]Laboratorio de Hidrobiología, ETSI Montes, Forestal y Medio Natural, Universidad Politécnica de Madrid, Camino de las Moreras s/n, Madrid 28040, Spain

[2]Institut d'Investigació per a la Gestió Integrada de Zones Costaneres (IGIC), Universitat Politècnica de València, C/ Paranimf 1, Grau de Gandia, Valencia 46730, Spain

[3]Fundación para la Investigación del Clima, C/ Tremps 11. Madrid 28040, Spain

Correspondence to: José María Santiago (jmsant@picos.com)

Abstract. Climate changes affect aquatic ecosystems by altering temperatures and precipitation patterns, and the rear edges of the distributions of cold-water species are especially sensitive to these effects. The main goal of this study was to predict in detail how changes in air temperature and precipitation will affect streamflow, the thermal habitat of a cold-water fish (the brown trout, *Salmo trutta*), and the synergistic relationships among these variables at the rear edge of the natural distribution of brown trout. Thirty-one sites in 14 mountain rivers and streams were studied in central Spain. Models of streamflow were built for several of these sites using M5 model trees, and a non-linear regression method was used to estimate stream temperatures. Nine global climate models simulations for Representative Concentration Pathways RCP4.5 and RCP8.5 scenarios were downscaled to the local level. Significant reductions in streamflow were predicted to occur in all of the basins (max. −49 %) by the year 2099, and seasonal differences were noted between the basins. The stream temperature models showed relationships between the model parameters, geology and hydrologic responses. Temperature was sensitive to streamflow in one set of streams, and summer reductions in streamflow contributed to additional stream temperature increases (max. 3.6 °C), although the sites that are most dependent on deep aquifers will likely resist warming to a greater degree. The predicted increases in water temperatures were as high as 4.0 °C. Temperature and streamflow changes will cause a shift in the rear edge of the distribution of this species. However, geology will affect the extent of this shift. Approaches like the one used herein have proven to be useful in planning the prevention and mitigation of the negative effects of climate change by differentiating areas based on the risk level and viability of fish populations.

1 Introduction

Water temperatures are a primary influence on the physical, chemical and biological processes in rivers and streams (Caissie, 2006; Webb et al., 2008) and, subsequently, the organisms that live completely or partially in the water. Temperature is a major feature of the ecological niche of poikilothermic species (e.g. Magnuson and Destasio, 1997; Angilletta, 2009) and a key factor in energy balance of fish. It affects the rate of food intake, metabolic rate and growth performance (Forseth et al., 2009; Elliott and Elliott, 2010; Elliott and Allonby, 2013). It is also involved in many other physiological functions, such as blood function and reproductive maturation (Jeffries et al., 2012), reproductive timing (Warren et al., 2012), gametogenesis (Lahnsteiner and Leitner, 2013), cardiac function (Vornanen et al., 2014), gene expression (White et al., 2012; Meshcheryakova et al., 2016), ecological relationships (Hein et al., 2013; Fey and Herren, 2014), and fish behaviour (Colchen et al., 2017).

Natural patterns of water temperature and streamflow are profoundly linked with climatic variables (Caissie, 2006; Webb et al., 2008). Therefore, stream temperature is strongly correlated with air temperature (Mohseni and Stefan, 1999), whereas streamflow has a complex relationship with precipitation (McCuen, 1998; Gordon et al., 2004). In addition, atmospheric temperature influences the type of precipitation (rain or snow) that occurs and the occurrence of snowmelt; conversely, river discharge is also a main explanatory factor of water temperature for some river systems (Neumann et al., 2003; van Vliet et al., 2011). Furthermore, geology affects surface water temperatures by means of groundwater discharge (Caissie, 2006; Loinaz et al., 2013), influenced by the aquifer depth (shallow or deep) and the water's residence time (Kurylyk et al., 2013; Snyder et al., 2015).

Climate change is already affecting aquatic ecosystems by altering water temperatures and precipitation patterns. Stream temperature increases have been documented over the last several decades over the whole globe, such as in Europe (e.g. Orr et al., 2015, documented a mean increase in stream temperature of 0.03 °C per year in England and Wales), Asia (e.g. Chen et al., 2016, documented a mean increase in stream temperature of 0.029–0.046 °C per year in the Yongan River, eastern China), America (e.g. Kaushal et al., 2010, documented mean increases in stream temperature of 0.009–0.077 °C per year) and Australia (e.g. Chessman, 2009, documented mean increases in stream temperature of 0.12 °C per year between macroinvertebrate sampling campaigns). Abundant information is also available regarding the impact of recent climate changes on streamflow regimes worldwide (e.g. Luce and Holden, 2009; Leppi et al., 2012) and, more specifically, in the Iberian Peninsula (e.g. Ceballos-Barbancho et al., 2008; Lorenzo-Lacruz et al., 2012; Morán-Tejeda et al., 2014). However, detailed predictions are uncommon (e.g. Thodsen, 2007). The predictions of the Intergovernmental Panel on Climate Change (IPCC, 2013) suggest that these alterations will continue throughout the 21st century, and they will have consequences for the distribution of freshwater fish (e.g. Comte et al., 2013; Ruiz-Navarro et al., 2016). These changes may have an especially strong effect on cold-water fish, which have been shown to be very sensitive to climate warming (Williams et al., 2015; Santiago et al., 2016). For example, among salmonids, DeWeber and Wagner (2015) found stream temperature to be the most important determinant of the probability of occurrence of brook trout, *Salvelinus fontinalis* (Mitchill, 1814).

The rear edge populations (sensu Hampe and Petit, 2005: "populations residing at the current low-latitude margins of species' distribution ranges") of a cold-water species are especially sensitive to changes in water temperature, in addition to reductions in the available habitable volume (i.e. streamflow). The rear edge is the eroding margin of the range where lineages mix, the genetic drift and local adaptations increase, and droughts put populations under stress. The impact of water temperatures on the distribution of salmonid fish is well documented (e.g. Beer and Anderson, 2013; Eby et al., 2014); however, the combined effects of rising stream temperatures and reductions in streamflow remain relatively unexamined, with some exceptions (e.g. Wenger et al., 2011; Muñoz-Mas et al., 2016). Jonsson and Jonsson (2009) predicted that the expected effects of climate change on water temperatures and streamflow will have implications for the migration, ontogeny, growth and life-history traits of Atlantic salmon, *Salmo salar* Linnaeus, 1758, and brown trout, *Salmo trutta* Linnaeus, 1758. Thus, investigation of these habitat variables in the context of several climate scenarios should help scientists to assess the magnitude of these changes on the suitable range and life history of these species.

The objective of this study is to predict how and to what extent the availability of suitable habitat for the brown trout, a sensitive cold-water species, will change within its current natural distribution under the new climate scenarios through a study of changes in streamflow and temperature and their interactions. Specifically, in this paper, we (i) assessed the effects of both streamflow and geology on stream temperature; (ii) predicted the changes in streamflow and stream temperature implied by the climate change scenarios used in the 5th Assessment Report (AR5) of the IPCC; and (iii) assessed the expected effects of these changes on trout habitat aptitude. To this end, hydrologic simulations with M5 model trees coupled with non-linear water temperature models at the daily time step were fed with high-resolution, downscaled versions of the air temperature and precipitation fields predicted using the most recent climate change scenarios (IPCC, 2013). The effects of basin geology on the stream temperature models and on the estimated changes in thermal regimes were studied. Finally, the changes in the thermal habitat of trout were assessed by studying the violation of the tolerable temperature thresholds of the brown trout.

2 Materials and methods

The logical framework followed is summarized in Fig. 1. First, the daily global climate models output presented by the IPCC were downscaled to the study area. Then, the obtained local climate models output were applied to generate simulations of streamflow and water temperature. The results are daily values that can be used for the assessment of fish habitat suitability and availability.

The procedure yielded results in the form of continuous time series, but they are presented for two time horizons: the year 2050 (H-2050) and the year 2099 (H-2099). The values for these horizons correspond to the average of the values of the different variables for the decades 2041–2050 and 2090–2099, respectively.

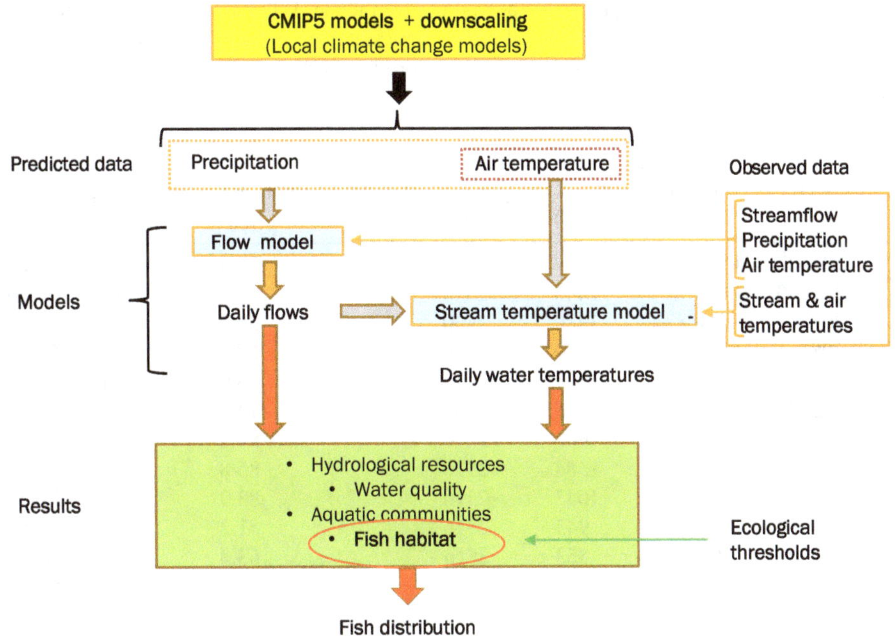

Figure 1. Logical framework of the study.

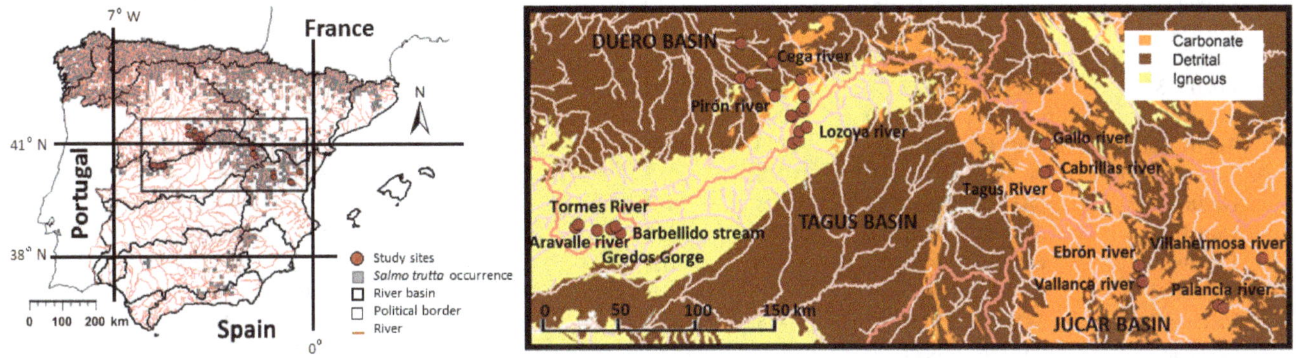

Figure 2. River network and location of the study sites (water temperature data loggers), with details regarding lithology. The grid depicts the actual occurrence of brown trout in Spain.

2.1 Study sites

In total, 31 sites in 14 mountain rivers and streams inhabited by brown trout were chosen with the aim of encompassing a diverse array of geological and hydrological conditions in the centre of Spain (between the latitudes of 39°53′ and 41°21′ N). Specifically, the investigated sites are located in the Tormes River and its tributaries, the Barbellido River, the Gredos Gorge and the Aravalle River (in the Duero Basin); the Cega River and the Pirón River (the Pirón River is a tributary of the Cega River in the larger Duero Basin); the Lozoya River, the Tagus River, the Gallo River, and the Cabrillas River (all four of which are in the Tagus Basin); the Ebrón River and the Vallanca River (the Vallanca River is a tributary of the Ebrón River in the Turia Basin); and the Palancia River and the Villahermosa River (Fig. 2). The major geological components that lithologically characterize the mountain sites in the Duero and Lozoya basins are igneous rocks; the altitudinally lower sites in the Duero Basin are underlain by Cenozoic detrital material, and the eastern basins (Tagus, Gallo, Cabrillas, Ebrón, Vallanca, Palancia and Villahermosa) are underlain primarily by Mesozoic carbonates. The distribution of geological materials was retrieved from the Lithological Map of Spain (IGME, 2015) (Table 1).

The land cover type is mainly pine forest in all of the studied basins (*Pinus sylvestris*, *P. nigra*, *P. pinea* and *P. pinaster*) (CORINE Land Cover 2006; European Environmental Agency, 2007). Only the lower basins of the downstream sites on the Cega and Pirón rivers are mosaics of forest and croplands, whereas the uppermost sites within the Tormes River basin (Barbellido and Gredos Gorge) lie above the current tree line. Territorial planning does not con-

Table 1. Description of the data logger (thermograph) sites, specifying given name, UTM coordinates (Europe WGS89), altitude (metres above sea level), code of the nearest temperature meteorological station with suitable time series for this study (AEMET: Spanish Meteorological Agency), orthogonal distance between the data logger and the meteorological station, number of recorded days for stream temperature and characteristic geological nature (lithology) of the data logger site (the last of which was obtained from IGME, 2015). Bold letters indicate sites associated with the gauging stations.

Sites	UTM-X	UTM-Y	Altitude (m a.s.l.)	AEMET code	Distance to AEMET station (km)	Recording days	Lithology
Aravalle	283623	4468847	1010	2440	76.4	1257	Igneous
Barbellido	311759	4465519	1440	2440	52.2	881	Igneous
Gredos Gorge	306363	4468087	1280	2440	55.7	644	Igneous
Tormes1	**308751**	**4469371**	**1270**	**2440**	**53.0**	**421**	**Igneous**
Tormes2	297543	4467191	1135	2440	64.0	537	Igneous
Tormes3	**285481**	**4470750**	**995**	**2440**	**74.1**	**588**	**Igneous**
Cega1	427627	4539806	1600	2516	84.5	544	Igneous
Cega2	429416	4541728	1384	2516	85.8	544	Igneous
Cega3	428892	4549370	1043	2516	83.9	544	Igneous
Cega4	**426932**	**4559076**	**943**	**2516**	**81.2**	**407**	**Quaternary detrital**
Cega5	**408504**	**4569772**	**853**	**2516**	**63.4**	**544**	**Quaternary detrital**
Cega6	389014	4581160	766	2516	47.9	501	Quaternary detrital
Pirón1	422082	4536456	1475	2516	80.1	544	Igneous
Pirón2	420660	4537094	1348	2516	78.6	483	Igneous
Pirón3	**409935**	**4549473**	**908**	**2516**	**65.2**	**544**	**Quaternary detrital**
Pirón4	394462	4556823	826	2516	48.9	544	Quaternary detrital
Pirón5	388615	4560166	815	2516	42.9	424	Quaternary detrital
Lozoya1	422060	4520319	1452	3104	7.3	2151	Igneous
Lozoya2	425445	4522314	1267	3104	4.6	1870	Igneous
Lozoya3	**425657**	**4527327**	**1142**	**3104**	**0.7**	**1776**	**Igneous**
Lozoya4	430740	4530050	1090	3104	6.4	2187	Igneous
Tagus-Peralejos	**590887**	**4494165**	**1149**	**3013**	**27.9**	**964**	**Carbonate**
Tagus-Poveda	582900	4502160	1028	3013	22.8	669	Carbonate
Gallo	**583771**	**4519743**	**998**	**3013**	**10.9**	**1019**	**Carbonate**
Cabrillas	**585619**	**4502986**	**1075**	**3013**	**20.8**	**1070**	**Carbonate**
Ebrón	**643551**	**4445027**	**879**	**8381B**	**9.5**	**592**	**Carbonate**
Vallanca1	644966	4435479	745	8381B	1.8	836	Carbonate
Vallanca2	645936	4435715	718	8381B	0.8	836	Carbonate
Palancia1	694348	4421176	760	8434A	10.4	334	Carbonate
Palancia2	697451	4419477	660	8434A	7.8	334	Carbonate
Villahermosa	722594	4449436	592	8478	13.5	334	Carbonate

sider significant changes in land use at mid-century; objectively, changes are not expected after that time because a high percentage of the territory is protected. The studied reaches are not effectively regulated (only small weirs or natural obstacles exist). One large dam lies on the Pirón River (the Torrecaballeros Dam, which has a capacity of 0.32 hm^3 and a maximum depth of 26 m and lies at an altitude of 1390 m a.s.l.), but it does not significantly alter the temporal pattern of streamflow (Santiago et al., 2013). In the Lozoya River, a large dam (the Pinilla Dam, which has a capacity of 38.1 hm^3 and a maximum depth of 30 m and lies at an altitude of 1060 m a.s.l.) exists that separates fish populations above and below the reservoir, although it lies downstream of the studied reach.

Hydrological data characterize the streamflow regimes as extreme winter/early spring (groups 13 and 14 in the classification of Haines et al., 1988). However, the hydrographs show a west-to-east smoothing gradient (Fig. 3). This smoothing is associated with the carbonate rocks, whereas greater seasonality is associated with the igneous and detrital geological materials.

2.2 Data collection

At each study site, water temperatures were recorded every 2 h throughout the year using 31 Hobo® Water Temperature Pro v2 (Onset®) and Vemco® Minilog data loggers located at several sites along the studied rivers and streams (Table 1). Loggers were tested for malfunctions before being deployed, and they were placed in areas not exposed to direct sunshine (Stamp et al., 2014). Meteorological data were obtained from nine thermometric and 15 pluviometric sta-

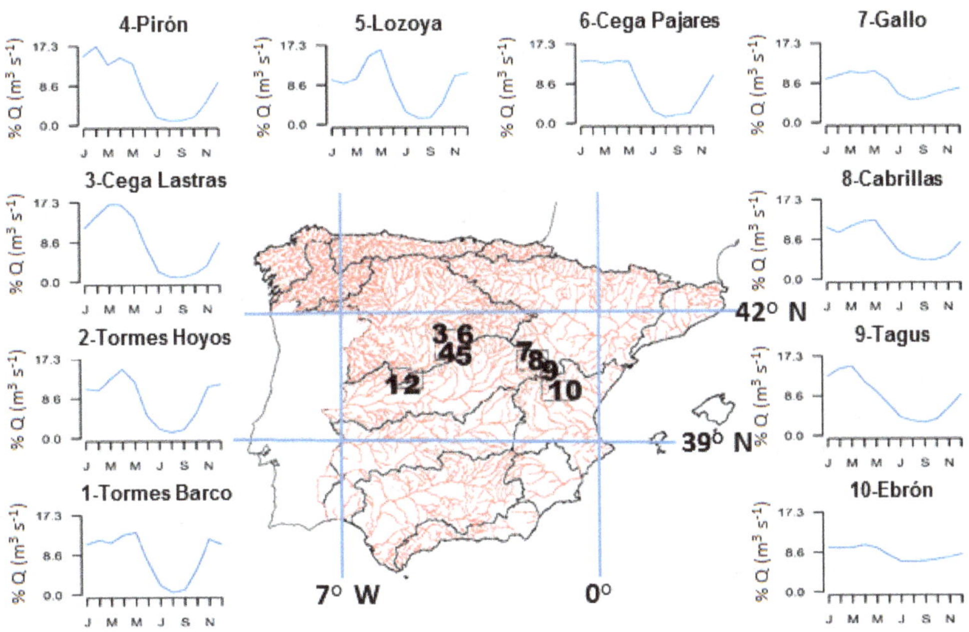

Figure 3. River regime patterns for the different gauging stations. The flows are expressed as percentages of the mean annual flow, and the months (horizontal axis) are ordered from January to December.

tions of the Spanish Meteorological Agency (AEMET) network, and data from 10 gauging stations (from the official network of the Spanish Water Administration) were obtained to model the streamflows. The AEMET thermometric stations that lie closest to the stream temperature monitoring sites and have at least 30 years of data between 1955 and the present were selected. The selected pluviometric stations were those located within the upstream river basin or near the corresponding gauging station (Table 2). The air temperature and precipitation data from AEMET were tested to assess their reliability by applying a homogeneity test. This test is based on a two-sample Kolmogorov–Smirnov test, and it marks years as possibly containing inhomogeneous data. In the second phase, the marked years are matched against the distribution of the entire series to determine if they contain true inhomogeneities, searching for possible dissimilarities between the empirical distribution functions. Only reliable series were used. The locations of the stations did not change in the studied period.

2.3 Climate change modelling and downscaling

Data from nine global climate models associated with the 5th Coupled Model Intercomparison Project (CMIP5) were used, namely BCC-CSM1-1, CanESM2, CNRM-CM5, GFDL-ESM2 M, HADGEM2-CC, MIROC-ESM-CHEM, MPI-ESM-MR, MRI-CGCM3, and NorESM1-M (Santiago et al., 2016). These models provided daily data to simulate future climate changes corresponding to the Representative Concentration Pathways RCP4.5 (a stable scenario) and RCP8.5 (a scenario including a pronounced increase in CO_2 concentrations) established in Taylor et al. (2009) and used in the AR5 of the IPCC (2013). An array of nine general climate models was used to avoid biases due to the particular assumptions and features of each particular model (Kurylyk et al., 2013). Historical simulations of the 20th century were used to control the quality of the procedure and to compare the magnitudes of the predicted changes.

Pourmokhtarian et al. (2016) note the importance of the use of fine downscaling techniques. Thus, a two-step analogue statistical method (Ribalaygua et al., 2013) was used to downscale the daily climatic data, specifically the maximum and minimum air temperatures and the precipitation for each station and for each day. For both air temperature and precipitation, the procedure begins with an analogue stratification (Zorita and von Storch, 1999) in which the n days most similar to each problem day to be downscaled are selected using four different meteorological large-scale fields as predictors, specifically (1) the speed and (2) direction of the geostrophic wind at 1000 hPa, as well as (3) the speed and (4) direction of the geostrophic wind at 500 hPa. In a second step, the temperature determination was obtained through multiple linear regression analysis using the selected n of the most analogous days. This was performed for the maximum and minimum air temperatures at each station and for each problem day. The linear regression uses forward and backward stepwise selections of the predictors to select only the relevant predictive variables for that particular case. For precipitation, a group of m problem days (the whole days of a month were used) were downscaled together, and the "preliminary precipitation quantity", or the average precipitation of the n most analogous days, was obtained for each prob-

Table 2. Official stations used (meteorological and hydrological), variables, length of time series used and geographical position. AEMET: Spanish Meteorological Agency; CHD: Water Administration of Duero Basin; CHT: Water Administration of Tagus Basin; and CHJ: Water Administration of Júcar Basin.

Institution	Code	Name	Variable	Series length used	UTM-X	UTM-Y	Altitude
AEMET	2180	Matabuena	pluviometry	1955–2013	436266	4549752	1154
AEMET	2186	Turégano	pluviometry	1955–2013	415346	4556596	935
AEMET	2196	Torreiglesias	pluviometry	1970–2013	413294	4550606	1053
AEMET	2199	Cantimpalos	pluviometry	1955–2013	402524	4547811	906
AEMET	2440	Aldea del Rey Niño	temperature	1955–2012	356059	4493201	1160
AEMET	2462	Puerto de Navacerrada	temperature and pluviometry	1967–2012	414745	4516276	1894
AEMET	2516	Ataquines	temperature	1970–2013	345716	4560666	802
AEMET	2813	Navacepeda de Tormes	pluviometry	1965–2012	308892	4470347	1340
AEMET	2828	El Barco de Ávila	temperature and pluviometry	1955–1983	285643	4470512	1007
AEMET	3009E	Orihuela del Tremedal	pluviometry	1986–2000	614383	4489759	1450
AEMET	3010	Ródenas	pluviometry	1968–2006	625505	4499963	1370
AEMET	3013	Molina de Aragón	temperature and pluviometry	1951–2010	594513	4521786	1056
AEMET	3015	Corduente	pluviometry	1961–2000	584125	4523281	1120
AEMET	3018E	Aragoncillo	pluviometry	1968–2010	580519	4531876	1263
AEMET	3104	Rascafría-El Paular	temperature and pluviometry	1967–2012	425165	4526895	1159
AEMET	8376B	Jabaloyas	pluviometry	1993–2006	635600	4456215	1430
AEMET	8381B	Ademuz-Agro	temperature and pluviometry	1989–2010	646722	4436034	740
AEMET	8434A	Viver	temperature	1971–2006	704704	4422256	562
AEMET	8478	Arañuel	temperature	1971–2006	714943	4438277	406
CHD	2006	Tormes-Hoyos del Espino	flow	1955–2012	314676	4467908	1377
CHD	2016	Cega-Pajares de Pedraza	flow	1955–2013	428296	4557678	938
CHD	2057	Pirón-Villovela de Pirón	flow	1972–2013	405596	4551929	869
CHD	2085	Tormes-El Barco de Ávila	flow	1955–2012	285173	4470362	992
CHD	2714	Cega-Lastras de Cuéllar	flow	2004–2013	403509	4571682	838
CHT	3001	Tagus-Peralejos de las Truchas	flow	1946–2010	590474	4494474	1143
CHT	3002	Lozoya-Rascafría (El Paular)	flow	1967–2013	425321	4522069	1270
CHT	3030	Gallo-Ventosa	flow	1946–2010	587349	4520522	1016
CHT	3268	Cabrillas-Taravilla	flow	1982–2010	587480	4503395	1107
CHJ	8104	Ebrón-Los Santos	flow	1989–2010	645963	4441366	750

lem day. Thus, the m problem days from the highest to the lowest "preliminary precipitation amount" could be sorted. To assign the final amount of precipitation, each precipitation amount of the $m \times n$ analogous days was taken. Then, those $m \times n$ amounts of precipitation were sorted, and then those amounts were clustered into m groups. Every quantity was then assigned in order to the m days previously sorted by the "preliminary precipitation amount". Further details of the methodology are described in Ribalaygua et al. (2013).

A systematic error is obtained when comparing the simulated data from the climate models with the observed data. Such errors are inherently associated with all downscaling methodologies and climate models, which usually introduce bias into their outputs. To eliminate this systematic error, the future climate projections were corrected according to a parametric quantile–quantile method (Monjo et al., 2014), which was performed by comparing the observed and simulated empirical cumulative distribution functions (ECDFs) and linking them using ECDFs obtained from the downscaled European Centre for Medium-Range Weather Forecasts ERA-40 reanalysis daily data (Uppala et al., 2005).

As a result, for each climate change scenario, the daily maximum and minimum air temperatures (which were used to infer the mean air temperature) and precipitation were obtained for each climate model, and the whole dataset were used as inputs to simulate the runoff and water temperatures under these climate change scenarios.

2.4 Hydrological modelling

Although process-based physical models are considered the standard hydrological models, flexible data-driven machine learning techniques are gaining popularity because they can be based solely on precipitation and temperature (Shortridge et al., 2016) and can be automatized to perform multiple simulations. Therefore, the prediction of the streamflows under the climate change scenarios was performed with data-driven hydrological models developed using the M5 algorithm (Quinlan, 1992). M5 has been shown to have skill in modelling daily streamflow (Solomatine and Dulal, 2003; Taghi Sattari et al., 2013), including in studies involving climate change (Muñoz-Mas et al., 2016), and it is sufficiently fast to deal proficiently with larger datasets (Quinlan, 2017)

Mathematically, M5 is a kind of decision tree that, instead of assigning a single value to each terminal node, assigns a multi-linear regression model (Quinlan, 1992). Therefore, the dataset is hierarchically divided into homogeneous parts and a multi-linear model is adjusted to every part (Hettiarachchi et al., 2005). In this regard, each node, and the corresponding multi-linear regression model, is specialized in particular areas of the data set, such as peak flows or base flows, to name the extremes (i.e. it is a piece-wise linear model with each part dedicated to a particular hydrologic condition) (Taghi Sattari et al., 2013). Based on the multi-linear models at the terminal nodes, M5 allows extrapolation, in contrast with other machine learning techniques that have demonstrated little or no extrapolation ability (e.g. random forest or multilayer perceptron) (Hettiarachchi et al., 2005; Shortridge et al., 2016).

The M5 hydrological models were developed in R (R Core Team, 2015) with the *Cubist* package (Kuhn et al., 2014). One single M5 model tree was trained for each gauging station (ten models were produced in total; Fig. 3 and Table 2), whereas the predictions were supported by the nearest observation (i.e. neighbours = 1) to avoid producing unreliable flows. Finally, M5 was allowed to determine the ultimate number of models (i.e. nodes or areas) into which the dataset is eventually divided (see Kuhn et al., 2014).

Following previous studies, the M5 hydrological models were trained by employing the daily, monthly and quarterly data lags of historical precipitation and air temperature collected at meteorological stations within or nearby the target river basins as input variables (Table 2) (Solomatine and Dulal, 2003; Taghi Sattari et al., 2013; Muñoz-Mas et al., 2016). These three groups of variables were intended to reflect the causes of peak, normal and base flows. The study encompassed several rivers and streams that may have different hydrologic behaviours; therefore, the starting set of input variables, which was afterwards subset, was larger than that used in other studies (Solomatine and Dulal, 2003; Taghi Sattari et al., 2013; Muñoz-Mas et al., 2016). The daily variables included the precipitation and air temperature from the current day to the 15th previous day (16 variables in total). The monthly variables were calculated using the moving average for the 12 previous months (12 variables in total), and the quarterly data were calculated from the moving average for the current month to the 24th previous month (8 variables in total). Consequently, the daily variables overlapped with the current month variable, and the first four quarterly variables overlapped with the monthly data. In the end, 72 variables were gathered, 36 each for air temperature and precipitation.

The whole set of input variables may be relevant for some river systems (Shortridge et al., 2016), although it may cause M5 to overfit the data in others (Schoups et al., 2008). Therefore, the ultimate variable subset was optimized following the forward stepwise approach (Kittler, 1978). This approach relies on iteratively adding input variables (one at a time) while the performance on the test data set improves and stopping (i.e. selecting a smaller subset of the input variables) as soon as the performance stagnates or degrades. However, the classical forward stepwise approach may cause consideration of unrelated variable sets (i.e. disjoint precipitation and air temperature variable lags). To address such potential inconsistencies, the optimization began by testing the precipitation-related variables and only tested the air temperature variables for lags coinciding with those precipitation-related variables that were already selected. No precautions were taken regarding correlations among inputs (D. P. Solomatine, personal communication, 2016), and the forward stepwise approach sought to maximize the Nash–Sutcliffe efficiency (NSE) index (which ranges from $-\infty$ to 1; Nash and Sutcliffe, 1970) in a fivefold cross-validation (i.e. for each combination of variables, five M5 model trees were trained on four parts and validated with the fifth part, which was held out) (Borra and Di Ciaccio, 2010; Bennett et al., 2013). Finally, in order to account for the uncertainty of the models (Bennett et al., 2013), the variance of the NSEs obtained during the cross-validation was inspected; large intervals led to alternative data partitions. Following previous studies (Fukuda et al., 2013; Muñoz-Mas et al., 2016), once the optimal variables set for each gauging station was determined, 10 M5 model trees (i.e. one per gauging station) were developed using the corresponding subset of variables, and they were used to predict the streamflows under the climate change scenarios.

The daily data were analysed monthly and seasonally using the following statistics: minimum flow (Q_{min}), the 10th percentile of flow (Q_{10}), the mean flow (Q_{mean}), and the maximum flow (Q_{max}). The annual runoff and days of zero flow were also examined.

To assess the significance of the streamflow trends throughout the century, Sen's slope was used (as implemented in the *Trend* package of R; Pohlert, 2016; p-value ≤ 0.05) with horizons H-2050 and H-2099.

Finally, the variation in the patterns of the monthly mean streamflow was studied by means of the Ward hierarchical clustering implemented in the *cluster* R package (Maechler, 2013) on the basis of the rate of change of the normalized monthly mean streamflows in H-2050 and H-2099 and the RCP4.5 and RCP8.5 scenarios. Performance of the obtained cluster was quantified by using the agglomerative coefficient (a.c.). This is a measure of the clustering structure of the dataset, as expressed by Kaufman and Rousseeuw (2005) and its value ranges between 0 (maximum dissimilarity) and 1 (minimum dissimilarity).

2.5 Stream temperature modelling

Stream temperature (T_s) at each thermal sampling site was simulated from air temperature (T_a) by means of a modified version of the bounded non-linear regression model described by Mohseni et al. (1998). A previous modification (Term 1 in Eq. 1; Santiago et al., 2016) served to improve the behaviour of the former model, permitting it to be used for daily inputs. In this study, the effect of the instream flow (Q) effect is incorporated. Thus, this model addresses daily mean stream temperature (DMST; T_s in Eq. 1) using the daily mean air temperature (DMAT, T_a in Eq. 1), the 1-day before variation in the daily mean air temperature (ΔT_a in Eq. 1), and the daily mean flow (Q_{mean}, Q in Eq. 1) as predictors. DMST was used because it better reflects the average conditions that fish (particularly trout) will experience for an extended period of time (Santiago et al., 2016), and it averages over daily fluctuations in the radiation and heat fluxes. The model is formulated as follows:

$$T_s = \mu + \underbrace{\frac{\alpha - \mu}{1 + e^{\gamma(\beta - T_a)}} + \lambda(\Delta T_a)}_{\text{Term 1}} + \underbrace{\frac{\omega}{1 + e^{\delta(\tau - Q)}}}_{\text{Term 2}}, \quad (1)$$

where μ is the minimum stream temperature (°C), α is the maximum stream temperature (°C), β represents the air temperature at which the rate of change of the stream temperature with respect to the air temperature is a maximum (°C), γ (°C^{-1}) is the value of the rate of change at β, and λ is a coefficient (dimensionless) that represents the resistance of DMST to change with respect to the 1-day variation in DMAT (ΔT_a). In the flow component (Term 2 in Eq. 1), ω is the maximum observable variation in stream temperature due to the flow difference (given in °C), τ represents the flow value at which the rate of change of the stream temperature with respect to the flow is a maximum (m^3 s^{-1}), and δ (m^{-3} s) is this maximum rate at τ. Negative values of λ are due to the resistance to stream temperature changes, and thus they must be subtracted from the expected temperature: the more resistant the stream is to temperature change, the closer λ will be to zero. The less resistant the stream is to change, the more negative λ is. The parameter μ was allowed to be less than zero in the modelling process, even though this is the freezing temperature. Thus, the function would truncate at the freezing point. The relationship between the thermal amplitude $\alpha - \mu$ and the indicator of thermal stability λ was studied using the Pearson correlation.

A blockwise non-parametric bootstrap regression (Liu and Singh, 1992) was used to estimate the parameters of both the modified Mohseni models (with and without streamflow), and residual normality and non-autocorrelation were checked with the Shapiro test and Durbin–Watson test. Moreover, the 7-day lag PACF (partial autocorrelation function) was obtained. These calculations were performed using R. A 95 % confidence interval was calculated for each parameter. Performance was quantified using two indicators: the residual standard error (RSE) and the Nash–Sutcliffe efficiency index (NSE). The Bayesian information criterion (BIC) and the Akaike information criterion (AIC) were used to test the eight-parameter models (Terms 1 and 2 of Eq. 1) against the five-parameter models (Term 1 of Eq. 1).

This model can be classified as semi-physically based. It has some advantages over machine learning methods, such as classification and regression trees (De'ath and Fabricius, 2000) or random forests (Breiman, 2001), because the model parameters imply a mechanistic interpretation of how process drivers act, yielding a higher transferability (Wenger and Olden, 2012). These features make of this model an advantageous option for our goals.

2.6 Effects of geology on stream temperature

Geology determines the residence time of deep groundwater in the aquifers underlying streams (Chilton, 1996), and residence times influence discharge temperatures. To explore the relationships between thermal regimes and geology, a stratified study of both the geology classes of the parameter values was completed by means of a t test with the Bonferroni correction (p-value < 0.05). In the same sense, increments of the annual averages of the daily mean (ΔT_{mean}), minimum (ΔT_{min}) and maximum (ΔT_{max}) stream temperatures were calculated and studied by lithological classes (Table 1).

The variation in the patterns of the monthly mean stream temperature was studied by means of cluster analysis of the temperature increases corresponding to H-2050 and H-2099 for the RCP4.5 and RCP8.5 scenarios (using Ward's hierarchical clustering as implemented in the *cluster* package of R; Maechler, 2013). As said above, agglomerative coefficient (a.c.) was used as a performance indicator.

2.7 Thermal habitat changes

Several tolerance temperatures and thermal niche limits have been described for brown trout (Table 3). The realized niche must reflect energetic efficiency: spending long periods above that threshold makes animals less efficient competitors, and their performance decreases critically (Magnuson et al., 1979; Verberk et al., 2016). Thus, we focused our study on the realized thermal niche. The elected threshold for this study was the occurrence of DMST values above 18.7 °C for seven or more consecutive days, because it has proven to be the most realistic value to represent the realized thermal niche (Santiago et al., 2016). The minimum period of seven consecutive days is usually the established time for determining thermal tolerance (Elliott and Elliott, 2010), and when this period is exceeded, the death risk (exclusion risk in our case) increases substantially. The chosen threshold was origi-

Table 3. Different classes of thermal thresholds for emerged trout classes found in the literature. The type of experiment differentiates the experiments with controlled (laboratory) and uncontrolled (wild) temperature. Latitude of the experiments' location is shown.

Variable	Temperature) (°C	Type of experiment	Latitude	Reference
maximum growth	13.1	laboratory	54° N	Elliott et al. (1995)
Maximum growth	16	laboratory	61° N	Forseth and Jonsson (1994)
Maximum growth	16.9	laboratory	43° N	Ojanguren et al. (2001)
Maximum growth	13.2	wild	43° N	Lobón-Cerviá and Rincón (1998)
Maximum growth	13	wild	41° S	Allen (1985)
Maximum growth	15.4–19.1	laboratory	59° N	Forseth et al. (2009)
Thermal optimum	14.2	wild	47° N	Hari et al. (2006)
Upper growth limit	19.5	wild	41° S	Allen (1985)
Upper thermal niche	20	wild	47° N	Hari et al. (2006)
Upper thermal niche*	18.1	wild	41° N	Santiago et al. (2016)
Upper thermal niche*	18.7	wild	41° N	Santiago et al. (2016)
Critical feeding temperature	19.4	laboratory	54° N	Elliott et al. (1995)
Critical feeding temperature	≥ 23	laboratory	59° N	Forseth et al. (2009)
Incipient lethal temperature*	24.7	laboratory	54° N	Elliott (1981)
Ultimate	27.8	laboratory	60° N	Grande and Andersen (1991)
Ultimate**	29.7	laboratory	54° N	Elliott (2000)

* Seven days; ** 10 min.

nally determined in one of the streams in this study (the Cega River).

Once DMST was modelled, the frequency of events of seven or more consecutive days above the threshold per year (times above the threshold, TAT ≥ 7), the total days above the threshold per year (DAT), and the maximum consecutive days above the threshold per year (MCDAT) were calculated for the whole period of 2015–2099.

To assess the general trend in thermal habitat alterations at the middle (H-2050) and the end of the century (H-2099), the TAT ≥ 7, DAT and MCDAT were calculated at each sampling site for each climate change scenario and compared with current conditions

2.8 Longitudinal interpolation and extrapolation

The number of sampling sites and their distribution in the Cega, Pirón and Lozoya rivers (Fig. 2, Table 1) permit the longitudinal interpolation and extrapolation of the predicted water temperatures to study the relationships between the annual average DMST and altitude (strong correlations were detected between these quantities; $R^2 = 0.986$, 0.985 and 0.881, respectively). A digital elevation model (DEM) with a resolution of 5 m made using lidar and obtained from the National Geographic Institute of the Spanish Government (IGN) was used to perform an altitudinal interpolation of the model parameters to determine the water temperature along the stream continuum to simulate the effects of the climate change scenarios and then to obtain the percentage of stream/river length that will be lost for trout. ArcGIS® 10.1 software (made by ESRI®) was used to manage the DEM.

All variables and abbreviations are summarized in the Appendix A. An overview of the uncertainty issue is given in Appendix B.

3 Results

3.1 Climate change

Under the climate change scenarios, all the meteorological stations will experience noticeable temperature (DMAT) increases through the century. As might be expected, this trend is steeper for the RCP8.5 scenario, especially in summer, though it is also noticeable in winter to a lesser extent (annual trends are shown in Fig. 4; the seasonal results are shown by location in Figs. S1 to S24 in the Supplement). The air temperature variations will run parallel to one another in the two scenarios until mid-century, when the RCP8.5 scenario predicts a similar trend and the increases decrease under the RCP4.5 scenario; the annual change in temperatures for RCP4.5 fluctuates between 2 and 2.5 °C at mid-century and between 2.5 and 3.5 °C at the end of the century (3–4 °C at mid-century and 3.5–4.5 °C at the end of the century in summer) The annual change for RCP8.5 is between 2 and 3 °C at mid-century and between 5 and 7 °C at the end of the century (3.5–4.5 °C at mid-century and 7–8 °C at the end of the century in summer).

The change in the annual precipitation (mm day^{-1}) will fluctuate around zero (Fig. 4), although seasonal values will vary (Fig. S1). RCP4.5 predicts a slight decrease (−7 %) by mid-century in total precipitation, which will return to current values by the end of the century. Conversely, RCP8.5

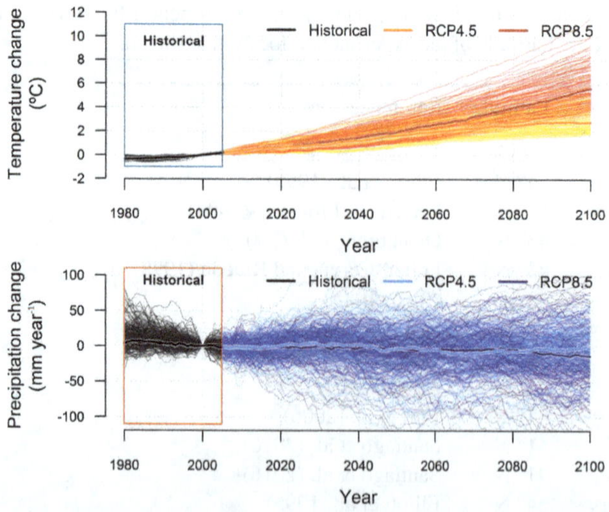

Figure 4. Changes in mean air temperature and total annual precipitation related to climate change for the nine general climate models and the two climate change scenarios for the all the studied meteorological stations.

predicts stable precipitation up to mid-century and a slight decrease (−10 %) by the end of the century. The most important changes appear to occur in autumn. Daily mean air temperatures of the ensemble members for each meteorological station are shown in the Supplement (Dataset S1).

3.2 Hydrological regimes

In general, decreases in flow will occur throughout the century, but the degree of change will vary among the sites. Stations located in the western (Tormes) and eastern (Ebrón) extremes of the study area will experience an increase in flow by 2099 after decreasing in the mid-21st century. Lozoya will suffer the most intense flow decreases, followed by Pirón and Cega-Lastras, Tagus and Gallo, and Cabrillas. These patterns of change in flow regimes are predicted to be linked to a west-to-east longitudinal gradient; climate change is expected to have less of an influence on discharge at the western stations and Ebrón (in the far eastern portion of the study area).

The hydrological models performed well; all of them achieved NSE values ≥ 0.7 when a number of assorted combinations of variables were selected (Table S1 in the Supplement). Figure 5 shows plots of the monthly Q_{mean} results of the simulations for the RCP4.5 and RCP8.5 scenarios in H-2050 and H-2099. Daily mean streamflow estimated from the climate change model ensemble is given in the Supplement (Dataset S2).

3.2.1 RCP4.5 scenario

Statistically significant ($p < 0.05$) shifts in the flow regime will be rare in H-2050 (Table 4, Fig. 5). In H-2099, these changes will be less pronounced, but significant changes become more frequent (Table 4, Fig. 5). Only two gauging stations (Lozoya and Tagus) exhibit significant reductions in annual discharge. By the end of the century (H-2099), annual discharge is expected to be significantly lower at seven gauging stations. The Tagus Basin will experience the greatest changes in annual discharge. Maximum, mean and minimum daily discharges (Q_{mean} and Q_{min}), as well as the Q_{10}, will become much lower in Tagus River basin. Only Cega-Lastras and Pirón (Duero River basin) will suffer a significant increase in the number of zero-flow days.

3.2.2 RCP8.5 scenario

According to the predictions, the most significant changes in flow regimes will occur at the gauging stations of Cega-Lastras and Lozoya in H-2050 (Table 4, Fig. 5). In H-2099, most sites will experience strong flow reductions, even in seasons where seasonal increases in flow are predicted (e.g. Ebrón and both stations in the Tormes River) (Table 4, Fig. 5). Significant annual runoff reductions in H-2050 will occur at five of the stations, increasing the occurrence of significant losses at 9 out of the 10 sites in H-2099 (i.e. all stations except Ebrón). The most important decreases in every variable and throughout the century were predicted for the stations in the middle Cega Basin and the Tagus Basin. A significant increase in the number of days with no flow was predicted for Cega-Lastras, Pirón and Gallo.

3.2.3 Geographical pattern

The cluster analysis of gauging stations based on seasonal variations in the flow regime revealed the importance of careful examinations at the local level, since hydrological behaviour is a consequence of both macroclimatic and mesoclimatic conditions. A geographical pattern is recognizable when the actual flow regime (2006–2015) is seasonally clustered (Fig. 6). Analysing the deviations in this geographical pattern by scenarios and horizons, the different gauging stations can grouped according to the seasonal behaviour of the flow changes (Fig. 7a). For the RCP4.5 scenario in H-2050 (agglomerative coefficients, a.c. = 0.73), the stations that differed most strongly from the remainder in terms of their deviations in the flow regime are those located at Cega-Lastras (winter), Pirón (autumn) and Ebrón (summer). For RCP4.5 in H-2099 (a.c. = 0.56), they are Cega-Pajares (spring), Tormes-Hoyos (summer) and Ebrón (autumn). For RCP8.5 in H-2050 (a.c. = 0.61), they are Pirón (spring, summer and autumn), Lozoya and Ebrón (both in winter). For RCP8.5 in H-2099 (a.c. = 0.72), they are Cega-Pajares (spring) and Ebrón (summer, autumn and winter).

Table 4. Seasonal significant changes of flow variables in percentage (DJF: winter; MAM: spring; JJA: summer; SON: autumn) in H-2050 and H-2099, and RCP4.5 and RCP8.5 scenarios.

2050 HORIZON

Scenario	Variable	Period	Tormes-Hoyos	Tormes-Barco	Cega-Pajares	Cega-Lastras	Pirón	Lozoya	Tagus	Gallo	Cabrillas	Ebrón
RCP4.5	Runoff	annual						−11.7	−13.5			
	$Q=0$	annual										
	Q_{min}	DJF				−49.5						
		MAM		−56.7		−34.6						
		JJA				−39.7		−93.8		−28.0		
		SON			−50.6	−41.4		−106.0				
	Q_{10}	DJF				−44.8						
		MAM				−25.4						
		JJA				−43.3		−92.9		−18.3		
		SON				−36.7		−98.1				
	Q_{mean}	DJF										
		MAM						−7.9	−22.0			
		JJA				−33.9		−48.7				
		SON	−32.9				51.6	−26.1		−13.0	−16.4	
	Q_{max}	DJF										
		MAM							−28.7			
		JJA				−30.3						19.6
		SON									−34.7	
RCP8.5	Runoff	annual				−13.6		−15.5	−25.8	−12.2	−12.7	
	$Q=0$	annual				147.6						
	Q_{min}	DJF				−32.9						
		MAM				−41.3		−39.4				−12.1
		JJA	−43.1			−42.7		−73.8		−33.9		
		SON	−88.0					−49.7				
	Q_{10}	DJF				−43.1			−23.5			
		MAM				−33.1		−31.9	−19.6			−8.8
		JJA	−40.6			−48.4		−89.1		−17.9		
		SON				−45.2		−94.6				
	Q_{mean}	DJF							−30.3		−17.1	
		MAM		−16.6				−15.0	−35.4		−9.7	
		JJA				−44.0	28.6	−47.2		−10.0	−8.5	
		SON			−34.7	−40.6		−31.1		−14.5		
	Q_{max}	DJF							−30.9			
		MAM						−9.4	−33.9			
		JJA				−33.6						
		SON				−40.4				−22.3		

2099 HORIZON

Scenario	Variable	Period	Tormes-Hoyos	Tormes-Barco	Cega-Pajares	Cega-Lastras	Pirón	Lozoya	Tagus	Gallo	Cabrillas	Ebrón
RCP4.5	Runoff	annual		−11.3		−8.6	−17.0	−11.5	−12.9	−6.5	−9.1	
	$Q=0$	annual				85.8	42.7					
	Q_{min}	DJF						−9.3		8.9		
		MAM		−28.8		−29.2		−22.7	−14.8	−13.0		−8.7
		JJA				−9.9		−85.0		−18.0		
		SON					−14.7	−76.8		−11.5		
	Q_{10}	DJF				−27.8						
		MAM	−10.5	−25.2		−19.4	−23.3	−19.2	−17.2	−8.2	−11.6	−6.2
		JJA				−23.8		−48.6		−15.2	−6.3	
		SON					−17.1	−86.9		−11.0		
	Q_{mean}	DJF									−11.6	6.8
		MAM	−10.5	−15.1			−28.1	−9.0	−20.9	−9.2		−5.3
		JJA				−37.3		−42.2	−6.1	−6.5	−8.4	5.7
		SON						−26.2		−7.1		
	Q_{max}	DJF									−15.9	
		MAM					−30.1		−20.9	−11.4		
		JJA				−27.8		−20.7	−18.5			
		SON						−17.0		−13.2		−8.7
RCP8.5	Runoff	annual	−7.7	−12.5	−12.5	−37.7	−48.6	−39.0	−32.1	−19.0	−17.5	
	$Q=0$	annual				332.2	116.9			203.5		
	Q_{min}	DJF			−30.7	−58.3	−69.4	−41.8	−29.3			
		MAM	−38.1	−38.2		−69.5	−31.2	−68.9	−32.2	−11.6	−18.3	−12.0
		JJA	66.2			−53.3	−14.2	−111.8		−36.2	−13.1	
		SON	24.7			−51.4	−16.0	−93.8		−30.5		
	Q_{10}	DJF			−24.0	−62.9	−62.7	−37.0	−37.1	−12.4	−16.2	
		MAM	−33.7	−36.2		−61.4	−33.3	−55.3	−34.7	−14.6	−19.2	14.8
		JJA	67.9	46.8		−50.4		−72.0		−28.0	−9.4	
		SON				−53.2	−18.9	−108.0		−19.9	−7.9	−6.2
	Q_{mean}	DJF		−23.4	−12.9	−40.0	−55.3	−21.1	−40.4	−20.3	−19.2	5.8
		MAM	−22.5	−21.4		−28.9	−49.3	−38.0	−38.8	−19.4	−16.7	−12.9
		JJA	27.5	28.5		−65.7		−80.2		−17.0	−11.2	12.6
		SON		−22.0	−37.4	−51.5	−50.6	−59.6	−21.5	−17.5	−18.0	−4.4
	Q_{max}	DJF		−26.4		−20.5	−44.3	−14.4	−32.5	−32.5	−27.2	
		MAM		−12.9			−50.1	−24.6	−24.5	−26.2		−11.1
		JJA				−62.8		−66.9	−21.2	−11.7	−22.8	26.2
		SON		−23.0	−40.2	−47.7	−47.7	−38.1	−45.7	−24.3	−24.6	

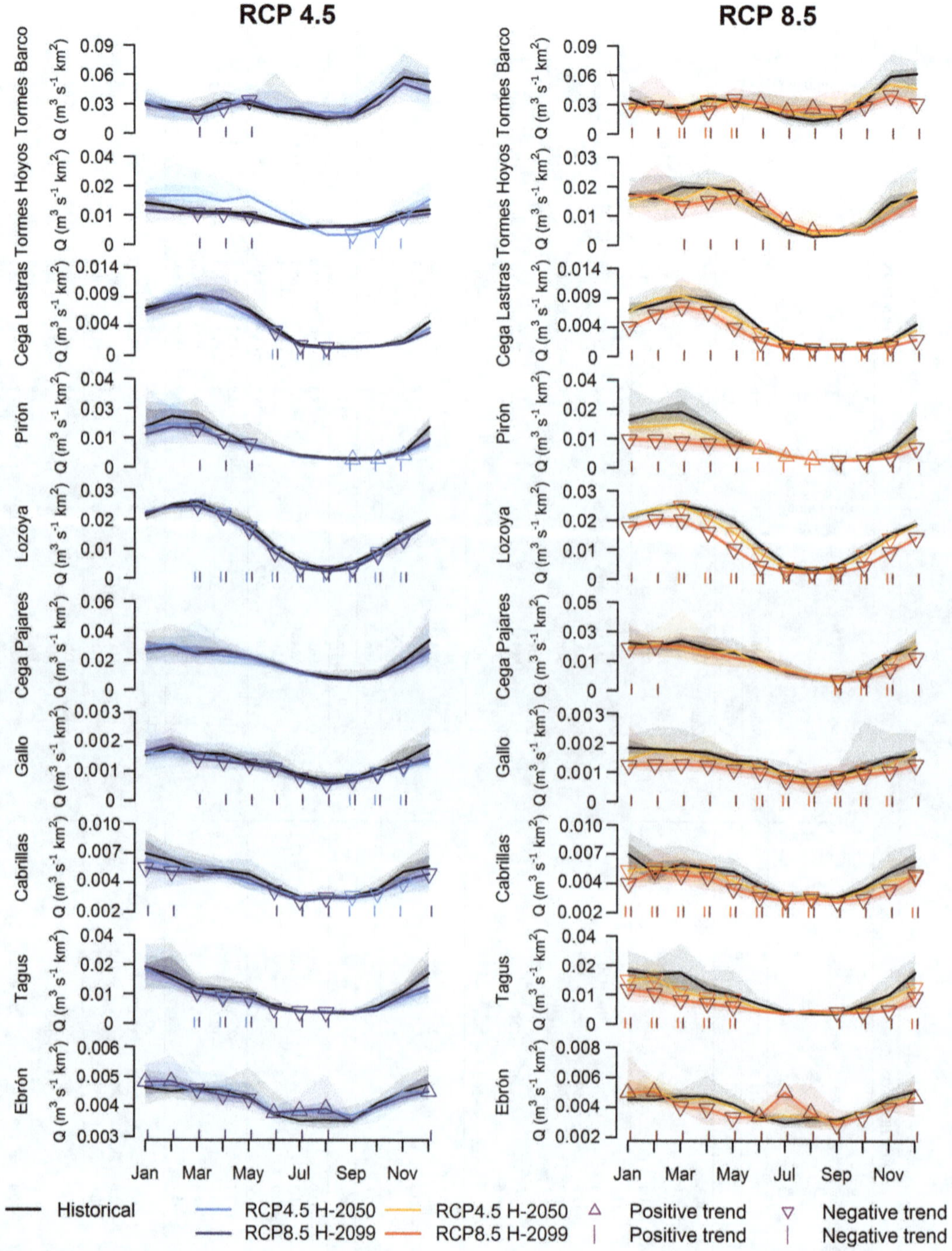

Figure 5. Predicted monthly mean specific flow in H-2050 and H-2099 for the RCP4.5 and RCP8.5 scenarios. Shaded areas indicate decadal fluctuations. Triangles show significant negative or positive trends (Sen's slope $p \leq 0.05$); the sign of each trend is indicated by the directions in which the triangles point.

3.3 Stream temperature

3.3.1 Model parameter behaviour and general trends

The inclusion of the streamflow component improves model performance at 12 out of the 28 study sites (Table 5). In the remaining 16 cases, either no convergence of values was observed in the regression process or the obtained values did not improve the results, as the streamflow component (Term 2 of the equation) is virtually zero at the other sites. The five-parameter model was used in these remaining 16 cases. The calculated parameters and the performance indicators (RSE and NSE) of the models are shown in Table 6, and daily mean stream temperatures estimated by the climate change models

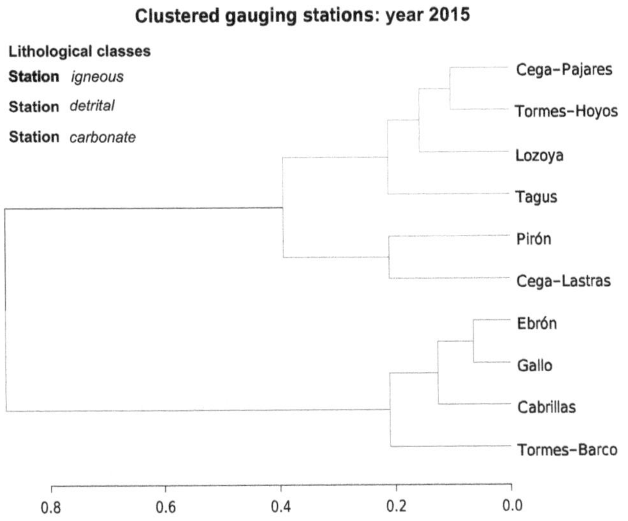

Figure 6. Gauging stations clustered by the current normalized seasonal streamflow regime (agglomerative coefficient, a.c. = 0.81). Stations are grouped by lithological classes.

are given in the Supplement (Dataset S3). Performance was high in all the cases except in Pirón 5 where NSE was low.

For the entire array of sites ($n = 31$), the Pearson correlation between thermal amplitude ($\alpha - \mu$) and λ is significant (Pearson's $r = -0.832$; $p < 0.0001$). As the thermal amplitude increases, λ becomes more negative (indicating less resistance). As the thermal amplitude decreases, λ approaches zero (indicating more resistance).

By the end of the 21st century, the predicted average increase in the mean annual stream temperature among the sites is 1.1 °C for the RCP4.5 scenario (range 0.3–1.6 °C) and 2.7 °C for RCP8.5 (range 0.8–4.0 °C). The average increases in maximum annual mean temperature are predicted to be 0.8 °C for RCP4.5 (range 0.1–1.5 °C) and 1.6 °C for RCP8.5 (range 0.2–3.0 °C), and the average increases in minimum annual mean temperature are predicted to be 1.0 °C (range 0.4–1.8 °C) and 2.7 °C (range 1.1–4.5 °C), respectively. The most important increases are predicted to occur in winter, with summer experiencing smaller increases.

3.3.2 Stream temperature and geological nature

The values of the model parameters showed different behaviours depending on the lithology found in each basin, which thus influences the thermal response to climate change. The thermal amplitude is greater at sites underlain by igneous bedrock ($\overline{\alpha - \mu} = 20.38$ °C) than at sites underlain by carbonate bedrock ($\overline{\alpha - \mu} = 13.07$ °C). β values are greater at sites underlain by igneous bedrock ($\overline{\beta} = 12.71$ °C) than at sites underlain by carbonate bedrock ($\overline{\beta} = 7.80$ °C), and λ is significantly greater ($\overline{\lambda} = -0.140$) at sites underlain by carbonate bedrock than at sites underlain by igneous bedrock ($\overline{\lambda} = -0.292$) and at sites underlain by Quaternary

Table 5. Bayesian (BIC) and Akaike (AIC) information criteria values for the stream-temperature models with five and eight parameters.

Site	BIC 5	BIC 8	AIC 5	AIC 8
Tormes2	2075.7	1911.8	2108.5	1837.9
Tormes3	2346.7	2274.2	2346.7	2234.8
Cega1	1814.8	1731.9	1789.1	1693.4
Pirón1	1725.9	1530.5	1700.2	1492.0
Lozoya1	5097.3	4924.6	5065.2	4876.4
Lozoya2	3979.3	3927.9	3948.6	3881.9
Lozoya3	3841.6	3673.6	3811.5	3628.5
Lozoya4	5076.6	4735.3	5044.8	4687.5
Cabrillas	2552.9	2172.2	2523.1	2127.4
Ebrón	624.8	169.8	598.5	130.3
Vallanca1	1438.9	1359.9	1410.6	1317.3
Vallanca2	1322.1	1279.7	1293.7	1237.1

detrital material ($\overline{\lambda} = -0.305$) (Fig. 8). All of these differences are significant (p-values < 0.001, as determined using t tests with the Bonferroni correction).

Among the eight-parameter models ($n = 12$), significant differences are also found among the lithological classes for ω and τ. For both ω and τ, the values were higher at sites underlain by carbonate bedrock ($\overline{\omega} = 0.96$ °C; $\overline{\tau} = 3.640\,m^3\,s^{-1}$) than at sites underlain by igneous bedrock ($\overline{\omega} = -2.12$ °C; $\overline{\tau} = 0.345\,m^3\,s^{-1}$). The differences in the δ values among the carbonate ($\overline{\delta} = 67.06\,m^{-3}\,s$) and igneous sites ($\overline{\delta} = 67.06\,m^{-3}\,s$) were only marginally significant ($p < 0.1$).

Under the RCP4.5 scenario, ΔT_{min} displays significantly different behaviour at sites underlain by Quaternary detrital material than at sites underlain by carbonate and igneous rocks. Under the RCP8.5 scenario, this difference is solely found between the sites underlain by Quaternary detrital material and those underlain by carbonate rocks. ΔT_{mean} exhibits significant differences between the sites underlain by all three lithologies in both scenarios. All of these results are common to H-2050 and H-2099. In terms of ΔT_{max}, in H-2050, significant differences are found between the sites underlain by carbonate and igneous rocks for both scenarios. These differences are also significant in H-2099 for RCP4.5 and RCP8.5 and between the sites underlain by carbonate rocks and Quaternary detrital material under the RCP8.5 scenario (Fig. 9).

The results of the cluster analysis of the monthly mean stream temperatures revealed a highly homogeneous aggregation of sites for the different combinations of horizons and scenarios, given that the thermal responses of the rivers and streams are tightly linked with lithology (Fig. 7b). The carbonate sites from the Cabrillas stream (in the east) and Pirón 3 (which is strongly influenced by a calcareous spring) form a group of sites that shows low thermal amplitude and in which λ is close to zero. At the other extreme, a group

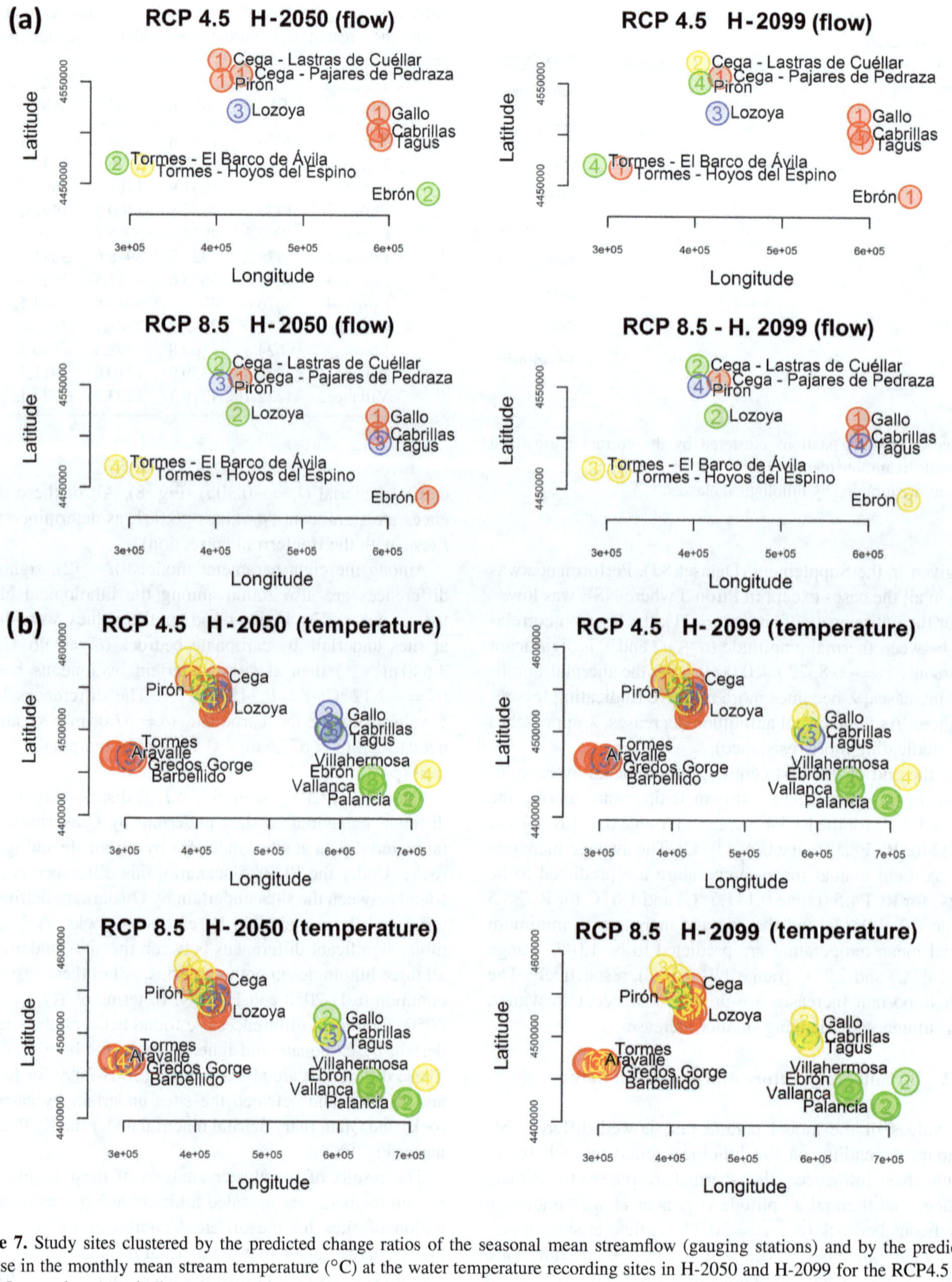

Figure 7. Study sites clustered by the predicted change ratios of the seasonal mean streamflow (gauging stations) and by the predicted increase in the monthly mean stream temperature (°C) at the water temperature recording sites in H-2050 and H-2099 for the RCP4.5 and RCP8.5 scenarios. Axes indicate geographic positions (UTM coordinates). The colours and numbers indicate the clusters.

that is made up mainly of sites underlain by igneous material (in the Lozoya and Tormes basins, in addition to several sites found in the detrital basin of Cega-Pirón) shows higher thermal amplitude and lower values of λ than the former group. The remaining sites have intermediate values of thermal amplitude and resistance.

Table 6. Parameter values of the stream temperature models for every thermograph site (λ is a dimensionless parameter) and values of the performance indicators residual standard error (RSE) and Nash–Sutcliffe efficiency index (NSE).

Site	μ (°C)	α (°C)	$\alpha-\mu$ (°C)	γ (°C^{-1})	β (°C)	λ	ω (°C)	δ (m^{-3} s)	τ (m^3 s^{-1})	RSE (°C)	NSE
Aravalle	0.1	23.3	23.2	0.14	10.79	−0.31				2.31	0.70
Barbellido	1.1	19.2	18.1	0.23	12.19	−0.30				3.29	0.81
Gredos Gorge	2.4	19.0	16.6	0.18	14.06	−0.29				3.68	0.71
Tormes1	−1.1	20.8	21.9	0.16	11.00	−0.35				1.97	0.83
Tormes2	3.4	24.6	21.3	0.14	11.82	−0.31	−3.98	61.43	0.37	1.56	0.78
Tormes3	−0.1	30.5	30.6	0.12	12.61	−0.39	−2.85	223.69	0.22	1.62	0.79
Pirón1	−1.4	19.1	20.5	0.09	14.33	−0.18	−2.08	72.82	0.15	0.97	0.91
Pirón2	0.6	15.5	14.9	0.22	12.11	−0.23				3.69	0.82
Pirón3	7.2	14.0	6.8	0.29	10.17	−0.10				1.16	0.77
Pirón4	−0.6	18.3	18.9	0.15	8.20	−0.29				1.26	0.85
Pirón5	−4.6	21.7	26.3	0.13	8.92	−0.40				1.57	0.39
Cega1	1.4	15.6	14.2	0.19	15.92	−0.22	−1.53	112.98	0.27	1.17	0.88
Cega2	−0.6	18.0	18.6	0.17	16.03	−0.31				1.81	0.85
Cega3	−2.0	24.7	26.6	0.14	15.35	−0.41				2.83	0.85
Cega4	−0.3	19.9	20.2	0.16	12.21	−0.35				1.65	0.87
Cega5	−2.4	18.1	20.5	0.13	7.85	−0.31				0.87	0.84
Cega6	0.7	22.4	21.7	0.13	13.84	−0.38				2.34	0.79
Lozoya1	0.4	19.5	19.1	0.18	11.90	−0.24	−1.33	11.93	0.41	1.13	0.90
Lozoya2	0.3	20.2	20.0	0.19	11.63	−0.28	−1.27	13.16	0.38	1.17	0.90
Lozoya3	1.1	21.0	19.9	0.19	10.62	−0.29	−1.74	23.44	0.49	1.23	0.89
Lozoya4	1.7	22.0	20.2	0.17	10.29	−0.27	−2.19	17.04	0.48	1.17	0.90
Tagus-Peralejos	1.1	21.0	19.9	0.11	11.01	−0.17				1.00	0.89
Tagus-Poveda	1.3	20.4	19.2	0.15	9.95	−0.38				1.26	0.83
Gallo	0.4	20.2	19.8	0.13	7.76	−0.18				0.90	0.92
Cabrillas	8.3	15.3	7.0	0.21	9.23	−0.04	−1.38	13.56	1.25	1.65	0.90
Ebrón	5.5	17.0	11.4	0.07	6.58	−0.06	1.73	−1.78	3.16	0.27	0.86
Vallanca1	−0.5	16.9	17.4	0.09	4.70	−0.12	1.95	−5.21	4.85	0.53	0.88
Vallanca2	1.4	16.8	15.4	0.10	5.36	−0.11	1.54	−11.29	5.29	0.50	0.89
Palancia1	11.7	15.3	3.6	0.19	13.92	−0.03				0.23	0.93
Palancia2	9.3	16.1	6.8	0.27	12.73	−0.11				0.59	0.88
Villahermosa	7.8	18.0	10.2	0.27	16.46	−0.20				1.02	0.85

3.3.3 Effect of streamflow reductions on stream temperature

The predicted flow reductions lead to notable increases in water temperature. The effect of streamflow variation on stream temperature is analysed at the following sites: Tormes 2, Tormes 3, Pirón 1, Cega 1, Lozoya 1 to 4, Cabrillas, Ebrón 1 and Vallanca 1 and 2. These are the sites at which the eight-parameter model improves upon the five-parameter model. In all cases, differences in stream temperature between the five- and eight-parameter models are found, and summer flow reductions lead to increases in stream temperature, increasing DAT, TAT ≥ 7 and MCDAT. Among these sites, the threshold is only surpassed at Lozoya and Tormes, increasing the thermal habitat loss. At Cega 1, Cabrillas and Ebrón, α is below the thermal threshold, and at Pirón 1, the stream temperature increase is not sufficient to exceed the threshold.

For all of the sites at which the influence of streamflow on stream temperature was revealed, the eight-parameter model estimates higher values of maximum annual DMST than the five-parameter model. The maximum annual DMST calculated by the eight-parameter model is 3.6 °C higher than that calculated by the five-parameter model at the Tormes 2 site. This difference is not so large at the other sites, and the minimum disagreement between the models (0.01 °C) is noted at the Ebrón and Cabrillas sites. In general, the maximum differences between the two models are noted in igneous catchments, whereas carbonate sites yield the lowest differences.

3.3.4 Effect of climate change on the thermal habitat of brown trout

The length of the thermal habitat of trout will undergo important reductions due to the rises in water temperatures and the increase in the extent of the warm period. In the predictions for H-2050, the 18.7 °C threshold (TAT ≥ 7) will be violated

Table 7. Maximum daily mean stream temperature (°C) in each site in the year 2015 and the horizons H-2050 and H-2099. Both scenarios (RCP4.5 and RCP8.5) are shown. Bold numbers: values > 18.7 °C.

	Maximum daily mean stream temperature (°C)					
	RCP4.5			RCP8.5		
Site	2015	H-2050	H-2099	2015	H-2050	H-2099
Aravalle	**19.8**	**20.4**	**21.0**	**19.8**	**20.7**	**22.0**
Barbellido	17.9	18.4	**18.7**	17.9	18.6	**19.3**
Gredos Gorge	16.5	17.1	17.6	16.5	17.4	18.5
Tormes1	18.1	18.6	**19.1**	18.0	**18.9**	**20.1**
Tormes2	**20.5**	**21.2**	**21.4**	**20.7**	**21.1**	**22.1**
Tormes3	**21.8**	**22.7**	**23.1**	**22.4**	**22.4**	**24.5**
Cega1	12.4	13.1	13.6	12.5	13.3	14.0
Cega2	15.2	15.9	16.3	15.2	16.1	17.3
Cega3	**19.8**	**20.7**	**21.4**	**19.8**	**21.0**	**22.8**
Cega4	18.1	18.5	**18.9**	18.1	**18.7**	**19.7**
Cega5	16.6	16.9	17.3	16.6	17.1	17.8
Cega6	**18.7**	**19.5**	**19.9**	**18.8**	**19.6**	**21.0**
Pirón1	12.9	13.8	14.2	13.2	13.9	15.6
Pirón2	14.9	15.1	15.4	14.9	15.3	15.7
Pirón3	14.1	14.1	14.2	14.0	14.2	14.3
Pirón4	17.2	17.5	17.8	17.2	17.7	18.3
Pirón5	**19.3**	**19.8**	**20.2**	**19.3**	**20.0**	**21.1**
Lozoya1	16.8	17.4	17.8	16.8	17.6	**18.8**
Lozoya2	17.6	18.1	18.6	17.5	18.4	**19.6**
Lozoya3	**19.0**	**19.5**	**19.9**	**18.9**	**19.7**	**20.8**
Lozoya4	**19.5**	**20.0**	**20.5**	**19.5**	**20.3**	**21.4**
Tagus-Peralejos	16.7	17.2	17.6	16.6	17.4	18.6
Tagus-Poveda	18.1	18.6	**19.0**	18.1	**18.8**	**19.9**
Gallo	17.9	18.3	18.6	17.9	18.4	**19.3**
Cabrillas	14.9	15.0	15.1	14.9	15.1	15.2
Ebrón	16.2	16.5	16.5	16.2	16.5	17.0
Vallanca1	16.8	17.1	17.3	16.8	17.2	17.9
Vallanca2	16.5	16.8	17.0	16.5	16.9	17.5
Palancia1	15.0	15.1	15.1	15.0	15.1	15.3
Palancia2	16.0	16.1	16.1	16.0	16.1	16.4
Vistahermosa	16.0	16.1	16.1	16.0	16.1	16.5

at eight sites under the RCP4.5 scenario and six sites under the RCP8.5 scenario. In H-2099, the threshold will be violated at eight sites under the RCP4.5 scenario and 13 sites under the RCP8.5 scenario.

By the end of the century (H-2099), the most notable increases in TAT ≥ 7 (Fig. 10) will be produced at Cega 6, Pirón 5 and Lozoya 3 under the RCP4.5 scenario and at Tormes 1, Cega 4, Cega 6, Lozoya 2, Gallo and Tagus-Poveda under the RCP8.5 scenario. The most significant increases in MCDAT (Fig. 10) will occur at low-altitude sites underlain by igneous rocks and detrital material. In general, the highest temperatures (maximum values of 24.5 °C, Table 7) are predicted to occur in the downstream reaches of the igneous and detrital river basins. In the carbonate basins, only two sites (Tagus-Poveda and Gallo) will exceed the thermal threshold. At mid-century (H-2050), the main changes under the RCP8.5 scenario are similar to those predicted for RCP4.5 at the end of the century (H-2099). RCP4.5 predicts a slower warming from mid-century onwards, whereas RCP8.5 predicts an acceleration of the warming during that period.

Continuous modelling of water temperature by means of the interpolation of model parameters along the Cega, Pirón and Lozoya rivers and the application of the model to DEM data predicts relevant losses of thermal habitat, which will affect up to 56, 11 and 66 % of the lengths of these streams, respectively. In the Cega and Pirón rivers, the habitat loss is expressed relative to the proportion of total stream length where trout currently dwell (98 and 77 km in the Cega and Pirón streams, respectively). In the Lozoya River, the loss is predicted to occur in the reach (20 km) immediately upstream of a large reservoir (the Pinilla reservoir), which produces a total disconnection of the stream. The losses in maximum usable habitat will shift the current downstream limit of

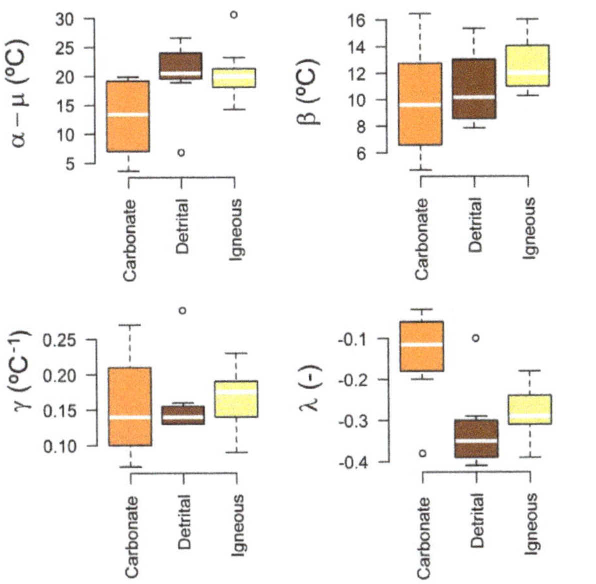

Figure 8. Distributions of the stream temperature model parameter values ($\alpha - \mu$, β, γ and λ) in relation to lithology. Differences were assessed using Student's *t* test with the Bonferroni correction ($p < 0.05$).

the trout distribution from 820 up to 831 m a.s.l in the Pirón River, from 730 up to 830 m a.s.l. in the Cega River, and from 1090 up to 1276 m a.s.l. in the Lozoya River. In the particular case of the Cega River, a window of usable thermal habitat is also predicted to occur upstream from this altitudinal range (from 913 up to 1050 m a.s.l.).

4 Discussion

4.1 Climate change

Our downscaled results predict greater air temperature increments than the original IPCC (2013) results. These higher temperatures may lead to increased ecological impacts (Magnuson and Destasio, 1997; Angilletta, 2009) caused by the combination of rising water temperatures and decreasing stream flows. The results from the AR5 of the IPCC and its annex, the Atlas of Global and Regional Climate Projections (IPCC, 2013), suggest that droughts are unlikely to increase in the near future for the Mediterranean area. However, air temperatures are expected to rise, subsequently increasing evapotranspiration. As a consequence, the available water in rivers and streams will be reduced. Regional studies have used coarser resolutions than ours, which may be appropriate for their goals (e.g. Thuiller et al., 2006). However, they may be insufficient when more local predictions are needed, as does our study, which treats geographically confined, stream-dwelling trout populations. Therefore, fine downscaling techniques like those applied in this study must be used when high-resolution, detailed predictions are needed.

4.2 Streamflow

This study predicts significant but diverse streamflow reductions during the present century. At the regional level, a reduction in water resources is expected in the Mediterranean area (IPCC, 2013). Milly et al. (2005) predicted a 10–30 % decrease in runoff in southern Europe in 2050. In another global-scale study, van Vliet et al. (2013) predicted a decrease in the mean flows of greater than 25 % in the Iberian Peninsula area by the end of the century (2071–2100), using averages for both the SRES A2 and B1 scenarios (Nakicenovic et al., 2000). Our results predict mean flows that are similar to that value (−23 %; range: 0–49 %), although the emissions scenarios in this study are more severe (that is, they involve greater increases in atmospheric CO_2) than those used in the aforementioned studies.

More specifically, the predictions for the RCP4.5 scenario show flow reductions that range from negligibly small to significant (up to 17 %). Under the RCP8.5 scenario, significant reductions become more widespread, ranging up to 49 % of the annual streamflow losses. Our results also predict a relevant increase in the number of days with zero flow for some stations in the detrital area under this scenario (RCP8.5). The predicted streamflow changes are compatible with those obtained in previous studies, although these studies were performed at larger scales (Milly et al., 2005; van Vliet et al., 2013). The apparent differences between the streamflow reductions estimated in this study and those obtained by Milly et al. (2005) and van Vliet et al. (2013) (who report lower flow reductions than those given in the present study) might be caused by the regional focus of their predictions (the entire Iberian Peninsula), whereas ours are focused on mountain reaches.

In terms of methods, process-based hydrological models are often preferred for climate change studies (Van Vliet et al., 2012). However, they can be overly complicated and require excessive data inputs, which may also lead to overfitting of the data (Zhuo et al., 2015). Constraining further predictions to within the training domain is a rule of thumb for machine learning studies (Fielding, 1999), although extrapolation is rather common (Elith and Leathwick, 2009). Therefore, taking into account the extrapolation that occurs towards lower flows, which are over-represented in the training dataset, we consider the magnitude of the extrapolation acceptable, and we consider the values, although they are not exempt from uncertainty, to be reliable.

4.3 Stream temperature

The model we present in this study showed good performance. Bustillo et al. (2013) recommended the assessment of the impacts of climate change on river temperatures using regression-based methods like ours that rely on logistic approximations of *equilibrium temperatures* (Edinger et al.,

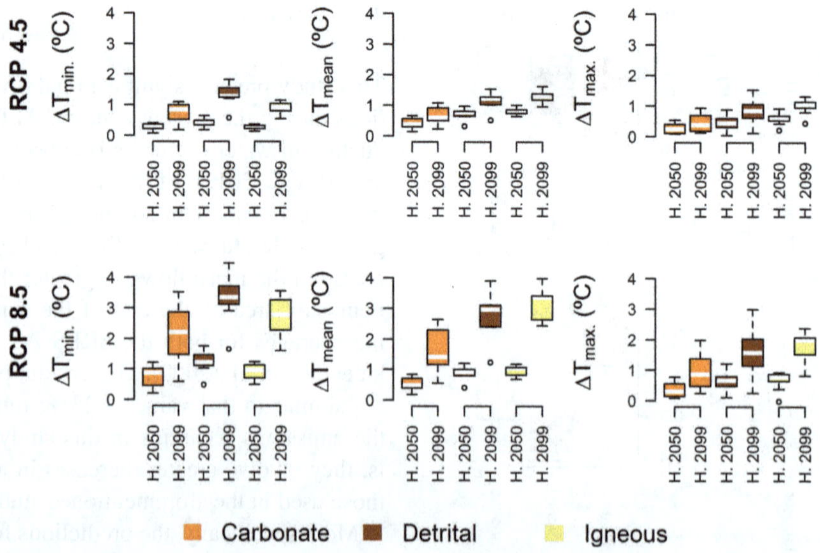

Figure 9. Distributions of $\Delta T_{\min}$, ΔT_{mean} and $\Delta T_{\max}$ in relation to lithology for the climate change scenarios RCP4.5 and RCP8.5 in H-2050 and H-2050. The reference period corresponds to the simulated period 2010–2019.

1968), which are at least as robust as the most refined classical heat balance models.

However, we also sought to identify relationships between thermal regime and other environmental variables besides air temperature and streamflow, such as geology. Bogan et al. (2003) showed that water temperatures were uniquely controlled by climate in only 26 % of 596 studied stream reaches. Groundwater, wastewater and reservoir releases influenced water temperatures in the remaining 74 % of the cases. Loinaz et al. (2013) quantified the influence of groundwater discharge on temperature variations in the Silver Creek Basin (Idaho, USA), and they concluded that a 10 % reduction in groundwater flow can cause increases of over 0.3 and 1.5 °C in the average and maximum stream temperatures, respectively. Our studied reaches were not influenced by wastewater or reservoir releases (with the exception of releases from the Torrecaballeros Dam on the Pirón River). Kurylyk et al. (2015) showed that the temperature of shallow groundwater influences the thermal regimes of groundwater-dominated streams and rivers. Since groundwater is strongly influenced by geology, we can expect it to be a good indicator of the thermal response, as shown here. The models used accurately described the thermal performance of the study sites, and we found significant relationships among the model parameters, the underlying lithologies and the hydrologic responses. Thermal amplitude ($\alpha - \mu$) and temperature at the maximum change rate (β) were lower, and the resistance parameter (λ) was closer to zero, in river basins that were highly influenced by aquifers (mainly carbonate) compared to the others, particularly compared with river basins underlain by carbonate rocks. Since DMST is a variable that is relevant for detecting departures from thermal niche, we can conclude that it is worthwhile to use the more complex eight-parameter model to predict the effects of global warming, especially in igneous catchments.

A wide range of models is described in the literature, and each such model has its strengths and weaknesses. Arismendi et al. (2014) hold that regression models based on air temperature can be inadequate for projecting future stream temperatures because they are only surrogates for air temperature, whereas Piccolroaz et al. (2016) argued that the adequacy depends on the hydrological regime, type of model and the timescale analysis. Their main objections to regressive methods arose when modelling reaches of regulated rivers, but this is not our case. In addition, our model improves the models that were tested in both studies (Arismendi et al., 2014; Piccolroaz et al., 2016). Performance indicators of our models produce good results, showing that the models are sufficiently competent. We show that our model implicitly integrates the effect of other factors, such as geology and flow regime by means of its parameters. A fine mechanistic solution to the modelling issue could need prohibitive methods (Kurylyk et al., 2015), losing the advantages that make attractive the model (input data easy to get). Therefore a compromise between improved precision and increased cost must be met.

The behaviour and dynamics of the parameters offer a promising research field. Their analysis may help to introduce new parametrization criteria to avoid the risk of ignoring the effect of climate warming on groundwater (subsurface water and deep water), for instance. The thermal sensitivity of shallow groundwater differs between short-term (e.g. seasonal) and long-term (e.g. multi-decadal) time horizons, and the relationship between air and water temperatures does not necessarily reflect this difference. This variability should be taken into account in order to avoid un-

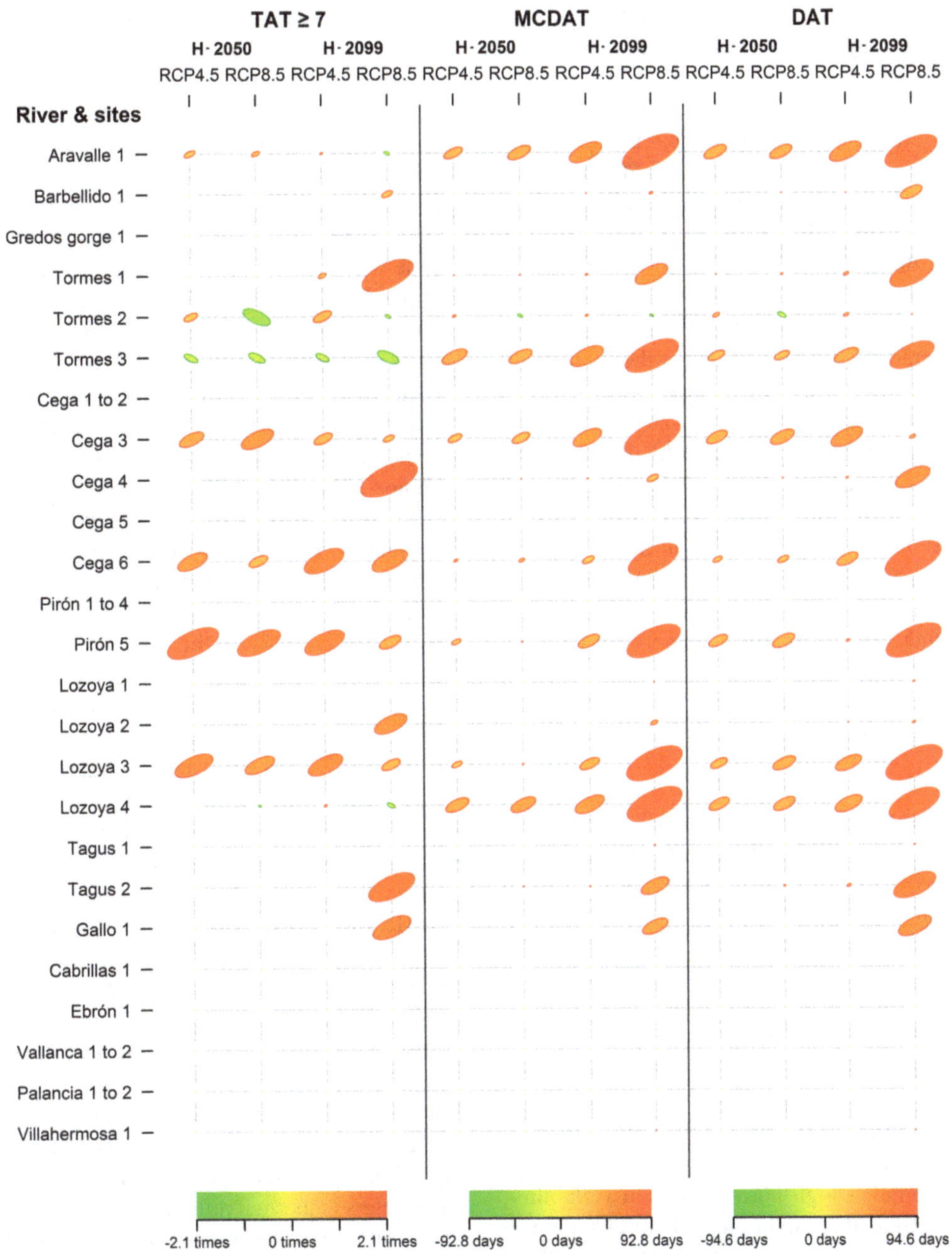

Figure 10. Increases in TAT ≥ 7 (time above the threshold during seven or more consecutive days), MCDAT (maximum consecutive days above the threshold) and DAT (days above the threshold per annum) from the present to H-2050 and H-2099 for RCP4.5 and RCP8.5.

derestimating the effects of climate warming (Kurylyk et al., 2015).

Regression models are substantially site-specific compared to deterministic approaches (Arismendi et al., 2014). However, the parameters of these regression approaches are still physically meaningful, and these models require fewer variables that can limit the applicability of more complex models in areas where data are scarce. Consequently, the value of this type of model is its applicability to a large number of sites where the only available data describe air temperatures (and precipitation and streamflow to a lesser extent). On the other hand, our results show that predictions can improve when streamflow is included in the water temperature model, although some streams show little or no sensitivity to the introduction of streamflow into the model. However, the lack of sensitivity is not necessarily due to the absence of the influence of flow on the water temperature but rather to its minor relevance compared to other sources of noise. Thus, when flow data are available, it may be recommended to use the more complex eight-parameter model to predict the effects of climate warming. This conclusion is especially applicable to lithologically sensitive basins, such as those underlain by igneous rocks.

The predicted increase in water temperature will be substantial at most of the study sites. The annual mean rates of change will increase with time. Stewart et al. (2015) predict an increase of 1–2 °C by mid-century in 80 % of the stream lengths in Wisconsin and by 1–3 °C by the latter part of the century in 99 % of the stream lengths, which corresponds to a significant loss in suitable areas for cold-water fish. The results of Stewart et al. (2015) do not differ from ours, except that we expect greater increases by the end of the studied period (up to 4 °C). Our results are also compatible with those of van Vliet et al. (2013), who predicted a water temperature increase > 2 °C in the Iberian Peninsula area by the end of the century; however, our results are more specific and precise. In this sense, Muñoz-Mas et al. (2016) also obtained similar results for H-2050 in a river reach in central Spain by mid-century (i.e. daily mean flow reductions between 20 and 29 % and daily mean stream temperature increases up to 0.8 °C). However, we predict that the minima are more sensitive to climate warming than the maxima.

4.4 Effects of climate change on brown trout populations

Brown trout are sensitive to changes in discharge patterns because high-intensity floods during the incubation and emergence periods may limit recruitment (Lobón-Cerviá and Rincón, 2004; Junker et al., 2015). In the Iberian Peninsula, the trout distribution is mainly concentrated in mountain streams, where extreme discharges during winter are expected to increase (Rojas et al., 2012). These extreme discharges will likely affect trout recruitment negatively. Thus, the predicted changes in the hydrological regime can subject brown trout populations to more variable conditions, which may occasionally present some populations with insuperable bottlenecks. Trout are polytypic and display an adaptable phenology and rather high intra-population variability in their life history traits that might allow them to show resilience to variations in habitat features (Gortázar et al., 2007; Larios-López et al., 2015), especially in the marginal ranges (Ayllón et al., 2016). However, despite these strong evolutionary responses, the current combination of warming and streamflow reduction scenarios is likely to exceed the capacity of many populations to adapt to new conditions (Ayllón et al., 2016). Consistent with regional predictions (Rojas et al., 2012; Garner et al., 2015), significant flow reductions are expected during summertime in most of the studied rivers and streams at the end of the century, and this may mean, in turn, the reduction in the suitable habitat (i.e. the available water volume) (Muñoz-Mas et al., 2016). Finally, the increase in extreme droughts, which involve absolute water depletion, in certain reaches of the streams may be critical for some trout populations.

The predicted increase in winter stream temperatures can affect the sessile phases (i.e. eggs and larvae) of trout development. These phases are very sensitive to temperature changes because it affects their physiology, and because their development is temperature dependent (e.g. Lobón-Cerviá and Mortensen, 2005; Lahnsteiner and Leitner, 2013). Thus, changes in the duration of incubation and yolk sac absorption can affect emergence times and, in turn, the sensitivity of these phases to hydrological regime alterations (Sánchez-Hernández and Nunn, 2016). An increase in stream temperature can also reduce hatchling survival (Elliott and Elliott, 2010). In accordance with the results presented herein, the predicted synergy of streamflow reductions and water temperature increases will cause substantial losses of suitable fish habitat, especially for cold-water fish such as brown trout (Muñoz-Mas et al., 2016).

The increases in threshold violations were important in our simulations. The duration of warm events (temperature above the threshold value) increased by up to 3 months at the end of the century in the most pessimistic scenario (RCP8.5). A continuous analysis of the whole-river response should be conducted to allow spatially explicit predictions and to identify reaches where thermal refugia are likely to occur. However, our results suggest that trout will not survive in these reaches because the persistence of thermal refugia is improbable or because their extents will be insufficient. In the Cega, Pirón and Lozoya rivers, important losses of thermal habitat will occur that could jeopardize the viability of the trout population. Behavioural thermoregulatory tactics are common in fish (Reynolds and Casterlin, 1979; Goyer et al., 2014); for instance, some species perform short excursions (< 60 min in experiments with brook char, *S. fontinalis*) that could be a common thermoregulatory behaviour adopted by cold freshwater fish species to sustain their body temperature below a critical temperature threshold, enabling them to exploit resources in an unfavourable thermal environment (Pépino et al., 2015). Brown trout can use pool bottoms during daylight hours to avoid the warmer and less oxygenated surface waters in thermal refugia (Elliott, 2000). Nevertheless, if the warm events became too long, the thermal refugia could become completely insufficient, thus compromising fish survival (Brewitt and Danner, 2014; Daigle et al., 2014).

4.5 The brown trout distribution

According to our results, streamflow reductions are able to synergistically contribute to the loss of thermal habitat by increasing daily mean stream temperatures. This effect is especially relevant in summer in the Mediterranean area, when the warmest temperatures and minimum flows usually occur. The existence of thermal refugia represents a possible means of fish survival, and the probability for a water body to become a thermal refugium is highly geologically dependent. In our simulations, the sites that are most dependent on deep aquifers (i.e. basins underlain by Mesozoic carbonate rocks) display improved resistance to warming. The habitat retraction at the rear edge of the actual distribution of brown trout is deduced to be geologically mediated.

The mountains of central and south-eastern Spain contain the rear edge of the distribution of native brown trout (Kottelat and Freyhof, 2007). Fragmentation and disconnection of populations by newly formed thermal barriers may aggravate the already significant losses of thermal habitat by reducing the viability of populations and increasing the extinction risk. Thus, the rear edge of the trout population in the Iberian Peninsula might shift to the northern mountains to varying extents depending on the presence of relevant mesological features, such as geology. The calcareous mountains of northern Spain could be a refuge for trout because they combine favourable geology and a relatively more humid climate. As a result of this differential response, the western portion of the Iberian range (which is plutonic and less buffered) will eventually experience more frequent local temperature-driven extinction events, thus producing a greater shift northward, than in the eastern Iberian end of this range, which is calcareous and highly buffered and will remain more resilient to these local extinction events. However, the predicted streamflow reductions may act synergistically, reducing the physical space, and this may jeopardize the less thermally exposed populations. In the Iberian Peninsula, stream temperatures will increase less in the central and northern mountains than in the central plateau, and the increases will be smaller in karstic than in granitic (igneous) mountains. At the same time, the side of the peninsula that faces the Mediterranean is expected to be more sensitive to warming and streamflow reductions than the side of the peninsula that faces the Atlantic. Thus, brown trout populations in the karstic mountains of northern Spain (the Cantabrian Mountains and the calcareous parts of the Pyrenees) are better able to resist the climate warming than the populations farther east in the granitic portion of the Pyrenees (Santiago, 2017). Similar patterns may occur in other parts of southern Europe. Most likely, the less pronounced thermal responses of rivers and streams in the karstic areas will allow for greater persistence of the brown trout population, although changes in streamflow regimes will likely also occur there.

In a study of the major basins of Europe, Lassalle and Rochard (2009) predicted that the brown trout would "lose all its suitable basins in the southern part of its distribution area ([the] Black Sea, the Mediterranean, the Iberian Peninsula and the South of France), but [would] likely to continue being abundant in [the] northern basins". Almodóvar et al. (2011) estimated that the brown trout will be eradicated over almost the entire stream length of the studied basins in northern Spain, and Filipe et al. (2013) estimated an expected loss of 57 % of the studied reaches in the Ebro Basin in north-eastern Spain. Our study shows important, yet not so dramatic, reductions in the thermal habitat of Iberian brown trout populations in mountainous areas. The number of general climate models used, the reliability of the downscaling procedure, the resolution of the stream temperature and streamflow models, and the method used to study the threshold imply a substantial improvement in detail (Santiago et al., 2016) over previous work. It is reasonable to infer that many mountain streams appear poised to become refugia for cold-water biodiversity during this century (Isaak et al., 2016).

5 Conclusions

The main findings of this study are as follows: (i) our downscaled results predict greater air temperature increments than the IPCC's averages, from which our estimations were made; (ii) significant but diverse streamflow reductions are predicted to occur during the present century; (iii) the models presented in this study have been shown to be useful for improving simulations; (iv) the predicted increases in water temperature will be influenced to varying degrees by the flow and geological features of rivers and streams; (v) the thermal habitat of brown trout, a cold-water species, will decrease as a consequence of the synergistic effects of flow reduction and water warming; and (vi) the peaks in water temperature and the complete depletion of the river channels will produce local extinctions, although the ultimate magnitude of the effect will be governed by the geological nature of the basins.

Our findings might be useful in planning the prevention and mitigation of the negative effects of climate change on freshwater fish species at the rear edge of their distributions. A differentiation of areas based on their risk level and viability is necessary to set standardized conservation goals. Our results show that trout conservation requires knowledge of both temperature and streamflow dynamics at fine spatial and temporal scales. Managers need easy-to-use tools to simulate the expected impacts and the management options to address them, and the methods and results we provide could provide key information in developing these tools and management options.

Appendix A: Table of symbols and abbreviations (alphabetically ordered)

Symbol/abbreviation	Meaning
a.c.	agglomerative coefficient
AEMET	Spanish Meteorological Agency
AIC	Akaike information criterion
α	maximum stream temperature
AR5	Fifth Assessment Report of the IPCC
β	air temperature at which the rate of change of the stream temperature with respect to the air temperature is a maximum
BIC	Bayesian information criterion
CMIP5	Fifth Coupled Model Intercomparison Project
DAT	total days above the threshold per year
δ	maximum rate at τ.
ΔTa	1-day variation in the daily mean air temperature
$\Delta T_{\max}$	increments of the annual averages of the daily maximum stream temperature
ΔT_{mean}	increments of the annual averages of the daily mean stream temperature
$\Delta T_{\min}$	increments of the annual averages of the daily minimum stream temperature
DEM	digital elevation model
DMAT	daily mean air temperature
DMST	daily mean steam temperature
ECDF	empirical cumulative distribution functions
γ	rate of change at β
H-2050	year horizon 2050
H-2099	year horizon 2099
IGN	National Geographic Institute of the Spanish Government
IPCC	Intergovernmental Panel on Climate Change
λ	represents the resistance of DMST to change with respect to the 1-day variation in DMAT (ΔT_a)
M5	machine learning technique
MCDAT	maximum consecutive days above the threshold per year
μ	minimum stream temperature
NSE	Nash–Sutcliffe efficiency
ω	maximum observable variation in stream temperature due to the flow difference
PACF	partial autocorrelation function
Q_{10}	10th percentile of flow
$Q_{\max}$	maximum flow
Q_{mean}	mean flow
$Q_{\min}$	minimum flow
RCP4.5	Representative Concentration Pathway 4.5
RCP8.5	Representative Concentration Pathway 8.5
RSE	residual standard error
T_a	air temperature
TAT ≥ 7	frequency of events of seven or more consecutive days above the threshold per year
τ	low value at which the rate of change of the stream temperature with respect to the flow is a maximum
T_s	stream temperature

Appendix B: Uncertainty

In science, and particularly in hydrological studies, the uncertainty is a matter which requires special attention, and it has led us to include a synthesis of our approach to this problem in this appendix. The uncertainty analysis is a necessary step in assessing the risk level in the applicability of a model (Pappenberger and Beven, 2006). Our aim was to study the viability of the brown trout populations, and we modelled the flow and the stream temperature for this purpose. From a conceptual point of view our approaches were consistent with it.

On data inputs to build the models, uncertainties and inconsistencies are a habitual issue (Juston et al., 2013). The meteorological and hydrological services subject data to their own quality controls but systematic error cannot always be completely controlled (Beven and Westerberg, 2011; McMillan et al., 2012). For this reason, in addition, we tested the input data seeking inconsistencies.

The modelling of the river reaches as one-dimensional elements implies a simplification of the fluvial ecosystem that is generally accepted at this scale (e.g. Viganò et al., 2015; Ahmed and Tsanis, 2016), especially for ecological purposes (e.g. Caiola et al., 2014). Nevertheless, the size of the rivers under study made little or nothing relevant the variations in width and depth (it was verified in the field).

Regarding the parameterization of the models, cross-validation was used to evaluate the uncertainty in these process, and indicators such as the NSE (for hydrological and thermal models), the deviance and the RSE (for thermal models) were calculated. In the case of the thermal model, the functions of distribution of the parameters of the model were built by non-parametric bootstrap, and the mean values were chosen as the most proficient estimators. As results show, parameters tell us about the functional behaviour of catchments (particularly on the effects of the catchments geology on the streams temperature) and this might improve predictions in ungauged basins by better controlling uncertainty (Juston et al., 2013).

Once the models were constructed, It was verified that the overlaps of the ranges of the model input variables and the ranges of the outputs were significant ($p<0.05$). The non-overlapping zones affected, on the one hand, infrequent events (great floods) and the extreme temperature zone (zone of extrapolation), being the last one the scope in which we expected to work. However, the weakness of the hydrological model in the flood zone should be considered for other applications and developments of the model. The hydrological model given us sufficient and relevant information since its possible weak points (extrapolation in the floods assessment) did not affect our goal: we focused on central trends and minimum values, and they were solidly represented in the samples and in the simulations.

As said, the inherent uncertainty of the climate predictions according to the scenarios RCP4.5 and RCP8.5 was attenuated by means of the ensemble technique, showing the dispersion of the results by mean of the percentiles in Figs. S1 to S24 (in the Supplement). Beven (2011) exposed his legitimate concerns on the credibility of climate models which fail when are compared with the control period and, consequently, we used ERA-40 reanalysis to control this source of bias with excellent results.

There was a time dependence between the errors of the model and the scope of the prediction, but these errors were only important in the zone of high temperature and low flow, as expected by the physical nature of the climatic variables. Moreover, this is the behaviour of the variables that was our intention to evaluate.

Competing interests. The authors declare that they have no conflict of interest.

Acknowledgements. We are grateful to the Consejería de Medio Ambiente y O.T. of the Government of Castilla y León, especially to Mariano Anchuelo, Fabián Mateo and the forest ranger team of Navafría. Also, we are indebted to the Sierra of Guadarrama National Park staff, and especially with Juan Bielva and Ángel Rubio for the temperature data of Lozoya stream. Juan Diego Alcaraz is the author of the temperature data of the Ebrón Basin. Valérie Ouellet and two anonymous referees provided valuable comments that substantially improved the original manuscript. The World Climate Research Programme's Working Group on Coupled Modelling is responsible for the 5th Coupled Model Intercomparison Project, and we thank the climate modelling groups for producing and making available their model output. Climate change study was partially funded by the Ministerio de Agricultura, Alimentación y Medio Ambiente of Spain through the Fundación para la Investigación del Clima (http://www.ficlima.org/); by the European Union, DG for Environment (DURERO project. C1.3913442); by the Ministerio de Economía y Competitividad of Spain (IMPADAPT project CGL2013-48424-C2-1-R); and by FEDER funds of the European Commission.

Edited by: Jan Seibert

References

Ahmed, S. and Tsanis, I.: Hydrologic and Hydraulic Impact of Climate Change on Lake Ontario Tributary, Am. J. Water Resour., 4, 1–15, https://doi.org/10.12691/ajwr-4-1-1, 2016.

Allen, K. R.: Comparison of the Growth Rate of Brown

Trout (Salmo trutta) in a New Zealand Stream with Experimental Fish in Britain, J. Anim. Ecol., 54, 487–495, https://doi.org/10.2307/4493, 1985.

Almodóvar, A., Nicola, G. G., Ayllón, D., and Elvira, B.: Global warming threatens the persistence of Mediterranean brown trout, Glob. Change Biol., 18, 1549–1560, https://doi.org/10.1111/j.1365-2486.2011.02608.x, 2011.

Angilletta Jr., M. J.: Thermal Adaptation: A Theoretical and Empirical Synthesis, Oxford University Press, New York, USA, 2009.

Arismendi, I., Safeeq, M., Dunham, J. B., and Johnson, S. L.: Can air temperature be used to project influences of climate change on stream temperature?, Environ. Res. Lett., 9, 084015, https://doi.org/10.1088/1748-9326/9/8/084015, 2014.

Ayllón, D., Railsback, S. F., Vincenzi, S., Groeneveld, J., Almodóvar, A., and Grimm, V.: InSTREAM-Gen: Modelling eco-evolutionary dynamics of trout populations under anthropogenic environmental change, Ecol. Model., 326, 36–53, https://doi.org/10.1016/j.ecolmodel.2015.07.026, 2016.

Beer, W. N. and Anderson, J. J.: Sensitivity of salmonid freshwater life history in western US streams to future climate conditions, Glob. Change Biol., 19, 2547–2556, https://doi.org/10.1111/gcb.12242, 2013.

Bennett, N. D., Croke, B. F. W., Guariso, G., Guillaume, J. H. A., Hamilton, S. H., Jakeman, A. J., Marsili-Libelli, S., Newham, L. T. H., Norton, J. P., Perrin, C., Pierce, S. A., Robson, B., Seppelt, R., Voinov, A. A., Fath, B. D., and Andreassian, V.: Characterising performance of environmental models, Environ. Modell. Softw., 40, 1–20, https://doi.org/10.1016/j.envsoft.2012.09.011, 2013.

Beven, K.: I believe in climate change but how precautionary do we need to be in planning for the future?, Hydrol. Process., 25, 1517–1520, https://doi.org/10.1002/hyp.7939, 2011.

Beven, K. and Westerberg, I.: On red herrings and real herrings: disinformation and information in hydrological inference, Hydrol. Process., 25, 1676–1680, https://doi.org/10.1002/hyp.7963, 2011.

Bogan, T., Mohseni, O., and Stefan, H. G.: Stream temperature-equilibrium temperature relationship, Water Resour. Res., 39, 1245, https://doi.org/10.1029/2003WR002034, 2003.

Borra, S. and Di Ciaccio, A.: Measuring the prediction error. A comparison of cross-validation, bootstrap and covariance penalty methods, Comput. Stat. Data An., 54, 2976–2989, https://doi.org/10.1016/j.csda.2010.03.004, 2010.

Breiman, L.: Random forests, Mach. Learn., 45, 5–32, https://doi.org/10.1023/A:1010933404324, 2001.

Brewitt, K. S. and Danner, E. M.: Spatio-temporal temperature variation influences juvenile steelhead (*Oncorhynchus mykiss*) use of thermal refuges, Ecosphere, 5, 92, https://doi.org/10.1890/ES14-00036.1, 2014.

Bustillo, V., Moatar, F., Ducharne, A., Thiéry, D., and Poirel, A.: A multimodel comparison for assessing water temperatures under changing climate conditions via the equilibrium temperature concept: case study of the Middle Loire River, France, Hydrol. Process., 28, 1507–1524, https://doi.org/10.1002/hyp.9683, 2013.

Caiola, N., Ibáñez, C., Verdú, J., and Munné, A.: Effects of flow regulation on the establishment of alien fish species: A community structure approach to biological validation of environmental flows, Ecol. Indic., 45, 598–604, https://doi.org/10.1016/j.ecolind.2014.05.012, 2014.

Caissie, D.: The thermal regime of rivers: a review, Freshwater Biol., 51, 1389–1406, https://doi.org/10.1111/j.1365-2427.2006.01597.x, 2006.

Ceballos-Barbancho, A., Morán-Tejeda, E., Luego-Ugidos, M. A., and Llorente-Pinto, J. M.: Water resources and environmental change in a Mediterranean environment: The south-west sector of the Duero river basin (Spain), J. Hydrol., 351, 126–138, https://doi.org/10.1016/j.jhydrol.2007.12.004, 2008.

Chen, D., Hu, M., Guo, Y., and Dahlgren, R.A.: Changes in river water temperature between 1980–2012 in Yongan watershed, eastern China: magnitude, drivers and models, J. Hydrol., 533, 191–199, https://doi.org/10.1016/j.jhydrol.2015.12.005, 2016.

Chessman, B. C.: Climatic changes and 13-year trends in stream macroinvertebrate assemblages in New South Wales, Australia, Glob. Change Biol., 15, 2791–2802, https://doi.org/10.1111/j.1365-2486.2008.01840.x, 2009.

Chilton, J.: Groundwater, in: Water quality assessments: A guide to the use of biota, sediments and water in environmental monitoring, 2nd ed., edited by: Chapman, D. E. and Spon, F. N., London, UK, 413–510, 1996.

Colchen, T., Teletchea, F., Fontaine, P., and Pasquet, A.: Temperature modifies activity, inter-individual relationships and group structure in fish, Curr. Zool., 63, 175–183, https://doi.org/10.1093/cz/zow048, 2017.

Comte, L., Buisson, L., Daufresne, M., and Grenouillet, G.: Climate-induced changes in the distribution of freshwater fish: observed and predicted trends, Freshwater Biol., 58, 625–639, https://doi.org/10.1111/fwb.12081, 2013.

Daigle, A., Jeong, D. I., and Lapointe, M. F.: Climate change and resilience of tributary thermal refugia for salmonids in eastern Canadian rivers, Hydrolog. Sci. J., 60, 1044–1063, https://doi.org/10.1080/02626667.2014.898121, 2014.

De'ath, G. and Fabricius, K. E.: Classification and regression trees: a powerful yet simple technique for ecological data analysis, Ecology, 81, 3178–3192, https://doi.org/10.2307/177409, 2000.

DeWeber, J. T. and Wagner, T.: Predicting Brook trout occurrence in stream reaches throughout their native range in the Eastern United States, T. Am. Fish. Soc., 144, 11–24, https://doi.org/10.1080/00028487.2014.963256, 2015.

Eby, L. A., Helmy, O., Holsinger, L. M., and Young, M. K.: Evidence of climate-induced range contractions in bull trout *Salvelinus confluentus* in a Rocky Mountain watershed, U.S.A., PLoS ONE, 9, e98812, https://doi.org/10.1371/journal.pone.0098812, 2014.

Edinger, J. E., Duttweiler, D. W., and Geyer, J. C.: The response of water temperatures to meteorological conditions, Water Resour. Res., 4, 1137–1143, https://doi.org/10.1029/WR004i005p01137, 1968.

Elith, J. and Leathwick, J. R.: Species distribution models: Ecological explanation and prediction across space and time, Annu. Rev. Ecol. Evol. S., 40, 677–697, https://doi.org/10.1146/annurev.ecolsys.110308.120159, 2009.

Elliott, J. M.: Some aspects of thermal stress on freshwater teleosts, in: Stress and Fish, edited by: Pickering, A. D., Academic Press, London, UK, 209–245, 1981.

Elliott, J. M.: Pools as refugia for brown trout during two summer droughts: trout responses to thermal and oxygen stress, J. Fish Biol., 56, 938–948, https://doi.org/10.1111/j.1095-8649.2000.tb00883.x, 2000.

Elliott, J. M. and Allonby, J. D.: An experimental study of ontogenetic and seasonal changes in the temperature preferences of unfed and fed brown trout, *Salmo trutta*, Freshwater Biol., 58, 1840–1848, https://doi.org/10.1111/fwb.12173, 2013.

Elliott, J. M. and Elliott, J. A.: Temperature requirements of Atlantic salmon *Salmo salar*, Brown trout *Salmo trutta* and Arctic charr *Salvelinus alpinus*: predicting the effects of climate change, J. Fish Biol., 77, 1793–1817, https://doi.org/10.1111/j.1095-8649.2010.02762.x, 2010.

Elliott, J. M., Hurley, M. A., and Fryer, J.: A new, improved growth model for brown trout, *Salmo trutta*, Funct. Ecol., 9, 290–298, https://doi.org/10.2307/2390576, 1995.

European Environment Agency: CLC2006 technical guidelines. Technical report No. 17/2007, Publications Office, Luxembourg, https://doi.org/10.2800/12134, 2007.

Fey, S. B. and Herren, C. M.: Temperature-mediated biotic interactions influence enemy release of non-native species in warming environments, Ecology, 95, 2246–2256, https://doi.org/10.1890/13-1799.1, 2014.

Fielding, A. H.: An introduction to machine learning methods. In: Machine Learning Methods for Ecological Applications, edited by: Fielding, A. H., Kluwer, Boston, USA, 1–35, 1999.

Filipe, A. F., Markovic, D., Pletterbauer, F., Tisseuil, C., De Wever, A., Schmutz, S., Bonada, N., and Freyhof, J: Forecasting fish distribution along stream networks: brown trout (*Salmo trutta*) in Europe, Divers. Distrib., 19, 1059–1071, https://doi.org/10.1111/ddi.12086, 2013.

Forseth, T. and Jonsson, B.: The growth and food ration of piscivorous brown trout (Salmo trutta), Funct. Ecol., 8, 171–177, https://doi.org/10.2307/2389900, 1994.

Forseth, T., Larsson, S., Jensen, A. J., Jonsson, B., Näslund, I., and Berglund, I.: Thermal growth performance of juvenile brown trout *Salmo trutta*?: no support for thermal adaptation hypotheses, J. Fish Biol., 74, 133–149, https://doi.org/10.1111/j.1095-8649.2008.02119.x, 2009.

Fukuda, S., De Baets, B., Waegeman, W., Verwaeren, J., and Mouton, A. M.: Habitat prediction and knowledge extraction for spawning European grayling (*Thymallus thymallus* L.) using a broad range of species distribution models, Environ. Modell. Softw., 47, 1–6, https://doi.org/10.1016/j.envsoft.2013.04.005, 2013.

Garner, G., Van Loon, A. F., Prudhomme, C., and Hannah, D. M.: Hydroclimatology of extreme river flows, Freshwater Biol., 60, 2461–2476, https://doi.org/10.1111/fwb.12667, 2015.

Gordon, N. D., McMahon, T. A., Finlayson, B. L., Gippel, C. J., and Nathan, R. J.: Stream hydrology. An introduction for ecologists, 2nd ed., John Wiley and Sons, Chichester, UK, 2004.

Gortázar, J., García de Jalón, D., Alonso-González, C., Vizcaíno, P., Baeza, D., and Marchamalo, M.: Spawning period of a southern brown trout population in a highly unpredictable stream, Ecol. Freshw. Fish, 16, 515–527, https://doi.org/10.1111/j.1600-0633.2007.00246.x, 2007.

Goyer, K., Bertolo, A., Pépino, M., and Magnan, P.: Effects of lake warming on behavioural thermoregulatory tactics in a cold-water stenothermic fish, PLoS ONE, 9, e92514, https://doi.org/10.1371/journal.pone.0092514, 2014.

Grande, M. and Andersen, S.: Critical Thermal Maxima for Young Salmonids, J. Freshwater Ecol., 6, 275–279, https://doi.org/10.1080/02705060.1991.9665304, 1991.

Haines, A. T., Finlayson, B. L., and McMahon, T. A.: A global classification of river regimes, Appl. Geogr., 8, 255–272, https://doi.org/10.1016/0143-6228(88)90035-5, 1988.

Hampe, A. and Petit, R. J.: Conserving biodiversity under climate change: the rear edge mattersm Ecol. Lett., 8, 461–467, https://doi.org/10.1111/j.1461-0248.2005.00739.x, 2005.

Hari, R. E., Livingstone, D. M., Siber, R., Burkhardt-Holm, P., and Guettinger, H.: Consequences of climatic change for water temperature and brown trout populations in Alpine rivers and streams, Glob. Change Biol., 12, 10–26, https://doi.org/10.1111/j.1365-2486.2005.001051.x, 2006.

Hein, C. L., Ohlund, G., and Englund, G.: Fish introductions reveal the temperature dependence of species interactions, P. Roy. Soc. B-Biol. Sci., 281, 20132641, https://doi.org/10.1098/rspb.2013.2641, 2013.

Hettiarachchi, P., Hall, M. J., and Minns, A. W.: The extrapolation of artificial neural networks for the modelling of rainfall–runoff relationships, J. Hydroinform., 7, 291–296, 2005.

IGME: Mapa de Litologías de España 1 : 1 000 000, Madrid, Spain, available at: http://mapas.igme.es/gis/rest/services/Cartografia_Geologica/IGME_Litologias_1M/MapServer (last access: 15 February 2016), 2015.

IPCC: Climate Change 2013. The Physical Science Basis. Working Group I Contribution to the Fifth Assessment Report of the Intergovernmental Panel on Climate Change, edited by: Stocker, T. F., Qin, D., Plattner, G.-K., Tignor, M. M. B., Allen, S. K., Boschung, J., Nauels, A., Xia, Y., Bex, V., and Midgley, P. M., Cambridge University Press, Cambridge, UK, 2013.

Isaak, D. J., Young, M. K., Luce, C. H., Hostetler S. W., Wenger, S. J., Peterson, E. E., Ver Hoef, J. M., Groce, M. C., Horan, D. L., and Nagel, D. E.: Slow climate velocities of mountain streams portend their role as refugia for cold-water biodiversity, P. Natl. Acad. Sci. USA, 113, 4374–4379, https://doi.org/10.1073/pnas.1522429113, 2016.

Jeffries, K. M., Hinch, S. G., Martins, E. G., Clark, T. D., Lotto, A. G., Patterson, D. A., Cooke, S. J., Farrell, A. P., and Miller, K. M.: Sex and proximity to reproductive maturity influence the survival, final maturation, and blood physiology of Pacific salmon when exposed to high temperature during a simulated migration, Physiol. Biochem. Zool., 85, 62–73, https://doi.org/10.1086/663770, 2012.

Jonsson, B. and Jonsson, N.: A review of the likely effects of climate change on anadromous Atlantic salmon *Salmo salar* and brown trout *Salmo trutta*, with particular reference to water temperature and flow, J. Fish Biol., 75, 2381–2447, https://doi.org/10.1111/j.1095-8649.2009.02380.x, 2009.

Junker, J., Heimann, F. U. M., Hauer, C., Turowski, J. M., Rickenmann, D., Zappa, M., and Peter, A.: Assessing the impact of climate change on brown trout (*Salmo trutta* fario) recruitment, Hydrobiologia, 751, 1–21, https://doi.org/10.1007/s10750-014-2073-4, 2015.

Juston, J. M., Kauffeldt, A., Montano, B. Q., Seibert, J., Beven, K. J., and Westerberg, I. K.: Smiling in the rain: Seven reasons to be positive about uncertainty in hydrological modelling, Hydrol. Process., 27, 1117–1122, https://doi.org/10.1002/hyp.9625, 2013.

Kaufman, L. and Rousseeuw, P. J.: Finding Groups in Data: An Introduction to Cluster Analysis. John Wiley & Sons, Hoboken, New Jersey, USA, https://doi.org/10.1002/9780470316801, 2005.

Kaushal, S. S., Likens, G. E., Jaworski, N. A., Pace, M. L., Sides, A. M., Seekell, D., Belt, K. T., Secor, D. H., and Wingate, R.: Rising stream and river temperatures in the United States, Front. Ecol. Environ., 8, 461–466, https://doi.org/10.1890/090037, 2010.

Kittler, J.: Feature set search algorithms, in: Pattern Recognition and Signal Processing, edited by: Chen, C. H., Sijthoff and Noordhoff, Alphen aan den Rijn, the Netherlands, 41–60, 1978.

Kottelat, M. and Freyhof, J.: Handbook of European freshwater fishes, Kottelat, Cornol, Switzerland and Freyhof, Berlin, Germany, 2007.

Kuhn, M., Weston, S., Keefer, C., and Coulter, N.: Cubist: Rule- and Instance-Based Regression Modeling – C code for Cubist by Ross Quinlan, R package, version 0.0.18, CRAN R-Project package, version 0.0.18, CRAN R-Project, available at: https://cran.r-project.org/package=Cubist (last access: 15 August 2015), 2014.

Kurylyk, B. L., Bourque, C. P.-A., and MacQuarrie, K. T. B.: Potential surface temperature and shallow groundwater temperature response to climate change: an example from a small forested catchment in east-central New Brunswick (Canada), Hydrol. Earth Syst. Sci., 17, 2701–2716, https://doi.org/10.5194/hess-17-2701-2013, 2013.

Kurylyk, B. L., MacQuarrie, K. T. B., Caissie, D., and McKenzie, J. M.: Shallow groundwater thermal sensitivity to climate change and land cover disturbances: derivation of analytical expressions and implications for stream temperature modeling, Hydrol. Earth Syst. Sci., 19, 2469–2489, https://doi.org/10.5194/hess-19-2469-2015, 2015.

Lahnsteiner, F. and Leitner, S.: Effect of temperature on gametogenesis and gamete quality in Brown trout, *Salmo trutta*, J. Exp. Zool. Part A, 319, 138–148, https://doi.org/10.1002/jez.1779, 2013.

Larios-López, J. E., Tierno de Figueroa, J. M., Galiana-García, M., Gortázar, J., and Alonso, C.: Extended spawning in brown trout (*Salmo trutta*) populations from the Southern Iberian Peninsula: the role of climate variability, J. Limnol., 74, 394–402, https://doi.org/10.4081/jlimnol.2015.1089, 2015.

Lassalle, G. and Rochard, E.: Impact of twenty-first century climate change on diadromous fish spread over Europe, North Africa and the Middle East, Glob. Change Biol., 15, 1072–1089, https://doi.org/10.1111/j.1365-2486.2008.01794.x, 2009.

Leppi, J. C., DeLuca, T. H., Harrar, S. W.. and Running, S. W.: Impacts of climate change on August stream discharge in the Central-Rocky Mountains, Climatic Change, 112, 997–1014, https://doi.org/10.1007/s10584-011-0235-1, 2012.

Liu, R. and Singh, K.: Moving blocks jackknife and bootstrap capture weak dependence, in: Exploring the Limits of Bootstrap, edited by: LePage, R. and Billard, L., John Wiley and Sons, New York, USA, 225–248, 1992.

Lobón-Cerviá, J. and Rincón, P. A.: Field assessment of the influence of temperature on growth rate in a brown trout population, T. Am. Fish. Soc., 127, 718–728, https://doi.org/10.1577/1548-8659(1998)127<0718:FAOTIO>2.0.CO;2, 1998.

Lobón-Cerviá, J. and Mortensen, E.: Population size in stream-living juveniles of lake-migratory brown trout *Salmo trutta* L.: the importance of stream discharge and temperature, Ecol. Freshw. Fish, 14, 394–401, https://doi.org/10.1111/j.1600-0633.2005.00111.x, 2005.

Lobón-Cerviá, J. and Rincón, P. A.: Environmental determinants of recruitment and their influence on the population dynamics of stream-living brown trout *Salmo trutta*, Oikos, 105, 641–646, https://doi.org/10.1111/j.0030-1299.2004.12989.x, 2004.

Loinaz, M. C., Davidsen, H. K., Butts, M., and Bauer-Gottwein, P.: Integrated flow and temperature modeling at the catchment scale, J. Hydrol., 495, 238–251, https://doi.org/10.1016/j.jhydrol.2013.04.039, 2013.

Lorenzo-Lacruz, J., Vicente-Serrano, S. M., López-Moreno, J. I., Morán-Tejeda, E., and Zabalza, J.: Recent trends in Iberian streamflows (1945–2005), J. Hydrol., 414–415, 463–475, https://doi.org/10.1016/j.jhydrol.2011.11.023, 2012.

Luce, C. H. and Holden, Z. A.: Declining annual streamflow distributions in the Pacific Northwest United States, 1948–2006, Geophys. Res. Lett., 36, L16401, https://doi.org/10.1029/2009GL039407, 2009.

Maechler, M.: Cluster analysis extended, Rousseeuw et al., R package, version 1.14.4, CRAN R-Project, available at: https://cran.r-project.org/package=cluster (last access: 15 August 2015), 2013.

Magnuson, J. J. and Destasio, B. T.: Thermal niche of fishes and global warming, in: Global Warming: Implications for Freshwater and Marine Fish, edited by: Wood, C. M. and McDonald, D. G., Cambridge University Press, Cambridge, UK, 377–407, 1997.

Magnuson, J. J., Crowder, L. B., and Medvick, P. A.: Temperature as an Ecological Resource, Am. Zool., 19, 331–343, https://doi.org/10.1093/icb/19.1.331, 1979.

McCuen, R. H.: Hydrologic analysis and design, 2nd ed., Prentice Hall, New Jersey, USA, 1998.

McMillan, H., Krueger, K., and Freer, J.: Benchmarking observational uncertainties for hydrology: Rainfall, river discharge and water quality, Hydrol. Process., 26, 4078–4111, https://doi.org/10.1002/hyp.9384, 2012.

Meshcheryakova, O. V., Churova, M. V., Veselov, A. E., and Nemova, N. N.: Activities of cytochrome c oxidase and mitochondrial lactate dehydrogenase isozymes and Cox1, Cox2, Cox4, and Cox6 gene subunit expression in cold adaptation of *Salmo trutta* L., Russ. J. Bioorganic Chem., 42, 162–169, https://doi.org/10.1134/S1068162016010106, 2016.

Milly, P. C. D., Dunne, K. A., and Vecchia, A. V.: Global pattern of trends in streamflow and water availability in a changing climate, Nature, 438, 347–350, https://doi.org/10.1038/nature04312, 2005.

Mohseni, O. and Stefan, H.: Stream temperature/air temperature relationship: a physical interpretation, J. Hydrol., 218, 128–141, https://doi.org/10.1016/S0022-1694(99)00034-7, 1999.

Mohseni, O., Stefan, H. G., and Eriksson, T. R.: A nonlinear regression model for weekly stream temperatures, Water Resour. Res., 34, 2685–2692, https://doi.org/10.1029/98WR01877, 1998.

Monjo, R., Caselles, V., and Chust, G.: Probabilistic correction of RCM precipitation in the Basque Country (Northern Spain), Theor. Appl. Climatol., 117, 317–329, https://doi.org/10.1007/s00704-013-1008-8, 2014.

Morán-Tejeda, E., Lorenzo-Lacruz, J., López-Moreno, J. I., Rahman, K., and Beniston, M.: Streamflow timing of mountain rivers in Spain: Recent changes and future projections, J. Hydrol., 517, 1114–1127, https://doi.org/10.1016/j.jhydrol.2014.06.053, 2014.

Muñoz-Mas, R., López-Nicolás, A., Martínez-Capel, F., and Pulido-Velázquez, M.: Shifts in the suitable habitat available for brown trout (*Salmo trutta* L.) under short-term climate change scenarios, Sci. Total Environ., 544, 686–700, https://doi.org/10.1016/j.scitotenv.2015.11.147, 2016.

Nakicenovic, N., Alcamo, J., Davis, G., de Vries, H. J. M., Fenhann, J., Gaffin, S., Gregory, K., Grubler, A., Jung, T. Y., Kram, T., La Rovere, E. L., Michaelis, L., Mori, S., Morita, T., Papper, W., Pitcher, H., Price, L., Riahi, K., Roehrl, A., Rogner, H.-H., Sankovski, A., Schlesinger, M., Shukla, P., Smith, S., Swart, R., van Rooijen, S., Victor, N., and Dadi, Z.: Emissions Scenarios: A Special Report of Working Group III of the Intergovernmental Panel on Climate Change, Cambridge University Press, Cambridge, UK, 2000.

Nash, J. E. and Sutcliffe, J. V.: River flow forecasting through conceptual models part I – A discussion of principles, J. Hydrol., 10, 282–290, https://doi.org/10.1016/0022-1694(70)90255-6, 1970.

Neumann, D. W., Rajagopalan, B., and Zagona, E. A.: Regression model for daily maximum stream temperature, J. Environ. Eng., 129, 667–674, https://doi.org/10.1061/(ASCE)0733-9372(2003)129:7(667), 2003.

Ojanguren, A. F., Reyes-Gavilán, F. G. and Braña, F.: Thermal sensitivity of growth, food intake and activity of juvenile brown trout, J. Therm. Biol., 26, 165–170, https://doi.org/10.1016/S0306-4565(00)00038-3, 2001.

Orr, H. G., Simpson, G. L., des Clers, S., Watts, G., Hughes, M., Hannaford, J., Dunbar, M. J., Laizé, C. L. R., Wilby, R. L., Battarbee, R. W., and Evans, R.: Detecting changing river temperatures in England and Wales, Hydrol. Process., 29, 752–766, https://doi.org/10.1002/hyp.10181, 2015.

Pappenberger, F. and Beven, K. J.: Ignorance is bliss: 7 reasons not to use uncertainty analysis, Water Resour. Res., 42, W05302, https://doi.org/10.1029/2005WR004820, 2006.

Pépino, M., Goyer, K., and Magnan, P.: Heat transfer in fish: are short excursions between habitats a thermoregulatory behaviour to exploit resources in an unfavourable thermal environment?, J. Exp. Biol., 218, 3461–3467, https://doi.org/10.1242/jeb.126466, 2015.

Piccolroaz, S., Calamita, E., Majone, B., Gallice, A., Siviglia, A., and Toffolon, M.: Prediction of river water temperature: a comparison between a new family of hybrid models and statistical approaches: Prediction of River Water Temperature, Hydrol. Process., 30, 3901–3917, https://doi.org/10.1002/hyp.10913, 2016

Pohlert, T.: Non-parametric trend tests and change-point detection, R package, version 0.1.0, CRAN R-Project, available at: https://cran.r-project.org/package=trend, last access: 10 June 2016.

Pourmokhtarian, A., Driscoll, C. T., Campbell, J. L., Hayhoe, K., and Stoner, A. M.: The effects of climate downscaling technique and observational dataset on modeled ecological responses, Ecol. Appl., 26, 1321–1337, https://doi.org/10.1890/15-0745, 2016.

Quinlan, J. R.: Learning with continuous classes, in: The 5th Australian Joint Conference on Artificial Intelligence, 16–18 November 1992, Hobart, Australia, 343–348, 1992.

Quinlan, J. R.: An overview of Cubist, available at: https://www.rulequest.com/cubist-win.html, last access: 8 June 2017.

R Core Team: R: A language and environment for statistical computing, R Foundation for Statistical Computing, Vienna, Austria, available at: http://www.R-project.org (last access: 20 May 2016), 2015.

Reynolds, W. W. and Casterlin, M. E.: Thermoregulatory behavior of brown trout, *Salmo trutta*, Hydrobiologia, 62, 79–80, https://doi.org/10.1007/BF00012567, 1979.

Ribalaygua, J., Torres, L., Pórtoles, J., Monjo, R., Gaitán, E., and Pino, M. R.: Description and validation of a two-step analogue/regression downscaling method, Theor. Appl. Climatol., 114, 253–269, https://doi.org/10.1007/s00704-013-0836-x, 2013.

Rojas, R., Feyen, L., Bianchi, A., and Dosio, A.: Assessment of future flood hazard in Europe using a large ensemble of bias-corrected regional climate simulations, J. Geophys. Res.-Atmos., 117, D17109, https://doi.org/10.1029/2012JD017461, 2012.

Ruiz-Navarro, A., Gillingham, P. K., and Britton, J. R.: Predicting shifts in the climate space of freshwater fishes in Great Britain due to climate change, Biol. Conserv., 203, 33–42, https://doi.org/10.1016/j.biocon.2016.08.021, 2016.

Sánchez-Hernández, J. and Nunn, A.D.: Environmental changes in a Mediterranean river: implications for the fish assemblage, Ecohydrology, 9, 1439–1451, https://doi.org/10.1002/eco.1737, 2016.

Santiago, J. M.: Thermal ecology of Brown trout and the climate change challenge, in: Tilapia and Trout: Harvesting, Prevalence and Benefits, edited by: Richardson, B., Nova Science Publishers, New York, USA, 79–119, 2017.

Santiago, J. M., García de Jalón, D., Alonso, C., and Solana, J.: Comportamiento térmico de dos tramos fluviales de cabecera del sistema central: impacto del embalse de Torrecaballeros (Segovia), in: III Jornadas del Ingeniería del Agua, Valencia, Vol. 1, Marcombo, Barcelona, Spain, 153–160, 2013.

Santiago, J. M., García de Jalón, D., Alonso, C., Solana, J., Ribalaygua, J., Pórtoles, J., and Monjo, R.: Brown trout thermal niche and climate change: expected changes in the distribution of cold-water fish in central Spain, Ecohydrology, 9, 514–528, https://doi.org/10.1002/eco.1653, 2016.

Santiago, J. M., Alonso, C., and García de Jalón, D.: Daily mean stream temperatures in Central Spain, PANGAEA, https://doi.org/10.1594/PANGAEA.879494, 2017.

Schoups, G., van de Giesen, N. C., and Savenije, H. H. G.: Model complexity control for hydrologic prediction, Water Resour. Res., 44, W00B03, https://doi.org/10.1029/2008WR006836, 2008.

Shortridge, J. E., Guikema, S. D., and Zaitchik, B. F.: Machine learning methods for empirical streamflow simulation: a comparison of model accuracy, interpretability, and uncertainty in seasonal watersheds, Hydrol. Earth Syst. Sci., 20, 2611–2628, https://doi.org/10.5194/hess-20-2611-2016, 2016.

Snyder, C. D., Hitt, N. P., and Young, J. A.: Accounting for groundwater in stream fish thermal habitat responses to climate change, Ecol. Appl., 25, 1397–1419, https://doi.org/10.1890/14-1354.1, 2015.

Solomatine, D. P. and Dulal, K. N.: Model trees as an alternative to neural networks in rainfall – runoff modelling, Hydrolog. Sci. J., 48, 399–411, https://doi.org/10.1623/hysj.48.3.399.45291, 2003.

Stamp, J., Hamilton, A., Craddock, M., Parker, L., Roy, A. H., Isaak, D. J., Holden, Z., Passmore, M., and Bierwagen, B. G.: Best practices for continuous monitoring of temperature and flow in wadeable streams. EPA/600/R-13/170F, U.S. Environmental Protection Agency, Washington, DC, USA, 2014.

Stewart, J. S., Westenbroek, S. M., Mitro, M. G., Lyons, J. D., Kammel, L. E., and Buchwald, C. A.: A model for evaluating stream temperature response to climate change in Wisconsin, Reston, Virginia, USA, U.S. Geological Survey Scientific Investigations Report 2014–5186, https://doi.org/10.3133/sir20145186, 2015.

Taghi Sattari, M., Pal, M., Apaydin, H., and Ozturk, F.: M5 model tree application in daily river flow forecasting in Sohu Stream, Turkey, Water Resour., 40, 233–242, https://doi.org/10.1134/S0097807813030123, 2013.

Taylor, K. E., Stouffer, R. J., and Meehl, G. A.: A summary of the CMIP5 experiment design, available at: http:

//cmip-pcmdi.llnl.gov/cmip5/docs/Taylor_CMIP5_design.pdf (last access: 1 February 2016), 2009.

Thodsen, H.: The influence of climate change on stream flow in Danish rivers, J. Hydrol., 333, 226–238, https://doi.org/10.1016/j.jhydrol.2006.08.012, 2007.

Thuiller, W., Lavorel, S., Sykes, M. T., and Araujo, M. B.: Using niche-based modelling to assess the impact of climate change on tree functional diversity in Europe, Divers. Distrib., 12, 49–60, https://doi.org/10.1111/j.1366-9516.2006.00216.x, 2006.

Uppala, S. M., KÅllberg, P. W., Simmons, A. J., Andrae, U., Bechtold, V. D. C., Fiorino, M., Gibson, J. K., Haseler, J., Hernandez, A., Kelly, G. A., Li, X., Onogi, K., Saarinen, S., Sokka, N., Allan, R. P., Andersson, E., Arpe, K., Balmaseda, M. A., Beljaars, A. C. M., Berg, L. V. D., Bidlot, J., Bormann, N., Caires, S., Chevallier, F., Dethof, A., Dragosavac, M., Fisher, M., Fuentes, M., Hagemann, S., Hólm, E., Hoskins, B. J., Isaksen, L., Janssen, P. A. E. M., Jenne, R., Mcnally, A. P., Mahfouf, J.-F., Morcrette, J.-J., Rayner, N. A., Saunders, R. W., Simon, P., Sterl, A., Trenberth, K. E., Untch, A., Vasiljevic, D., Viterbo, P., and Woollen, J.: The ERA-40 re-analysis, Q. J. Roy. Meteor. Soc., 131, 2961–3012, https://doi.org/10.1256/qj.04.176, 2005.

van Vliet, M. T. H., Ludwig, F., Zwolsman, J. J. G., Weedon, G. P., and Kabat, P.: Global river temperatures and sensitivity to atmospheric warming and changes in river flow, Water Resour. Res., 47, W02544, https://doi.org/10.1029/2010WR009198, 2011.

van Vliet, M. T. H., Yearsley, J. R., Franssen, W. H. P., Ludwig, F., Haddeland, I., Lettenmaier, D. P., and Kabat, P.: Coupled daily streamflow and water temperature modelling in large river basins, Hydrol. Earth Syst. Sci., 16, 4303–4321, https://doi.org/10.5194/hess-16-4303-2012, 2012.

van Vliet, M. T. H., Franssen, W. H. P., Yearsley, J. R., Ludwig, F., Haddeland, I., Lettenmaier, D. P., and Kabat, P.: Global river discharge and water temperature under climate change, Glob. Environ. Change, 23, 450–464, https://doi.org/10.1016/j.gloenvcha.2012.11.002, 2013.

Verberk, W. C. E. P., Durance, I., Vaughan, I. P., and Ormerod, S. J.: Field and laboratory studies reveal interacting effects of stream oxygenation and warming on aquatic ectotherms, Glob. Change Biol., 22, 1769–1778, https://doi.org/10.1111/gcb.13240, 2016.

Viganò, G., Confortola, G., Fornaroli, R., Cabrini, R., Canobbio, S., Mezzanotte, V., and Bocchiola, D.: Effects of future climate change on a river habitat in an Italian alpine catchment, J. Hydrol. Eng., 21, 04015063, https://doi.org/10.1061/(ASCE)HE.1943-5584.0001293, 2015.

Vornanen, M., Haverinen, J., and Egginton, S.: Acute heat tolerance of cardiac excitation in the brown trout (*Salmo trutta* fario), J. Exp. Biol., 217, 299–309, https://doi.org/10.1242/jeb.091272, 2014.

Warren, D. R., Robinson, J. M., Josephson, D. C., Sheldon, D. R., and Kraft, C. E.: Elevated summer temperatures delay spawning and reduce redd construction for resident brook trout (*Salvelinus fontinalis*), Glob. Change Biol., 18, 1804–1811, https://doi.org/10.1111/j.1365-2486.2012.02670.x, 2012.

Webb, B. W., Hannah, D. M., Moore, R. D., Brown, L. E., and Nobilis, F.: Recent advances in stream and river temperature research, Hydrol. Process., 22, 902–918, https://doi.org/10.1002/hyp.6994, 2008.

Wenger, S. J. and Olden, J. D.: Assessing transferability of ecological models: an underappreciated aspect of statistical validation, Methods Ecol. Evol., 3, 260–267, https://doi.org/10.1111/j.2041-210X.2011.00170.x, 2012.

Wenger, S. J., Isaak, D. J., Luce, C. H., Neville, H. M., Fausch, K. D., Dunham, J. B., Dauwalter, D. C., Young, M. K., Elsner, M. M., Rieman, B. E., Hamlet, A. F., and Williams, J. E.: Flow regime, temperature, and biotic interactions drive differential declines of trout species under climate change, P. Natl. Acad. Sci. USA, 108, 14175–14180, https://doi.org/10.1073/pnas.1103097108, 2011.

White, C. R., Alton, L. A., and Frappell, P. B.: Metabolic cold adaptation in fishes occurs at the level of whole animal, mitochondria and enzyme, P. Roy. Soc. B-Biol. Sci., 279, 1740–1747, https://doi.org/10.1098/rspb.2011.2060, 2012.

Williams, J. E., Isaak, D. J., Imhof, J., Hendrickson, D. A., and McMillan, J. R.: Cold-water fishes and climate change in North America, in: Reference Module in Earth Systems and Environmental Sciences 2015, https://doi.org/10.1016/B978-0-12-409548-9.09505-1, 2015.

Zhuo, L., Dai, Q., and Han, D.: Meta-analysis of flow modeling performances – to build a matching system between catchment complexity and model types, Hydrol. Process., 29, 2463–2477, https://doi.org/10.1002/hyp.10371, 2015.

Zorita, E. and Von Storch, H.: The analog method as a simple statistical downscaling technique: comparison with more complicated methods, J. Climate, 12, 2474–2489, https://doi.org/10.1175/1520-0442(1999)012<2474:TAMAAS>2.0.CO;2, 1999.

11

A thermodynamic formulation of root water uptake

Anke Hildebrandt[1,2], **Axel Kleidon**[2], **and Marcel Bechmann**[1,2]

[1]Friedrich-Schiller-Universität, Jena, Germany
[2]Max-Planck-Institut für Biogeochemie, Jena, Germany

Correspondence to: Anke Hildebrandt (anke.hildebrandt@uni-jena.de)

Abstract. By extracting bound water from the soil and lifting it to the canopy, root systems of vegetation perform work. Here we describe how root water uptake can be evaluated thermodynamically and demonstrate that this evaluation provides additional insights into the factors that impede root water uptake. We derive an expression that relates the energy export at the base of the root system to a sum of terms that reflect all fluxes and storage changes along the flow path in thermodynamic terms. We illustrate this thermodynamic formulation using an idealized setup of scenarios with a simple model. In these scenarios, we demonstrate why heterogeneity in soil water distribution and rooting properties affect the impediment of water flow even though the mean soil water content and rooting properties are the same across the scenarios. The effects of heterogeneity can clearly be identified in the thermodynamics of the system in terms of differences in dissipative losses and hydraulic energy, resulting in an earlier start of water limitation in the drying cycle. We conclude that this thermodynamic evaluation of root water uptake conveniently provides insights into the impediments of different processes along the entire flow path, which goes beyond resistances and also accounts for the role of heterogeneity in soil water distribution.

1 Introduction

Root water uptake is an important process, determining the transport of water between soil and atmosphere and influencing plant productivity and crop yield. A wealth of studies using both models and observations deal therefore with understanding root water uptake, that is, to learn where plants take up water (Doussan et al., 2006; Javaux et al., 2008; Schneider et al., 2010), how root length and hydraulic properties affect uptake (Zwieniecki et al., 2003; Bechmann et al., 2014), how plant communities exploit heterogeneously distributed soil water (Lhomme, 1998; Couvreur et al., 2012; Guswa, 2012), how to identify efficient rooting depth (Guswa, 2010), how soil water storage is shared between plants (Ivanov et al., 2012; Hildebrandt and Eltahir, 2007), how plants may optimize water flow in order to prevent cavitation (Sperry et al., 1998; Johnson et al., 2014), and about relations between root water uptake and stomatal control (Tuzet et al., 2003; Janott et al., 2011), as well as crop yield (Hammer et al., 2009).

In order to evaluate the efficiency of root water uptake and learning how plants may regulate it, we require some understanding of the impediment for water flow and how it is distributed along the soil–plant–atmosphere continuum, especially whether it lies within the plant or the soil compartment (Draye et al., 2010; Vadez et al., 2014). Much of our process understanding on the spatial distribution of water uptake and its evolution in drying soil is based on physically based models of the root system (Dunbabin et al., 2013). Relying on the electrical analogue of water flow and mass balance (van den Honert, 1948; Lhomme, 1998), they mimic the flow of water over a chain of resistances along continuously dropping water potentials from the soil to the root, further up within the root xylem, and sometimes up the canopy (Janott et al., 2011). At the same time, root water uptake depletes the soil reservoir, leading to more negative soil hydraulic potentials which need to be overcome in order to maintain the necessary gradient between soil and atmosphere to allow for flow. Both processes (flow over a resistance network and increasing soil water retention) impede transpiration, but comparing their mutual contribution in form of resistances is not suitable, since the change of soil water retention per water removed has no proper resistance analogue.

In this paper we show that additional information about the system can be obtained from a thermodynamic perspective, specifically by combining the hydraulic potentials with mass fluxes, yielding fluxes of energy. This approach has the advantage that different processes, such as the change of soil water potential with decreasing soil water content as well as the transport of water over a resistance, can be expressed in the same currency of energy fluxes and dissipation, with units of joules per second ($J s^{-1}$).

While thermodynamics is most commonly associated with heat, its formulation is much more general and can be used to express the constraints and directions of energy conversions of any form (Kondepudi and Prigogine, 1998; Kleidon, 2012). As soil water movement and uptake by plants involve changes in binding and gravitational energy, as expressed by the respective matric and gravitational potentials, the fluxes of water in the soil–vegetation–atmosphere system are associated with fluxes of energy, and we can compare which one of the processes in the uptake chain requires most energy, as well as quantifying the total energy expense of the uptake. Thus, the thermodynamic perspective allows us to evaluate the efficiency of different temporal dynamics of root water uptake and differentiate more efficient from less efficient root systems.

As will be shown in this paper, the thermodynamic formulations are comparatively simple and straightforward to implement in models. Since the hydraulic potential is just the specific energy per mass (or volume), that is, the derivative of the Gibbs free energy to mass (or volume), the related soil energy content can be obtained by integration. The thermodynamic representation has, however, several advantages that are currently not well explored by the hydrological community. One of these advantages is, for example, related to describing the effects of soil heterogeneity. While soil water potential is an intensive property (i.e., a property that does not depend on the size of the system) that cannot meaningfully be averaged, the associated energy content is an extensive property (i.e., a property that depends on the size of the system) and therefore is additive, and the total energy content in heterogenous soil can be calculated. As will be shown, the total energy content offers insights into the role of soil heterogeneity that cannot be derived when focussing only on the potential or the soil water content alone.

In the following, we will derive formulations for the energy contained in unsaturated soil as well as for the dissipation of energy for fluxes in unsaturated soil and along the root system. In order to illustrate how these fluxes can be interpreted to evaluate impediments to root water uptake and the role of soil water heterogeneity, we illustrate them in a simplified process model, which is a conceptual four-box model for root water uptake.

2 Thermodynamics and soil hydrology

2.1 Thermodynamic background

Thermodynamics is a general theory of physics that describes the rules for energy conversions. The first law of thermodynamics ensures energy conservation and formulates that the internal changes in energy are balanced with external additions or removals and internal conversions between different forms. The second law describes that with every conversion of energy, energy is increasingly dispersed, which is described by entropy as a physical quantity. It is the second law that sets the natural direction of processes to deplete their driving gradients, and that is, for instance, reflected in soil water movement depleting gradients in soil water potential. The state of thermodynamic equilibrium is then described as a state of maximum entropy and represents a state in which no driving gradients are present within the system.

To describe soil water movement in thermodynamic terms, it needs to be formulated in terms of the energies involved, and it needs to be associated with entropy. The energies involved consist of the binding energies associated with capillary and adhesive forces, gravitational energy, and heat. The first two forms of energy are directly relevant to soil water movement. Their formulation in energetic terms is straightforward as these are directly related to the matric and gravitational potentials. These potentials are formally defined as chemical potentials (Edlefsen and Anderson, 1943; Kondepudi and Prigogine, 1998), i.e., defined as the change in Gibbs free energy resulting from an incremental change in mass.

The use of heat is important as it is required to ensure energy conservation within the soil when the other forms of energy change, and because heat is directly linked to the entropy of the system. When water is redistributed within the soil due to gradients in soil water potential, this results in a reduction of the binding and gravitational energy, with the reduced energy being released as heat of immersion (see also below). The state of thermodynamic equilibrium is reached when there is no gradient in soil water potential. This state corresponds to a state of minimum Gibbs free energy; i.e., the binding and gravitational energy is minimized for a given amount of stored water. As the remaining energy is converted into heat, this reduction to a minimum of Gibbs free energy corresponds to a maximum conversion into heat and thus a maximization of entropy that can be achieved by soil water redistribution. This is despite the fact that the actual amounts of heat involved are rather small compared to the heat fluxes involved in heat diffusion in the soil.

Next, we describe how these forms of energies are determined quantitatively from their respective potentials and how these forms of energy change during root water uptake and soil water redistribution. We state equations for discrete bulk soil compartments $V_{s,i}$ of sufficiently large size such that the porosity and the soil water retention curve are defined at that scale. Soil properties are considered homogenous within but may vary between soil compartments. We thus use sums to integrate over the soil space in the following. The variable names used in this manuscript are summarized in Table 2.

2.2 Forms of energy associated with soil water content

Two types of energy are relevant for describing soil water states. We will refer to the sum of these as the total hydraulic energy (U_w, J) contained in a soil volume, which consists of the binding energy, U_{wb} (J), and the gravitational energy, U_{wg} (J), defined in each soil compartment i (with $i = 1..n$):

$$U_{w,i} = U_{wb,i} + U_{wg,i}. \tag{1}$$

For the total soil volume we have

$$U_w = \sum_{i=1}^{n} U_{w,i}. \tag{2}$$

The gravitational energy ($U_{wg,i}$, J) relates to the energy necessary to lift the water from a reference level up to the point where it is stored in the soil:

$$U_{wg,i} = V_{s,i} \cdot \rho_w \cdot g \cdot \theta_i \cdot (z_i - z_r), \tag{3}$$

where ρ_w is the density of water ($\rho_w = 1000\,\mathrm{kg\,m^{-3}}$), g is the gravitational acceleration ($g = 9.81\,\mathrm{m\,s^{-1}}$), z_r is the elevation of the reference level, z_i the elevation of compartment i, and θ_i refers to the volumetric water content. This formulation neglects variations of gravitational energy within the compartment i. More rigorously, the gravitational energy should be integrated over depth from the bottom to the top of the compartment. This requires information on the vertical distribution of soil water within the compartment, which is typically not available in model applications. If compartments are chosen to be shallow, the error associated with this simplification is small.

The binding energy ($U_{wb,i}$, J) relates to the capillary forces in the soil pores. With the soil matric potential being the change of Gibbs free energy per change of mass, the related energy can be found by integration. We obtain it here by integration of soil water volume of each compartment i:

$$U_{wb,i} = U_{wb,i}(\theta_i) = V_{s,i} \cdot \rho_w \cdot g \cdot \int_{\theta_{min}}^{\theta_i} \psi_M(\theta') \cdot d\theta'. \tag{4}$$

Essentially, $U_{wb,i}$ is the integral of the water retention curve ($\psi_M(\theta)$). The multiplication with $\rho_w \cdot g$ serves to convert the units of the matric potential (ψ_M) from meter water to joule. An example for both $\psi_M(\theta)$ and the related $U_{wb}(\theta)$ is depicted in Fig. 1 for a sandy loam using the van Genuchten parameterization with parameters given in Table 1. The lower integration point (θ_{min}) should refer to completely dry soil. However, some analytic formulations for the water retention function are not well defined in the very dry range (Rossi and Nimmo, 1994; Ciocca et al., 2014), and we therefore chose $\theta_{min} > 0$. This has a great influence on the absolute values of binding energy in the soil, but, as will be shown below, the relative (i.e., temporal) differences in $U_{wb,i}$ are of relevance. Therefore, the exact choice of θ_{min} does not affect the results. We propose choosing a value just below the water content at the permanent wilting point, or another suitable value smaller than the water contents that will be reached in the desired application.

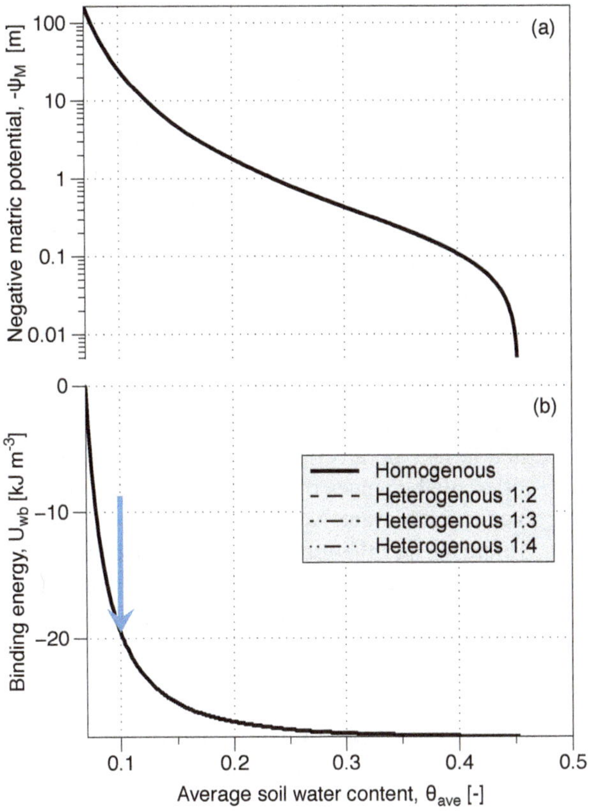

Figure 1. Example of the hydraulic and thermodynamic states of a sample soil (sandy loam; Rawls et al., 1993): **(a)** water retention curve with logarithmic y axis; parameters are given in Table B1. **(b)** For the same soil, binding energy, $U_{wb} = \sum_{i=1}^{n} U_{wb,i}$, as a function of soil water content, for homogenous and heterogenous soil water distribution in a total soil volume $1\,\mathrm{m^3}$. The ratio indicated in the legend corresponds to the ratio of soil water contents in two compartments of equal size but different soil water content. The blue arrow indicates how much energy is available for driving fluxes to equalize the gradients in water potentials between compartments.

Figure 1b shows the binding energy of the soil water as a function of the volumetric soil water content both for homogeneously and heterogeneously wetted soil. Like the soil matric potential, U_{wb} is negative, reaching the lowest values at soil saturation. The negative sign relates to the fact that energy is released (in form of a very small amount of heat) when water attaches to the pore walls ("heat of immersion"; Edlefsen and Anderson, 1943; Hillel, 1998). The same amount of energy has to be transferred to the soil when water is removed from the pores, and hence the bond between the water and the pore wall is broken. Thus, decreasing the water content via root water uptake constitutes an export of negative energy along with the mass export of water from the soil system to the plant (with details described below).

When soil water potential is distributed heterogeneously, the binding energy increases (is less negative). Technically, this results from the strongly nonlinear water retention function. From a process perspective, this additional energy will drive water fluxes for equalizing the soil water gradients between compartments and will during this process eventually dissipate this amount of energy by conversion into heat.

During root water uptake, a given amount of energy has to be invested to take up a certain volume of water over time. Hence differential changes of binding energy per change in water content are relevant. Note that the slopes on the curves in Fig. 1b are steeper the greater the soil water heterogeneity. This illustrates that more energy has to be invested per decrease in total soil water content in heterogeneously compared to homogeneously wetted soils if water is to be extracted at an equal rate from the compartments.

2.3 Dissipation and energy export associated with soil water movement and root water uptake

Soil water fluxes lead to dissipation (D, $\mathrm{J\,s^{-1}}$) of total hydraulic energy. Generally, dissipation of energy occurs when water flows over a resistance, and it is expressed as the product of the driving gradient and the water flow. Those fluxes may occur within the soil between compartments during redistribution of bulk soil water or at the small scale due to root water uptake, or further, within the plant tissue. The representation of fluxes in models differs according to model complexity, yet the dissipative nature of these fluxes should remain unaffected by how these are represented.

Energy dissipation due to soil water flow between compartments is written as

$$D_{\mathrm{f},k} = \rho_\mathrm{w} \cdot g \cdot (h_i - h_j) J_{\mathrm{w},k} \tag{5}$$

and for the entire soil volume as

$$D_\mathrm{f} = \sum_{k=1}^{l} D_{\mathrm{f,k}}, \tag{6}$$

where h refers is the soil hydraulic potential ($h_i = \psi_\mathrm{M} + (z_i - z_\mathrm{r})$, m water) of the neighboring compartments i and j; $J_{\mathrm{w},k}$ refers to the water flux between two neighboring compartments i and j; $D_{\mathrm{f},k}$ ($\mathrm{J\,s^{-1}}$) refers to the respective dissipation of energy over the boundary k between those compartments; and D_f ($\mathrm{J\,s^{-1}}$) refers to the total dissipation due to water fluxes within the total soil volume, i.e., over all interfaces between compartments (l). The dissipation is always negative, since it indicates a loss of hydraulic energy from the system, which is released in the form of a very small quantity of thermal energy.

The same applies to the dissipation of energy due to the small-scale radial root water uptake ($D_{\mathrm{u},i}$, $\mathrm{J\,s^{-1}}$). The formulation depends a great deal on the complexity of the applied flow model. The flow may be represented over several steps, i.e., from bulk soil to the root surface, from root surface to the xylem, and within the xylem. Here, we represent the flow in a simple form, corresponding to the water uptake ($J_{\mathrm{wr},i}$) of the water flow model presented in the next section (Eq. 15). Note that this may easily be adapted to more comprehensive formulations of water uptake. We write dissipation due to microscopic radial water uptake as

$$D_{\mathrm{u},i} = \rho_\mathrm{w} \cdot g \cdot (\psi_{\mathrm{M},i} - \psi_\mathrm{x}) \cdot J_{\mathrm{wr},i}, \tag{7}$$

with $D_{\mathrm{u},i}$ ($\mathrm{J\,s^{-1}}$) being the dissipation due to root water uptake in each reservoir i and ψ_x being the root xylem potential (m). The total dissipation D_u becomes

$$D_\mathrm{u} = \sum_{i=1}^{n} D_{\mathrm{u},i}. \tag{8}$$

Lastly, the root water uptake constitutes an export of energy ($J_\mathrm{E,exp}$, $\mathrm{J\,s^{-1}}$) from the soil root system which is generally defined as the product of the potential at and the flow across the boundary. In our case, water leaves at the top of the root collar as total transpiration at the xylem water potential:

$$J_\mathrm{E,exp} = \rho_\mathrm{w} \cdot g \cdot \psi_\mathrm{x} \cdot J_\mathrm{wu}. \tag{9}$$

The sign of $J_\mathrm{E,exp}$ is positive since in our case water leaves the system (a negative flux) over the root collar at negative hydraulic potential. Correspondingly, this increases the total hydraulic energy as the soil dries (compare Fig. 1). $J_\mathrm{E,exp}$ would be negative should water enter the system via the roots.

Although the dissipation terms (D_f, D_u) and energy fluxes ($J_\mathrm{E,exp}$) carry the same units, their difference is noteworthy. Dissipative fluxes refer to internal processes within the thermodynamic system. They are irreversible. In our example they reflect the heat dissipated when water fluxes degrade the gradients in soil water potential. On the other hand, $J_\mathrm{E,exp}$ is an energy flux (energy transported) across the system boundary. Note, however, that in general this flux also depletes a gradient (between the soil and the atmosphere), but this gradient is not described in our simple soil–root model explicitly.

2.4 Energy balance equation

The energy balance for the soil–root system can be written as the sum of the changes in total hydraulic energy over all compartments, the dissipation terms, and the energy export:

$$\sum_{i=1}^{n} \frac{\mathrm{d}U_{\mathrm{w},i}}{\mathrm{d}t} = \sum_{k=1}^{l} D_{\mathrm{f},k} + \sum_{i=1}^{n} D_{\mathrm{u},i} + J_\mathrm{E,exp}. \tag{10}$$

Some properties of this equation are noteworthy. First, rearranging Eq. (10) yields an expression that relates the characteristics at the outlet of the system to a series of internal processes:

$$\begin{aligned} J_{\mathrm{E,exp}} &= \rho_{\mathrm{w}} \cdot g \cdot \psi_{\mathrm{x}} \cdot J_{\mathrm{wu}} \\ &= \sum_{i=1}^{n} \frac{\mathrm{d}U_{\mathrm{w},i}}{\mathrm{d}t} - \sum_{k=1}^{l} D_{\mathrm{f},k} - \sum_{i=1}^{n} D_{\mathrm{u},i}. \end{aligned} \tag{11}$$

The units in all terms of Eq. (11) are joules per second ($\mathrm{J\,s^{-1}}$), as they all indicate rates of energy flux and changes of energy content with time. More practically, $J_{\mathrm{E,exp}}$, as the product of root collar xylem potential and transpirational flux, is influenced by several processes, and Eq. (11) shows that they act as a sum (remember that all dissipative terms have negative signs). For a given water flux, Eq. (11) shows that the collar xylem potential has to be more negative when water has to be moved within the soil (thus decreasing D_{f}), when water is taken up in drier soil at more negative soil water potentials, and also when soil water potentials are more heterogeneously distributed (both increasing U_{wb}, as shown in Fig. 1). For model applications, comparison of the magnitude of the separate terms of the sum in Eq. (11) provides a tool to assess which of the successive pathways involved in root water uptake most strongly impedes water flow.

3 Conceptual root water uptake model

The thermodynamic evaluation introduced in the last section is meant to be applied to a water flow model (process model). We illustrate the application using a simple model system as shown in Fig. 2. The system consists of four soil water reservoirs, from which water is extracted by root uptake. No water is added during the simulation. All soil reservoirs are assumed to be of equal volume, $V_{\mathrm{s},i}$ ($\mathrm{m^3}$), and their water storage is described by the variables $W_i = \theta_i \cdot V_{\mathrm{s},i}$ (in $\mathrm{m^3}$), with θ_i being the volumetric soil water content (–) of the reservoir i. Soil water is extracted from the soil reservoirs by root water uptake. In the model scenarios presented below, we will assume that soil compartments are isolated; hence no flow between compartments takes place. However, in order to formulate all terms in the energy balance, we also state the corresponding flow equations for bulk water flow between compartments for reference.

The mass balances of the reservoirs describing the temporal changes in W_i in terms of the root water uptake fluxes, $J_{\mathrm{wr},i}$, ($\mathrm{m^3\,s^{-1}}$) and soil water flow between the reservoirs, $J_{\mathrm{w},k}$ ($\mathrm{m^3\,s^{-1}}$), are expressed as

$$\frac{\mathrm{d}W_i}{\mathrm{d}t} = J_{\mathrm{wr},i} + \sum_{k=1}^{m} J_{\mathrm{w},k}, \tag{12}$$

with both fluxes carrying negative signs when directed outward of the reservoir. Here, m describes the total number

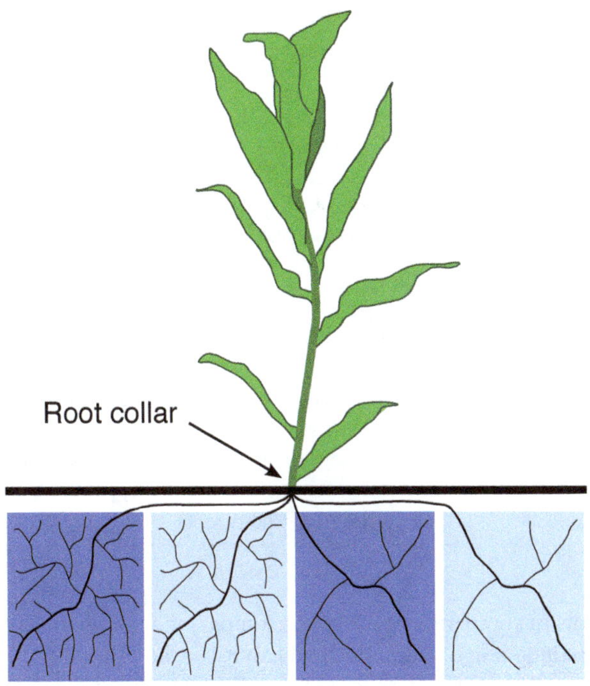

Figure 2. Schematic of the numerical split-root experiment. One plant has access to four soil compartments, two densely rooted (left) and two sparely rooted (right). Color shading of the containers indicates high (dark color) and low (bright color) initial soil water content. The average initial soil water content is the same in all simulations. In the same way, the average water content over the two left (densely rooted) and two right (sparsely rooted) containers is the same in all simulations.

of neighbors of cell i. The water flux ($J_{\mathrm{w},k}$, $\mathrm{m^3\,s^{-1}}$) from the neighboring reservoirs is expressed by Darcy's law, being proportional to the difference in hydraulic potentials ($h_i = \psi_{\mathrm{M}} + (z_i - z_{\mathrm{r}})$, m water). For any two neighboring cells i and j, it is given as

$$J_{\mathrm{w},k} = -K_{ij}(h_i - h_j), \tag{13}$$

where K_{ij} ($\mathrm{m^2\,s^{-1}}$) is the effective unsaturated soil hydraulic conductivity between adjacent compartments i and j. For convenience the spatial scale is factored into the effective conductivity K_{ij}.

The total root water uptake (J_{wu}, $\mathrm{m^3\,s^{-1}}$) is the sum of the uptake fluxes from each compartment:

$$J_{\mathrm{wu}} = \sum_{i} J_{\mathrm{wr},i}, \tag{14}$$

which are described in analogy to Darcy's law:

$$J_{\mathrm{wr},i} = -K_{\mathrm{r},i}(\psi_{\mathrm{M},i} - \psi_{\mathrm{x}}), \tag{15}$$

where ψ_{x} (m) is the xylem water potential which is taken to be equal throughout the entire root system and the index i runs over the number of compartments. The conductivities $K_{\mathrm{r},i}$ ($\mathrm{m^2\,s^{-1}}$) are effective radial conductivities of active

Table 1. Parameters and initial conditions applied for each of the scenarios in the conceptual model for the compartments ($i = 1 \ldots 4$). Given are the differences between scenarios in words and the corresponding manipulations in initial states and parameters.

Scenario	Variable	$i=1$	$i=2$	$i=3$	$i=4$
All	$V_{s,i}$ (m^3)	0.125	0.125	0.125	0.125
(1) optimal case: soil and roots homogenous	Initial soil water $\theta_{init,i}$ (–)	Average 0.18	Average 0.18	Average 0.18	Average 0.18
	Root abundance $K_{r,i}$ ($m^2\,s^{-1}$)	Average 5.0×10^{-6}	Average 5.0×10^{-6}	Average 5.0×10^{-6}	Average 5.0×10^{-6}
Heterogenous soil cases	Initial soil water $\theta_{init,i}$ (–)	Dry 0.15	Dry 0.15	Wet 0.21	Wet 0.21
(2) homogenous roots	Root abundance $K_{r,i}$ ($m^2\,s^{-1}$)	Average 5.0×10^{-6}	Average 5.0×10^{-6}	Average 5.0×10^{-6}	Average 5.0×10^{-6}
(3) heterogenous roots	Root abundance $K_{r,i}$ ($m^2\,s^{-1}$)	Many 7.5×10^{-6}	Few 2.5×10^{-6}	Many 7.5×10^{-6}	Few 2.5×10^{-6}
(4) strongly heterogenous roots	$K_{r,i}$ ($m^2\,s^{-1}$)	9.0×10^{-6}	1.0×10^{-6}	9.0×10^{-6}	1.0×10^{-6}

roots in compartment i. They encompass the notion of active root length and hydraulic conductivity of the flow path from the bulk soil into the root xylem, all of which are positively related to K_r. In our conceptual model, we will change the proportion of $K_{r,i}/K_{r,j}$ to create heterogenous root water uptake from the different reservoirs (see below). As mentioned above, root water uptake may be represented more comprehensively. We keep it simple here to better support the purpose of demonstrating the thermodynamic diagnostics.

For the purpose of demonstration, we keep the model simulations deliberately simple. All soil compartments are arranged horizontally, so that differences in gravitational energy do not play a role and all changes in hydraulic energy will be due to changes in water content. Also, we model a split-root experiment, where no water flow between compartments is possible and all changes in soil water content are due to root water uptake. This enables us to increase the heterogeneity in the soil water content and demonstrate its effect on the energy balance, xylem potential, and uptake dynamics. In this simplified setup we solve Eqs. (14) and (15) with a prescribed boundary condition (total transpiration, J_{wu}) for the unknown xylem water potential ψ_x as follows:

$$\psi_x = \frac{J_{wu} + \sum_{i=1}^{n} (\psi_{M,i} \cdot K_{r,i})}{\sum_{i=1}^{n} (K_{r,i})}. \quad (16)$$

For each time step (Δt), Eq. (16) yields the xylem water potential, based on the current water contents in the compartments. Next, we obtain the water uptake for the same time step using Eqs. (12) and (15). Water contents are then updated for the next time step based on the root water uptake. Initial soil water contents and root conductivities are applied as shown in Table 1. For purposes of simplicity, we run the model until soil water limits uptake. We somewhat arbitrarily assume that soil water becomes limiting when the root xylem potential falls somewhat below the permanent wilting point (−150 m). When this point is reached, we fix $\psi_x = -150$ m.

Additional scenarios with diurnal fluctuations of transpirational forcing and other soil hydraulic properties are given as a reference in the Supplement. They yield similar results.

4 Scenarios

We run the model for four scenarios, as shown in Table 1. In the scenarios we vary the distribution of initial soil water content and the implied root length by changing to compartment root conductivity to impose increasingly heterogenous conditions while keeping the average constant.

The first scenario is completely homogenous with a uniform initial soil water content and root conductivity across compartments. Three additional scenarios are initialized with heterogenous initial soil water and differ with regard to the heterogeneity of root conductivity. In all simulations the average initial soil water content is the same. In the same way, the effective root conductivities ($K_{r,i}$) were either homogeneously distributed or heterogenous with two compartments having more roots and two less than average. Working with four compartments allows us to combine the manipulation such that average root conductivity is equal between the dry and wet compartments and between all scenarios (see Table 1, Fig. 2).

The model is representative of a plant having access to a soil volume of 0.5 m^2, consisting of a soil monolith of 0.5 m depth and a surface area of 1 m^2. Each of the compartments is the same volume (0.125 m^3). The transpiration rate is indicative of a hot summer day in Germany, with 6 mm d^{-1}. Effective root conductivities correspond to roots with radial conductivity of 3×10^{-6} m s^{-1} MPa^{-1}, which is on the upper end of the values summarized in Draye et al. (2010), and total root length densities in the compartments varying between 1 cm cm^{-3} (most densely rooted), 0.5 cm cm^{-3} (average), and 0.1 cm cm^{-3} (least densely rooted). This is

Table 2. Variables used in this study.

Symbol	Variable	Units
D_f	Dissipation due to soil water flow	$J\,s^{-1}$ or W
D_u	Dissipation due to root water uptake	$J\,s^{-1}$ or W
g	Gravitational acceleration	$9.81\,m\,s^{-2}$
h	Soil hydraulic potential	m
i, j	Soil compartment indices	–
$J_{E,exp}$	Export of energy from the soil–root system through the root collar	$J\,s^{-1}$ or W
J_{wu}	Total root water uptake	$m^3\,s^{-1}$
$J_{w,k}$	Soil water redistribution between compartments over the interface k	$m^3\,s^{-1}$
$J_{wr,i}$	Water flux from bulk soil into the root	$m^3\,s^{-1}$
k	Index for interfaces between compartments	–
l	Number of interfaces between compartments	–
K_{ij}	Soil hydraulic conductivity between compartments i and j	$m^2\,s^{-1}$
$K_{r,i}$	Effective radial conductivity of the active roots in compartment i	$m^2\,s^{-1}$
n	Number of soil compartments	4
t	Time	s
U_w	Total hydraulic energy	J
U_{wb}	Binding energy	J
U_{wg}	Gravitational energy	J
W_i	Total soil water storage	m^3
V	Volume	m^3
$V_{s,i}$	Volume of the model compartments	m^3
z	Elevation	m
z_r	Elevation of the reference level	m
ψ_M	Soil matric potential	m
ψ_x	Xylem water potential	m
ρ_w	Density of water	$1000\,kg\,m^{-3}$
θ	Volumetric soil water content	–
θ_{ave}	Average volumetric soil water content	–
θ_{min}	Lower integration boundary for U_{wb}	–

within the range of observed root length densities for maize (Kuchenbuch et al., 2009).

The soil hydraulic properties are equal in all compartments and derived using van Genuchten (1980) with parameters for a sandy loam (Rawls et al., 1993) given in Table B1. Starting from the initial condition, we model a dry-down event until limiting soil water xylem potential is reached. In each simulation, first root water uptake and the resulting temporal evolution of soil water content were explicitly solved for each compartment. Second, the thermodynamic evaluation was applied a posteriori based on the results of the root water uptake model.

5 Results

Figure 3 shows the evolution of the root collar potential obtained from the process model over the course of the drying cycle and the associated creation of heterogeneity of soil water contents (as reflected in the coefficient of variation). Shown are the results of all scenarios given in Table 1. Remember that the difference between scenarios is only with regard to the prescribed heterogeneity. The average initial water contents, root conductivities, and root water uptake are the same in all simulations. The scenario called "optimal" is one where both initial soil water content and root distribution are homogenous. It can be seen as the optimal scenario from a plant's point of view, as it minimizes dissipation, which keeps the xylem potentials less negative and delays time to water stress. It is obvious from the evolution of the root collar potential that, despite everything relating to the overall water balance being the same in all scenarios, the homogenous (optimal) scenario is the one where limiting xylem potentials are reached at the lowest average soil water content and longest time after the beginning of the experiment. The limiting xylem potential is reached earlier the more heterogenous the distribution of root water uptake and soil water contents is. Also, Appendix A shows analytically that uptake from homogeneously distributed soil water minimizes (i.e., optimizes from the plant's point of view) the dissipative losses due to root water uptake in a situation in which soil hydraulic properties are homogeneous.

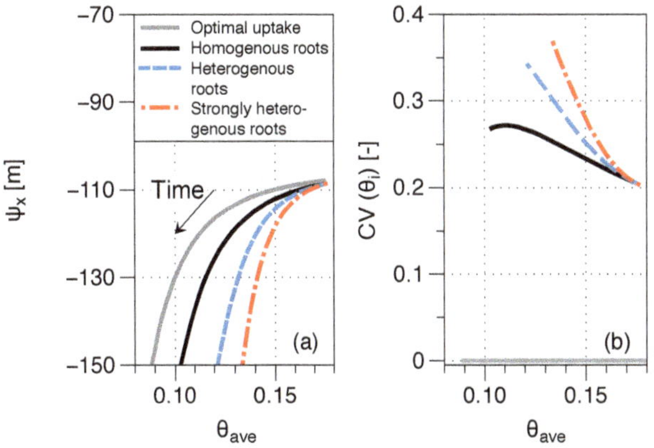

Figure 3. Model results of the simple model: **(a)** evolution of xylem potential over the course of root water uptake, **(b)** evolution of the coefficient of variation of soil water content during the simulation. Average initial soil water content is the same in all simulations. Only the unstressed uptake is shown. The time axis has been replaced by average volumetric soil water, which evolves parallel with time in this constant-flux experiment.

Based on the output of the root water uptake model, we applied Eq. (11) to diagnose the impediments to root water uptake. The individual terms of Eq. (11) (except dissipation to soil water flow, which was not modeled) are plotted separately in Fig. 4: on the left the total export of energy ($J_{E,exp}$), which proves to be composed of the change of binding energy in the soil (dU_{wb}/dt, middle) and dissipation due to root water uptake ($-D_u$, right). All individual terms (dU_{wb}/dt, D_u and $J_{E,exp}$) were calculated separately, applying Eqs. (4),

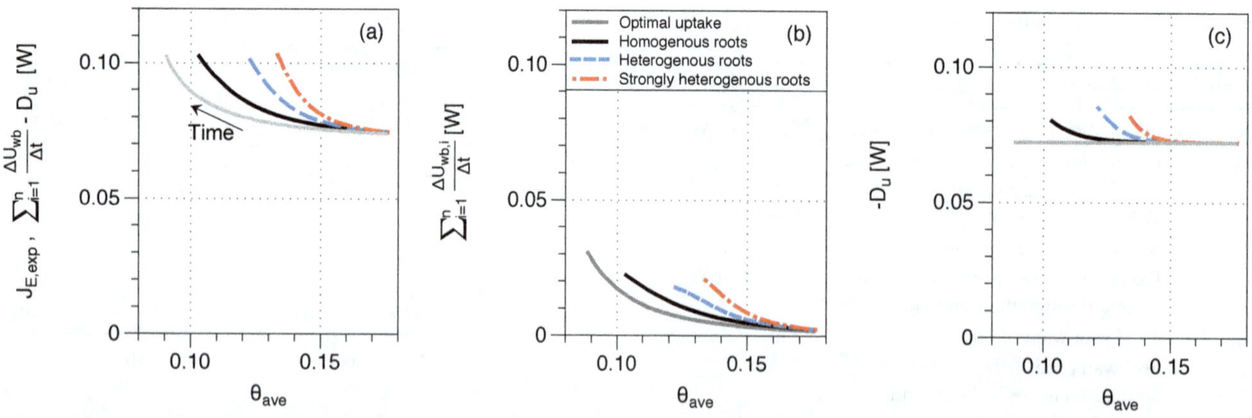

Figure 4. Components of the energy balance (Eq. 11) for the soil–plant system over the course of a drying experiment and different root water uptake scenarios. As in Fig. 3, the time axis was replaced by the average soil water content. **(a)** Total energy exported from the system at the root collar. It is the sum of the two components given in the other subplots. **(b)** Component due to decrease of soil binding energy, which is due to both soil drying and enhanced heterogeneity (compare Fig. 3). **(c)** Component due to energy dissipation by water flow from the soil into the root.

(7), and (9). Thus, Fig. 4 provides a proof of concept for the correct derivation of Eq. (11) because all the terms balance.

The energy export ($J_{E,exp}$) corresponds closely to the evolution of the root xylem potential (Fig. 3, left), because the transpirational flux is prescribed as constant. $J_{E,exp}$ increases continuously as the soil dries in order to maintain the constant rate of uptake. The decomposition of the energy export provides information about the impediments to root water uptake in the different scenarios. For example, the greatest contribution to $J_{E,exp}$ in wet soil in our setup originates from the dissipation when water flows from the soil into the root, which constitutes about 97 % of the energy export. When the soil dries out, it becomes increasingly more costly to detach water from the soil matrix, and the change of the binding energy makes up a somewhat more substantial proportion of the total energy exported from the system (17–22 %, depending on the scenario).

The optimal case (grey solid line) is the one with the least possible expenditure in $\mathrm{d}U_{wb}/\mathrm{d}t$, and the difference between the solid grey curve and the other curves illustrates the impact of soil water heterogeneity on the water uptake at each time step. At the same average soil water content, differences in $\mathrm{d}U_{wb}/\mathrm{d}t$ between our scenarios are entirely due to heterogenous soil water distribution. When comparing the optimal scenario and the one with strongly heterogenous roots at $\theta_{ave} = 0.15$, we observe that less than half of the investment in detaching water is due to soil drying and the remaining part is due to the heterogenous distribution of the soil water. The effect of soil heterogeneity increases further after this point.

At the same time, in heterogenous soils the impediment to uptake due to water flow over the root resistance increases, since uptake occurs preferentially in a limited part of the root system (the compartment with greatest root length that was initialized as wet, data not shown). However, this dissipation effect is less dynamic over time than the one related to soil drying in this modeling exercise.

6 Discussion

We used thermodynamics to evaluate the dynamics of a very simple process model for root water uptake to demonstrate that besides fluxes and potentials there is more relevant information in the system that relates to change of hydraulic energy and dissipative losses of water uptake. The main contribution of the paper lies in providing a tool for assessing where the impediments to root water uptake lie along the flow path between soil and atmosphere. For this the thermodynamic formulations are applied a posteriori to water fluxes and changes of soil water contents calculated with the hydrological model. The relative contribution of each of the impediments can then be quantified, by evaluation of the relative contribution of each process to the total energy export. At the same time, the calculations with the simple model serve as a proof of concept: the energy balance is closed; i.e., the sum of change in hydraulic energy and dissipation equal the energy export.

In our thermodynamic description of the soil–plant system, we have not considered the changes of soil temperature, which should be induced particularly when heat is generated as water attaches to the soil. We have done this because the related changes of temperature are so small that they would not affect the water flow and generally small compared to changes of temperature due to radiative soil heating.

Also, we have assumed in this derivation that the soil water retention function is known and is non-hysteretic. The latter may have considerable influence on the resulting trajectory of $\mathrm{d}U_{wb}/\mathrm{d}t$. Generally, hysteresis can be included in the framework to investigate this effect further in the future.

Finally, we have also deliberately limited our model scenarios to situations where roots do not grow and where root length does not depend on water availability, and we have not allowed for redistribution of water between compartments. This way, we artificially maintained heterogeneity, which was done in order to demonstrate in the separate scenarios how heterogeneity alone affects uptake and its thermodynamics.

An important advantage of evaluating the process model output in the energy domain lies in the possibility for evaluating the role of heterogenous soil water potentials. The water potentials, the derivative of the Gibbs free energy per mass, are an intensive property of the system, and in heterogenous systems they cannot be meaningfully averaged. The Gibbs free energy itself is an extensive property, can be averaged, and hence allows states in heterogenous systems to be efficiently described. An additional advantage of working in the "energy domain" is the possibility to consider the influence of the water retention function, the heterogeneous soil water distribution, and the various resistances along the flow path in the same realm and using the same units. In particular, heterogeneity of soil water increases the total hydraulic energy, which necessarily implies that xylem water potentials have to be more negative to transpire at the same rate and same average soil water content if root systems are equally distributed. Thus, with everything else being equal and independent of soil water potential distribution, plants rooted in heterogeneously wetted soils are expected to reach water limitation earlier. This phenomenon has already been observed in models dealing with spatially heterogenous infiltration patterns caused by forest canopies (Guswa and Spence, 2011). At the same time heterogeneous soil water retention properties may induce root growth that alleviates water stress, and root systems are likely adapted to such conditions. This reasoning may be further extended to understanding horizontal and vertical distribution of root systems and uptake in adapting to their environment (Adiku et al., 2000; van der Ploeg et al., 2008) also in terms of reducing their dissipative losses.

We have given equations for our simple system, but the concept can easily be extended to more complex systems, for example three-dimensional models of root water uptake (Doussan et al., 2006; Javaux et al., 2008; Kalbacher et al., 2011) – which include more process details, particularly more complex description of water flow within the root system – or any other process models describing root water uptake. Application of thermodynamics as proposed in this paper may help to identify and understand the effects of heterogeneity in more realistic models of root water uptake. Furthermore, a thermodynamic evaluation may be applied to investigate the effects of dynamic root growth, aquaporin and stomatal regulation (Aroca et al., 2012; Vadez et al., 2014), or mucilage (Carminati et al., 2011) on the impediments of the whole plant root water uptake. Bechmann et al. (2014) have applied thermodynamics to root water uptake studies in order to differentiate efficient root parameterizations from less efficient ones by minimizing the time average of $J_{E,exp}$. More practically, measurements of leaf water potential and transpiration are used to assess plant water relations, and Eq. (11) elucidates the processes involved. Thus, when information on potentials and flux along the flow path are available, the formulations can also be implemented in experimental studies, while imposed system boundaries can be adapted to fit the specific setup.

At the more general level, this study adds to the thermodynamic formulation of hydrologic processes and the application of thermodynamic optimality approaches (Kleidon and Schymanski, 2008; Porada et al., 2011; Kleidon et al., 2013; Zehe et al., 2013). What we described here is more targeted towards reduction in dissipative losses, rather than the maximization of dissipation, or entropy production, as suggested by some previous studies (Kleidon and Schymanski, 2008; Zehe et al., 2013). This is, however, not a contradiction. A reduction of dissipative losses in a system allows greater fluxes for the same forcing gradient to be maintained, which may then result in a greater depletion of the driving gradients, thus maintaining low xylem potentials and delaying the onset of water stress or cavitation. In our study, we did not consider this effect on the driving gradients, which in the case of root systems are the difference in chemical potential between soil moisture and the water vapor in the near-surface atmosphere. The minimization of internal dissipation has already been applied in hydrology for characterization of river network structure (Rinaldo et al., 1996, 2014) and vertical root water uptake (Adiku et al., 2000). Notably, it has also been used in vascular networks as the starting point to derive scaling laws and the fractal nature of plant branching systems (West et al., 1997). It would seem that our study fits very well into the scope of this previous study and extends it to include the transport of soil water towards the vascular network of the rooting system. In a further step, this transport would need to be linked into the whole soil–vegetation–atmosphere system along with its driving gradient to fully explore the thermodynamic implications of an optimized root system. Such extensions could form the scope of future research. The thermodynamic formulation of root water uptake as described here provides the necessary basis upon which to test the applicability of thermodynamic optimality approaches to root system functioning.

7 Summary and conclusions

Systems approaches and modeling will certainly be tools to investigate plant water relations and efficient rooting strategies in the future (Lobet et al., 2014). In this paper we give a description of how root water uptake relates to changes of total energy in the system, which can be used to quantify the contribution of individual processes to impeding root water uptake. It also sheds new light on some impediments not yet accounted for, like heterogeneity in soil water. This is a slightly different and potentially complementary approach to describing flow resistances over potential gradients. Our derivation shows that the product of xylem water potential and transpiration flux carries a great deal of information, as it can be partitioned into the sum of individual processes impeding water flow in the soil–plant system. Particularly in process models on root water uptake (Doussan et al., 2006; Kalbacher et al., 2011; Couvreur et al., 2012), the changes of total hydraulic energy and energy dissipation provide the opportunity to evaluate which processes dominate the impedance to root water uptake at given times and shed light on whether those are of biotic (within the plant) or abiotic (within the soil) origin.

Appendix A: Analytical derivations

It can be shown analytically that a homogeneous soil water distribution results in the least dissipation associated with root water uptake (as shown in Fig. 4c). Such a minimization of dissipation then results in a lower decrease dU_w/dt and/or in a lower export $J_{E,exp}$, as expressed in Eq. (11). To show this minimum analytically, we consider a simplified setup of only two reservoirs, a and b, yet use the same formulations as in the main text and the same boundary condition of a prescribed flux of root water uptake, J_{wu}.

We consider the case of a uniform root system (i.e., $K_{r,a} = K_{r,b} = K_r$) that takes up water from the two soil reservoirs. The distribution of soil water is described by matric potentials $\psi_{M,a} = \psi_M - \Delta\psi$ and $\psi_{M,b} = \psi_M + \Delta\psi$. When $\Delta\psi$ is relatively small, then the water retention curve is approximately linear with the soil water content, so that this formulation represents a case in which the total soil water of the two reservoirs is the same, and it is only the distribution across the two reservoirs that differs, as described by $\Delta\psi$. With this formulation, the prescribed boundary condition in terms of the root water uptake J_{wu} results in a constraint of the form

$$\begin{aligned} J_{wu} &= -K_r(\psi_M - \Delta\psi_M - \psi_x) - K_r(\psi_M + \Delta\psi_M - \psi_x) \\ &= -2K_r(\psi_M - \psi_x) \end{aligned} \quad \text{(A1)}$$

so that $\psi_x = \psi_M + J_{wu}/(2K_r)$. The dissipation, D_u, associated with root water uptake then becomes

$$D_u = D_{u,a} + D_{u,b} = -\rho_w g \cdot K_r \left[2\left(\frac{J_{wu}}{2K_r}\right)^2 + 2\Delta\psi_M^2 \right]. \quad \text{(A2)}$$

It is easy to see in this expression that the minimum is reached when $\Delta\psi_M = 0$ (which can also be derived analytically by $\partial D_u / \partial \Delta\psi_M = 0$). In other words, for a uniform root system, dissipation associated with root water uptake is at a minimum when moisture is distributed homogeneously in the soil.

In principle, one can also show that a uniform root system results in a minimum of dissipation. This requires an integration over time, which makes an analytical treatment more complex so that it is more easily illustrated by the numerical simulations done in the main text.

Appendix B: Soil parameters

Table B1. Parameters used for calculation of soil hydraulic properties using van Genuchten (1980).

Symbol	Description	Value
n	Shape parameter	1.38
m	Shape parameter, $m = 1 + \frac{1}{n}$	0.275
α	Shape parameter	0.068 cm^{-1}
θ_{min}	Lower integration boundary in Eq. (4)	0.07
θ_r	Residual soil water content	0.041
θ_s	Porosity	0.453

Acknowledgements. Anke Hildebrandt acknowledges support from the Collaborative Research Centre "AquaDiva" (SFB 1076/1, B02) funded by the German Science Foundation (DFG). Marcel Bechmann has been supported by the Jena School for Microbial Communication (JSMC). Axel Kleidon acknowledges financial support from the "Catchments As Organized Systems (CAOS)" research group funded by DFG through grant KL 2168/2-1. We thank Stefan Kollet for discussion and helpful references; Thomas Wutzler; and the reviewers Gerrit de Rooij and Uwe Ehret as well as the editors Erwin Zehe and Alberto Guadagnini for critical comments during the manuscript discussion.

Edited by: A. Guadagnini

References

Adiku, S. G. K., Rose, C. W., Braddock, R. D., and Ozier-Lafontaine, H.: On the simulation of root water extraction: examination of a minimum energy hypothesis, Soil Sci., 165, 226–236, doi:10.1097/00010694-200003000-00005, 2000.

Aroca, R., Porcel, R., and Ruiz-Lozano, J. M.: Regulation of root water uptake under abiotic stress conditions, J. Exp. Bot., 63, 43–57, doi:10.1093/jxb/err266, 2012.

Bechmann, M., Schneider, C., Carminati, A., Vetterlein, D., Attinger, S., and Hildebrandt, A.: Effect of parameter choice in root water uptake models – the arrangement of root hydraulic properties within the root architecture affects dynamics and efficiency of root water uptake, Hydrol. Earth Syst. Sci., 18, 4189–4206, doi:10.5194/hess-18-4189-2014, 2014.

Carminati, A., Schneider, C. L., Moradi, A. B., Zarebanadkouki, M., Vetterlein, D., Vogel, H.-J., Hildebrandt, A., Weller, U., Schüler, L., and Oswald, S. E.: How the Rhizosphere May Favor Water Availability to Roots, Vadose Zone J., 10, 988–998, doi:10.2136/vzj2010.0113, 2011.

Ciocca, F., Lunati, I., and Parlange, M. B.: Effects of the water retention curve on evaporation from arid soils, Geophys. Res. Lett., 41, 3110–3116, doi:10.1002/2014GL059827, 2014.

Couvreur, V., Vanderborght, J., and Javaux, M.: A simple three-dimensional macroscopic root water uptake model based on the hydraulic architecture approach, Hydrol. Earth Syst. Sci., 16, 2957–2971, doi:10.5194/hess-16-2957-2012, 2012.

Doussan, C., Pierret, A., Garrigues, E., and Pagès, L.: Water uptake by plant roots: II – Modelling of water transfer in the soil root-system with explicit account of flow within the root system – comparison with experiments, Plant Soil, 283, 99–117, doi:10.1007/s11104-004-7904-z, 2006.

Draye, X., Kim, Y., Lobet, G., and Javaux, M.: Model-assisted integration of physiological and environmental constraints affecting the dynamic and spatial patterns of root water uptake from soils, J. Exp. Bot., 61, 2145–2155, doi:10.1093/jxb/erq077, 2010.

Dunbabin, V. M., Postma, J. A., Schnepf, A., Pagès, L., Javaux, M., Wu, L., Leitner, D., Chen, Y. L., Rengel, Z., and Diggle, A. J.: Modelling root – soil interactions using three – dimensional models of root growth, architecture and function, Plant Soil, 372, 93–124, doi:10.1007/s11104-013-1769-y, 2013.

Edlefsen, N. E. and Anderson, A. B. C.: Thermodynamics of soil moisture, Hilgardia, 15, 31–298, doi:10.3733/hilg.v15n02p031, 1943.

Guswa, A. J.: Effect of plant uptake strategy on the wateroptimal root depth, Water Resour. Res., 46, 1–5, doi:10.1029/2010WR009122, 2010.

Guswa, A. J.: Canopy vs. roots: Production and destruction of variability in soil moisture and hydrologic fluxes, Vadose Zone J., 11, 13 pp., doi:10.2136/vzj2011.0159, 2012.

Guswa, A. J. and Spence, C. M.: Effect of throughfall variability on recharge: application to hemlock and deciduous forests in western Massachusetts, Ecohydrology, 5, 563–574, doi:10.1002/eco.281, 2011.

Hammer, G. L., Dong, Z., McLean, G., Doherty, A., Messina, C., Schussler, J., Zinselmeier, C., Paszkiewicz, S., and Cooper, M.: Can changes in canopy and/or root system architecture explain historical maize yield trends in the U.S. corn belt?, Crop Sci., 49, 299–312, doi:10.2135/cropsci2008.03.0152, 2009.

Hildebrandt, A. and Eltahir, E. A. B.: Ecohydrology of a seasonal cloud forest in Dhofar: 2. Role of clouds, soil type, and rooting depth in tree-grass competition, Water Resour. Res., 43, 1–13, doi:10.1029/2006WR005262, 2007.

Hillel, D.: Environmental Soil Physics, Academic Press, Boston, 1998.

Ivanov, V. Y., Hutyra, L. R., Wofsy, S. C., Munger, J. W., Saleska, S. R., de Oliveira, R. C., and de Camargo, P. B.: Root niche separation can explain avoidance of seasonal drought stress and vulnerability of overstory trees to extended drought in a mature Amazonian forest, Water Resour. Res., 48, W12507, doi:10.1029/2012WR011972, 2012.

Janott, M., Gayler, S., Gessler, A., Javaux, M., Klier, C., and Priesack, E.: A one-dimensional model of water flow in soil-plant systems based on plant architecture, Plant Soil, 341, 233–256, doi:10.1007/s11104-010-0639-0, 2011.

Javaux, M., Schroder, T., Vanderborght, J., and Vereecken, H.: Use of a three-dimensional detailed modeling approach for predicting root water uptake, Vadose Zone J., 7, 1079–1088, 2008.

Johnson, D. M., Sherrard, M. E., Domec, J.-C., and Jackson, R. B.: Role of aquaporin activity in regulating deep and shallow root hydraulic conductance during extreme drought, Trees, 28, 1323–1331, doi:10.1007/s00468-014-1036-8, 2014.

Kalbacher, T., Schneider, C. L., Wang, W., Hildebrandt, A., Attinger, S., and Kolditz, O.: Modeling soil-coupled water uptake of multiple root systems with automatic time stepping, Vadose Zone J., 10, 727, doi:10.2136/vzj2010.0099, 2011.

Kleidon, A.: How does the earth system generate and maintain thermodynamic disequilibrium and what does it imply for the future of the planet?, Philos. T. R. Soc. A, 370, 1012–1040, doi:10.1098/rsta.2011.0316, 2012.

Kleidon, A. and Schymanski, S.: Thermodynamics and optimality of the water budget on land: A review, Geophys. Res. Lett., 35, L20404, doi:10.1029/2008GL035393, 2008.

Kleidon, A., Zehe, E., Ehret, U., and Scherer, U.: Thermodynamics, maximum power, and the dynamics of preferential river flow structures at the continental scale, Hydrol. Earth Syst. Sci., 17, 225–251, doi:10.5194/hess-17-225-2013, 2013.

Kondepudi, D. and Prigogine, I.: Modern thermodynamics – from heat engines to dissipative structures, Wiley, Chichester, 1998.

Kuchenbuch, R. O., Gerke, H. H., and Buczko, U.: Spatial distribution of maize roots by complete 3D soil monolith sampling, Plant Soil, 315, 297–314, doi:10.1007/s11104-008-9752-8, 2009.

Lhomme, J.-P.: Formulation of root water uptake in a multi-layer soil-plant model: does van den Honert's equation hold?, Hydrol. Earth Syst. Sci., 2, 31–39, doi:10.5194/hess-2-31-1998, 1998.

Lobet, G., Couvreur, V., Meunier, F., Javaux, M., and Draye, X.: Plant Water Uptake in Drying Soils, Plant Physiol., 164, 1619–1627, doi:10.1104/pp.113.233486, 2014.

Porada, P., Kleidon, A., and Schymanski, S. J.: Entropy production of soil hydrological processes and its maximisation, Earth Syst. Dynam., 2, 179–190, doi:10.5194/esd-2-179-2011, 2011.

Rawls, W. J., Ahuja, L. A., Brakensiek, D. L., and Shirmohammadi, A.: Infiltration and soil water movement, in: Handbook of Hydrology, edited by: Maidment, D. R., chap. 5, 5.1–5.51, MacGraw-Hill, New York, 1993.

Rinaldo, A., Maritan, A., Colaiori, F., Flammini, A., Rigon, R., Rodríguez-Iturbe, I., Banavar, J. J. J., Rodriguez-Iturbe, I., and Banavar, J. J. J.: Thermodynamics of Fractal Networks, Phys. Rev. Lett., 76, 3364–3367, doi:10.1103/PhysRevLett.76.3364, 1996.

Rinaldo, A., Rigon, R., Banavar, J. R., Maritan, A., and Rodriguez-Iturbe, I.: Evolution and selection of river networks: Statics, dynamics, and complexity, P. Natl. Acad. Sci. USA, 111, 2417–2424, doi:10.1073/pnas.1322700111, 2014.

Rossi, C. and Nimmo, J. R.: Modeling of soil water retention from saturation to oven dryness, Water Resour. Res., 30, 701–708, doi:10.1029/93WR03238, 1994.

Schneider, C. L., Attinger, S., Delfs, J.-O., and Hildebrandt, A.: Implementing small scale processes at the soil-plant interface – the role of root architectures for calculating root water uptake profiles, Hydrol. Earth Syst. Sci., 14, 279–289, doi:10.5194/hess-14-279-2010, 2010.

Sperry, J. S., Adler, F. R., Campbell, G. S., and Comstock, J. P. C. N.: Limitation of plant water use by rizosphere and xylem conductance: results of a model, Plant Cell Environ., 21, 347–359, 1998.

Tuzet, A., Perrier, A., and Leuning, R.: A coupled model of stomatal conductance , photosynthesis, Plant Cell Environ., 26, 1097–1116, 2003.

Vadez, V., Kholova, J., Medina, S., Kakkera, A., and Anderberg, H.: Transpiration efficiency: New insights into an old story, J. Exp. Bot., 65, 6141–6153, doi:10.1093/jxb/eru040, 2014.

van den Honert, T. H.: Water transport as a catenary process, Discuss. Faraday Soc., 3, 146–153, doi:10.1039/DF9480300146, 1948.

van der Ploeg, M. J., Gooren, H. P. A., Bakker, G., and de Rooij, G. H.: Matric potential measurements by polymer tensiometers in cropped lysimeters under water-stressed conditions, Vadose Zone J., 7, 1048–1054, doi:10.2136/vzj2007.0104, 2008.

van Genuchten, M. T.: A closed-form equation for predicting the hydraulic conductivity of unsaturated soils, Soil Sci. Soc. Am. J., 44, 892–898, doi:10.2136/sssaj1980.03615995004400050002x, 1980.

West, G. B., Brown, J. H., and Enquist, B. J.: A general model for the origin of allometric scaling laws in biology, Science, 276, 122–126, doi:10.1126/science.276.5309.122, 1997.

Zehe, E., Ehret, U., Blume, T., Kleidon, A., Scherer, U., and Westhoff, M.: A thermodynamic approach to link self-organization, preferential flow and rainfall-runoff behaviour, Hydrol. Earth Syst. Sci., 17, 4297–4322, doi:10.5194/hess-17-4297-2013, 2013.

Zwieniecki, M. A., Thompson, M. V., and Holbrook, N. M.: Understanding the hydraulics of porous pipes: Tradeoffs between water uptake and root length utilization, J. Plant Growth Regul., 21, 315–323, doi:10.1007/s00344-003-0008-9, 2003.

12

Seasonal drought prediction for semiarid northeastern Brazil: verification of six hydro-meteorological forecast products

José Miguel Delgado[1], **Sebastian Voss**[1], **Gerd Bürger**[1], **Klaus Vormoor**[1], **Aline Murawski**[2], **José Marcelo Rodrigues Pereira**[3], **Eduardo Martins**[3], **Francisco Vasconcelos Júnior**[3], **and Till Francke**[1]

[1]Institute of Earth and Environmental Sciences, University of Potsdam, Potsdam, Germany
[2]German Research Centre of Geosciences GFZ Potsdam, Potsdam, Germany
[3]Research Institute for Meteorology and Water Resources – FUNCEME, Fortaleza, Brazil

Correspondence: José Miguel Delgado (martinsd@uni-potsdam.de)

Abstract. A set of seasonal drought forecast models was assessed and verified for the Jaguaribe River in semiarid northeastern Brazil. Meteorological seasonal forecasts were provided by the operational forecasting system used at FUNCEME (Ceará's research foundation for meteorology) and by the European Centre for Medium-Range Weather Forecasts (ECMWF). Three downscaling approaches (empirical quantile mapping, extended downscaling and weather pattern classification) were tested and combined with the models in hindcast mode for the period 1981 to 2014. The forecast issue time was January and the forecast period was January to June. Hydrological drought indices were obtained by fitting a multivariate linear regression to observations. In short, it was possible to obtain forecasts for (a) monthly precipitation, (b) meteorological drought indices, and (c) hydrological drought indices.

The skill of the forecasting systems was evaluated with regard to root mean square error (RMSE), the Brier skill score (BSS) and the relative operating characteristic skill score (ROCSS). The tested forecasting products showed similar performance in the analyzed metrics. Forecasts of monthly precipitation had little or no skill considering RMSE and mostly no skill with BSS. A similar picture was seen when forecasting meteorological drought indices: low skill regarding RMSE and BSS and significant skill when discriminating hit rate and false alarm rate given by the ROCSS (forecasting drought events of, e.g., $SPEI_1$ showed a ROCSS of around 0.5). Regarding the temporal variation of the forecast skill of the meteorological indices, it was greatest for April, when compared to the remaining months of the rainy season, while the skill of reservoir volume forecasts decreased with lead time.

This work showed that a multi-model ensemble can forecast drought events of timescales relevant to water managers in northeastern Brazil with skill. But no or little skill could be found in the forecasts of monthly precipitation or drought indices of lower scales, like SPI_1. Both this work and those here revisited showed that major steps forward are needed in forecasting the rainy season in northeastern Brazil.

1 Introduction

Northeastern Brazil has historically been the epicenter of major drought events. Fioreze et al. (2012) identified 100 severe droughts since the 16th century in this region, while Marengo et al. (2017) identified 68 major events for the same period. Within this region, the state of Ceará has been in the frontline of the fight against this natural hazard. This has been due to both the impacts suffered in the past and the measures taken to improve its resilience.

Droughts in Ceará reflect a meteorological anomaly over the tropical Atlantic Ocean. Dry years are generally related to a positive sea surface temperature (SST) anomaly in the tropical North Atlantic, associated with a negative anomaly in the tropical South Atlantic and over the Equator. This forces a northward shift of the intertropical convergence zone, taking the rainbelt to northern latitudes. The causes of this anomaly are linked to the occurrence of the El Niño–Southern Oscillation and to the North Atlantic Oscillation (Hastenrath, 2012).

Past famines and mass migrations triggered large investments in infrastructure in recent decades. These investments brought hundreds of strategic reservoirs and thousands of small dams to a semi-arid landscape, which are being managed according to a transparent water allocation process (Formiga-Johnsson and Kemper, 2005). In order to support water allocation and management, the state runs a seasonal drought forecasting system and issues annual quantitative and qualitative forecasts of the magnitude of the rainy season. These predictions can support decisions ranging from agricultural management (choice of crop, planning of seeding time) to water distribution and reservoir operation.

Currently, the forecasting system in Ceará is based on the ECHAM4.6 general circulation model (Roeckner et al., 1992). It runs from January to August on persisted SSTs (observed SST anomalies which are assumed invariant), covering each year's rainy season (February to April). The forecasts produced by this model are generally downscaled with the NCEP regional spectral model (Juang et al., 1997), in order to resolve the spatial variability of Ceará. Verification and the current forecast can be retrieved under http://www3.funceme.br/previsao-climatica/ (last access: 24 September 2018). For downscaled forecasts check http://seca-vista.geo.uni-potsdam.de/ (last access: 24 September 2018).

In this study we intend to evaluate and extend this prediction system by employing

1. an additional underlying GCM,
2. a statistical approach based on the classification of weather patterns,
3. empirical–statistical downscaling methods to increase the spatial resolution and temporal fidelity of the predictions, and
4. drought indices as powerful integrative descriptors for the description of drought severity.

By these means, we aim to address the following question.

> What skill do the seasonal *meteorological drought* forecasts have?

While the term *meteorological drought* focuses on the atmospheric forcing causing water shortage, its effective implications for society are more specifically accounted for by the term *hydrological drought* (de Araújo and Bronstert, 2016). Since the aim of the prediction system is to support water management, we sharpen the previous question in this regard.

> Can we forecast *hydrological* droughts in Ceará based on these seasonal meteorological forecasts?

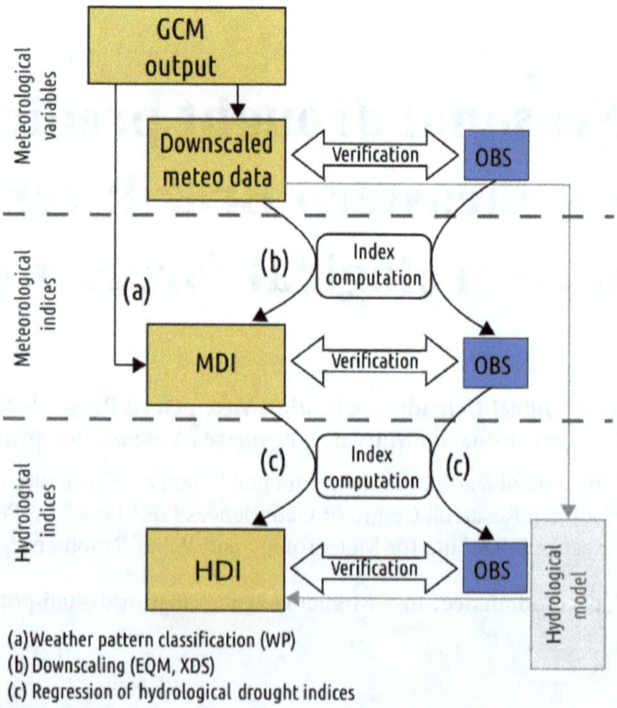

Figure 1. Flowchart explaining the methodology used for predicting meteorological data, meteorological drought indices (MDIs) and hydrological drought indices (HDIs).

2 Methods

2.1 General approach

This work employed a cascade of models and algorithms ranging from two general circulation models (one atmospheric and one coupled) at the top to hydrological indices at the bottom (Fig. 1). Each step involved different types of target variables being forecasted: the *meteorological forecasts* (Fig. 1, top) refer to meteorological variables ("meteo data") from GCM forecasts and the subsequent downscaling and bias correction to match the spatial and temporal resolution. The *meteorological indices* (same figure, center) refer to the indices that were used to describe the magnitude of the forecasted meteorological drought. Finally, the *hydrological indices* (same figure, bottom) were calculated based on meteorological indices in an attempt to infer the magnitude of a hydrological drought characterized by meteorological and hydrological properties. To allow for the comparison with observations, we use results of GCM hindcast, i.e., a model that has been run with data only known until the specified time in the past. As these are supposed to represent and technically resemble true forecasts, they are referred to as "forecasts" henceforward. All results and computations after the statistical downscaling have a monthly time step. Similarly, all results and computations here presented were aggregated to selected subbasins (Fig. 2).

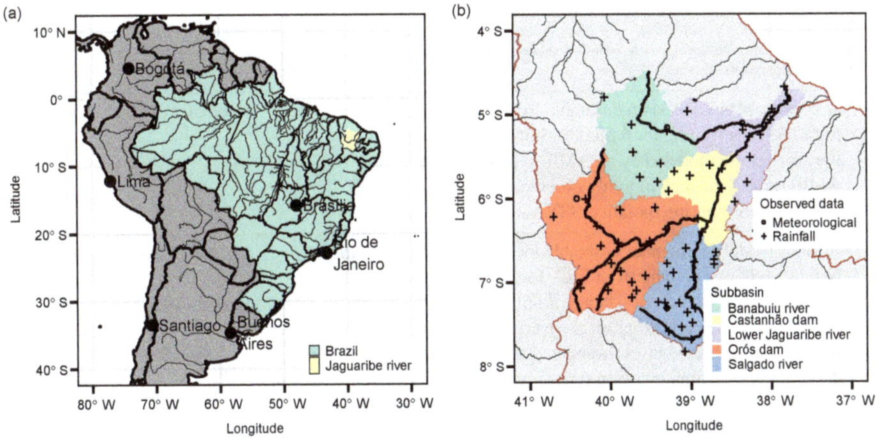

Figure 2. Left panel shows the location of the Jaguaribe River basin in South America. Right panel shows the Jaguaribe River together with its main tributaries, division into subcatchments used in this work, and meteorological and rainfall observation stations.

2.2 Study area

The spatial domain chosen for this analysis is the Jaguaribe River basin. Due to the river's regional importance, a lot has been written about its hydrology and development (see, e.g., de Araújo and Bronstert, 2016; de Aragão Araújo, 1990). The Jaguaribe is the most important river in Ceará. Its catchment has an area of 70 000 km^2 and is home to about 2.7 million people (IPECE, 2017). Annual precipitation amounts to 755 mm, of which about 90 % falls in the months January to June. The rainfall season comprising the months January to May is often considered key in securing water reserves for the whole year. June contributes with considerably less rainfall.

Average potential evapotranspiration is estimated to 2100 mm. Due to its dominant geology composed of a crystalline complex, aquifers in the region are unproductive. Runoff is practically the only source of drinking water for people and animals as well as irrigation. To that end, most of the water is stored in thousands of reservoirs of all scales across the watershed.

The main tributaries are the Banabuiú River in central Ceará and the Salgado River in southern Ceará. We aggregated the results of this research into five subcatchments: the aforementioned tributaries Banabuiú and Salgado, and the upper (upstream of Orós Reservoir), middle (upstream of Castanhão Reservoir) and lower (downstream of Castanhão Reservoir) Jaguaribe. An overview of the state and location of these catchments and tributaries is given in Fig. 2.

2.3 Seasonal forecast models ("GCM output")

To address the first research question we employed different combinations of dynamical and statistical models and a weather pattern classification methodology to produce meteorological drought indices. The dynamical seasonal forecast models were provided by FUNCEME and ECMWF in the form of hindcasts for the period 1981 to 2014. Details like resolution, reference and a short description are given in Table 1.

ECMWF operational seasonal forecasting system S4 has 51 ensemble members and 6 months' lead time. It is a fully coupled atmosphere–ocean model. The system has been systematically verified (Vitart, 2013; Molteni et al., 2011; Richardson et al., 2012). The hindcast version of the system has the same specifications of the operational model but only 15 ensemble members. It is available for academic purposes and is here employed as a benchmark for the verification of the regional forecasting system in operation in Ceará.

The seasonal forecasting system implemented at FUNCEME (Ceará's hydro-meteorological agency) is based on the ECHAM4.6 general circulation model. Details on this model can be found in Table 1. The operational and hindcast versions have 20 ensemble members and are run on initial sea surface temperature (SST) anomalies that persisted during the forecasting period (8 months). The initial state represents a typical (but random) realization of late December as derived from AMIP-type runs (Gates et al., 1999). The AMIP run starts in 1961 and is forced by monthly observed SSTs (NOAA Optimum Interpolation SST V2). Therefore, potential forecast skill is solely based on oceanic memory. The forecasting system of FUNCEME is in operational use and seasonal forecasts are released monthly.

2.4 Downscaling of GCM output

In order to predict precipitation over particular locations it is necessary to downscale the GCM forecasts. Three statistical downscaling approaches were employed: expanded downscaling (XDS), empirical quantile mapping (EQM) and weather pattern classification (WP; see Table 1 for details and references). To differentiate between two fundamentally different downscaling approaches, weather pattern classifica-

Table 1. Output variables of each prediction model used in this paper.

Model/method	Short description	Reference	Spatial scale
FUNCEME seasonal forecast system	A 20-member ECHAM4.6 ensemble. Atmospheric circulation model, initial SSTs persisting for 6 months. Initial conditions of the atmosphere modeled by an AMIP-type run (starting in 1961). AMIP run is forced by monthly observed SSTs (NOAA Optimum Interpolation SST V2).	Roeckner et al. (1992)	approx. 2.8° longitude/latitude
ECMWF seasonal forecast system	A multi-model 15-member ensemble including ocean circulation. Initial conditions coming from ERA-Interim	e.g., Stockdale et al. (1998)	approx. 0.7° latitude/longitude
expanded downscaling	Simulates local events consistent with prevailing atmospheric circulation while preserving local covariability	Bürger (1996)	network of monitoring stations
empirical quantile mapping	Improves systematic biases throughout the statistical distribution by mapping the empirical cumulative distributions of the observed and modeled variables	e.g., Wetterhall et al. (2012)	network of monitoring stations
weather pattern classification	Including pre-selection of variables, variable combinations and spatial domain	e.g., Murawski et al. (2016), Philipp et al. (2007)	network of monitoring stations

tion will not be referred to as downscaling approach/method throughout the text.

The downscaling approaches used here yielded a full set of meteorological variables distributed across the catchment at points where observations were available (daily mean temperature, relative humidity, wind speed and daily total precipitation and radiation). The forecasting products obtained from the combinations of GCMs and downscaling will be named after their components: *XDS:ECHAM*, *XDS:ECMWF*, *EQM:ECHAM*, *EQM:ECMWF*, *WP:ECHAM*, and *WP:ECMWF*.

Weather patterns were classified using the SANDRA methodology described in Philipp et al. (2016). The selection of the *optimal* classification was done visually in respect to the explained variation of the observed meteorological drought indices. The classification itself was independent of the MDIs, so that no artificial skill was to be expected from forecasting the stations. Only MDI scales of 1, 12 and 36 months were calculated.

2.5 Drought quantification using drought indices

Meteorological droughts were quantified in magnitude and temporal scale using meteorological drought indices (MDIs). After careful appraisal regarding data demand and current conventions, the following indices were selected: SPI_1, SPI_3, SPI_6 (WMO, 2012b; McKee et al., 1993), $SPEI_1$, $SPEI_3$ and $SPEI_6$ (Vicente-Serrano et al., 2009). The subscripted numbers (e.g., SPI_1) refer to the temporal scale in months for which the index was computed.

The forecast is generated at the beginning of January for the period from January until June. Indices obtained by downscaling forecasts with a temporal scale greater than the lead time of the forecast will include values from the observation set. SPI_6, for example, will contain 5 months of measured precipitation in the January forecast. In June, the same index will be calculated exclusively with forecasted precipitation. The skill of a SPI_6 forecast for some months is therefore expected to be greater than the skill of a SPI_1 forecast beforehand. This feature does not apply to WP classification.

Timescales greater than 6 months are of no value for the verification of the forecasting system in terms of meteorology, as rainfall in the preceding dry season is usually negligible. However, the hydrology of Ceará is characterized by long-term memory introduced by a vast network of reservoirs. Additionally, drought events in this region are known to be long and creeping phenomena that must be quantified on large temporal scales. MDIs with long temporal scales will therefore have to be considered when designing the hydrological drought index (HDI) forecast model in the next section. To that end, we will employ shorter timescales for MDI verification, but keep longer timescales (greater than 6 months) in the regression of hydrological drought indices (HDIs), since they provide a better fit for the forecast model.

Regarding hydrological droughts, various HDIs were reviewed and two were considered suitable for this work. All other indices either (a) require consumptive data for water use, which is impractical for the given settings, or (b) focus on streamflow, which misses the most important features (ephemeral rivers, role of reservoirs) of the hydrological system of Ceará and many other semi-arid regions. The only index chosen from the literature was the surface water supply index (SWSI) as formulated in Doesken et al. (1991) with a weight of 0.5 for precipitation within the reservoir catchment and 0.5 for reservoir volume:

$$\mathrm{SWSI} = \frac{0.5P(\mathrm{rs}) + 0.5P(\mathrm{pr}) - 50}{12}, \tag{1}$$

where $P(x)$ is the non-exceedance probability of x based on available historical records of x, rs is mean monthly reservoir storage in the respective catchment and pr is the monthly precipitation averaged for the respective catchment. The second index, V, was defined as the regional reservoir volume at the end of each month relative to the total regional reservoir storage capacity.

In terms of event prediction, the event considered for the meteorological drought indices in use in this work is "dry spells of moderate to extreme magnitude", translated by values lower than or equal to -1 in the SPI/SPEI scale. For precipitation a threshold based on the 30th percentile of the series of observed monthly precipitation was used. The threshold for defining HDI drought events was based on the 30th percentile of the series of observed monthly HDIs. The reason for using the 30th percentile was the classification used by the regional agencies to separate between a "dry", a "wet" (above the 70th percentile) and a "normal" year.

2.6 Regression of hydrological drought indices

Forecasts of hydrological drought indices were obtained by searching and fitting a multivariate regression model to observations of hydrological drought indices and reservoir volume changes. As candidate predictors, meteorological drought indices of all temporal scales were used.

For predicting SWSI the multivariate linear regression was fit directly to the hydrological index. For the regional reservoir volume, V, two different approaches were followed. With the first approach, M1, the multivariate linear regression was fit directly to the values of V, analogous to SWSI. With M2, the second approach, the multivariate linear regression was first fit to the monthly changes in V. Then the predicted value of V was calculated by adding its predicted monthly changes to the most recent measured value in December. The regional reservoir volumes predicted by the two regression models M1 and M2 are denoted V_{M1} and V_{M2}, respectively.

Model parsimony was enforced by predictor selection comprising a heuristic search for the best Akaike information criterion (AIC) under the constraint of checking the predictors for multi-collinearity. To eliminate multi-collinearity between predictors, correlated predictors were replaced by their ratios.

Possible forms of multilinear regression include predictors as denominators of fractions. This implies that these predictors must not take the value zero, in order to exclude division by zero. To enforce this condition, the MDIs in question were removed from the time series when approaching zero, in particular values in $]-0.1, 0.1[$.

2.7 Forecast verification

At each level of Fig. 1, a verification of the forecast was performed. Three metrics were employed: the root mean square error (RMSE), the relative operating characteristic skill score (ROCSS) and the Brier skill score (BSS) (Wilks, 2005). Root mean square error is a scalar accuracy measure applied to the realizations of the ensemble forecast. The Brier score is also a scalar accuracy measure, though for verification of probabilistic forecasts of predefined events. The relative operating characteristic is a discrimination-based verification metric for forecasts of defined events. For more information on these metrics, we recommend chap. 8 of Wilks (2005).

RMSE was computed for each member, ensemble mean and climatology, i.e., the long-term mean annual cycle. Climatology was considered the reference forecast. The mean square error was computed for monthly values in the forecast period (1981–2014, January–June) and averaged over the entire period. The square root of this measure is the RMSE. It shows the capability of the model to correctly forecast monthly values, but it does not quantify the skill to predict particular events of water scarcity. January to June precipitation represents over 90 % of the annual precipitation in the Jaguaribe basin.

Another important metric employed was the BSS. The Brier score can be seen as the sum of three terms: *reliability*, *resolution* and *uncertainty*. The term *reliability* measures the differences between forecast probabilities and relative frequencies of the observed event. Thus, low values of this score correspond to high reliability. *Resolution* measures the ability of the forecast to discern periods in which observed frequencies depart from average. Finally the term *uncertainty* quantifies the variability of the observations: when the event being forecasted almost never or almost always happens, the uncertainty of the forecast is small. The Brier score is here understood as in Wilks (2005) as

$$\mathrm{BS} = \frac{1}{n}\sum_{k=1}^{n}(y_k - o_k)^2, \tag{2}$$

where BS is the Brier score and k denotes the index of the n forecast-event pairs. y_k is the forecast probability for each forecast-event pair k and belongs to $[0, 1]$. The forecast probability is calculated as the number of members of the ensemble that forecast an event divided by their total count. o_k is the observation for each forecast-event pair, which can take the value 1 for an event and 0 when no event is observed in k.

The BSS is computed with respect to the Brier score of the reference forecast ($\mathrm{BS}_{\mathrm{ref}}$),

$$\mathrm{BSS} = 1 - \frac{\mathrm{BS}}{\mathrm{BS}_{\mathrm{ref}}}, \tag{3}$$

and it can take any value lower than or equal to one. A forecast is said to have skill if its BSS is greater than zero.

The reference forecast was considered to be the climatological relative frequency of the predefined event. For example, the climatological relative frequency for February is the number of times that a February observation, e.g., of precipitation, is considered an event divided by the total number of years in the hindcast period.

The last metric employed was the ROCSS. The relative operating characteristic describes the ability to discriminate between true positives and false positives when forecasting a given event. It is normally calculated for a set of forecast

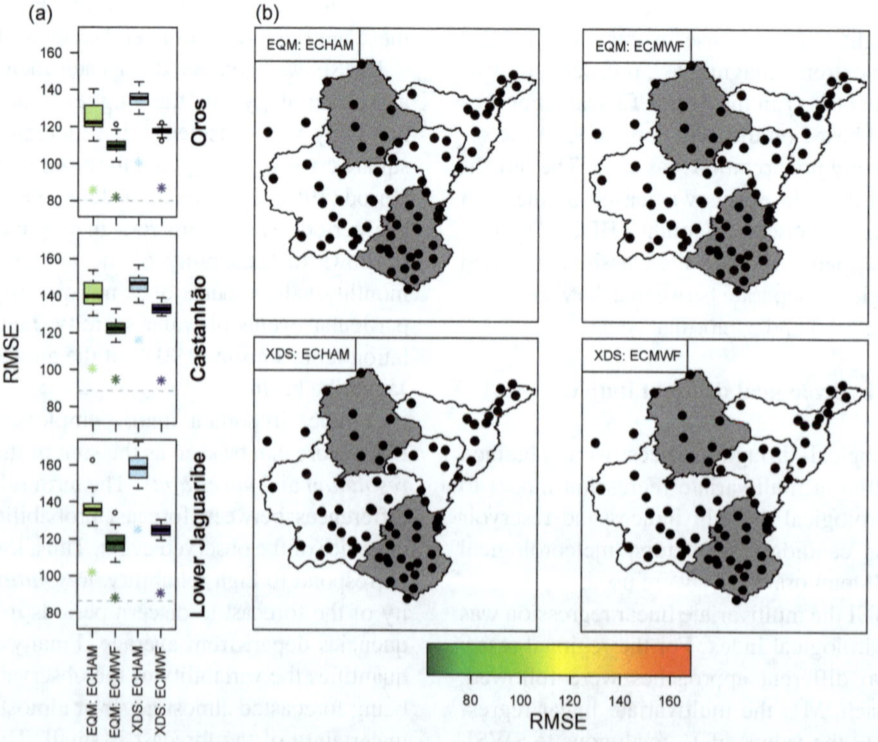

Figure 3. Root mean square error of the forecast of monthly precipitation. Panels **(a)**: box plots show the spread of the RMSE of each model. The asterisks (*) show the RMSE of the ensemble mean. The RMSE of using climatology as a forecasting product is given by the grey dashed line. The four panels **(b)** show the RMSE for each individual station for each model. Note that in general the ensemble mean ranks better than the best of the ensemble members.

probability bins, thereby having great importance for decision makers. ROCSS was calculated as

$$ROCSS = 2 \cdot AUC - 1 \quad (4)$$

as in Wilks (2005), where AUC is the area under the relative operating characteristic curve. The ROCSS can have values between -1 and 1, where anything below zero means *no skill*. A ROCSS of 0 corresponds to the skill of a reference random forecast.

3 Results and discussion

3.1 Forecasting precipitation

The RMSEs of the precipitation forecast are presented in Fig. 3. ECMWF ranks better than ECHAM, while EQM:ECMWF results in the lowest RMSEs and XDS:ECHAM in the greatest. Still, the best results in terms of RMSE are comparable to the climatology, meaning that there is limited skill in forecasting monthly precipitation. The spatial distribution of RMSE of the ensemble mean in April shows a concentration of high RMSEs in the lower Jaguaribe catchment for EQM and in the Salgado catchment for XDS.

The ensemble mean of the forecast, shown by the asterisks in Fig. 3 as well as in other figures below, always displayed a lower RMSE than *any of the ensemble members*. This happens because the ensemble mean "smoothes out unpredictable detail and presents the more predictable elements of the forecast" (WMO, 2012a). Despite its usefulness, the ensemble mean is not entirely appropriate for predicting drought events. Ensemble means do not provide any information on the probability of an extreme event.

Unlike RMSE, which does not provide any information on the skill of event forecasts, the BSS is explicitly suited for that purpose and is shown in Fig. 4. One remarkable observation is to be made regarding the BSS: skill is mostly absent when forecasting drought events based on precipitation and its thresholds. The only exception is the forecast for April, where the multi-model ensemble shows limited skill in the three regions considered. These results will be discussed further in light of the greater skill shown when forecasting drought events based on MDIs.

Still, all forecasting systems here presented show skill in discriminating events against false alarm forecasts. This is expressed by the ROC curve shown in Fig. 5. The variation of the ROCSS over time can be attributed to lead time (skill decreasing with increasing lead time) and to low or no precipitation in the months before and after the rainy season.

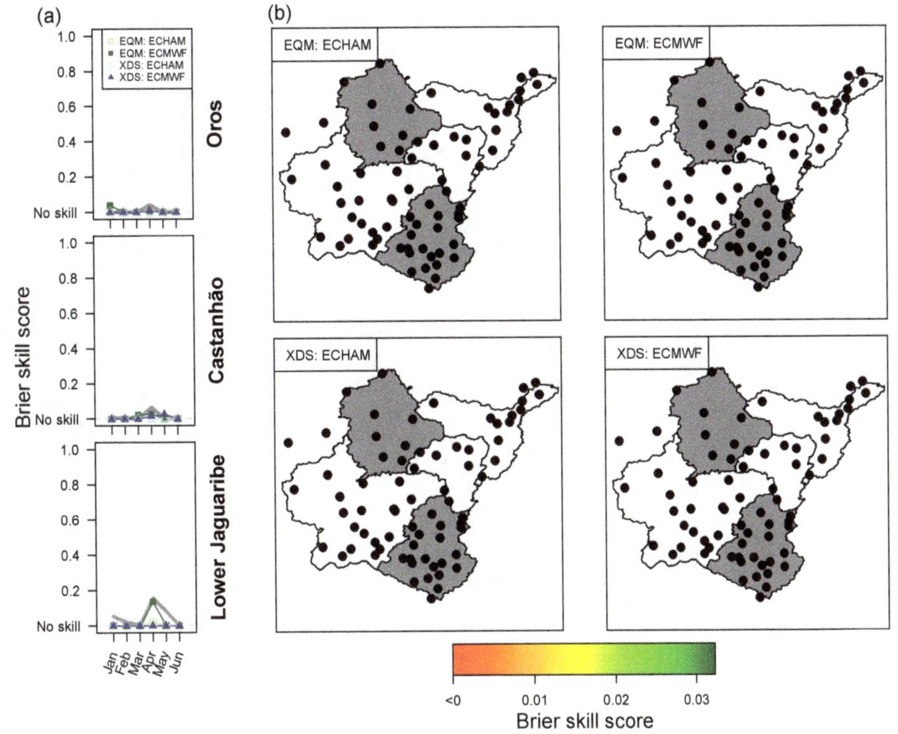

Figure 4. BSS of the forecast of a monthly low precipitation event. **(a)** The BSS is shown for each model/downscaling combination and for the forecasting months averaged over the respective subcatchments. BSS values below zero were assigned a "no skill" category in order to improve readability. The grey line is the BSS of the multi-model ensemble. Panels **(b)** show BSS averaged over all forecasting months for each station. Note that in most cases the forecast of monthly precipitation has no skill.

Months of typically low precipitation showed poor ROCSS (Fig. 5: January, May, June). When comparing downscaling techniques and GCMs, EQM mostly outperformed XDS, while the skill was less affected by using different GCMs.

To put our results into context, we could find three reports with a statement of verification concerning precipitation forecast in Ceará. de Castro et al. (2013) present a RMSE of between 120 and 130 mm for the *Sertão Interior de Inhamuns*, using an empirical model with forecasts issued in January for the period February to June. Moura and Hastenrath (2004), with a forecast issued at the end of February for the period of March to June, i.e., with a shorter lead time than our work, show a RMSE of 50 to 70 mm (Hastenrath and Greischar (1993) obtained similar results).

3.2 Forecasting meteorological drought indices

A time series of seasonal MDI forecasts was plotted to illustrate the forecast spread given by model EQM:ECMWF (Fig. 6). The improvement provided by the ensemble mean, when compared to each member, is clearly visible. Also visible are several observed events of moderate to severe drought (below the dashed grey line). The ensemble mean is able to forecast at least a few of these events.

A measure of the general agreement (for all kinds of conditions, dry, average or wet) between forecasted and observed MDIs is given by the RMSE in Fig. 7. The relationship between forecast probability and relative frequency of a drought event (i.e., the BSS) is provided in Fig. 8, whereas the balance between hit rate and false alarm rate for the same event can be seen in the form of ROCSS in Fig. 9 below.

The RMSE of MDI forecasts is shown in Fig. 7. With the exception of the predictions produced by the WP approach for SPI_1, the general ranking of the approaches is quite consistent among the three subbasins. As with precipitation, the RMSE of SPI_1 and $SPEI_1$ generally does not differ from that of the climatology and is greatest for ECHAM and EQM. EQM:ECMWF and XDS:ECMWF show the consistently lowest RMSE and XDS:ECMWF performs better than the climatology. Interestingly, ECMWF consistently outperforms ECHAM on all scales.

RMSE reflects the prediction skill for the whole range of the indices, including wet spells and dry spells/droughts. When aiming primarily at forecasting drought events, this verification may be misleading. Nevertheless, this metric shows which models are most appropriate for this domain and confirms the plausibility of the forecasting system also for wet years.

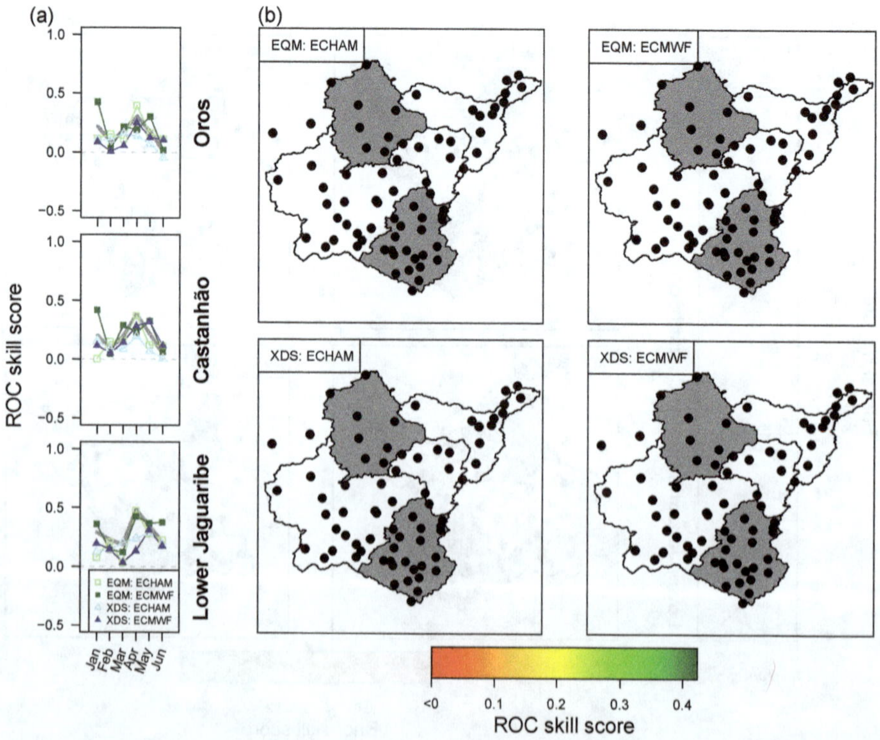

Figure 5. ROCSS of the forecast of a monthly low precipitation event. **(a)** The ROCSS is shown for each model/downscaling combination and for the forecasting months averaged over the respective subcatchments. Panels **(b)** show ROCSS averaged over all forecasting months for each station.

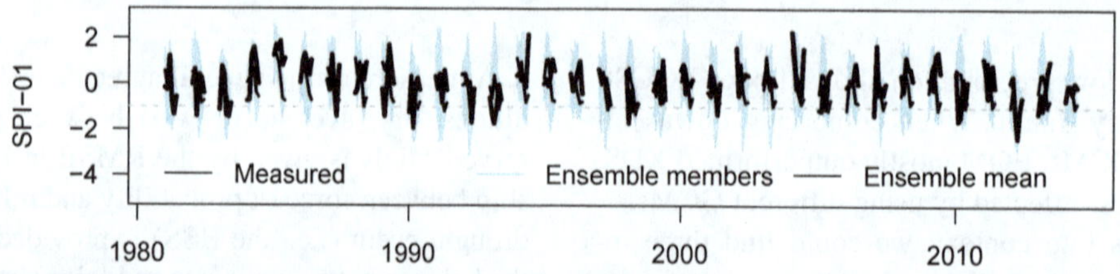

Figure 6. Time series of the seasonal forecast of SPI_1 in the Castanhão subcatchment given by ECMWF:EQM. Only periods from January to June are shown. The threshold "moderate drought event" is given by the grey dashed line.

As for the BSS, Fig. 8 shows this indicator of skill for timescales of 1, 3 and 6 months in three regions of the Jaguaribe River. For the 1-month timescale, it is noteworthy that the first 3 months of the forecast display the lowest skill. In particular the March forecast shows no skill in most models, March being a key contributing month in the rainy season. The second half of the rainy season, April/May/June, has generally more skill. The same BSS minimum can be observed in the SPI_3 and SPI_6 panels, but this time with slightly greater value than for SPI_1, since these indices entail some measured data. Another interesting observation is that, contrary to RMSE, here no product can be considered a clear winner.

For the ECHAM model a possible explanation for lower skill at the onset of the rainy season may lie in its initial conditions. Since the initial conditions for each model run are provided by the output of an AMIP-type run (Gates et al., 1999), they may depart considerably from actual atmospheric conditions. According to this hypothesis, the model would come closer to atmospheric conditions only through the SST forcing, which could explain a certain lag in the forecasting skill. Still, this explanation can only account for ECHAM and not the ECMWF model, which is fully coupled and whose initial conditions are derived from ERA-Interim.

The ROCSS for the different months of the forecasting period shows a slightly different picture than the RMSE and BSS previously presented. Figure 9 shows ROCSS for timescales of 1, 3 and 6 months in three regions of the Jaguaribe River. There is no clear pattern concerning the relationship between lead time and skill for any of the fore-

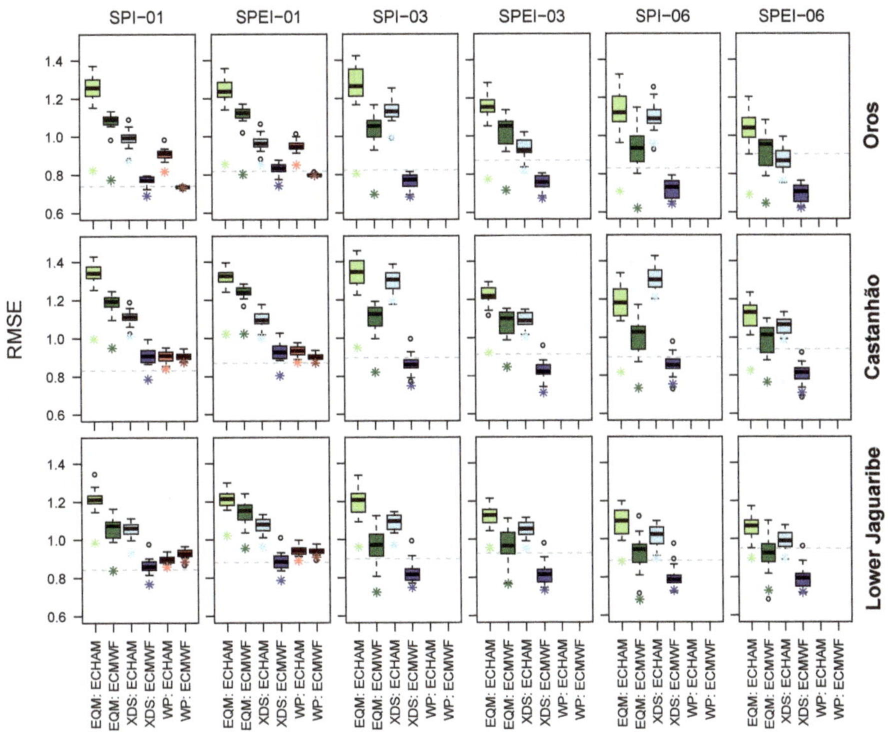

Figure 7. Box plots of the root mean square error of a forecasted meteorological drought index. The asterisk (*) shows the RMSE of the ensemble mean and box plots show the spread of the individual members. Note that in general the ensemble mean ranks better than the best of the ensemble members.

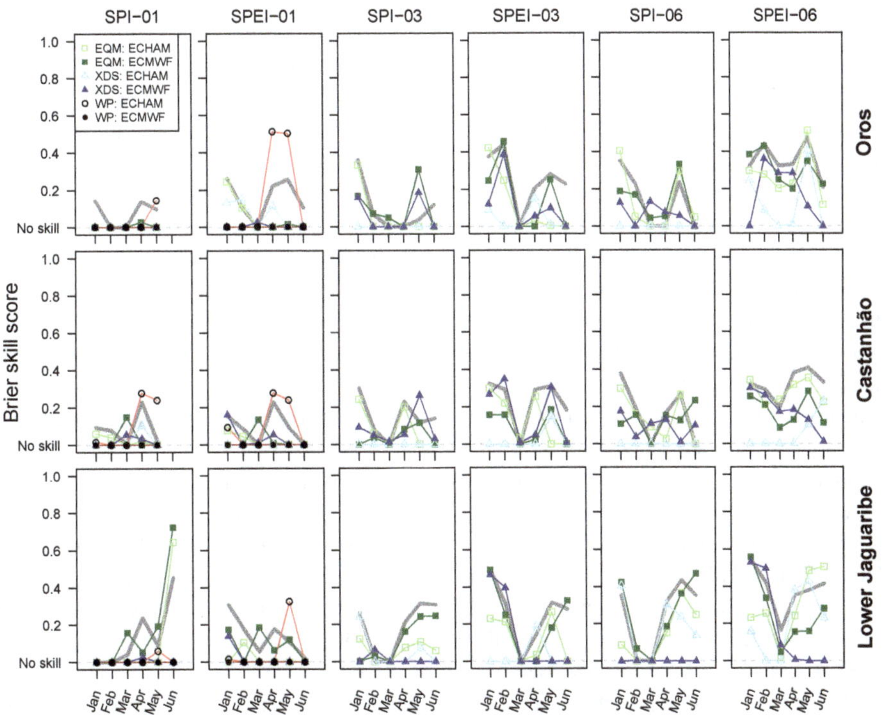

Figure 8. BSS of a forecasted meteorological drought event based on an event of index lower than −1. The grey line shows the result of the multi-model ensemble mean.

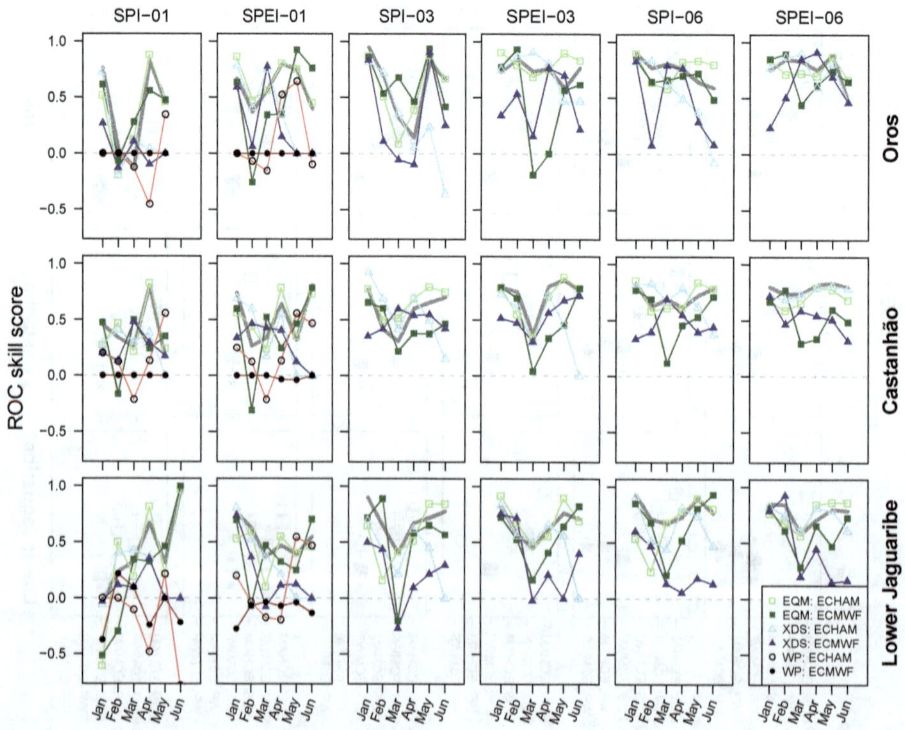

Figure 9. ROCSS of a forecasted meteorological drought event based on an event of an index lower than −1. The grey line shows the result of the multi-model ensemble mean.

casting models. As in previous plots, the forecasting skills for different MDIs tend to display a minimum in March.

Contrary to the results for RMSE, ECHAM shows comparably good ROCSS and BSS in forecasting MDI drought events of all scales in all three regions. Still, the comparably low skill of the March forecast is problematic, March being the month of greatest precipitation in most of the catchment. WP:ECHAM features the best BSSs for $SPI_1/SPEI_1$ in April and May, whereas EQM:ECHAM features generally the highest ROCSS in April for the same scale.

It is worth looking at the BSS of $SPI_6/SPEI_6$, even if they partly encompass measured values. BSS in June in particular is a valuable indicator of the ability of the models in forecasting the whole rainy season. Here, most products display some skill in forecasting a drought event. XDS:ECMWF is the only one displaying no skill for all three regions in $SPEI_6$. Generally the skills are higher with $SPEI_6$ than with SPI_6. Regarding $SPEI_6$, EQM:ECHAM and EQM:ECMWF display skill in all three regions. In the important region of Castanhão, where the largest reservoir and most infrastructure is located, EQM:ECHAM and XDS:ECHAM perform best in forecasting $SPEI_6$ for June, although with a low value of BSS.

The multi-model ensemble skill shown by the grey line is generally close to the upper envelope formed by that of the individual models. For $SPEI_1$ in the months January to May (rainy season) the ROCSS of the multi-model ensemble is always positive and oscillates around 0.5. An interesting result is the improvement in skill when SPI_1 is replaced by $SPEI_1$. The grey line, which shows the ROCSS/BSS for the multi-model ensemble, has an increase in skill at all scales and regions.

A similar forecast assessment has been reported by, e.g., Dutra et al. (2013). Events were defined by a SPI_3 lower than −1, with a lead time of 3 months. ROCSS obtained were on the order of 0.6 for the Blue Nile basin, which is comparable with the results presented in this paper, but much lower for other river basins, e.g., the Zambezi.

3.3 Forecasting hydrological drought indices

The multivariate regression model equations obtained and their respective R^2 are shown in the Appendix, Table A1. Long-scale MDIs (like SPI_{12} or SPI_{36}) prevail as predictors of reservoir volume, whereas short-scale MDIs like SPI_1 are mostly present as predictors of reservoir volume change. This reflects the timescale of reservoir storage variations. At a given moment in time, the reservoir storage results from several months of inflow. Similarly, the effect of a month of high inflow on the reservoir storage level is likely to be only residual.

The forecast of the three HDIs shows notable differences between downscaling techniques EQM/XDS and the WP classification (Fig. 10). WP classification has a lower RMSE

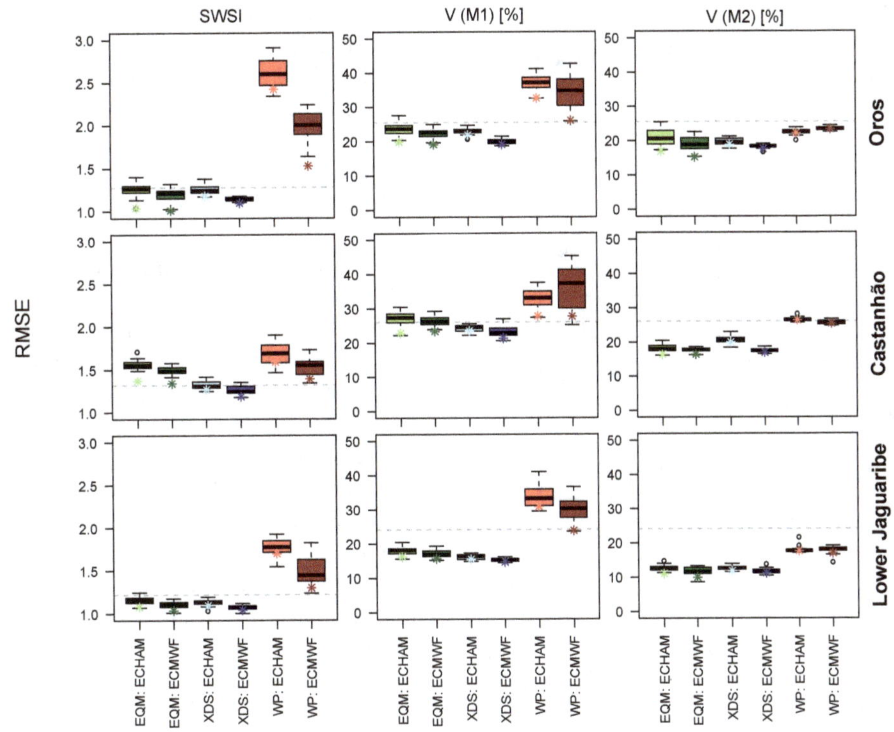

Figure 10. Root mean square error of the forecast of SWSI, V predicted with M1 and V predicted with M2 (based on month-to-month variation). The forecast period is January to June. Three regions are presented: lower Jaguaribe, Orós and Castanhão. The horizontal grey dashed line shows the RMSE of the climatology.

than EQM/XDS when predicting SWSI or V_{M1}. For V_{M2}, the difference between WP and EQM/XDS is much smaller. The ensemble spread of WP classification shrinks considerably from V_{M1} to V_{M2}. All methods show a decrease in RMSE from V_{M1} to V_{M2}.

Again, WP classification considers by design only a range of discrete MDIs, which can affect RMSE. MDIs were limited to nine values, of which −0.75, 0 and 0.75 are the closest to zero. The continuous values of MDIs derived by the other products are problematic, because the multilinear regression also considers division by the meteorological drought index. When the MDIs are close to zero, outliers arise and skew the RMSE. These datapoints were therefore removed from the verification metrics.

Regarding the prediction of HDI drought events, Fig. 11 clearly points out that prediction performs best when targeting reservoir volume with model M2 (adding predicted monthly value to the December observed regional reservoir volume). Here, all products show reasonable performance for most regions, but a decreasing skill with increasing lead time. Another important observation is that WPs do not display skill in forecasting HDI events as shown in Fig. 11.

Contrary to the MDIs, the BSSs of the HDIs do not feature a minimum in March. A slight tendency of lower skills towards the end of the rainy season is observable in V_{M1} forecasts. V_{M2} shows comparably good results for all

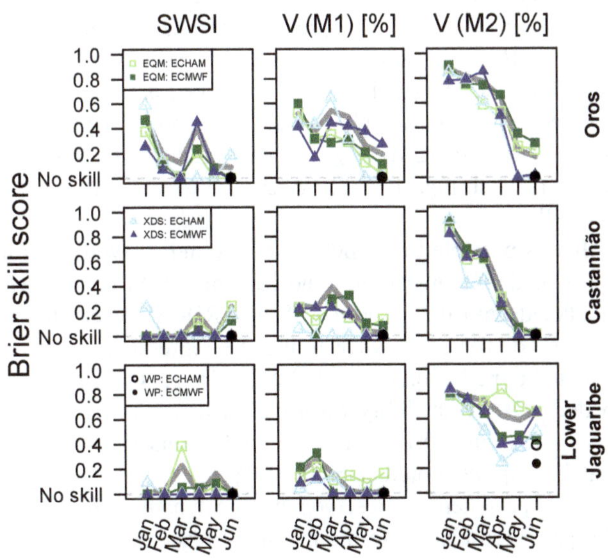

Figure 11. BSS of the forecast of drought events as defined by SWSI, V predicted with M1 and V predicted with M2 (based on month-to-month variation). An event is defined when an index is lower than the 30th percentile of the observations. The forecast period is January to June. Three regions are presented: lower Jaguaribe, Orós and Castanhão.

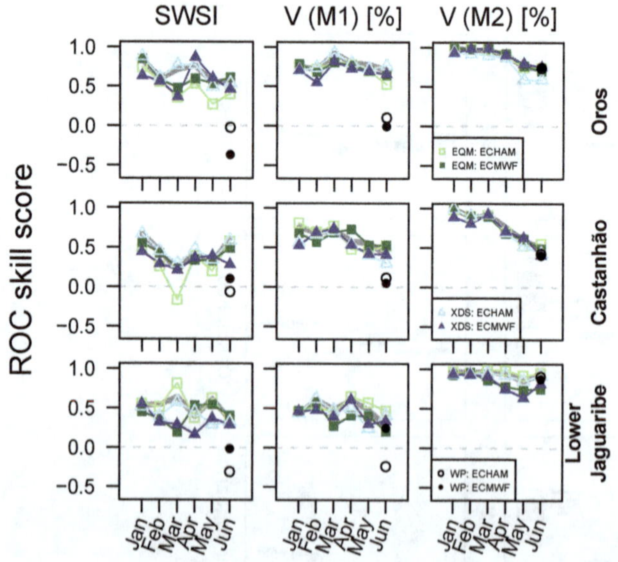

Figure 12. ROCSS of the forecast of drought events as defined by SWSI, *V* predicted with M1 and *V* predicted with M2 (based on month-to-month variation). An event is defined when an index is lower than the 30th percentile of the observations. The forecast period is January to June. Three regions are presented: lower Jaguaribe, Orós and Castanhão.

GCM/downscaling combinations in predicting HDI events, confirming the results in Fig. 10.

The ROCSS shows small differences between GCMs or downscaling methods (Fig. 12). V_{M2} features the highest ROCSS of the different indices used and very little variability among downscaling approaches and GCMs employed. As with BSS, the ROCSS of V_{M2} decreases with increasing lead time. The results of SWSI and V_{M1} are very similar, with SWSI showing higher variability among downscaling approaches and GCMs. V_{M2} could be predicted by WP classification with high ROCSS, whereas V_{M1} and SWSI show no skill.

It was possible to predict any of the indices with skill in most modeling approaches and catchments. Still, V_{M2} was predicted with the greatest BSS and ROCSS, even though it showed worse R^2 when fitting the regression model on which the prediction is based (Table A1). This result hints at better HDI predictability when the predictant is a change in reservoir volume than the reservoir volume itself. One reason for the improved predictability of V_{M2} is surely the importance of persistence in reservoir storage dynamics. By adding the predicted change to the measured reservoir volume we are providing valuable measured information to the forecast model that SWSI and V_{M1} do not have.

We could not find reports on streamflow/reservoir forecasting systems for the region of Ceará stating BSS, ROCSS, RMSE or another verification measure. Still, for other semi-arid regions of the world, similar skill values could be found in the literature. Trambauer et al. (2015) forecasted events of a standardized runoff index of 6 months lower than −0.5 with variable lead times. Their best catchment points to a ROCSS of 0.7 with a lead time of 5 months. Seibert et al. (2017) forecasted events with a standardized streamflow index (Vicente-Serrano et al., 2012) below −0.5, reporting a ROCSS of 0 at the outlet of a large river (the Limpopo in southern Africa) to close to 1 in its headwater catchments.

3.4 Multi-model ensemble forecast

Finally, we present the skill score of the multi-model ensemble forecast in Table 2. Each type of index considered (precipitation, meteorological drought index and hydrological drought index) is presented. Results of the WP classification were excluded from the multi-model ensemble, because they did not cover all the indices addressed in this work.

The BSS of forecasts of low precipitation events (given in column *P*), as well as that of the forecasts of drought defined by the SPI_1, show either very low or no skill. Forecasts of SPEI generally display greater skill than the forecasts of SPI. This points to a possible bias in the forecasting that is compensated for by introducing temperature into the equation of SPEI.

The best skill obtained by the multi-model ensemble was forecasting drought events related to reservoir storage in the lower Jaguaribe region. The good skill of the reservoir storage forecast is likely related to the long memory of the reservoir system. The forecasted precipitation will affect the reservoir only marginally, since most of its storage is accumulated throughout several years. Most importantly, BSS increases when V_{M2} is used instead of V_{M1}, i.e., when reservoir volume is forecasted by adding forecasted reservoir volume change to measured December reservoir volume.

Table 2 reveals an interesting pattern in this work: additional information to the forecast model tends to increase forecast skill. $SPEI_1$ is based on temperature and precipitation data and was forecasted with greater skill than SPI_1, which is only based on precipitation. Similarly, $SPEI_6$, which combines forecasted and measured precipitation and temperature from months prior to the forecasting period, has more skill than $SPEI_1$ forecasts. The greatest BSS is given by V_{M2}, a HDI that requires measured initial reservoir volumes as well as a combination of several MDIs. This last point stresses the importance of assimilating prior hydrological conditions into the forecast products.

4 Conclusions

The plausibility and skill of a set of drought forecasting models were presented. Different types of drought events were considered: a rainfall anomaly during the rainy season, standardized precipitation indices below a given threshold and anomalies in regional reservoir storage. The forecast products considered were combinations of two models, ECHAM

Table 2. BSS of January–May multi-model ensemble forecast. The ensemble includes ECHAM and the ECMWF seasonal forecast model, as well as the EQM and XDS downscaling techniques. The BSSs are averaged over each region. Columns show different indices used for the forecast: P is seasonal precipitation, SPI_1 and $SPEI_1$ are standardized precipitation indices with scale 1 month, and *Reservoir volume* stands for regional reservoir volume in percentage of regional storage capacity. The BSS refers to meteorological and hydrological drought events described in Sect. 2.

Region	BSS					
	P	SPI_1	$SPEI_1$	$SPEI_6$	Reservoir volume (M1)	Reservoir volume (M2)
Orós	no skill	no skill	0.11	0.38	0.43	0.66
Castanhão	no skill	0.08	0.11	0.32	0.23	0.52
Lower Jaguaribe	0.06	no skill	0.16	0.37	0.12	0.71

and the ECMWF seasonal forecast, two downscaling techniques, XDS and EQM, and a weather pattern classification approach.

Each model provided an ensemble of predictions, so deterministic and probabilistic measures of skill could be used. The deterministic measure allowed us to see the significant improvement introduced by the ensemble mean: the ensemble mean had in most cases a lower root mean square error than the climatology. The RMSE of the ensemble mean however was comparable to the climatology and in some cases greater. Still, no approach had a RMSE that significantly departed from the RMSE of the climatology.

A multi-model ensemble forecast was obtained by binding all members of all models into one product. The skill of this forecast is given in Figs. 4, 8, and 11, and Table 2. Multi-model ensembles can be considered to be our best guess of a probabilistic drought forecast, since they are consistently among the best forecast skills provided by the individual models. Individual members surpassed the multi-model ensemble skill only occasionally, for particular combinations of regions, months and indices.

The skill of the hydrological drought forecast, namely the relative reservoir storage V_{M2}, was 0.66, 0.52 and 0.71 for the regions of Orós, Castanhão and lower Jaguaribe, respectively. The skill obtained for the hydrological drought forecast is likely inflated by the long memory of the reservoir system and the use of observed reservoir volume to define the conditions prior to each forecast. Still, the R^2 of the regression that provides the reservoir variation underlying V_{M2} was lower than that of V_{M1}, indicating that a regression might be a poor prediction of reservoir inflow. Improvements are expected by coupling a process-based hydrological model to the seasonal forecasting system.

This work showed that a multi-model ensemble can forecast drought events of timescales relevant to water managers in northeastern Brazil with skill. But no or little skill could be found in the forecasts of monthly precipitation or drought indices of smaller temporal scales, like SPI_1. Both this work and others here revisited showed that major steps forward are needed in forecasting the rainy season in northeastern Brazil.

Appendix A: Multivariate regression of regional reservoir volume

Table A1. Regression used for predicting regional reservoir volume and regional reservoir volume change using a set of MDIs as predictors. Regional reservoir volume was taken at the end of each month relative to the total regional reservoir storage capacity. Regional reservoir volume change refers to the difference between the given and previous months.

Region	Predictant	Formula	R^2
Orós	Reservoir volume	$59.0 - 22.9\,\mathrm{SPEI}_{36} + 6.67\,\mathrm{SPI}_{12} + 45.4\,\mathrm{SPI}_{36} + 6.00\mathrm{SPI}_{12}\mathrm{SPEI}_{1} - 5.30\mathrm{SPEI}_{36}\mathrm{SPI}_{36}$	0.64
Orós	Reservoir volume change	$0.416 + 2.43\,\mathrm{SPEI}_1\,\mathrm{SPI}_1 + 2.23\,\mathrm{SPI}_1\,\mathrm{SPI}_{12} - 0.173\,\mathrm{SPI}_{36}\frac{\mathrm{SPEI}_{36}}{\mathrm{SPEI}_{12}} + 4.20\mathrm{SPEI}_1\frac{\mathrm{SPI}_{12}}{\mathrm{SPEI}_{12}} - 0.00334\frac{\mathrm{SPEI}_{36}}{\mathrm{SPEI}_{12}}\frac{\mathrm{SPI}_{36}}{\mathrm{SPI}_{12}}$	0.36
Castanhão	Reservoir volume	$55.4 + 12.5\mathrm{SPEI}_{36} + 12.4\frac{\mathrm{SPI}_{36}}{\mathrm{SPEI}_{36}} - 3.12\mathrm{SPEI}_{12}\frac{\mathrm{SPI}_{12}}{\mathrm{SPEI}_{36}} - 3.19\frac{\mathrm{SPI}_{12}}{\mathrm{SPEI}_{36}}\frac{\mathrm{SPI}_{12}}{\mathrm{SPEI}_{12}} + 10.19\frac{\mathrm{SPI}_{36}}{\mathrm{SPEI}_{36}}\mathrm{SPEI}_{12}$	0.41
Castanhão	Reservoir volume change	$2.95 + 3.47\mathrm{SPI}_1 - 1.15\frac{\mathrm{SPI}_1}{\mathrm{SPEI}_1} - 1.27\frac{\mathrm{SPI}_1}{\mathrm{SPEI}_1}\mathrm{SPEI}_{36} - 0.791\mathrm{SPI}_1\frac{\mathrm{SPI}_{36}}{\mathrm{SPEI}_{12}} + 1.41\mathrm{SPEI}_{12}\frac{\mathrm{SPI}_{36}}{\mathrm{SPEI}_{36}}$	0.21
Lower Jaguaribe	Reservoir volume	$33.4 + 16.3\,\mathrm{SPEI}_{36} + 16.6\,\mathrm{SPI}_{12} + 5.65\,\mathrm{SPEI}_{36}\,\mathrm{SPI}_{12} - 13.5\,\mathrm{SPI}_{12}\frac{\mathrm{SPI}_{12}}{\mathrm{SPEI}_{12}} + 0.877\,\mathrm{SPI}_{12}\frac{\mathrm{SPI}_{12}}{\mathrm{SPEI}_{36}}$	0.60
Lower Jaguaribe	Reservoir volume change	$0.689 + 2.22\mathrm{SPI}_1 + 0.0353\frac{\mathrm{SPI}_{36}}{\mathrm{SPI}_{12}} + 2.12\mathrm{SPI}_1\mathrm{SPI}_{12} + 1.08\,\mathrm{SPI}_{12}\frac{\mathrm{SPI}_{12}}{\mathrm{SPEI}_{12}} + 0.286\,\mathrm{SPI}_1\frac{\mathrm{SPI}_{12}}{\mathrm{SPEI}_{36}}$	0.38

Author contributions. Conceived and designed the experiments: TF. Performed the experiments: SV, JMRP, GB, KV, AM, JMD, FVJ, EM. Analyzed the data: JMD, SV, TF. Wrote the paper: JMD, TF, SV.

Competing interests. The authors declare that they have no conflict of interest.

Acknowledgements. This work was funded by the Federal Ministry of Education and Research of Germany under grant number 01DN14013. The first author was also supported by the German Research Foundation under grant number BR1731/18-1. One of the hindcast datasets was kindly provided by the European Centre for Medium-Range Weather Forecasts. The Open Access Publication Fund of the University of Potsdam supported the publication of this research paper.

Edited by: Carlo De Michele

References

Bürger, G.: Expanded downscaling for generating local weather scenarios, Clim. Res., 7, 111–128, https://doi.org/10.3354/cr007111, 1996.

de Aragão Araújo, J. A.: Barragens no Nordeste do Brasil, DNOCS, Fortaleza, 1990.

de Araújo, J. C. and Bronstert, A.: A method to assess hydrological drought in semi-arid environments and its application to the Jaguaribe River basin, Brazil, Water Int., 41, 213–230, https://doi.org/10.1080/02508060.2015.1113077, 2016.

de Castro, T. N., Souza, F., Alves, J. M. B., Pontes, R. S. T., dos Reis, L. L. N., and Daher, S.: Neo-fuzzy neuron model for seasonal rainfall forecast: A case study of Ceara's eight homogenous regions, J. Intell. Fuzzy Syst., 25, 389–394, https://doi.org/10.3233/IFS-2012-0645, 2013.

Doesken, N., McKee, T., and Kleist, J.: Development of a Surface Water Supply Index for the Western United States, Tech. Rep. Climatology Report 91-3, Colorado Climate Center, Department of Atmospheric Science, Colorado State University, available at: http://climate.colostate.edu/pdfs/climo_rpt_91-3.pdf (last access: 24 September 2018), 1991.

Dutra, E., Di Giuseppe, F., Wetterhall, F., and Pappenberger, F.: Seasonal forecasts of droughts in African basins using the Standardized Precipitation Index, Hydrol. Earth Syst. Sci., 17, 2359–2373, https://doi.org/10.5194/hess-17-2359-2013, 2013.

Fioreze, A. P., Bubel, A. P. M., Callou, A. É. P., Mendonça, B. C. d. S., Nunes, C. M., Pinto, C. G., Viana, C. F. G., Junior, D. S. R., Martins, E. S. P. R., Rodrigues, F. S. F., Filho, F. d. A. d. S., Teixeira, F. J. C., Viana, F. L., Nascentes, J. C. d. M., Filho, J. G. C. G., Júnior, J. A. d. L., Campos, J. N. B., Carvalho, J. O. d., Gonçalves, J. Y. d. B., Burte, J., Silva, L. M. C. d., Azevedo, L. G. T. d., Bursztyn, M., Cerqueira, M. R. S., Coimbra, T. P., Nobre, P., Vieira, R. F., Alves, R. F. F., Chacon, S. S., and Paulino, W. D.: A questão da Água no Nordeste, Ministério da Ciência e Tecnologia (MCT), available at: http://livroaberto.ibict.br/handle/1/669 (last access: 24 September 2018), 2012.

Formiga-Johnsson, R. M. and Kemper, K.: Institutional and Policy Analysis of River Basin Management: The Jaguaribe River Basin, Ceara, Brazil, SSRN Scholarly Paper ID 757424, Social Science Research Network, Rochester, NY, available at: https://papers.ssrn.com/abstract=757424 (last access: 24 September 2018), 2005.

Gates, W. L., Boyle, J. S., Covey, C., Dease, C. G., Doutriaux, C. M., Drach, R. S., Fiorino, M., Gleckler, P. J., Hnilo, J. J., Marlais, S. M., Phillips, T. J., Potter, G. L., Santer, B. D., Sperber, K. R., Taylor, K. E., and Williams, D. N.: An Overview of the Results of the Atmospheric Model Intercomparison Project (AMIP I), B. Am. Meteorol. Soc., 80, 29–56, https://doi.org/10.1175/1520-0477(1999)080<0029:AOOTRO>2.0.CO;2, 1999.

Hastenrath, S.: Exploring the climate problems of Brazil's Nordeste: a review, Climatic Change, 112, 243–251, https://doi.org/10.1007/s10584-011-0227-1, 2012.

Hastenrath, S. and Greischar, L.: Further Work on the Prediction of Northeast Brazil Rainfall Anomalies, J. Climate, 6, 743–758, https://doi.org/10.1175/1520-0442(1993)006<0743:FWOTPO>2.0.CO;2, 1993.

IPECE: Anuário Estatístico do Ceará, available at: http://www.ipece.ce.gov.br/index.php/anuario-estatistico-do-ceara (last access: 24 September 2018), 2017.

Juang, H.-M. H., Hong, S.-Y., and Kanamitsu, M.: The NCEP Regional Spectral Model: An Update, B. Am. Meteorol. Soc., 78, 2125–2143, https://doi.org/10.1175/1520-0477(1997)078<2125:TNRSMA>2.0.CO;2, 1997.

Marengo, J. A., Torres, R. R., and Alves, L. M.: Drought in Northeast Brazil – past, present, and future, Theor. Appl. Climatol., 129, 1189–1200, https://doi.org/10.1007/s00704-016-1840-8, 2017.

McKee, T. B., Doesken, N. J., and Kleist, J.: The Relationship of Drought Frequency and Duration of Time Scales, Eighth Conference on Applied Climatology, Anaheim, California, American Meteorological Society, available at: http://clima.cptec.inpe.br/~rclima1/pdf/paper_spi.pdf (last access: 24 September 2018), 1993.

Molteni, F., Stockdale, T., Balmaseda, M., Balsamo, G., Buizza, R., Ferranti, L., Magnusson, L., Mogensen, K., Palmer, T., and Vitart, F.: The new ECMWF seasonal forecast system (System 4), ECMWF Research Department, Technical Memorandum No. 656, available at: https://www.ecmwf.int/sites/default/files/elibrary/2011/11209-new-ecmwf-seasonal-forecast-system-system-4.pdf (last access: 24 September 2018), 2011.

Moura, A. D. and Hastenrath, S.: Climate Prediction for Brazil's Nordeste: Performance of Empirical and Numerical Modeling Methods, J. Climate, 17, 2667–2672, https://doi.org/10.1175/1520-0442(2004)017<2667:CPFBNP>2.0.CO;2, 2004.

Murawski, A., Bürger, G., Vorogushyn, S., and Merz, B.: Can local climate variability be explained by weather patterns? A multi-station evaluation for the Rhine basin, Hydrol. Earth Syst. Sci., 20, 4283–4306, https://doi.org/10.5194/hess-20-4283-2016, 2016.

Philipp, A., Della-Marta, P. M., Jacobeit, J., Fereday, D. R., Jones, P. D., Moberg, A., and Wanner, H.: Long-Term Variability of Daily North Atlantic–European Pressure Patterns since 1850 Classified by Simulated Annealing Clustering, J. Climate, 20, 4065–4095, https://doi.org/10.1175/JCLI4175.1, 2007.

Philipp, A., Beck, C., Huth, R., and Jacobeit, J.: Development and comparison of circulation type classifications using the COST 733 dataset and software, Int. J. Climatol., 36, 2673–2691, https://doi.org/10.1002/joc.3920, 2016.

Richardson, D. S., Bidlot, J., Ferranti, L., Ghelli, A., Haiden, T., Hewson, T., Janousek, M., Prates, F., and Vitart, F.: Verification statistics and evaluations of ECMWF forecasts in 2011–2012, ECMWF Research Department, Technical Memorandum No. 688, available at: https://www.ecmwf.int/sites/default/files/elibrary/2012/11917-verification-statistics-and-evaluations-ecmwf-forecasts (last access: 24 September 2018), 2012.

Roeckner, E., Arpe, K., Bengtsson, L., Brinkop, S., Dümenil, L., Esch, M., Kirk, E., Lunkeit, F., Ponater, M., Rockel, B., Sausen, R., Schlese, U., Schubert, S., and Windelband, M.: Simulation of the present-day climate with the ECHAM-3 model: impact of model physics and resolution, Report, Max-Planck-Institut für Meteorologie, 93, 1992.

Seibert, M., Merz, B., and Apel, H.: Seasonal forecasting of hydrological drought in the Limpopo Basin: a comparison of statistical methods, Hydrol. Earth Syst. Sci., 21, 1611–1629, https://doi.org/10.5194/hess-21-1611-2017, 2017.

Stockdale, T. N., Anderson, D. L. T., Alves, J. O. S., and Balmaseda, M. A.: Global seasonal rainfall forecasts using a coupled ocean–atmosphere model, Nature, 392, 370–373, https://doi.org/10.1038/32861, 1998.

Trambauer, P., Werner, M., Winsemius, H. C., Maskey, S., Dutra, E., and Uhlenbrook, S.: Hydrological drought forecasting and skill assessment for the Limpopo River basin, southern Africa, Hydrol. Earth Syst. Sci., 19, 1695–1711, https://doi.org/10.5194/hess-19-1695-2015, 2015.

Vicente-Serrano, S. M., Beguería, S., and López-Moreno, J. I.: A Multiscalar Drought Index Sensitive to Globa Warming: The Standardized Precipitation Evapotranspiration Index, J. Climate, 23, 1696–1718, https://doi.org/10.1175/2009JCLI2909.1, 2009.

Vicente-Serrano, S. M., López-Moreno, J. I., Beguería, S., Lorenzo-Lacruz, J., Azorin-Molina, C., and Morán-Tejeda, E.: Accurate Computation of a Streamflow Drought Index, J. Hydrol. Eng., 17, 318–332, https://doi.org/10.1061/(ASCE)HE.1943-5584.0000433, 2012.

Vitart, F.: Evolution of ECMWF sub-seasonal forecast skill scores over the past 10 years, ECMWF Research Department, Technical Memorandum Nr. 694, available at: https://www.ecmwf.int/sites/default/files/elibrary/2013/12932-evolution-ecmwf-sub-seasonal-forecast-skill-scores (last access: 24 September 2018), 2013.

Wetterhall, F., Pappenberger, F., He, Y., Freer, J., and Cloke, H. L.: Conditioning model output statistics of regional climate model precipitation on circulation patterns, Nonlin. Processes Geophys., 19, 623–633, https://doi.org/10.5194/npg-19-623-2012, 2012.

Wilks, D. S.: Statistical Methods in the Atmospheric Sciences, Academic Press, Burlington, MA,, 2005.

World Meteorological Organization (WMO): Guidelines on Ensemble Prediction Systems and Forecasting, WMO-No. 1091, Geneva, available at: http://library.wmo.int/pmb_ged/wmo_1091_en.pdf (last access: 24 September 2018), 2012a.

World Meteorological Organization (WMO): Standardized Precipitation Index User Guide (M. Svoboda, M. Hayes and D. Wood), WMO-No. 1090, Geneva, available at: https://library.wmo.int/pmb_ged/wmo_1090_en.pdf, (last access: 24 September 2018), 2012b.

13

Evaluation of uncertainties in mean and extreme precipitation under climate change for northwestern Mediterranean watersheds from high-resolution Med and Euro-CORDEX ensembles

Antoine Colmet-Daage[1,2,3], Emilia Sanchez-Gomez[1], Sophie Ricci[1], Cécile Llovel[2], Valérie Borrell Estupina[3], Pere Quintana-Seguí[4], Maria Carmen Llasat[5], and Eric Servat[6]

[1]CECI, CERFACS – CNRS TOULOUSE, Toulouse, France
[2]WSP France, Toulouse, France
[3]Hydrosciences Montpellier, Univ. Montpellier, Montpellier, France
[4]Observatori de l'Ebre Fundació, Tarragona, Spain
[5]Universitat de Barcelona, Barcelona, Spain
[6]Institut Montpelliérain de l'Eau et de l'Environnement – IRD, Montpellier, France

Correspondence: Antoine Colmet-Daage (colmet@cerfacs.fr)

Abstract. The climate change impact on mean and extreme precipitation events in the northern Mediterranean region is assessed using high-resolution EuroCORDEX and Med-CORDEX simulations. The focus is made on three regions, Lez and Aude located in France, and Muga located in northeastern Spain, and eight pairs of global and regional climate models are analyzed with respect to the SAFRAN product. First the model skills are evaluated in terms of bias for the precipitation annual cycle over historical period. Then future changes in extreme precipitation, under two emission scenarios, are estimated through the computation of past/future change coefficients of quantile-ranked model precipitation outputs. Over the 1981–2010 period, the cumulative precipitation is overestimated for most models over the mountainous regions and underestimated over the coastal regions in autumn and higher-order quantile. The ensemble mean and the spread for future period remain unchanged under RCP4.5 scenario and decrease under RCP8.5 scenario. Extreme precipitation events are intensified over the three catchments with a smaller ensemble spread under RCP8.5 revealing more evident changes, especially in the later part of the 21st century.

1 Introduction

The IPCC SRES report (IPCC, 2012) concludes with an increase in the frequency of heavy precipitation episodes over most areas of the globe at the end of 21st century. In particular, northwestern Mediterranean regions are often affected by extreme precipitation events that generate flash floods and cause serious damage (Ricard et al., 2012; Gaume et al., 2016). This climatically homogeneous region (Metzger et al., 2005) has been identified as a hotspot of climate change (CC) in the form of possible amplification of extreme precipitation associated with a decrease in total precipitation (Gao et al., 2006; Giorgi, 2006; Giorgi and Lionello, 2008; Milano et al., 2013). Assessing the impacts of regional climate change on precipitation is necessary in order to help and support policy makers develop strategies for future hydrological vulnerabilities like flash floods, but it constitutes a major challenge.

Global climate models (GCMs) are powerful tools to assess global-scale climate variability and change. GCMs have allowed for better understanding of interactions between the different components of the climate system (e.g., atmosphere, ocean, sea ice, continents). However GCMs generally operate at coarse horizontal resolutions (100–250 km in the atmospheric component); hence, they are not appropriate for investigating the hydrological impacts of future extreme precipitation events at local scale, such as over hydrological watersheds that are as small as hundreds of kilometers.

Interest in better representing climate variability and change at local scales has driven the development of regional climate models (RCMs), which are currently able to perform dynamical downscaling of GCM at very high horizontal resolutions (~ 10 km). RCMs run on limited area domains thereby allowing increased spatial resolution, and thus enabling better representation of surface heterogeneities and mesoscale atmospheric processes like convection (Fowler et al., 2007a). At European scale, collaborative research projects such as MERCURE (Hagemann et al., 2004), PRUDENCE (Christensen et al., 2007), NARCCAP (Paulsen et al., 2009) and ENSEMBLES (van der Linden and Mitchell, 2009) have contributed to further development and improvement of regional modeling. More recently, the international CORDEX (Coordinated Regional Climate Downscaling Experiment) initiative (Giorgi et al., 2009) has provided multimodel regional climate simulations at very high spatial resolution over different regions in the world. In particular, for the northwestern Mediterranean region, the EuroCORDEX and MedCORDEX sub-projects (EMCORDEX hereinafter) have produced present and future climate simulations at 12 km resolution.

The merits of increased spatial resolution in RCMs have largely been assessed in the literature. Comprehensive evaluations of RCMs have been undertaken over the Euro-Mediterranean region by applying evaluation metrics to mean values of precipitation (Déqué and Somot, 2010; Fisher et al., 2012; Jacob et al., 2007; Kjellström et al., 2010; Kotlarski et al., 2005) as well as focusing on extreme precipitation associated with hydrological floods (Frei et al., 2006; Fowler et al., 2007b; Herrera et al., 2010; Kysel et al., 2012; Maraun et al., 2012). For recent EMCORDEX models, initial evaluations over past periods have been conducted over Europe (Drobinski et al., 2016; Katragkou et al., 2015; Kotlarski et al., 2014). The latter evaluations mainly focused on mean and extreme precipitation over the whole EMCORDEX domain or in large regional boxes (e.g., France, the Alps, Mediterranean coastal regions, Morocco) using sparse observation data sets. Prein et al. (2016) highlighted that the added value of high-resolution models (12.5 km versus 50 km) for the simulated mean and extreme precipitation due to improved representation of orography and large-scale convection. Indeed, Drobinski et al. (2016) show the higher ability of high-resolution RCMs to reproduce the Clausius–Clapeyron relation and thus precipitation-related processes. In the context of climate change, Jacob et al. (2013) show that future climate projections performed by high-resolution (12.5 km) scenario under the Representative Concentration Pathways 4.5 and 8.5 (RCP4.5 and RCP8.5, respectively) project higher daily precipitation intensities than GCMs, in particular for RCP8.5. These results are consistent with the conclusions of Giorgi et al. (2016) over the Alps. Even though both GCM and RCM scenario experiments project a reduction of summer precipitation over the Alps, increased convective rainfall due to enhanced potential instability related to a finer representation of the orography over the Alps is found in RCMs.

Together with increasing model resolution, high-resolution observation-based products have also been recently developed over different Mediterranean countries. Moving from CRU product at 50 km resolution (Harris et al., 2014) to E-OBS product at 25 km resolution (Haylock et al., 2008), reanalysis such as SAFRAN (Système d'Analyse Fournissant des Renseignements Atmosphériques à la Neige; Durand et al., 1993; Quintana-Seguí et al., 2008) or interpolated products such as SPAIN02 (Herrera et al., 2012, 2016) now provide precipitation products at a resolution comparable to those of the RCMs. In particular, the SAFRAN data set available over France (Quintana-Seguí et al., 2008; Vidal et al., 2010) and Spain (Quintana-Seguí et al., 2016a, b) comprises a much larger observed data network than E-Obs or ERA-Interim (Dee et al., 2011), which were previously used for EMCORDEX model assessment (Cavicchia et al., 2016; Kotlarski et al., 2014). The SAFRAN-France data set was used together with downscaled products issued from CMIP5 (Coupled Model Intercomparison Phase 5) to assess future hydrological changes over France (Dayon, 2015; Quintana-Seguí et al., 2010, 2011). Themeßl et al. (2011) provide a review of downscaling methods. This implies systematic bias correction of model precipitation before being used as input for hydrological models, for instance in the framework of future flash flood simulation. Harader (2015) used the regional model ALADIN5.2 outputs at 12 km resolution from CORDEX as well as SAFRAN-France product to describe future flash flood events over the Lez catchment using a "futurization" method described in the following.

The present study focuses on extreme precipitation over mesoscale northwestern Mediterranean watersheds with complex orography. Three watersheds of various sizes are investigated here: Lez and Aude located in southern France, and Muga, located in northeastern Spain. The goals of this study are as follows:

- to assess RCM skills from the EMCORDEX multimodel ensemble in terms of mean and extreme precipitation values over past periods;
- to assess the influence of GCMs' lateral boundaries condition on the RCMs' skills;
- to evaluate future changes in precipitation extremes for further simulations of flash floods with an event-based hydrological model over future periods.

For this purpose, the futurization approach proposed by Harader (2015) is used. This method utilizes the computation of a past/future change coefficients of quantile-ranked RCM precipitation outputs. Each multiplicative coefficient is then applied to each quantile-ranked short-term observed precipitation event. It should be noted that the quantile-ranked observed precipitation is computed from SAFRAN daily precipitation that have generated flash floods. The futurization method is applied for each RCM in the EMCORDEX ensembles, forced by two emission scenarios (RCP4.5 and RCP8.5). We thus explicitly take into account climate model-related uncertainty. In further work, the "futurized" precipitation events will be used with different hydrological models, so that we explicitly take into account hydrology model-related uncertainty.

The paper is organized as follows, Sect. 2 includes a brief presentation of the EMCORDEX simulations, the reference data sets and the statistical metrics applied to seasonal mean and extreme precipitation values. Section 3 presents the RCM evaluation in terms of mean and extreme precipitation values over the present period when the global scale is prescribed by ERA-Interim. Section 4 analyzes present climate simulations to understand the role of the GCMs in driving the RCMs. The impacts of climate change on precipitation are then examined in Sect. 5. Conclusions and perspectives are finally given in Sect. 6.

2 Data and methods

2.1 The river catchments

In the current study, the futurization approach is applied over three Mediterranean watersheds with different characteristics and external influences. The Lez, Aude and Muga catchments displayed in Fig. 1 are frequently subject to flash floods that cause considerable damage to surrounding areas and cities.

The upstream part of the Lez watershed, shown in red in Fig. 1, is located 15 km north of the city of Montpellier and covers 114 km^2. The landscape is dominated by garrigue vegetation, very common in the Mediterranean countries. The spring of the Lez River is the resurgence of a karstic aquifer of about 380 km^2. The karst aquifer plays an important role in water resources in the basin, and the karst outcrops actively participate in flash flood dynamics (Raynaud et al., 2016). Cumulative annual rainfall is around 909 mm, which, on average, falls 60 days per year (Coustau, 2011; Harader, 2015). The Lez catchment is frequently subject to flash floods caused by extreme precipitation episodes regionally known as the "Cévenols" events (Ducrocq et al., 2008; Nuissier et al., 2008, 2011). Cumulative extreme precipitation can locally reach 600 mm in 24 h within the river catchment (Boudevillain et al., 2011).

The Aude watershed, shown in brown in Fig. 1, covers more than 5000 km^2 upstream the nearby city of Narbonne. The Aude River springs at the Pyrenees and flows along the catchment for 223 km before entering the Mediterranean Sea. The Orbieu and the Fresquel are its major tributaries. The Aude catchment is surrounded by several mountain chains such as the Cevennes massif to the north and the Pyrenees to the south. The catchment is mainly dominated by a Mediterranean climate, but large climate contrasts can be found over its sub-watersheds. A severe flash flood episode occurred in November 1999 (more than 200 mm of rain in 24 h over a major part of the catchment) and caused severe damage over an extended region (Aude, Tarn, Pyrénées Orientales, Hérault; Estupina, 2004; Bechtold and Bazille, 2001; Ducrocq et al., 2003; Gaume et al., 2004).

Finally, the Muga catchment (shown in purple in Fig. 1) located in northeastern Spain over Catalonia covers 854 km^2. The Muga River is about 58 km long, between the Pyrenees (maximum altitude of the catchment, 1214 m) and the Gulf of Roses. This catchment is usually affected by heavy precipitation associated with convective events, with annual precipitation average between 700 mm in the upper part and 530 mm at the mouth, and daily precipitation of 200 mm for return period of 10 years (Llasat et al., 2009, 2014). The previously mentioned November 1999 flash flood event led to a historical peak discharge recorded near the mouth of the Muga River, with 925 $m^3\,s^{-1}$ compared to an average value of 3.34 $m^3\,s^{-1}$. The Muga catchment was affected by 26 severe flood events between 1982 and 2010 (Llasat et al., 2014).

It should be noted that, in this paper, statistical analysis is carried out over the Lez, Aude and Muga catchment areas as well as over more extended regional boxes also displayed in Fig. 1.

2.2 RCM Euro-Mediterranean CORDEX simulations

The set of EMCORDEX simulations used in this study are summarized in Table 1. The five RCMs used are presented with the main reference papers (in particular with respect to boundary layer and convection schemes). The EMCORDEX community has provided three types of RCM simulations with driving conditions (also detailed in Table 1) issued from either ERA-Interim or GCMs simulations over past and future periods:

- The evaluation simulations (EVAL hereinafter). The lateral boundary conditions (LBCs) are driven by ERA-Interim reanalysis (Dee et al., 2011) over 1981–2010. These simulations are used to evaluate the RCM intrinsic biases.

- The historical simulations (HIST hereinafter). The LBCs are issued from numerical experiments performed with four different GCMs and extracted from the CMIP5 historical archive. These historical simulations represent the climate conditions over 1976–2005.

- The future climate scenario simulations (RCP hereinafter). In this case, the LBCs correspond to four future climate scenarios (from four different GCMs from

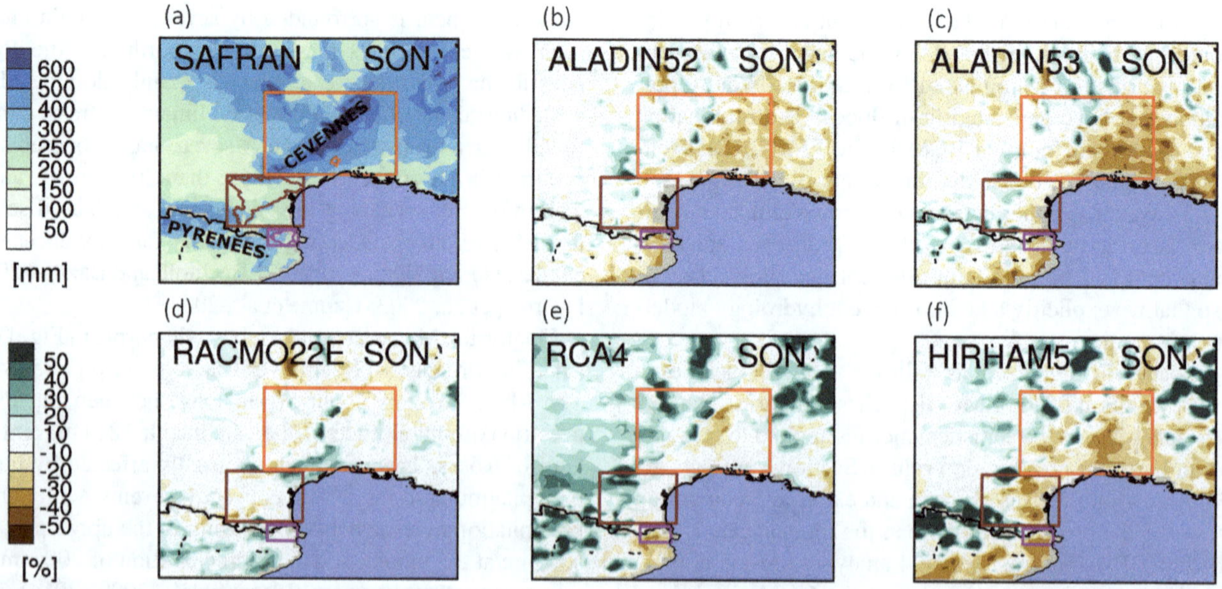

Figure 1. Mean relative seasonal precipitation bias (%) in all the RCMs forced with ERA-Interim for the period 1981–2010. The top left panel shows the horizontal pattern of seasonal precipitation provided by the SAFRAN reference data set (mm season $^{-1}$). Only the September–October–November (SON) season is shown here. On the SAFRAN map **(a)**, the colored rectangles represent the regional boxes, while the outline represents the hydrological catchment. The Cevennes box and the Lez catchment are in red, the Aude box and catchment in brown, and the Muga box and catchment in purple.

Table 1. The parameterization and physics scheme related to the RCM (0.11°) used in the present study are presented in the five first lines. The lateral boundary conditions (LBCs) for each type of simulations are presented with four different GCMs of CMIP5. The last row lists the name used for each GCM–RCM pair throughout this study.

	RCM	ALADIN5.2	ALADIN5.3	RCA4				HIRHAM5	RAMO22E
	Institution	Météo-France	Météo-France	SMHI				DMI	KNMI
	Boundary layer scheme	Ricard and Royer (1993)	Ricard and Royer (1993)	Cuxart et al. (2000)				Louis (1979)	Lenderink and Holtslag (2004); Siebesma et al. (2007)
	Convection	Mass flux Bougeault (1985)	Mass flux Bougeault (1985)	Kain and Fritsch (1990, 1993); Kain (2004); Jones and Sanchez (2002)				Tiedtke (1989)	Tiedtke (1989); Nordeng (1994); Neggers et al. (2009)
	Main references	Colin et al. (2010); Hermann et al. (2011)	Colin et al. (2010); Hermann et al. (2011)	Samuelsson et al. (2011); Kupiainen et al. (2011)				Christensen et al. (2008)	Meijgaard van et al. (2012)
Lateral boundary conditions	Evaluation simulation	ERA-INTERIM (reanalysis)							
	Historical simulation								
	RCP4.5 simulation	CNRM-CERFACS-CNRM-CM5			MOHC-Had	MPI-M-MPI-	ICHEC-EC-EARTH		
	RCP8.5 simulation				GEM2-ES	ESM-LR			
	GCM members	r8i1p1	r1i1p1	r1i1p1	r1i1p1	r1i1p1	r12i1p1	r3i1p1	r1i1p1
	Name of pair GCM_ RCM	CNRM-CM5 _ ALADIN52	CNRM-CM5 _ ALADIN53	CNRM-CM5 _ RCA4	MOHC_ RCA4	MPI_ RCA4	ICHEC_ RCA4	ICHEC_ HIRHAM5	ICHEC_ RACMO22E

CMIP5) under RCP4.5 and RCP8.5 (Clarke et al., 2007; Riahi et al., 2007; Meinshausen et al., 2011; van Vuuren et al., 2011) over 2011–2040, 2041–2070 and 2071–2100. These simulations cover 30-year time slice periods like the HIST simulation. Thus they can be statistically compared to assess future changes.

This results in eight pairs of RCM–GCM simulations analyzed in this study, for HIST and RCP as indicated in the last row of Table 1. Only simulations at 12 km for which EVAL, HIST and both RCP experiments were available at the beginning of the study were considered.

2.3 The reference data set: SAFRAN

SAFRAN reanalysis provides daily precipitation data for the period 1958–2008 over France and Spain on an 8 and 5 km grid, respectively (Vidal et al., 2010, Quintana-Seguí et al., 2016a, b). SAFRAN-France was built by using data from 3675 selected rain gauges that were gridded through an optimal interpolation algorithm described in Quintana-Seguí et al. (2008). The great number of rain gauges considered in SAFRAN and its high spatial resolution produce more accurate precipitation analyses over France and Spain than those proposed by other products commonly used for model assessments, such as CRU (Harris et al., 2014) and E-OBS (Haylock et al., 2008). A recent study shows that, for precipitation, the performance of SAFRAN is very similar to that of Spain02 (Quintana-Seguí et al., 2016b). SAFRAN data set has been evaluated using the Météo-France and Spanish State Meteorological Agency (AEMET) gauging station network as well as independent data (Quintana-Seguí et al., 2008, 2016a, b; Vidal et al., 2010).

SAFRAN product is used here as a reference data set to evaluate the simulated precipitation from EVAL and HIST ensembles. For that purpose, the 12 km RCM outputs and also SAFRAN Spain were regridded on the 8 km SAFRAN-France grid using the ESMF (Earth System Modeling Framework) bilinear regridding method. The impact of interpolation has been largely investigated in the literature, for instance in Diaconescu et al. (2015), and it should be quantified before interpretation of interpolated fields. In the present study, the impact of a bilinear scheme for the interpolation of precipitation fields from 12 km simulation grid to 8 km SAFRAN grid was assessed from an analytical test case based on the cosine function of longitude and latitude. The comparison is achieved by mapping the analytical field onto the 12 km fields and interpolated it onto an 8 km grid with the analytical field mapped onto the 8 km. The quadratic error reaches 5.10^{-5}; the impact of interpolation was thus neglected in the following.

2.4 Statistical metrics to evaluate RCM performance

Given the small size of the Lez, Aude and Muga catchments (3, 84 and 11 SAFRAN grid points, respectively), and in order to allow for proper statistical analysis, the evaluation of mean precipitation (seasonal and annual cycle) was achieved on larger regional boxes (excluding sea grid points) shown in Fig. 1. These regional boxes have been selected according to regions of homogeneous climate conditions. In the particular case of the Lez catchment, the RCM precipitation is evaluated over the entire Cevennes region. However for all three catchments, extreme precipitation metrics are computed over grid points that are strictly inside the catchments. The precipitation extremes are analyzed with respect to the 90th to 99.9th quantiles of the daily precipitation distribution discretized as follows:

- one point per 1 quantile rank from 90 to 95th;
- one point per 0.5 quantile rank from 95 to 98th;
- one point per 0.2 quantile rank from 98 to 99th;
- one point per 0.1 quantile rank from 99 to 99.9th.

Short-term observed precipitation events are thereby completely covered and evenly distributed with this quantile discretization (not shown). Quantiles are computed considering all the days (rainy days and dry days), thus allowing for a comparison of precipitation quantiles between the different RCM–GCM pairs (Giorgi et al., 2016; Schär et al., 2016). However, it should be noted that when precipitation is below 0.1 mm day^{-1}, the precipitation is set to 0 so that the cumulative precipitation is not affected by overrepresentation of light rainy days in the models (Harader et al., 2015; Tramblay et al., 2013, Paxian et al., 2015).

In addition to the classical metrics (spatial bias, annual cycle bias, quantile–quantile plot), two original metrics are used in this study. First, assuming additivity between GCM and RCM errors, the impact of the GCM bias on the RCM solution can be diagnosed by computing the difference ΔB between the HIST and the EVAL precipitation bias with respect to SAFRAN:

$$\Delta B = \frac{\text{HIST} - \text{SAFRAN}}{\text{SAFRAN}} - \frac{\text{EVAL} - \text{SAFRAN}}{\text{SAFRAN}}. \tag{1}$$

The ΔB criteria corresponds to the bias in the annual cycle of precipitation simulated with the RCMs that is strictly related to the influence of the lateral boundary condition imposed by the GCM. A high positive value indicates overestimation of the total monthly precipitation, and a negative value indicates underestimation of the total monthly precipitation.

Secondly, in the present study, change coefficients between the past (HIST) and future precipitation (RCP) quantile distributions are computed. For that purpose, precipitation from HIST and RCPs is quantile-ranked, and for each quantile rank "qi", a change coefficient "Aqi" for precipitation intensities between HIST and RCP is computed as follows:

$$\text{Aqi} = \frac{\text{Pqi(RCP)}}{\text{Pqi(HIST)}}, \tag{2}$$

where Pqi(RCP) and Pqi(HIST) represent the values of the quantiles for HIST and RCP, respectively.

All metrics are computed on a seasonal basis, considering hereinafter four seasons: autumn (September, October and November), winter (December, January and February), spring (March, April and May) and summer (June, July and August). Only autumn and spring results are presented in the following section.

3 Analysis of RCM evaluation simulations

In this section the RCM precipitation biases are diagnosed through the comparison between the EVAL simulations and SAFRAN. The spatial pattern for the mean cumulative precipitation, the annual cycle and the extreme values are investigated.

3.1 Spatial bias pattern

Figure 1 shows the spatial distribution of the cumulative precipitation normalized difference between each RCM from the EVAL ensemble and SAFRAN, averaged over the previous 30 years in autumn. The top-left panel in Fig. 1a displays the mean cumulative precipitation for SAFRAN-France and SAFRAN-Spain reference data sets. Large values of the cumulative precipitation (greater than 400 mm season^{-1}) are observed over mountainous regions, in particular the Cevennes and the Pyrenees chains. Lower cumulative precipitation values are observed in the valleys (Garonne, Aude) and over the coastal regions.

In general, the cumulative precipitation is overestimated for most RCMs over the mountainous regions (+30 %) and underestimated over the Mediterranean coastal region (−30 %), as shown in Fig. 1b to f, most likely because of imperfect representation of the orography as well as the parameterization of the convection scheme. Note that the RACMO22E pattern of precipitation bias differs from the other RCM patterns. Indeed, slight overestimation is observed over the south of the Pyrenees and the Cevennes (+20 %), and almost no bias is observed over the valleys.

We focus now on the precipitation bias over river catchments that are analyzed in the present study: for Cevennes, the cumulative precipitation is overestimated by 20 % in the southwest mountainous region and underestimated by 30 % in the northeast valley region for all RCMs (Fig. 1b, c, e, f) except for RACMO22E (Fig. 1d).

For the Aude region, no relevant bias is presented by ALADIN52 (Fig. 1b), ALADIN53 (Fig. 1c) and RACMO22E (Fig. 1d). A positive bias (+40 %) is presented by RCA4 (Fig. 1e) in the western region under continental influence, while no relevant bias is presented in the eastern region under Mediterranean influence. HIRHAM5 (Fig. 1f) displays a strong positive bias (+50 %) in the high-elevation areas

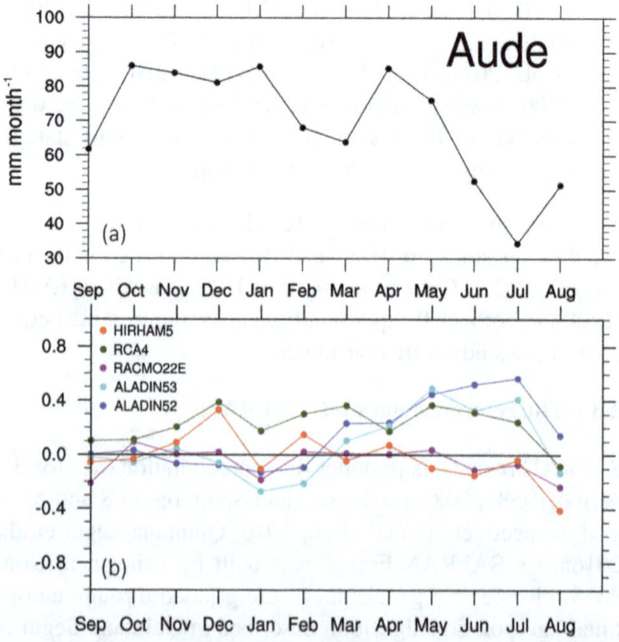

Figure 2. (a) Annual cycle of precipitation over Aude box with SAFRAN data set **(b)**. Bias of the annual cycle of precipitation simulated by the five RCMs with respect to SAFRAN's annual cycle of precipitation for the period 1981–2010 (no units), over the Aude box. The colored lines with different markers represent the biases of RCM-simulated precipitation with respect to SAFRAN data set (black line).

(Pyrenees in the southwest, and Black mountain in the northwest) and a strong negative bias elsewhere.

The mean cumulative precipitation is underestimated (−30 %) by all RCMs over the Muga region, except for RACMO22E.

For other seasons, the mean cumulative precipitation pattern tends to be overestimated over all three regions in spring (not shown), and a strong positive bias (+50 %) is presented by ALADIN52 and ALADIN53 over the Pyrenees in summer (not shown).

3.2 Annual cycle of precipitation bias

The 30-year climatology for the monthly cumulative precipitation is spatially averaged over each region of interest and normalized by the SAFRAN monthly climatology. Figure 2a displays the annual cycle of precipitation over the Aude box for SAFRAN data set. Figure 2b displays the bias of the annual cycle over the Aude region for each RCM. Globally speaking, the RCM bias and the inter-model spread are smaller in autumn than in summer. This can be explained by the influence of the large-scale atmospheric circulation being weaker in summer, the control exerted by the LBCs on the RCM is reduced in this season and the RCM solution has more degrees of freedom to deviate from the large-scale forcing (Déqué et al., 2012; Lucas-Picher et al., 2008).

ALADIN52 and ALADIN53 clearly overestimate the cumulative monthly precipitation in late spring and summer. This is likely due to a large presence of low precipitation days in the RCM precipitation distribution (0 to 10 mm day^{-1}) compared to SAFRAN distribution (not shown). This behavior is also observed over the Cevennes and Muga regions (not shown). Consistent with Fig. 1d, RACMO22E displays a low bias in the annual cycle. Similar behavior is observed for HIRHAM5, suggesting an error compensation between positive and negative spatial biases displayed in Fig. 1f. For the Muga and the Cevennes boxes (not shown), HIRHAM5 presents a negative bias. Finally, RCA4 simulates a large positive bias (+30 %) for the entire annual cycle that tends to intensify over the Alps and the western Pyrenees.

3.3 Extreme precipitation bias

The ability of the EVAL simulations to represent the extreme of precipitation is investigated through a quantile–quantile analysis with respect to SAFRAN. This analysis is performed for each grid point within the Lez, Aude and Muga catchments. The highest-order quantiles (95 to 99th) are displayed in Fig. 3 for each RCM and SAFRAN in autumn.

For the Lez, Aude and Muga catchments, the RCMs underestimate the higher-order quantiles. This underestimation is more important for the Muga catchment, where the extreme precipitation intensities are stronger.

ALADIN52 underestimates the upper precipitation quantiles; for instance, the intensity of the 99th quantiles only reaches 70 mm day^{-1} compared to 140 mm day^{-1} for SAFRAN. ALADIN53 performance is slightly improved with respect to ALADIN52, especially for the Aude and Muga catchments. RACMO22E and RCA4 underestimate the extreme precipitation above the 99.5th quantiles. HIRHAM5 provides a satisfactory description of all extreme quantiles for the three catchments.

3.4 Summary of means and extremes analysis

The mean and extreme precipitation is investigated over the Aude, Lez and Muga catchments. Largest biases in mean precipitation simulated by the regional climate models are located over the mountainous (positive bias) and coastal (negative bias) regions. These biases are stronger during the summer season, when the control exerted by the LBCs on the RCM is weaker, due to a reduction of the large-scale circulation and North Atlantic inflow. Overall, while the climate variability is covered by the spread of EMCORDEX ensemble of RCMs, each RCM simulates plausible precipitation.

These results are coherent with specific studies for mean and extreme precipitation as cited hereafter. ALADIN52 precipitation biases are consistent with those obtained by Harader (2015), and slightly reduced in ALADIN53. The RCA4 overestimation of mean precipitation is in accordance with the results showed in Prein et al. (2016). The underestimation of extreme precipitation diagnosed in our study is possibly even more severe than our reference, SAFRAN, found to underestimate observed extreme precipitation (Quintana-Seguí et al., 2008a). This last point was confirmed through a comparison between pluviometer data and SAFRAN over our catchments of interest.

Parameterization and physical processes that are involved in the generation of low precipitation differ from those involved in stronger precipitation events. Thus the precipitation bias varies according to the precipitation intensity. Quantile classification is thereby relevant for the determination of past/future precipitation change.

4 Analysis of RCM historical simulations

RCM intrinsic biases, determined from the EVAL ensemble, have been characterized and described in the previous section. Here we analyze the historical simulations (HIST), performed with the GCM–RCM pairs, as described in Sect. 2.2. In this case, GCM biases are expected to affect the RCM simulation. A difficult challenge is to understand how GCM and RCM biases are interacting (Dequé et al., 2012). In this study we make the assumption of additivity between the impact of the GCM biases and the RCM intrinsic biases. In this study, four different GCM forcings are considered: CNRM-CM5, ICHEC, MOHC and MPI (see Table 1).

4.1 Annual cycle of precipitation bias

Figure 4 displays the annual cycle of ΔB over the Aude watershed for each GCM–RCM pairs. The color code refers to the GCM (for instance for CNRM-CM5 forcing), while the markers refer to the RCM (for instance star for ALADIN53). From Fig. 4 it is evident that CNRM-CM5 forcing leads to systematic overestimation of summer precipitation, hence proving that the positive precipitation bias identified in EVAL (Sect. 3, Fig. 2 for ALADIN52, ALADIN53 and RCA4) is enhanced in HIST simulations.

ICHEC large-scale conditions induce no significant changes on RACMO22E and RCA4 errors, while they lead to strong overestimation of the HIRHAM5 precipitation. Except when forced by CNRM-CM5, the bias of GCM–RCA4 pairs is similar to the intrinsic bias of RCA4, as displayed for all three, MPI, MOHC and HIRHAM5. Illustration is provided for the Aude box in Fig. 4, but similar results are obtained for the Cevennes and the Muga boxes.

It should be noted that previous results might also be explained by bias compensation between GCM impacts and RCM intrinsic bias. Despite the GCM–RCM deficiencies shown here, all the GCM–RCM pairs display plausible precipitation and are then considered in the following for future change analysis over mean precipitation annual cycle.

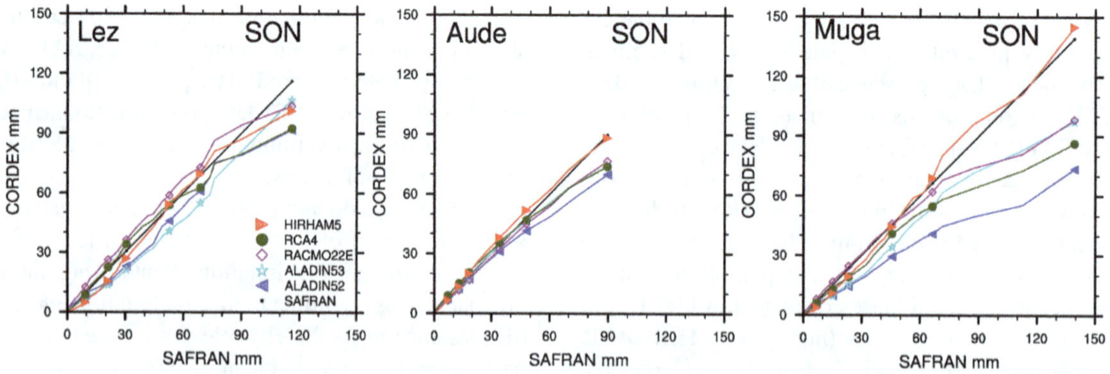

Figure 3. Quantile–quantile diagram of daily precipitation in the cells of the three catchments for the period 1981–2010. EVAL-simulated precipitation (colored lines with different markers) is compared to SAFRAN (black line). The x axis represents the precipitation quantile values with respect to the SAFRAN reference data. The y axis represents the precipitation intensity simulated by the EMCORDEX models for the same quantiles. If a colored line is above/below the black line, the corresponding RCM over/underestimates quantile intensities with respect to SAFRAN. Units are in mm day^{-1}. The colored dots represent, from left to right, the 90, 95, 97, 99, 99.5 and 99.9th quantiles.

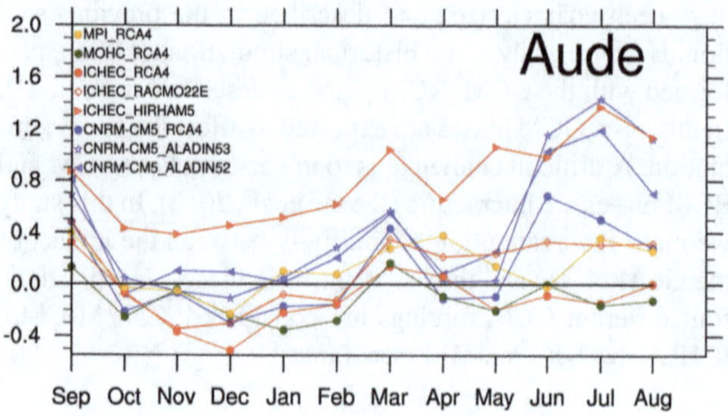

Figure 4. Bias on the annual cycle RCM-simulated precipitation induced by GCMs lateral boundary conditions. Aude regional box is shown here for the period 1976–2005. This bias is estimated through the computation of the difference between HIST and EVAL simulated precipitation with respect to SAFRAN. ΔB (no units) represent this bias with $\Delta B = \frac{\text{HIST}-\text{SAFRAN}}{\text{SAFRAN}} - \frac{\text{EVAL}-\text{SAFRAN}}{\text{SAFRAN}}$. Colored lines refer to the GCMs, and the markers refer to the RCMs driven. If a colored line is above (below) the black line, the corresponding GCM induce overestimation (underestimation) of the precipitation simulated by the RCM.

4.2 Extreme precipitation bias

Figure 5 is the counterpart of Fig. 3; it displays quantile–quantile diagrams of precipitation in autumn for the HIST ensemble over the Lez, Aude and Muga catchments. First, the spread amongst the HIST simulations displayed in Fig. 5 is larger, except for the Muga catchment, larger than the one generated by the EVAL ensemble (Fig. 3) that the EVAL ones (Fig. 3), and extreme quantiles tend to be systematically underestimated over all three catchments of interest.

Generally, CNRM-CM5 forcing leads to underestimation of SAFRAN quantiles, thus enhancing the RCMs' intrinsic biases displayed in Fig. 3. MPI and ICHEC provide good quantile distribution for all RCMs except beyond 99.5th quantile. In more detail, ICHEC-RCA4 extreme quantiles are slightly overestimated over the Aude catchment, and in good agreement with SAFRAN over the Muga catchment. MOHC-RCA4 provides good statistics over Aude and Muga catchments, while it overestimates all quantiles over the Lez catchment.

To conclude, the impact of the GCM on RCMs tends to systematically intensify the underestimation of extreme precipitation values. However, the HIST ensemble remains consistent with SAFRAN statistics and thus will be used in the following step, in order to estimate future changes over extreme precipitation over the different catchments.

5 Effect of climate change on precipitation

5.1 Annual cycle of precipitation

Figure 6 presents the precipitation annual cycle simulated for HIST computed over 1976–2005 period and the scenario RCP4.5 and RCP8.5 computed over 2071–2100 period, for the Cevennes, Aude and Muga boxes.

The results reveal stronger mean precipitation change between RCP4.5 and RCP8.5 than for the 2011–2040 and 2041–2070 periods. The annual cycle of the RCP4.5 ensemble is similar to HIST one in terms of mean and spread, suggesting that radiative forcing in RCP45 seems to have a weak impact on monthly averaged precipitation. In contrast, the annual cycle of the RCP8.5 ensemble displays a general decrease in mean precipitation from April to October for the three river catchments. These results are consistent with the conclusions from the study of Jacob et al. 2013 over EMCORDEX domain. As previously mentioned the spread of the ensemble is larger in summer when the LBCs exert a weaker control in the RCM domain. It should be noted

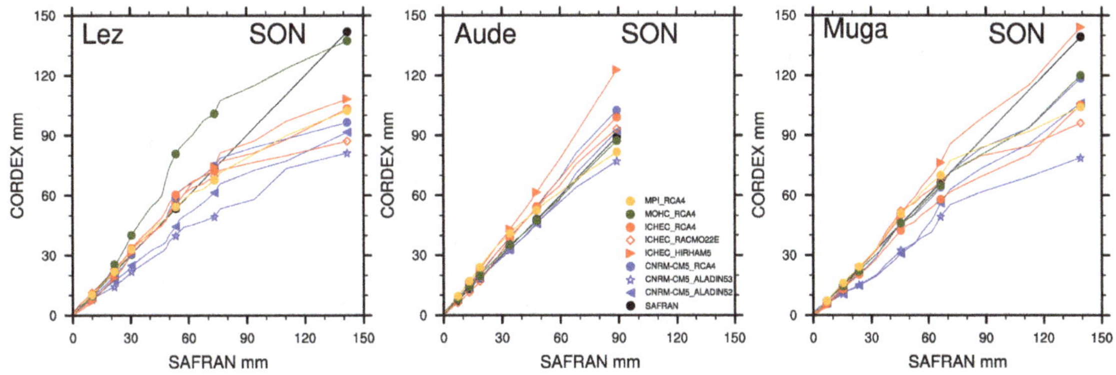

Figure 5. Quantile–quantile diagram of daily precipitation in the cells of the three catchments for the period 1981–2010. HIST-simulated precipitation (colored lines) is compared to SAFRAN (black line). The x axis represents the precipitation quantile values with respect to the SAFRAN reference data. The y axis represents the precipitation intensity simulated by the EMCORDEX models for the same quantiles. If a colored line is above/below the black line, the corresponding RCM over/underestimates quantile intensities with respect to SAFRAN. Units are in mm day^{-1}. The colored dots represent, from left to right, the 90th, 95th, 97th, 99th, 99.5th and 99.9th quantiles.

that these results hold for the three boxes of interest that are of significantly different sizes. This suggests that the high-resolution EMCORDEX simulations can be confidently used to investigate precipitation at local scale.

As mentioned in the previous section, the analysis of the annual cycle of HIST simulations emphasized that the association of GCM and RCM models induces a different precipitation bias that the intrinsic RCM biases (comparison between black and blue curve in Fig. 6). Nevertheless, the precipitation field issued from HIST is not bias corrected here. They are used, together with the RCP simulations, to estimate the change coefficient between past and future values of the quantile of the precipitation distribution.

5.2 Change coefficients for extreme precipitation

The change coefficients between quantile-ranked precipitation presented in Eq. (2) are displayed in Fig. 7. They allow the estimation of the changes between the past (HIST) and future precipitation (RCP) quantile distributions

Following Eq. (2), a change/transfer coefficient greater than 1 for a given quantile indicates an increase of the future precipitation value associated with this quantile. On the other hand, Aqi < 1 means a decrease in RCP precipitation with respect to HIST.

The Aqi coefficients are computed for each GCM–RCM pair over the Lez, Aude and Muga catchments. The multi-model approach adopted here allows for estimating an uncertainty in the values of the change coefficients. The Aqi coefficients are represented in Fig. 7 over the upper level quantile range (90–99.9) for RCP4.5 and RCP8.5 in autumn for the three river catchments and for 2041–2070 and 2071–2100 future time periods. The ensemble mean for Aqi computed amongst the GCM–RCM pairs is compared to the associated ensemble spread (standard deviation) in Fig. 7.

The interesting result here is that while RCP simulations tend to decrease mean precipitation with respect to HIST simulation (Fig. 6), extreme precipitation events are intensified for both time periods over the three catchments. Globally speaking, the mean change coefficient for RCP4.5 and RCP8.5 is similar except for the Lez (2041–2070) and the Muga (2071–2100) catchments, where RCP8.5 displays biggest changes. The ensemble spread from RCP8.5 is smaller than the one from RCP4.5, meaning that the change in extreme precipitation displays higher level of certainty under RCP8.5 scenario, which can be explained by stronger radiative forcing in RCP85 compared to RCP45. Few models indicate a decrease in extreme precipitation for RCP4.5 over 2041–2070 period. In contrast, for 2071–2100, all the GCM–RCM members agree on an increase in extreme precipitation for both RCP4.5 and RCP8.5 scenarios. For instance, the mean change coefficient over 2071–2100 reaches 1.15 for the 99.5th quantile over the Aude catchment for both RCPs, while RCP4.5 spread is about 0.3, versus 0.1 for RCP8.5. In terms of precipitation, a change coefficient of 1.35 over the Lez catchment for the 99.9th quantile of RCP8.5 represents an increase from 140 to 189 mm day^{-1}.

6 Discussion

The present study assessed the intensification of extreme precipitation events under climate change on small Mediterranean catchments using high-resolution RCM simulations (~ 12 km) from the EMCORDEX exercise. It was shown that over the past period (1976–2005), EVAL simulations (RCM driven by ERAI) and HIST simulations (RCM is driven by a GCM) underestimate extreme events with respect to SAFRAN data set.

Despite SAFRAN being the best gridded observation data set available covering the region of interest as explained in

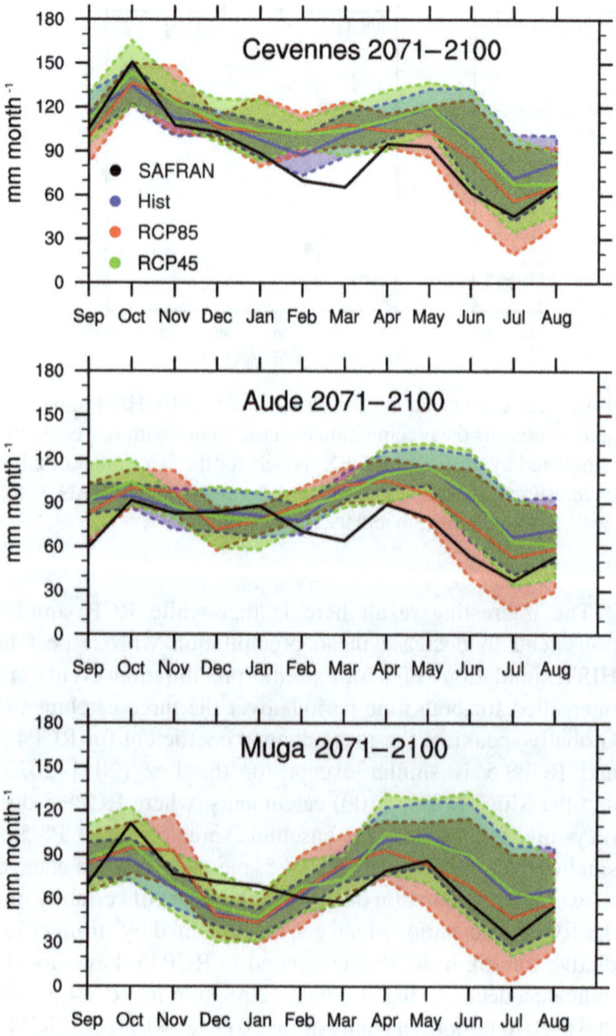

Figure 6. Annual cycle of mean precipitation (mm month^{-1}) simulated in the three regional boxes (Cevennes, Aude, Muga; see Fig. 1). Solid lines represent the ensemble means computed amongst the eight GCM–RCM pairs. The shaded areas represent the ensemble spreads characterized by the standard deviation. HIST ensemble mean (1976–2005) is plotted in blue. RCP4.5 and RCP8.5 ensemble means (2071–2100) are respectively plotted in green and red. SAFRAN annual cycle computed over 1981–2010 is plotted in black.

Sect. 2.3, the underestimation highlighted here is potentially more severe as SAFRAN is itself deemed to underestimate observed extreme precipitation events (Quintana-Seguí et al., 2008a). Indeed, complementary analysis (not shown here) on the Aude and Lez catchments highlighted that SAFRAN underestimates precipitation quantiles beyond the 95th quantile by up to 30 % with respect to pluviometer data from the Météo-France network. It should be noted that comparison between the SAFRAN gridded product and the pluviometer sparse product is challenging and that these products are not independent. However, more representative computation of observed quantiles from pluviometer data set in place of SAFRAN ones is one potential gain from the present study. This paves the way towards a time-varying quantile classification with an intra-daily cycle, for instance by using 3 h pluviometer data. Such an intra-daily cycle could also be represented in the past/future change coefficients using 3 h EMCORDEX precipitation outputs. Working on higher temporal resolution data (observation and simulation) would lead to a futurization method more consistent with flash flood timescales than the daily current method.

Future changes in precipitation are assessed through a multi-model approach focusing on the mean and standard deviations of the EMCORDEX ensemble. In order to reduce the model uncertainty, different classifications and clustering methods, based on advanced ensemble statistics, have been tested to exclude the outlier models from a specific ensemble, or to apply weightings procedures in the ensemble mean computation (Boberg and Christensen, 2012). However, as discussed in Reifen and Toumi (2009) and Knutti et al. (2010), model performances in the past do not necessarily relate to model performances in the future and the choice of the criteria for multi-model studies should be further investigated.

As previously explained in Fig. 6, the monthly averaged precipitation simulated by the RCP8.5 ensemble decreases with respect to the HIST monthly averaged precipitation over the past period, especially from April to October for the Lez, Aude and Muga catchments. For the Aude and Muga boxes, there is a shift in the annual cycle; the peak occurs earlier in spring (in April instead of May). This suggests that a change in precipitation amplitude and in temporal seasonality should be expected. Llasat and Puigcerver (1997) explained that intense precipitation events are mostly due to convective rainfall in autumn and to global circulation in spring. The maximum number of convective days and ratio between convective and total precipitation is recorded between August and September in la Boadella, and a positive trend (5 % error level) in the annual number of convective days has been founded in the Muga catchment (Llasat et al., 2016). The change in spring precipitation is thus most likely due to changes in global fluxes that have different impacts on western coastal regions (Aude and Muga) and on southern coastal regions (Cevennes). Indeed, Nuissier et al. (2011) explained that strong precipitation events over the west coast of the Mediterranean are mostly correlated with easterly fluxes, while strong precipitation events in the northern coastal region are correlated with south to southeasterly fluxes. Cassou et al. (2016) reported that climate change will have a greater impact on easterly fluxes; this could explain the change in the annual cycle observed in Aude and Muga regions only. This hypothesis should be validated with an analysis of changes in geopotential fields.

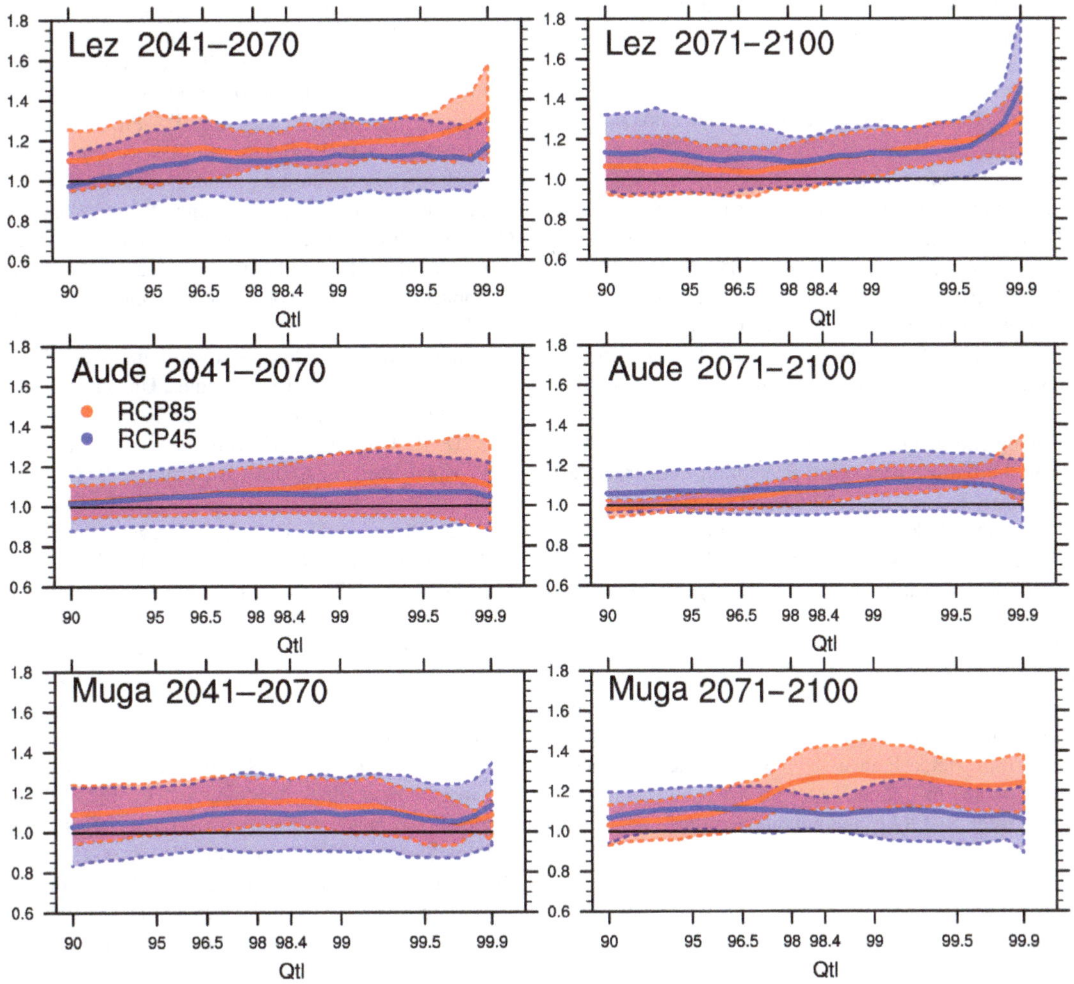

Figure 7. Change coefficients (Aqi) over the 90–99.9 quantile range computed for each GCM–RCM pairs over the Lez, Aude and Muga catchments during the autumn season (SON). The no change line (ai = 1) is also displayed in solid black. The thick solid lines represent the ensemble mean for (Aqi), and the associated ensemble spread is presented by the shaded areas (standard deviation). RCP4.5 and RCP8.5 are respectively plotted in blue and red. For clarity purposes, the scale of the x axis was distorted according to the quantile discretization.

7 Conclusions

A futurization approach is presented in this article; it consists of the computation of a past/future change coefficients applied to quantile-ranked RCM precipitation outputs. We use the EMCORDEX ensemble to estimate the model and scenario uncertainty. The study focuses over the Lez, Aude and Muga Mediterranean catchments. As a first step, EMCORDEX models' skills are evaluated in terms of mean and extreme precipitation over the present climate period 1976–2005. This analysis legitimizes the use of the EMCORDEX models for past/future changes assessment.

It has been shown that cumulative precipitation is overestimated over the mountainous regions and underestimated over the coastal regions in autumn. Extreme events beyond 95th quantiles are underestimated. GCM forcing tends to enhance RCM underestimation of extreme precipitation events over the three catchments. Climate change impact is investigated from the RCP4.5 and RCP8.5 EMCORDEX simulations. In comparison with the present period, the monthly averaged precipitation decreases in spring and summer for RCP8.5. Past/future change coefficients computed from the EMCORDEX ensemble display an increase in extreme precipitation event intensity (beyond 90th quantile). This result is stronger over the end of 21st century, for RCP8.5, for all catchments of interest: all models within the ensemble agree on change coefficients larger than 1. The multi-model approach developed here allows for quantifying the uncertainty related to the past/future change coefficients. As a major conclusion of this study, we have shown with a high degree of confidence that all RCM models in EMCORDEX ensemble forecast an increase in extreme precipitation events.

In the future, change coefficients could be used to provide a kind of futurized extreme precipitation event that occurred in the past. In further studies, the hydrological impact of this futurized precipitation will be assessed using rainfall–

runoff models over the Lez, Aude and Muga catchments. This generic method could also be applied to other catchments.

Competing interests. The authors declare that they have no conflict of interest.

Acknowledgements. This work is an initiative of WSP-France. The financial support for this work has been provided by WSP-France under the CIFRE contract 2015/005. This paper was written under the framework of the International HYMEX project and the Spanish HOPE (CGL2014-52571-R) project. We acknowledge the World Climate Research Programme's Working Group on Regional Climate, and the Working Group on Coupled Modelling, former coordinating body of CORDEX and responsible panel for CMIP5. We also thank the climate modeling groups (listed in Table 1 of this paper) for producing and making available their model output. We also acknowledge the Earth System Grid Federation infrastructure an international effort led by the U.S. Department of Energy's Program for Climate Model Diagnosis and Intercomparison, the European Network for Earth System Modelling and other partners in the Global Organisation for Earth System Science Portals (GO-ESSP). The authors wish to especially acknowledge Samuel Somot (CNRM) for his fruitful discussions about regional modeling and the MedCORDEX database. Thanks are given to Elisabeth Harader for some methodological ideas examined in their PhD thesis. Special thanks are also given to Jean-Marc Lacave and Gwenaëlle Hello from Météo-France for providing the daily and 3 h pluviometer data, and to Céline Vargel for her useful work on the comparison between SAFRAN and pluviometer data.

Edited by: Shraddhanand Shukla

References

Bechtold, P. and Bazile, E.: The 12–13 November 1999 flash flood in southern France, Atmos. Res., 56, 171–189, https://doi.org/10.1016/S0169-8095(00)00097-1, 2001.

Boberg, F. and Christensen, J. H.: Overestimation of Mediterranean summer temperature projections due to model deficiencies, Nat. Clim. Change, 2, 433, https://doi.org/10.1038/NCLIMATE1454, 2012.

Boudevillain, B., Delrieu, G., Galabertier, B., Bonnifait, L., Bouilloud, L., Kirstetter, P.-E., and Mosini, M.-L.: The Cévennes-Vivarais Mediterranean Hydrometeorological Observatory database, Water Resour. Res., 47, doi:10.1029/2010WR010353, 2011.

Bougeault, P.: A simple parameterization of the large-scale effects of cumulus convection, Mon. Weather Rev., 113, 2108–2121, https://doi.org/10.1175/1520-0493(1985)113<2108:ASPOTL>2.0.CO;2, 1985.

Cassou, C. and Cattiaux, J.: Disruption of the European climate seasonal clock in a warming world, Nat. Clim. Change, https://doi.org/10.1038/NCLIMATE2969, 2016.

Cavicchia, L., Scoccimarro, E., Gualdi, S., Marson, P., Ahrens, B., Berthou, S., Conte, D., Dell'Aquila, A., Drobinski, P., Djurdjevic, V., Dubois, C., Gallardo, C., Li, L., Oddo, P., Sanna, A., and Torma, C.: Mediterranean extreme precipitation: a multi-model assessment, Clim. Dynam., 1–13, https://doi.org/10.1007/s00382-016-3245-x, 2016.

Christensen, J. H., Carter, T. R., Rummukainen, M., and Amanatidis, G.: Evaluating the performance and utility of regional climate models: the PRUDENCE project, Clim. Change, 81, 1–6, https://doi.org/10.1007/s10584-006-9211-6, 2007.

Clarke, L., Edmonds, J., Jacoby, H., Pitcher, H., Reilly, J., and Richels, R.: CCSP synthesis and assessment product 2.1, Part A: scenarios of greenhouse gas emissions and atmospheric concentrations, US Government Printing Office, Washington DC, 2007.

Coustau, M.: Contribution à la prévision des crues sur le bassin du Lez: modélisation de la relation pluie-débit en zone karstique et impact de l'assimilation de débits, Université de Montpellier II, 2011.

Dayon, G.: Evolution du cycle hydrologique continental en France au cours des prochaines décennies, Université de Toulouse, Université Toulouse III-Paul Sabatier, 2015.

Dee, D. P., Uppala, S. M., Simmons, A. J., Berrisford, P., Poli, P., Kobayashi, S., Andrae, U., Balmaseda, M. A., Balsamo, G., Bauer, P., Bechtold, P., Beljaars, A. C. M., van de Berg, L., Bidlot, J., Bormann, N., Delsol, C., Dragani, R., Fuentes, M., Geer, A. J., Haimberger, L., Healy, S. B., Hersbach, H., Hólm, E. V., Isaksen, L., Kållberg, P., Köhler, M., Matricardi, M., McNally, A. P., Monge-Sanz, B. M., Morcrette, J.-J., Park, B.-K., Peubey, C., de Rosnay, P., Tavolato, C., Thépaut, J.-N., and Vitart, F.: The ERA-Interim reanalysis: configuration and performance of the data assimilation system, Q. J. R. Meteorol. Soc., 137, 553–597, https://doi.org/10.1002/qj.828, 2011.

Déqué, M. and Somot, S.: Weighted frequency distributions express modelling uncertainties in the ENSEMBLES regional climate experiments, Clim. Res., 44, 195–209, https://doi.org/10.3354/cr00866, 2010.

Déqué, M., Somot, S., Sanchez-Gomez, E., Goodess, C. M., Jacob, D., Lenderink, G., and Christensen, O. B.: The spread amongst ENSEMBLES regional scenarios: regional climate models, driving general circulation models and interannual variability, Clim. Dynam., 38, 951–964, https://doi.org/10.1007/s00382-011-1053-x, 2012.

Diaconescu, E. P., Gachon, P., and Laprise, R.: On the remapping procedure of daily precipitation statistics and indices used in regional climate model evaluation, J. Hydrometeorol., 16, 2301–2310, https://doi.org/10.1175/JHM-D-15-0025.1, 2015.

Drobinski, P., Da Silva, N., Panthou, G., Bastin, S., Muller, C., Ahrens, B., Borga, M., Conte, D., Fosser, G., Giorgi, F., Güttler, I., Kotroni, V., Li, L., Morin, E., Önol, B., Quintana-Segui, P., Romera, R., and Torma, C. Z.: Scaling precipitation extremes with temperature in the Mediterranean: past climate assessment and projection in anthropogenic scenarios, Clim. Dynam., 1–21, https://doi.org/10.1007/s00382-016-3083-x, 2016.

Ducrocq, V., Aullo, G., and Santurette, P.: Les précipitations intenses et les inondations des 12 et 13 novembre 1999 sur le sud de la France, https://doi.org/10.4267/2042/36293, 2003.

Ducrocq, V., Nuissier, O., Ricard, D., Lebeaupin, C., and Thouvenin, T.: A numerical study of three catastrophic precipitating events over southern France, II: Mesoscale triggering and stationarity factors, Q. J. Roy. Meteorol. Soc., 134, 131–145, https://doi.org/10.1002/qj.199, 2008.

Durand, Y., Brun, E., Mérindol, L., Guyomarc'h, G., Lesaffre, B., and Martin, E.: A meteorological estimation of relevant parameters for snow models, Ann. Glaciol., 18, 65–71, https://doi.org/10.3198/1993AoG18-1-65-71, 1993.

Estupina, V.: Vers une modélisation hydrologique adaptée à la prévision opérationnelle des crues éclair: application à de petits bassins versants du sud de la France, Toulouse, INPT, 2004.

Fowler, H. J., Ekström, M., Blenkinsop, S., and Smith, A. P.: Estimating change in extreme European precipitation using a multimodel ensemble, J. Geophys. Res., 112, D18104, https://doi.org/10.1029/2007JD008619, 2007a.

Fowler, H. J., Blenkinsop, S., and Tebaldi, C.: Linking climate change modelling to impacts studies: recent advances in downscaling techniques for hydrological modelling, Int. J. Climatol., 27, 1547–1578, https://doi.org/10.1002/joc.1556, 2007b.

Frei, C., Schöll, R., Fukutome, S., Schmidli, J., and Vidale, P. L.: Future change of precipitation extremes in Europe: Intercomparison of scenarios from regional climate models, J. Geophys. Res.-Atmos., (1984–2012), 111, https://doi.org/10.1029/2007JD008619, 2006.

Gao, X., Pal, J. S., and Giorgi, F.: Projected changes in mean and extreme precipitation over the Mediterranean region from a high resolution double nested RCM simulation, Geophys. Res. Lett., 33, https://doi.org/10.1029/2005GL024954, 2006.

Gaume, E., Livet, M., Desbordes, M., and Villeneuve, J.-P.: Hydrological analysis of the river Aude, France, flash flood on 12 and 13 November 1999, J. Hydrol., 286, 135–154, https://doi.org/10.1016/j.jhydrol.2003.09.015, 2004.

Gaume, E., Borga, M., Llasat, M. C., Maouche, S., Lang, M., and Diakakis, M.: Mediterranean extreme floods and flash floods. Into Hydro-meteorological extremes, chapter 3, The Mediterranean Region under Climate Change, A Scientific Update (coordinated by All Envi), 133–144, 2016.

Giorgi, F.: Climate change hot-spots, Geophys. Res. Lett., 33, https://doi.org/10.1029/2006GL025734, 2006.

Giorgi, F. and Lionello, P.: Climate change projections for the Mediterranean region, Glob. Planet. Change, 63, 90–104, https://doi.org/10.1016/j.gloplacha.2007.09.005, 2008.

Giorgi, F., Jones, C., Asrar, G. R.: Addressing climate information needs at the regional level: the CORDEX framework, World Meteorological Organization (WMO) Bulletin, 58, 175, 2009.

Giorgi, F., Torma, C., Coppola, E., Ban, N., Schär, C., and Somot, S.: Enhanced summer convective rainfall at Alpine high elevations in response to climate warming, Nat. Geosci., 9, 584–589, https://doi.org/10.1038/ngeo2761, 2016.

Hagemann, S., Machenhauer, B., Jones, R., Christensen, O. B., Déqué, M., Jacob, D., and Vidale, P. L.: Evaluation of water and energy budgets in regional climate models applied over Europe, Clim. Dynam., 23, 547–567, https://doi.org/10.1007/s00382-004-0444-7, 2004.

Harader, E.: L'impact du changement climatique sur les événements hydrologiques extrêmes des petits bassins versants méditerranéens: le cas du bassin versant du Lez, Université de Toulouse, Université Toulouse III-Paul Sabatier, 2015.

Harris, I., Jones, P. D., Osborn, T. J., and Lister, D. H.: Updated high-resolution grids of monthly climatic observations – the CRU TS3. 10 Dataset, Int. J. Climatol., 34, 623–642, https://doi.org/10.1002/joc.3711, 2014.

Haylock, M. R., Hofstra, N., Klein Tank, A. M. G., Klok, E. J., Jones, P. D., and New, M.: A European daily high-resolution gridded data set of surface temperature and precipitation for 1950–2006, J. Geophys. Res.-Atmos., 113, https://doi.org/10.1029/2008JD010201, 2008.

Herrera, S., Gutiérrez, J. M., Ancell, R., Pons, M. R., Frías, M. D., and Fernández, J.: Development and analysis of a 50-year high-resolution daily gridded precipitation dataset over Spain (Spain02), Int. J. Climatol., 32, 74–85, https://doi.org/10.1002/joc.2256, 2012.

Herrera, S., Fernández, J., and Gutiérrez, J. M.: Update of the Spain02 gridded observational dataset for EURO-CORDEX evaluation: assessing the effect of the interpolation methodology, Int. J. Climatol., 36, 900–908, https://doi.org/10.1002/joc.4391, 2016.

Jacob, D., Bärring, L., Christensen, O. B., Christensen, J. H., de Castro, M., Deque, M., Giorgi, F., Hagemann, S., Hirschi, M., Jones, R., Kjellström, E., Lenderink, G., Rockel, B., Sánchez, E., Schär, C., Seneviratne, S. I., Somot, S., van Ulden, A., and van den Hurk, B.: An inter-comparison of regional climate models for Europe: model performance in present-day climate, Clim. Change, 81, 31–52, https://doi.org/10.1007/s10584-006-9213-4, 2007.

Jacob, D., Petersen, J., Eggert, B., Alias, A., Christensen, O. B., Bouwer, L. M., Braun, A., Colette, A., Déqué, M., Georgievski, G., Georgopoulou, E., Gobiet, A., Menut, L., Nikulin, G., Haensler, A., Hempelmann, N., Jones, C., Keuler, K., Kovats, S., Kröner, N., Kotlarski, S., Kriegsmann, A., Martin, E., Meijgaard, E. van, Moseley, C., Pfeifer, S., Preuschmann, S., Radermacher, C., Radtke, K., Rechid, D., Rounsevell, M., Samuelsson, P., Somot, S., Soussana, J.-F., Teichmann, C., Valentini, R., Vautard, R., Weber, B., and Yiou, P.: EURO-CORDEX: new high-resolution climate change projections for European impact research, Reg. Environ. Change, 14, 563–578, https://doi.org/10.1007/s10113-013-0499-2, 2013.

Katragkou, E., García-Díez, M., Vautard, R., Sobolowski, S., Zanis, P., Alexandri, G., Cardoso, R. M., Colette, A., Fernandez, J., Gobiet, A., Goergen, K., Karacostas, T., Knist, S., Mayer, S., Soares, P. M. M., Pytharoulis, I., Tegoulias, I., Tsikerdekis, A., and Jacob, D.: Regional climate hindcast simulations within EURO-CORDEX: evaluation of a WRF multi-physics ensemble, Geosci. Model Dev., 8, 603–618, https://doi.org/10.5194/gmd-8-603-2015, 2015.

Kjellström, E., Boberg, F., Castro, M., Christensen, J. H., Nikulin, G., and Sánchez, E.: Daily and monthly temperature and precipitation statistics as performance indicators for regional climate models, Climate research (Open Access for articles 4 years old and older), 44, 135, https://doi.org/10.3354/cr00932, 2010.

Knutti, R., Furrer, R., Tebaldi, C., Cermak, J., and Meehl, G. A.: Challenges in combining projections from multiple climate models, Journal of Climate, 23(10), 2739–2758, https://doi.org/10.1175/2009JCLI3361.1, 2010.

Kotlarski, S., Block, A., Böhm, U., Jacob, D., Keuler, K., Knoche, R., Rechid, D., and Walter, A.: Regional climate model simulations as input for hydrological applications: evaluation of uncertainties, Adv. Geosci., 5, 119–125, 2005.

Kotlarski, S., Keuler, K., Christensen, O. B., Colette, A., Déqué, M., Gobiet, A., Goergen, K., Jacob, D., Lüthi, D., van Meijgaard, E., Nikulin, G., Schär, C., Teichmann, C., Vautard, R., Warrach-Sagi, K., and Wulfmeyer, V.: Regional climate modeling on European scales: a joint standard evaluation of the EURO-CORDEX RCM ensemble, Geosci. Model Dev., 7, 1297–1333, https://doi.org/10.5194/gmd-7-1297-2014, 2014.

Kysel, J., Beguería, S., Beranová, R., Gaál, L., and López-Moreno, J. I.: Different patterns of climate change scenarios for short-term and multi-day precipitation extremes in the Mediterranean, Glob. Planet. Change, 98, 63–72, https://doi.org/10.1016/j.gloplacha.2012.06.010, 2012.

Llasat, M. C., Llasat-Botija, M., and López, L.: A press database on natural risks and its application in the study of floods in Northeastern Spain, Nat. Hazards Earth Syst. Sci., 9, 2049–2061, https://doi.org/10.5194/nhess-9-2049-2009, 2009.

Llasat, M. C., Marcos, R., Llasat-Botija, M., Gilabert, J., Turco, M., and Quintana-Seguí, P.: Flash flood evolution in North-Western Mediterranean, Atmos. Res., 149, 230–243, https://doi.org/10.1016/j.atmosres.2014.05.024, 2014.

Llasat, M. C., Marcos, R., Turco, M., Gilabert, J., and Llasat-Botija, M.: Trends in flash flood events versus convective precipitation in the Mediterranean region: The case of Catalonia, J. Hydrol., 541, 24–37, https://doi.org/10.1016/j.jhydrol.2016.05.040, 2016.

Lucas-Picher, P., Caya, D., de Elía, R., and Laprise, R.: Investigation of regional climate models' internal variability with a ten-member ensemble of 10-year simulations over a large domain, Clim. Dynam., 31, 927–940, https://doi.org/10.1007/s00382-008-0384-8, 2008.

Maraun, D., Osborn, T. J., and Rust, H. W.: The influence of synoptic airflow on UK daily precipitation extremes, Part II, regional climate model and E-OBS data validation, Clim. Dynam., 39, 287–301, https://doi.org/10.1007/s00382-011-1176-0, 2012.

Meinshausen, M., Smith, S. J., Calvin, K., Daniel, J. S., Kainuma, M. L. T., Lamarque, J. F., Matsumoto, K., Montzka, S. A., Raper, S. C. B., Riahi, K., Thomson, A., Velders, G. J. M., and van Vuuren D. P. P.: The RCP greenhouse gas concentrations and their extensions from 1765 to 2300, Clim. Change, 109, 213–241, https://doi.org/10.1007/s10584-011-0156-z, 2011.

Metzger, M. J., Bunce, R. G. H., Jongman, R. H. G., Mücher, C. A., and Watkins, J. W.: A climatic stratification of the environment of Europe, Glob. Ecol. Biogeogr., 14, 549–563, https://doi.org/10.1111/j.1466-822X.2005.00190.x, 2005.

Milano, M., Ruelland, D., Fernandez, S., Dezetter, A., Fabre, J., Servat, E., Fritsch, J.-M., Ardoin-Bardin, S., and Thivet, G.: Current state of Mediterranean water resources and future trends under climatic and anthropogenic changes, Hydrol. Sci. J., 58, 498–518, https://doi.org/10.1080/02626667.2013.774458, 2013.

Nuissier, O., Ducrocq, V., Ricard, D., Lebeaupin, C., and Anquetin, S.: A numerical study of three catastrophic precipitating events over southern France, I: Numerical framework and synoptic ingredients, Q. J. Roy. Meteorol. Soc., 134, 111–130, https://doi.org/10.1002/qj.200, 2008.

Nuissier, O., Joly, B., Joly, A., Ducrocq, V., and Arbogast, P.: A statistical downscaling to identify the large-scale circulation patterns associated with heavy precipitation events over southern France, Q. J. Roy. Meteorol. Soc., 137, 1812–1827, https://doi.org/10.1002/qj.866, 2011.

Mearns, L. O., Gutowski, W., Jones, R., Leung, R., Macginnis, S., Nunes, A., and Qian, Y.: A regional climate change assessment program for North America, Eos, 90, https://doi.org/10.1029/2009EO360002, 2009.

Paxian, A., Hertig, E., Seubert, S., Vogt, G., Jacobeit, J., and Paeth, H.: Present-day and future mediterranean precipitation extremes assessed by different statistical approaches, Clim. Dynam., 44, 845–860, https://doi.org/10.1007/s00382-014-2428-6, 2015.

Prein, A. F., Gobiet, A., Truhetz, H., Keuler, K., Goergen, K., Teichmann, C., Maule, C. F., Van Meijgaard, E., Déqué, M., Nikulin, G., and others: Precipitation in the EURO-CORDEX 0.11° and 0.44° simulations: high resolution, high benefits, Clim. Dynam., 46, 383–412, https://doi.org/10.1007/s00382-015-2589-y, 2016.

Quintana Seguí, P., Ribes, A., Martin, E., Habets, F., and Boé, J.: Comparison of three downscaling methods in simulating the impact of climate change on the hydrology of Mediterranean basins, J. Hydrol., 383, 111–124, https://doi.org/10.1016/j.jhydrol.2009.09.050, 2010.

Quintana-Seguí, P., Le Moigne, P., Durand, Y., Martin, E., Habets, F., Baillon, M., Canellas, C., Franchisteguy, L., and Morel, S.: Analysis of Near-Surface Atmospheric Variables: Validation of the SAFRAN Analysis over France, J. Appl. Meteor. Climatol., 47, 92–107, https://doi.org/10.1175/2007JAMC1636.1, 2008.

Quintana-Seguí, P., Habets, F., and Martin, E.: Comparison of past and future Mediterranean high and low extremes of precipitation and river flow projected using different statistical downscaling methods, Nat. Hazards Earth Syst. Sci., 11, 1411–1432, https://doi.org/10.5194/nhess-11-1411-2011, 2011.

Quintana-Seguí, P., Peral, C., Turco, M., Llasat, M. C., and Martin, E.: Meteorological Analysis Systems in North-East Spain: Validation of SAFRAN and SPAN, J. Environ. Inf., 27, https://doi.org/10.3808/jei.201600335, 2016a.

Quintana-Seguí, P., Turco, M., Herrera, S., and Miguez-Macho, G.: Validation of a new SAFRAN-based gridded precipitation product for Spain and comparisons to Spain02 and ERA-Interim, https://doi.org/10.5194/hess-2016-349, 21, 2187, 2017.

Ramis, C., Llasat, M. C., Genovés, A., and Jansà, A.: The October-1987 floods in Catalonia: Synoptic and mesoscale mechanisms, Meteorol. Appl., 1, 337–350, https://doi.org/10.1002/met.5060010404, 1994.

Reifen, C. and Toumi, R.: Climate projections: Past performance no guarantee of future skill?, Geophys. Res. Lett., 36, https://doi.org/10.1029/2009GL038082, 2009.

Raynaud, F., Borrell-Estupina, V., Pistre, S., Van-Exter, S., Bourgeois, N., Dezetter, A., and Servat, E.: Combining hydraulic model, hydrogeomorphological observations and chemical analyses of surface waters to improve knowledge on karst flash floods genesis, Proc. Int. Assoc. Hydrol. Sci., 369, 55–60, 2015.

Riahi, K., Grübler, A., and Nakicenovic, N.: Scenarios of long-term socio-economic and environmental development under climate stabilization, Technol. Forecast. Soc. Change, 74, 887–935, https://doi.org/10.1016/j.techfore.2006.05.026, 2007.

Ricard, D., Ducrocq, V., and Auger, L.: A climatology of the mesoscale environment associated with heavily precipitating events over a northwestern Mediterranean area, J. Appl. Meteorol. Climatol., 51, 468–488, https://doi.org/10.1175/JAMC-D-11-017.1, 2012.

Schär, C., Ban, N., Fischer, E. M., Rajczak, J., Schmidli, J., Frei, C., Giorgi, F., Karl, T. R., Kendon, E. J., Tank, A. M. K., O'Gorman, P. A., Sillmannn, J., Zhang, X., and Zwiers, F. W.: Percentile indices for assessing changes in heavy precipitation events, Clim. Change, 137, 201–216, https://doi.org/10.1007/s10584-016-1669-2, 2016.

Themeßl, M. J., Gobiet, A., and Heinrich, G.: Empirical-statistical downscaling and error correction of regional climate models and its impact on the climate change signal, Clim. Change, 112, 449–468, https://doi.org/10.1002/joc.2168, 2012.

Tramblay, Y., Ruelland, D., Somot, S., Bouaicha, R., and Servat, E.: High-resolution Med-CORDEX regional climate model simulations for hydrological impact studies: a first evaluation of the ALADIN-Climate model in Morocco, Hydrol. Earth Syst. Sci., 17, 3721–3739, https://doi.org/10.5194/hess-17-3721-2013, 2013.

Trapero, L., Bech, J., Duffourg, F., Esteban, P., and Lorente, J.: Mesoscale numerical analysis of the historical November 1982 heavy precipitation event over Andorra (Eastern Pyrenees), Nat. Hazards Earth Syst. Sci., 13, 2969–2990, https://doi.org/10.5194/nhess-13-2969-2013, 2013.

van der Linden P. and J. F. B. Mitchell (eds.): ENSEMBLES: Climate Change and its Impacts: Summary of research and results from the ENSEMBLES project. Met Office Hadley Centre, FitzRoy Road, Exeter EX1 3PB, Uk, 160 pp, 2009.

Van Vuuren, D. P., Edmonds, J., Kainuma, M., Riahi, K., Thomson, A., Hibbard, K., Hurtt, G. C., Kram, T., Krey, V., Lamarque, J.-F., Masui, T., Meinshausen, M., Nakicenovic, N., Smith, S. J., and Rose, S. K.: The representative concentration pathways: an overview, Clim. Change, 109, 5–31, https://doi.org/10.1007/s10584-011-0148-z, 2011.

Vidal, J.-P., Martin, E., Franchistéguy, L., Baillon, M., and Soubeyroux, J.-M.: A 50-year high-resolution atmospheric reanalysis over France with the Safran system, Int. J. Climatol., 30, 1627–1644, https://doi.org/10.1002/joc.2003, 2010.

PERMISSIONS

All chapters in this book were first published in HESS, by Copernicus Publications; hereby published with permission under the Creative Commons Attribution License or equivalent. Every chapter published in this book has been scrutinized by our experts. Their significance has been extensively debated. The topics covered herein carry significant findings which will fuel the growth of the discipline. They may even be implemented as practical applications or may be referred to as a beginning point for another development.

The contributors of this book come from diverse backgrounds, making this book a truly international effort. This book will bring forth new frontiers with its revolutionizing research information and detailed analysis of the nascent developments around the world.

We would like to thank all the contributing authors for lending their expertise to make the book truly unique. They have played a crucial role in the development of this book. Without their invaluable contributions this book wouldn't have been possible. They have made vital efforts to compile up to date information on the varied aspects of this subject to make this book a valuable addition to the collection of many professionals and students.

This book was conceptualized with the vision of imparting up-to-date information and advanced data in this field. To ensure the same, a matchless editorial board was set up. Every individual on the board went through rigorous rounds of assessment to prove their worth. After which they invested a large part of their time researching and compiling the most relevant data for our readers.

The editorial board has been involved in producing this book since its inception. They have spent rigorous hours researching and exploring the diverse topics which have resulted in the successful publishing of this book. They have passed on their knowledge of decades through this book. To expedite this challenging task, the publisher supported the team at every step. A small team of assistant editors was also appointed to further simplify the editing procedure and attain best results for the readers.

Apart from the editorial board, the designing team has also invested a significant amount of their time in understanding the subject and creating the most relevant covers. They scrutinized every image to scout for the most suitable representation of the subject and create an appropriate cover for the book.

The publishing team has been an ardent support to the editorial, designing and production team. Their endless efforts to recruit the best for this project, has resulted in the accomplishment of this book. They are a veteran in the field of academics and their pool of knowledge is as vast as their experience in printing. Their expertise and guidance has proved useful at every step. Their uncompromising quality standards have made this book an exceptional effort. Their encouragement from time to time has been an inspiration for everyone.

The publisher and the editorial board hope that this book will prove to be a valuable piece of knowledge for researchers, students, practitioners and scholars across the globe.

LIST OF CONTRIBUTORS

Z. K. Tesemma, Y. Wei, M. C. Peel and A. W. Western
Department of Infrastructure Engineering, The University of Melbourne, Parkville, Victoria, 3010, Australia

Chuanhao Wu, Peng Wang and Kai Xu
Institute of Groundwater and Earth Sciences, Jinan University, Guangzhou 510632, China

Bill X. Hu
Institute of Groundwater and Earth Sciences, Jinan University, Guangzhou 510632, China
Department of Earth, Ocean and Atmospheric Sciences, Florida State University, Tallahassee, FL 32306, USA

Guoru Huang
School of Civil Engineering and Transportation, South China University of Technology, Guangzhou 510640, China
State Key Laboratory of Subtropical Building Science, South China University of Technology, Guangzhou 510640, China

Tingting Ning and Wenzhao Liu
State Key Laboratory of Soil Erosion and Dryland Farming on the Loess Plateau, Institute of Soil and Water Conservation, CAS & MWR, Yangling, Shaanxi 712100, China
University of the Chinese Academy of Sciences, Beijing 100049, China

Zhi Li
College of Natural Resources and Environment, Northwest A&F University, Yangling, Shaanxi 712100, China

Sangchul Lee
Department of Environmental Science and Technology, University of Maryland, College Park, MD 20742, USA
US Department of Agriculture-Agricultural Research Service, Hydrology and Remote Sensing Laboratory, Beltsville, MD 20705, USA

In-Young Yeo
School of Engineering, the University of Newcastle, Callaghan NSW 2308, Australia
Department of Geographical Sciences, University of Maryland, College Park, MD 20742, USA

Ali M. Sadeghi and Gregory W. McCarty
US Department of Agriculture-Agricultural Research Service, Hydrology and Remote Sensing Laboratory, Beltsville, MD 20705, USA

Wells D. Hively
US Geological Survey, Eastern Geographic Science Center, Reston, VA 20192, USA

Megan W. Lang
Department of Geographical Sciences, University of Maryland, College Park, MD 20742, USA
US Fish & Wildlife Service, National Wetlands Inventory, Falls Church, VA 22041, USA

Amir Sharifi
Department of Energy and Environment, Government of the District of Columbia, Washington, DC 20002, USA

Javier Fluixá-Sanmartín and Javier García-Hernández
Centre de Recherche sur l'Environnement Alpin (CREALP), 1951 Sion, Switzerland

Deng Pan and Jijun Xu
Changjiang River Scientific Research Institute, Changjiang Water Resources Commission, Wuhan Hubei, 430010, China

Luzia Fischer and Christoph Haemmig
GEOTEST AG, 3052 Zollikofen, Switzerland

Boris Orlowsky
climate-babel, 8047 Zurich, Switzerland

Frédéric Jordan
Hydrique Ingénieurs, 1052 Le Mont-sur-Lausanne, Switzerland

Fangwei Zhang
Bureau of Hydrology, Changjiang Water Resources Commission, Wuhan Hubei, 430017, China

Hjalte Jomo Danielsen Sørup
Technical University of Denmark, Global Decision Support Initiative, Lyngby, Denmark
Technical University of Denmark, Department of Environmental Engineering, Lyngby, Denmark

Stylianos Georgiadis
Technical University of Denmark, Global Decision Support Initiative, Lyngby, Denmark
Technical University of Denmark, Department of Applied Mathematics and Computer Science, Lyngby, Denmark

Ida Bülow Gregersen
Ramboll Danmark A/S, Department of Climate Adaptation and Green Infrastructure, Copenhagen, Denmark

Karsten Arnbjerg-Nielsen
Technical University of Denmark, Global Decision Support Initiative, Lyngby, Denmark
Technical University of Denmark, Department of Environmental Engineering, Lyngby, Denmark

Katrina E. Bennett, Kurt Solander, Chonggang Xu and Richard S. Middleton
Earth and Environmental Sciences, Los Alamos National Lab, Los Alamos, NM 87545, USA

Theodore J. Bohn
Julie Ann Wrigley Global Institute of Sustainability, Arizona State University, Tempe, AZ 85287, USA
School of Earth and Space Exploration, Arizona State University, Tempe, AZ 85287, USA

Nathan G. McDowell
Earth and Environmental Sciences, Los Alamos National Lab, Los Alamos, NM 87545, USA
Pacific Northwest National Laboratory, Richland, WA 99354, USA

Enrique Vivoni
School of Earth and Space Exploration, Arizona State University, Tempe, AZ 85287, USA
School of Sustainable Engineering and the Built Environment, Arizona State University, Tempe, AZ 85287, USA

Enrica Perra
Dipartimento di Ingegneria Civile, Ambientale ed Architettura, Università degli Studi di Cagliari, Cagliari, Italy
Centre Eau Terre Environnement, Institut National de la Recherche Scientifique, Quebec City, Canada

Monica Piras and Roberto Deidda
Dipartimento di Ingegneria Civile, Ambientale ed Architettura, Università degli Studi di Cagliari, Cagliari, Italy
Consorzio Interuniversitario nazionale per la Fisica dell'Atmosfere e dell'Idrosfere, Tolentino, Italy

Claudio Paniconi
Centre Eau Terre Environnement, Institut National de la Recherche Scientifique, Quebec City, Canada

Giuseppe Mascaro
Consorzio Interuniversitario nazionale per la Fisica dell'Atmosfere e dell'Idrosfere, Tolentino, Italy
School of Sustainable Engineering and the Built Environment and School of Earth and Space Exploration, Arizona State University, Tempe, AZ, USA

Enrique R. Vivoni
School of Sustainable Engineering and the Built Environment and School of Earth and Space Exploration, Arizona State University, Tempe, AZ, USA

Pierluigi Cau and Pier Andrea Marras
Centro di Ricerca, Sviluppo e Studi Superiori in Sardegna, Pula, Cagliari, Italy

Ralf Ludwig
Physical Geography and Environmental Modeling, Department of Geography, Ludwig-Maximilians-Universität, Munich, Germany

Swen Meyer
Physical Geography and Environmental Modeling, Department of Geography, Ludwig-Maximilians-Universität, Munich, Germany
Leibniz-Institut für Gemüse- und Zierpflanzenbau Großbeeren/Erfurt e.V., Grossbeeren, Germany

José María Santiago, Joaquín Solana-Gutiérrez, Diego García de Jalón and Carlos Alonso
Laboratorio de Hidrobiología, ETSI Montes, Forestal y Medio Natural, Universidad Politécnica de Madrid, Camino de las Moreras s/n, Madrid 28040, Spain

Rafael Muñoz-Mas and Francisco Martínez-Capel
Institut d'Investigació per a la Gestió Integrada de Zones Costaneres (IGIC), Universitat Politècnica de València, C/ Paranimf 1, Grau de Gandia, Valencia 46730, Spain

Javier Pórtoles, Robert Monjo and Jaime Ribalaygua
Fundación para la Investigación del Clima, C/ Tremps 11. Madrid 28040, Spain

Axel Kleidon
Max-Planck-Institut für Biogeochemie, Jena, Germany

Anke Hildebrandt and Marcel Bechmann
Friedrich-Schiller-Universität, Jena, Germany
Max-Planck-Institut für Biogeochemie, Jena, Germany

José Miguel Delgado, Sebastian Voss, Gerd Bürger, Klaus Vormoor and Till Francke
Institute of Earth and Environmental Sciences, University of Potsdam, Potsdam, Germany

Aline Murawski
German Research Centre of Geosciences GFZ Potsdam, Potsdam, Germany

José Marcelo Rodrigues Pereira, Eduardo Martins and Francisco Vasconcelos Júnior
Research Institute for Meteorology and Water Resources – FUNCEME, Fortaleza, Brazil

Antoine Colmet-Daage
CECI, CERFACS - CNRS TOULOUSE, Toulouse, France
WSP France, Toulouse, France
Hydrosciences Montpellier, Univ. Montpellier, Montpellier, France

Emilia Sanchez-Gomez and Sophie Ricci
CECI, CERFACS - CNRS TOULOUSE, Toulouse, France

Cécile Llovel
WSP France, Toulouse, France

Valérie Borrell Estupina
Hydrosciences Montpellier, Univ. Montpellier, Montpellier, France

Pere Quintana-Seguí
Observatori de l'Ebre Fundació, Tarragona, Spain

Maria Carmen Llasat
Universitat de Barcelona, Barcelona, Spain

Eric Servat
Institut Montpelliérain de l'Eau et de l'Environnement - IRD, Montpellier, France

Index

www.ingramcontent.com/pod-product-compliance
Lightning Source LLC
LaVergne TN
LVHW060548210926
848485LV00005B/1278

9781647401108